HADRON
PHYSICS

Previous Proceedings in the Series of Workshops on Hadron Physics

Year	Held in	Publisher	ISBN
1st 1999	Coimbra, Portugal	AIP Conference Proceedings vol. 508	1-56396-927-0

Other Related Titles from AIP Conference Proceedings

642 High Intensity and High Brightness Hadron Beams: 20th ICFA Advanced Beam Dynamics Workshop on High Intensity and High Brightness Hadron Beams; ICFA-HB2002
Edited by Weiren Chou, Yoshiharu Mori, David Neuffer, and Jean-François Ostiguy, November 2002, 0-7354-0097-0

631 New States of Matter in Hadronic Interactions: Pan-American Advanced Study Institute
Edited by H.-Thomas Elze, Erasmo M. Ferreira, Takeshi Kodama, Jean Letessier, Johann Rafelski, and Robert L. Thews, October 2002, 0-7354-0086-5

623 Particles and Fields: Eighth Mexican Workshop
Edited by J. L. Diaz-Cruz, J. Engelfried, M. Kirchbach, and M. Mondragon, July 2002, 0-7354-0072-5

619 Hadron Spectroscopy: Ninth International Conference on Hadron Spectroscopy; HADRON2001
Edited by Dmitry Amelin and Alexander M. Zaitsev, June 2002, 0-7354-0067-9

618 Heavy Flavor Physics: Ninth International Symposium on Heavy Flavor Physics
Edited by Anders Ryd and Frank C. Porter, May 2002, 0-7354-0064-4

603 Mesons and Light Nuclei: 8th Conference
Edited by Jiri Adam, Petr Bydzovsky, and Jiri Mares, December 2001, 0-7354-0047-4

602 QCD@Work: Int'l. Workshop on Quantum Chromodynamics: Theory and Experiment
Edited by Pietro Colangelo and Giuseppe Nardulli, December 2001, 0-7354-0046-6

594 Hadrons and Nuclei: First International Symposium
Edited by Il-Tong Cheon, Taekeun Choi, Seung-Woo Hong, and Su Houng Lee, November 2001, 0-7354-0037-7

588 Physics with an Electron Polarized Light-Ion Collider: Second Workshop, EPIC 2000
Edited by Richard G. Milner, October 2001, 0-7354-0028-8

561 Tours Symposium on Nuclear Physics IV: Tours 2000
Edited by M. Arnould, M. Lewitowicz, Yu. Ts. Oganessian, H. Akimune, M. Ohta, H. Utsunomiya, T. Wada, and T. Yamagata, April 2001, 1-56396-996-3

549 Intersections of Particle and Nuclear Physics: 7th Conference, CIPANP2000
Edited by Zohreh Parsa and William J. Marciano, December 2000, 1-56396-978-5

To learn more about these titles, or the AIP Conference Proceedings Series, please visit the webpage **http://proceedings.aip.org/proceedings**

HADRON PHYSICS

Effective Theories of Low Energy QCD
Second International Workshop on Hadron Physics

Coimbra, Portugal   25-29 September 2002

SPONSORING ORGANIZATIONS
Fundação para a Ciência e a Tecnologia
Fundo de Apoio à Comunidade Científica
  Programa Operacional Ciência, Tecnologia
  Inovação do Quadro Comunitário de Apoio III
Fundação Calouste Gulbenkian
Grupo Teórico de Altas Energias
Universidade de Coimbra
  Centro de Física Teórica
  Faculdade de Ciências e Tecnologia

EDITORS
A. H. Blin
B. Hiller
A. A. Osipov
M. C. Ruivo
E. van Beveren
Universidade de Coimbra
Portugal

AMERICAN INSTITUTE OF PHYSICS

Melville, New York, 2003
AIP CONFERENCE PROCEEDINGS ■ VOLUME 660

Editors:

A. H. Blin
B. Hiller
A. A. Osipov
M. C. Ruivo
E. van Beveren

Centro de Física Teórica
Universidade de Coimbra
P-3004-516 Coimbra
PORTUGAL

E-mail: alex@teor.fis.uc.pt
brigitte@teor.fis.uc.pt
alexguest@teor.fis.uc.pt
maria@teor.fis.uc.pt
eef@teor.fis.uc.pt

L.C. Catalog Card No. 2003102223
ISBN 0-7354-0120-9
ISSN 0094-243X
Printed in the United States of America

CONTENTS

J/Ψ PHYSICS

QUARK MODELS

Preface

The present volume contains selected topics of current interest in Hadron Physics, with focus on effective theories, models and methods for low energy QCD, hadronic matter under extreme conditions and physics involving heavy quarks. The first meeting of this kind (Coimbra 1999) also addressed the subject of Hadron Physics and was published in AIP Conference Proceedings 508.

It is a great pleasure to thank all the speakers for their successful efforts in presenting comprehensive and stimulating talks, as well as for the interesting discussions they have triggered. This workshop offered the opportunity to discuss at leisure problems of mutual interest. We are very grateful to Georges Ripka for going through the trouble of delivering the Summary Talk.

This conference has been made possible through the financial support of the following organizations and grants: Fundação para a Ciência e a Tecnologia, Fundo de Apoio à Comunidade Científica, Programa Operacional Ciência, Tecnologia, Inovação do Quadro Comunitário de Apoio III, Fundação Calouste Gulbenkian, Grupo Teórico de Altas Energias, Centro de Física Teórica da Universidade of Coimbra, Faculdade de Ciências e Tecnologia da Universidade of Coimbra, Coimbra Tourism Office, POCTI Grants FIS/-35304/2000 and FIS/35308/2000, and CERN Grants P/FIS/40119/2000 and FIS/43697/-2001.

The valuable secretarial assistance of Miss Sandra Correia and Mrs. Natália Cristina Silva is gratefully acknowledged.

The Editors
January 10, 2003

Organizing Committee

A.H. Blin (Coimbra)
B. Hiller (Coimbra)
A.A. Osipov (Coimbra and JINR, Dubna)
M.C. Ruivo (Coimbra)
E. van Beveren (Coimbra)

Secretaries

Sandra Correia
Natália Cristina Silva

EFFECTIVE FIELD THEORY METHODS AND APPLICATIONS

Mesons at Large N_C

A. Pich

Departament de Física Teòrica, IFIC, Universitat de València – CSIC
Edifici d'Instituts de Paterna, Apt. 22085, E-46071 València, Spain

Abstract. The $N_C \to \infty$ limit of QCD gives a useful approximation scheme to the physical hadronic world. A brief overview of the mesonic sector is presented. The large–N_C constraints on the low-energy couplings of Chiral Perturbation Theory are summarized and the role of unitarity corrections is discussed.[1]

INTRODUCTION

The limit of an infinite number of quark colours turns out to be a very useful starting point to understand many features of the strong interaction [2, 3]. The $SU(N_C)$ gauge theory simplifies considerably at $N_C \to \infty$, while keeping the most essential properties of QCD. Choosing the coupling constant g_s to be of $O\left(1/\sqrt{N_C}\right)$, *i.e.*, taking the large–N_C limit with $\alpha_s N_C$ fixed, there exists a systematic expansion in powers of $1/N_C$, which for $N_C = 3$ provides a good quantitative approximation scheme to the hadronic world [4]. The combinatorics of Feynman diagrams at large N_C results in simple counting rules, which characterize the $1/N_C$ expansion:

1. Dominance of planar diagrams with an arbitrary number of gluon exchanges (and a single quark loop at the edge for matrix elements of quark bilinears).
2. Non-planar diagrams are suppressed by factors of $1/N_C^2$.
3. Internal quark loops are suppressed by factors of $1/N_C$.

The summation of the leading planar diagrams is a very formidable task, which we are still unable to perform. Nevertheless, making the very plausible assumption that colour confinement persists at $N_C \to \infty$, a very successful picture of the meson world emerges.

Let us consider a generic n-point function of local quark bilinears $J = \bar{q}\Gamma q$:

$$\langle T\left(J_1 \cdots J_n\right)\rangle \sim O(N_C). \tag{1}$$

A simple diagrammatic analysis shows that at large N_C the only singularities are one-meson poles [3]. For instance, the two-point function takes the form:

$$\langle J(k)J(-k)\rangle = \sum_n \frac{f_n^2}{k^2 - M_n^2}. \tag{2}$$

---

[1] An expanded discussion with applications to electroweak interactions can be found in ref. [1]

CP660, *Hadron Physics: Effective Theories of Low Energy QCD*, edited by A. H. Blin et al.
© 2003 American Institute of Physics 0-7354-0120-9/03/$20.00

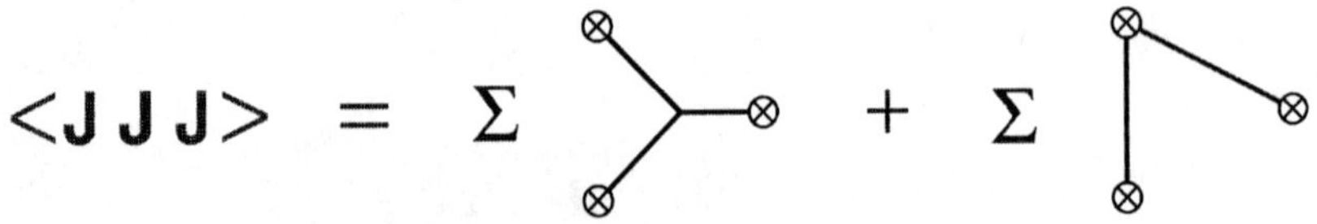

$$\langle J\,J\,J\rangle \;=\; \Sigma \quad \otimes + \quad \Sigma$$

FIGURE 1. 3-point function at large N_C

Thus:

i) $f_n = \langle 0|J|n\rangle \sim O(\sqrt{N_C})$ and $M_n \sim O(1)$.

ii) There are an infinite number of meson states, since $\langle J(k)J(-k)\rangle$ behaves logarithmically for large k^2.

iii) Mesons are free, stable and non-interacting.

At $N_C \to \infty$, the n-point functions are given by sums of tree diagrams with free meson propagators and effective local interaction vertices among m mesons, which scale as $V_m \sim O(N_C^{1-m/2})$. Moreover, $\langle 0|J|M_1 \cdots M_m\rangle \sim O(N_C^{1-m/2})$. Each additional meson coupled to the current J or to an interaction vertex brings then a suppression factor $1/\sqrt{N_C}$.

Including gauge-invariant gluon operators, such as $J_G = \mathrm{Tr}\left(G_{\mu\nu}G^{\mu\nu}\right)$, the diagrammatic analysis can be easily extended to glue states [3]. Since $\left\langle T\left(J_{G_1} \cdots J_{G_n}\right)\right\rangle \sim O(N_C^2)$, one derives the large–$N_C$ counting rules $\langle 0|J_G|G_1 \cdots G_m\rangle \sim O(N_C^{2-m})$ and $V[G_1, \cdots, G_m] \sim O(N_C^{2-m})$. Thus, at $N_C \to \infty$, glueballs are also free, stable, non-interacting and infinite in number. From the mixed correlators $\left\langle T\left(J_1 \cdots J_n J_{G_1} \cdots J_{G_m}\right)\right\rangle \sim O(N_C)$, one gets $V[M_1, \cdots, M_p; G_1, \cdots, G_q] \sim O(N_C^{1-q-p/2})$. Therefore, glueballs and mesons decouple at large N_C, their mixing being suppressed by a factor $1/\sqrt{N_C}$.

Many known phenomenological features of the hadronic world are easily understood at lowest order in the $1/N_C$ expansion: suppression of the $\bar{q}q$ sea (exotics), quark model spectroscopy, Zweig's rule, light SU(3) meson nonets, narrow resonances, multiparticle decays dominated by resonant two-body final states, etc. In some cases, the large–N_C limit is in fact the only known theoretical explanation that is sufficiently general. Clearly, the expansion in powers of $1/N_C$ appears to be a sensible physical approximation at $N_C = 3$.

The large–N_C limit provides a weak coupling regime to perform quantitative QCD studies. At leading order in $1/N_C$, the scattering amplitudes are given by sums of tree diagrams with physical hadrons exchanged. Crossing and unitarity imply that this sum is the tree approximation to some local effective Lagrangian. Higher-order corrections correspond to hadronic loop diagrams.

CHIRAL SYMMETRY

With n_f massless quark flavours, the QCD Lagrangian $[\bar{q} = (\bar{u}, \bar{d}, \dots)]$

$$\mathscr{L}^0_{QCD} = -\frac{1}{4} G^a_{\mu\nu} G^{\mu\nu}_a + i\bar{q}_L \gamma^\mu D_\mu q_L + i\bar{q}_R \gamma^\mu D_\mu q_R \qquad (3)$$

is invariant under global $U(n_f)_L \otimes U(n_f)_R$ transformations of the left- and right-handed quarks in flavour space: $q_{L,R} \to g_{L,R} \, q_{L,R}$, $g_{L,R} \in U(n_f)_{L,R}$. Under very general assumptions it has been shown that, at $N_C \to \infty$, the symmetry group must spontaneously break down to the diagonal $U(n_f)_{L+R}$ [5]. According to Goldstone's theorem [6], n_f^2 pseudoscalar massless bosons appear in the theory, which for $n_f = 3$ can be identified with the $U(3)$ multiplet

$$\Phi = \begin{pmatrix} \frac{1}{\sqrt{2}}\pi^0 + \frac{1}{\sqrt{6}}\eta_8 + \frac{1}{\sqrt{3}}\eta_1 & \pi^+ & K^+ \\ \pi^- & -\frac{1}{\sqrt{2}}\pi^0 + \frac{1}{\sqrt{6}}\eta_8 + \frac{1}{\sqrt{3}}\eta_1 & K^0 \\ K^- & \bar{K}^0 & -\frac{2}{\sqrt{6}}\eta_8 + \frac{1}{\sqrt{3}}\eta_1 \end{pmatrix}. \qquad (4)$$

The unitary matrix

$$U(\phi) = u(\phi)^2 = \exp\left\{ i\sqrt{2}\Phi/f \right\} \qquad (5)$$

gives a very convenient parameterization of the Goldstone fields. Under the chiral group it transforms as $U(\phi) \to g_R U(\phi) g_L^\dagger$.

The Goldstone nature of the pseudoscalar mesons implies strong constraints on their interactions, which can be most easily analyzed on the basis of an effective Lagrangian [7]. Since there is a mass gap separating the pseudoscalar nonet from the rest of the hadronic spectrum, we can build an effective field theory [8] (EFT) containing only the Goldstone modes. Moreover, the low-energy effective Lagrangian can be organized in terms of increasing powers of momenta (derivatives).

Let us consider an extended QCD Lagrangian, with quark couplings to external Hermitian matrix-valued fields l_μ, r_μ, s, p:

$$\mathscr{L}_{QCD} = \mathscr{L}^0_{QCD} + \bar{q}_L \gamma^\mu l_\mu q_L + \bar{q}_R \gamma^\mu r_\mu q_R - \bar{q}_L(s - ip)q_R - \bar{q}_R(s + ip)q_L. \qquad (6)$$

The external fields can be used to incorporate the electromagnetic and semileptonic weak interactions, and the explicit breaking of chiral symmetry through the quark masses:

$$s = \mathscr{M} + \dots, \qquad \mathscr{M} = \mathrm{diag}(m_u, m_d, m_s). \qquad (7)$$

At lowest order in derivatives and quark masses, the most general effective Lagrangian consistent with chiral symmetry has the form [9]:

$$\mathscr{L}_2 = \frac{f^2}{4} \langle D_\mu U^\dagger D^\mu U + U^\dagger \chi + \chi^\dagger U \rangle, \qquad \chi \equiv 2B_0(s + ip), \qquad (8)$$

where $D_\mu U = \partial_\mu U - ir_\mu U + iU l_\mu$, $\langle A \rangle$ denotes the flavour trace of the matrix A and B_0 is a constant, which, like f, is not fixed by symmetry requirements alone. Taking functional derivatives with respect to the appropriate external fields, one finds that f equals the pion decay constant (at lowest order) $f = f_\pi = 92.4$ MeV, while B_0 is related to the quark condensate:

$$B_0 = -\frac{\langle \bar{q}q \rangle}{f^2} = \frac{M_\pi^2}{m_u + m_d} = \frac{M_{K^0}^2}{m_s + m_d} = \frac{M_{K^\pm}^2}{m_s + m_u}. \tag{9}$$

Formally, the chiral Lagrangian could be computed (non-perturbatively) from the QCD generating functional. The leading-order terms in $1/N_C$ should be of $O(N_C)$, like the corresponding correlation functions of fermion bilinears. Moreover, they should have a single flavour trace since diagrams with n quark loops have n flavour traces and are of $O(N_C^{2-n})$. The Lagrangian $\mathscr{L}_2$ obeys the correct N_C counting rules: $f^2 \sim O(N_C)$, $B_0 \sim M_\phi^2 \sim U(\phi) \sim O(1)$. The $U(\phi)$ matrix generates an expansion in powers of ϕ/f, giving the required $1/\sqrt{N_C}$ suppression for each additional meson field. Clearly, interaction vertices with n mesons scale as $V_n \sim f^{2-n} \sim O(N_C^{1-n/2})$. Since $\mathscr{L}_2$ has an overall factor of N_C and U is N_C-independent, the $1/N_C$ expansion is equivalent to a semiclassical expansion. Quantum corrections computed with the chiral Lagrangian will have a $1/N_C$ suppression for each loop.

At $O(p^4)$, the conventional $SU(3)_L \otimes SU(3)_R$-invariant chiral Lagrangian is usually written as [9]:

$$
\begin{aligned}
\mathscr{L}_4 = \ & L_1 \langle D_\mu U^\dagger D^\mu U \rangle^2 + L_2 \langle D_\mu U^\dagger D_\nu U \rangle \langle D^\mu U^\dagger D^\nu U \rangle \\
& + L_3 \langle D_\mu U^\dagger D^\mu U D_\nu U^\dagger D^\nu U \rangle + L_4 \langle D_\mu U^\dagger D^\mu U \rangle \langle U^\dagger \chi + \chi^\dagger U \rangle \\
& + L_5 \langle D_\mu U^\dagger D^\mu U \left(U^\dagger \chi + \chi^\dagger U \right) \rangle + L_6 \langle U^\dagger \chi + \chi^\dagger U \rangle^2 \\
& + L_7 \langle U^\dagger \chi - \chi^\dagger U \rangle^2 + L_8 \langle \chi^\dagger U \chi^\dagger U + U^\dagger \chi U^\dagger \chi \rangle \\
& - i L_9 \langle F_R^{\mu\nu} D_\mu U D_\nu U^\dagger + F_L^{\mu\nu} D_\mu U^\dagger D_\nu U \rangle + L_{10} \langle U^\dagger F_R^{\mu\nu} U F_{L\mu\nu} \rangle,
\end{aligned}
\tag{10}
$$

where $F_{L,R}^{\mu\nu}$ are field-strength tensors of the l^μ and r^μ flavour fields.

Thus, at $O(p^4)$ we need ten additional coupling constants L_i to determine the low-energy behaviour of the Green functions. Terms with a single trace are of $O(N_C)$, while those with two traces should be of $O(1)$. However, a 3×3 matrix relation has been used to eliminate the additional structure $c \langle D_\mu U^\dagger D_\nu U D^\mu U^\dagger D^\nu U \rangle$ with the result $2\delta L_1 = \delta L_2 = -\frac{1}{2}\delta L_3 = c \sim O(N_C)$. As shown in Table 1, the phenomenologically determined values [10, 11] of those couplings follow the pattern suggested by the $1/N_C$ counting rules. Moreover, their average order of magnitude, $L_i \sim f^2/(4\Lambda_\chi^2) \sim 2 \times 10^{-3}$, suggests a chiral symmetry-breaking scale $\Lambda_\chi \sim 1$ GeV.

One-loop graphs with the lowest-order Lagrangian $\mathscr{L}_2$ contribute also at $O(p^4)$ in the chiral expansion, but they are suppressed by a factor of $1/N_C$. Their divergent parts are renormalized by the $\mathscr{L}_4$ couplings:

$$L_i = L_i^r(\mu) + \Gamma_i \frac{\mu^{D-4}}{32\pi^2} \left\{ \frac{2}{D-4} + \gamma_E - \log(4\pi) - 1 \right\}. \tag{11}$$

TABLE 1. Phenomenological values of the renormalized couplings $L_i^r(M_\rho)$ in units of 10^{-3}. The fourth column shows the source used to get this information. The large-N_C predictions obtained within the single-resonance approximation are given in the last column

i	$L_i^r(M_\rho)$	$O(N_C)$	Source	Γ_i	$L_i^{N_C \to \infty}$
$2L_1 - L_2$	-0.6 ± 0.6	$O(1)$	K_{e4}, $\pi\pi \to \pi\pi$	$3/16$	0
L_2	1.4 ± 0.3	$O(N_C)$	K_{e4}, $\pi\pi \to \pi\pi$	$3/16$	1.8
L_3	-3.5 ± 1.1	$O(N_C)$	K_{e4}, $\pi\pi \to \pi\pi$	0	-4.3
L_4	-0.3 ± 0.5	$O(1)$	Zweig rule	$1/8$	0
L_5	1.4 ± 0.5	$O(N_C)$	$F_K : F_\pi$	$3/8$	2.1
L_6	-0.2 ± 0.3	$O(1)$	Zweig rule	$11/144$	0
L_7	-0.4 ± 0.2	$O(1)$	GMO, L_5, L_8	0	-0.3
L_8	0.9 ± 0.3	$O(N_C)$	M_ϕ, L_5	$5/48$	0.8
L_9	6.9 ± 0.7	$O(N_C)$	$\langle r^2 \rangle_V^\pi$	$1/4$	7.1
L_{10}	-5.5 ± 0.7	$O(N_C)$	$\pi \to e\nu\gamma$	$-1/4$	-5.4

This introduces a renormalization scale dependence,

$$L_i^r(\mu_2) = L_i^r(\mu_1) + \frac{\Gamma_i}{(4\pi)^2} \log\left(\frac{\mu_1}{\mu_2}\right), \tag{12}$$

which is subleading in $1/N_C$. The phenomenological couplings given in Table 1 have been normalized at $\mu = M_\rho$.

The chiral loops generate non-polynomial contributions, with logarithms and threshold factors as required by unitarity, which are completely predicted as functions of f and the Goldstone masses. Although they are suppressed by a factor of $1/N_C$, the chiral logarithms can be numerically important since $\frac{1}{N_C} \log(\Lambda_\chi^2/M_\pi^2) \sim 4/3$.

Anomalies

Since chiral symmetry is explicitly violated by fermion anomalies at the fundamental QCD level [12], we need to add a functional $Z_\mathscr{A}$ with the property that its change under chiral transformations reproduces the anomalous change of the QCD generating functional. For the non-Abelian anomalies associated with the external sources l_μ and r_μ, such a functional was first constructed by Wess and Zumino [13], and reformulated in a nice geometrical way by Witten [14]. It is an $O(p^4)$ effect, which is completely calculable with no free parameters. This contribution is of $O(N_C)$, because it is generated by a triangle quark loop coupled to external sources.

Much more subtle is the $U(1)_A$ gluonic anomaly which breaks the conservation of the singlet axial quark current in the chiral limit:

$$\partial_\mu(\bar{q}\gamma^\mu\gamma_5 q) = 2n_f\omega \quad ; \quad \omega = \frac{\alpha_s}{16\pi}\varepsilon^{\mu\nu\rho\sigma}G_{\mu\nu}G_{\rho\sigma}. \tag{13}$$

The corresponding anomalous change of the QCD generating functional can be accounted for by adding a term $\Delta\mathscr{L}_{QCD} = -\theta\,\omega$ with the appropriate chiral transformation for the so-called vacuum angle $\theta(x)$ [15]. Notice that in the large–N_C limit the $U(1)_A$ anomaly is absent [16].

To lowest non-trivial order in $1/N_C$, the chiral symmetry breaking effect induced by the $U(1)_A$ anomaly can be taken into account in the effective low-energy theory, through the term [17]

$$\mathscr{L}_{U(1)_A} = -\frac{f^2}{4}\frac{a}{N_C}\left\{\theta - \frac{i}{2}\left[\log\left(\det U\right) - \log\left(\det U^\dagger\right)\right]\right\}^2, \tag{14}$$

which breaks $U(3)_L \otimes U(3)_R$ but preserves $SU(3)_L \otimes SU(3)_R \otimes U(1)_V$.

The parameter a has dimensions of mass squared and, with the factor $1/N_C$ pulled out, is booked to be of $O(1)$ in the large–N_C counting rules. Its value is not fixed by symmetry requirements alone; it depends crucially on the dynamics of instantons. In the presence of the term (14), the η_1 field becomes massive even in the chiral limit:

$$M_{\eta_1}^2 = 3\frac{a}{N_C} + O(\mathscr{M}). \tag{15}$$

Owing to the large mass of the η', the effect of the $U(1)_A$ anomaly cannot be treated as a small perturbation. Rather, one should keep the term (14) together with the lowest-order Lagrangian (8). It is possible to build a consistent combined expansion in powers of momenta, quark masses and $1/N_C$, by counting the relative magnitude of these parameters as [18]:

$$O(p^2) \sim O(\mathscr{M}) \sim O(1/N_C). \tag{16}$$

This expansion has been already analyzed at the next-to-leading order [15, 19].

RESONANCE CHIRAL THEORY

Let us consider a chiral-invariant Lagrangian $\mathscr{L}(U,V,A,S,P)$, describing the couplings of resonance nonet multiplets of the type $V(1^{--})$, $A(1^{++})$, $S(0^{++})$ and $P(0^{-+})$ to the Goldstone bosons [20]:

$$
\begin{aligned}
\mathscr{L}_2[V(1^{--})] &= \sum_i \left\{ \frac{F_{V_i}}{2\sqrt{2}}\langle V_i^{\mu\nu}f_{+\mu\nu}\rangle + \frac{iG_{V_i}}{\sqrt{2}}\langle V_i^{\mu\nu}u_\mu u_\nu\rangle \right\}, \\
\mathscr{L}_2[A(1^{++})] &= \sum_i \frac{F_{A_i}}{2\sqrt{2}}\langle A_i^{\mu\nu}f_{-\mu\nu}\rangle, \\
\mathscr{L}_2[S(0^{++})] &= \sum_i \left\{ c_{d_i}\langle S_i u^\mu u_\mu\rangle + c_{m_i}\langle S_i\chi_+\rangle \right\}, \\
\mathscr{L}_2[P(0^{-+})] &= \sum_i id_{m_i}\langle P_i\chi_-\rangle,
\end{aligned} \tag{17}
$$

where $u_\mu \equiv iu^\dagger D_\mu U u^\dagger$, $f_\pm^{\mu\nu} \equiv uF_L^{\mu\nu}u^\dagger \pm u^\dagger F_R^{\mu\nu}u$ and $\chi_\pm \equiv u^\dagger\chi u^\dagger \pm u\chi^\dagger u$. The resonance couplings $F_{V_i}, G_{V_i}, F_{A_i}, c_{d_i}, c_{m_i}$ and d_{m_i} are of $O\left(\sqrt{N_C}\right)$.

The lightest resonances have an important impact on the low-energy dynamics of the pseudoscalar bosons. Below the resonance mass scale, the singularity associated with the pole of a resonance propagator is replaced by the corresponding momentum expansion; therefore, the exchange of virtual resonances generates derivative Goldstone couplings proportional to powers of $1/M_R^2$. At lowest order in derivatives, this gives the large–N_C predictions for the $O(p^4)$ couplings of chiral perturbation theory (χPT) [20]:

$$2L_1 = L_2 = \sum_i \frac{G_{V_i}^2}{4M_{V_i}^2}, \qquad L_3 = \sum_i \left\{ -\frac{3G_{V_i}^2}{4M_{V_i}^2} + \frac{c_{d_i}^2}{2M_{S_i}^2} \right\},$$

$$L_5 = \sum_i \frac{c_{d_i} c_{m_i}}{M_{S_i}^2}, \qquad L_8 = \sum_i \left\{ \frac{c_{m_i}^2}{2M_{S_i}^2} - \frac{d_{m_i}^2}{2M_{P_i}^2} \right\}, \qquad (18)$$

$$L_9 = \sum_i \frac{F_{V_i} G_{V_i}}{2M_{V_i}^2}, \qquad L_{10} = \sum_i \left\{ \frac{F_{A_i}^2}{4M_{A_i}^2} - \frac{F_{V_i}^2}{4M_{V_i}^2} \right\}.$$

All these couplings are of $O(N_C)$, in agreement with the counting made in Table 1, while for the couplings of $O(1)$ we get $2L_1 - L_2 = L_4 = L_6 = L_7 = 0$.

Owing to the $U(1)_A$ anomaly, the η_1 field is massive and it is often integrated out from the low-energy chiral theory. In that case, the $SU(3)_L \otimes SU(3)_R$ chiral coupling L_7 gets a contribution from η_1 exchange [9, 20]:

$$L_7 = -\frac{\tilde{d}_m^2}{2M_{\eta_1}^2}, \qquad \tilde{d}_m = -\frac{f}{\sqrt{24}}. \qquad (19)$$

Since, $M_{\eta_1}^2 \sim O\left(1/N_C, \mathcal{M}\right)$, the coupling L_7 could then [9] be considered of $O(N_C^2)$. However, the large–N_C counting is no longer consistent if one takes the limit of a heavy η_1 mass (N_C small) while keeping m_s small [21].

Short-Distance Constraints

The short-distance properties of the underlying QCD dynamics impose some constraints on the low-energy EFT parameters [22]:

1. *Vector Form Factor.* At leading order in $1/N_C$, the two-Goldstone matrix element of the vector current, is characterized by

$$F_V(t) = 1 + \sum_i \frac{F_{V_i} G_{V_i}}{f^2} \frac{t}{M_{V_i}^2 - t}. \qquad (20)$$

Since the vector form factor $F_V(t)$ should vanish at infinite momentum transfer t, the resonance couplings should satisfy

$$\sum_i F_{V_i} G_{V_i} = f^2. \qquad (21)$$

2. *Axial Form Factor.* The matrix element of the axial current between one Goldstone and one photon is parameterized by the axial form factor. From the resonance Lagrangian (17), one gets

$$G_A(t) = \sum_i \left\{ \frac{2F_{V_i} G_{V_i} - F_{V_i}^2}{M_{V_i}^2} + \frac{F_{A_i}^2}{M_{A_i}^2 - t} \right\}, \tag{22}$$

which vanishes at $t \to \infty$ provided that

$$\sum_i \frac{2F_{V_i} G_{V_i} - F_{V_i}^2}{M_{V_i}^2} = 0. \tag{23}$$

3. *Weinberg Sum Rules.* The two-point function built from a left-handed and a right-handed vector quark current defines the correlator

$$\Pi_{LR}(t) = \frac{f^2}{t} + \sum_i \frac{F_{V_i}^2}{M_{V_i}^2 - t} - \sum_i \frac{F_{A_i}^2}{M_{A_i}^2 - t}. \tag{24}$$

Since gluonic interactions preserve chirality, $\Pi_{LR}(t)$ satifies an unsubtracted dispersion relation. Moreover, [23] in the chiral limit it vanishes faster than $1/t^2$ when $t \to \infty$. This implies the well-known conditions [24]:

$$\sum_i \left(F_{V_i}^2 - F_{A_i}^2 \right) = f^2, \qquad \sum_i \left(M_{V_i}^2 F_{V_i}^2 - M_{A_i}^2 F_{A_i}^2 \right) = 0. \tag{25}$$

The second relation is correct up to very small quark-mass contributions.

4. *Scalar Form Factor.* The two-pseudoscalar matrix element of the scalar quark current contains another dynamical form factor, which for the $K\pi$ case takes the form [25]:

$$F_{K\pi}^S(t) = 1 + \sum_i \frac{4c_{m_i}}{f^2} \left\{ c_{d_i} + \left(c_{m_i} - c_{d_i} \right) \frac{M_K^2 - M_\pi^2}{M_{S_i}^2} \right\} \frac{t}{M_{S_i}^2 - t}. \tag{26}$$

Requiring $F^S(t)$ to vanish at $t \to \infty$, one gets the constraints [25]:

$$4 \sum_i c_{d_i} c_{m_i} = f^2, \qquad \sum_i \frac{c_{m_i}}{M_{S_i}^2} \left(c_{m_i} - c_{d_i} \right) = 0. \tag{27}$$

5. *SS − PP Sum Rules.* The two-point correlation functions of two scalar or two pseudoscalar currents would be equal if chirality was absolutely preserved. Their difference is easily computed in the hadronic EFT:

$$\Pi_{SS-PP}(t) = 16B_0^2 \left\{ \sum_i \frac{c_{m_i}^2}{M_{S_i}^2 - t} - \sum_i \frac{d_{m_i}^2}{M_{P_i}^2 - t} + \frac{f^2}{8t} \right\}. \tag{28}$$

For massless quarks, $\Pi_{SS-PP}(t)$ vanishes as $1/t^2$ when $t \to \infty$, with a coefficient proportional to $\alpha_s \langle \bar{q}\Gamma q \bar{q}\Gamma q \rangle$ [26]. The vacuum four-quark condensate provides a non-perturbative breaking of chiral symmetry. In the large–N_C limit, it factorizes as $\alpha_s \langle \bar{q}q \rangle^2 \sim \alpha_s B_0^2$. Imposing this behaviour on (28), one gets [27]:

$$8 \sum_i \left(c_{m_i}^2 - d_{m_i}^2 \right) = f^2, \qquad \sum_i \left(c_{m_i}^2 M_{S_i}^2 - d_{m_i}^2 M_{P_i}^2 \right) = \frac{3\pi\alpha_s}{4} f^4. \qquad (29)$$

Single-Resonance Approximation

Let us approximate each infinite resonance sum with the contribution from the first meson nonet with the given quantum numbers. This is meaningful at low energies where the contributions from higher-mass states are suppressed by their corresponding propagators. The single-resonance approximation (SRA) corresponds to work with a low-energy EFT below the scale of the second resonance multiplets. The resulting short-distance constraints are nothing else than the matching conditions between this EFT and the underlying QCD dynamics. Thus, we are assuming that the short-distance operator product expansion provides an acceptable description at energies above 1.5 GeV.

Within the SRA, Eqs. (21), (23) and (25) determine the vector and axial-vector couplings in terms of M_V and f [22]:

$$F_V = 2G_V = \sqrt{2}F_A = \sqrt{2}f, \qquad M_A = \sqrt{2}M_V. \qquad (30)$$

The scalar [25] and pseudoscalar parameters are obtained from (27) and (29):

$$c_m = c_d = \sqrt{2}d_m = f/2, \qquad M_P = \sqrt{2}M_S (1-\delta)^{1/2}. \qquad (31)$$

The last relation involves a small correction $\delta \approx 3\pi\alpha_s f^2/M_S^2 \sim 0.08\,\alpha_s$, which we can neglect together with the tiny effects from light quark masses.

Inserting these predictions into Eqs. (18), one finally gets all $O(N_C p^4)$ χPT couplings, in terms of M_V, M_S and f:

$$2L_1 = L_2 = \frac{1}{4}L_9 = -\frac{1}{3}L_{10} = \frac{f^2}{8M_V^2}, \qquad (32)$$

$$L_3 = -\frac{3f^2}{8M_V^2} + \frac{f^2}{8M_S^2}, \qquad L_5 = \frac{f^2}{4M_S^2}, \qquad L_8 = \frac{3f^2}{32M_S^2}. \qquad (33)$$

The last column in Table 1 shows the results obtained with $M_V = 0.77$ GeV, $M_S = 1.0$ GeV and $f = 92$ MeV. Also shown is the L_7 prediction in (19), taking $M_{\eta_1} = 0.80$ GeV. The agreement with the measured values is a clear success of the large–N_C approximation. It demonstrates that the lightest resonance multiplets give indeed the dominant effects at low energies.

The study of other Green functions provides further matching conditions between the hadronic and fundamental QCD descriptions. Clearly, it is not possible to satisfy

11

all of them within the SRA. A useful generalization is the so-called *Minimal Hadronic Ansatz*, which consists of keeping the minimum number of resonances compatible with all known short-distance constraints for the problem at hand [28]. Some $O(p^6)$ χPT couplings have been already analyzed in this way, by studying an appropriate set of three-point functions [29].

UNITARITY CORRECTIONS

The χPT loops incorporate the unitarity field theory constraints in a perturbative way, order by order in the chiral expansion. Although subleading in the $1/N_C$ counting, these corrections may be enhanced by infrared logarithms. Their effect appears to be crucial for a correct understanding of some observables, in particular in the scalar sector, because the S–wave rescattering of two pseudoscalars is very strong. The combined constraints of analyticity and unitarity make possible to perform appropriate resummations of chiral logarithms, which describe the leading $1/N_C$ corrections in the resonance region.

A simple example is provided by the Omnès [30] exponentiation of the pion form factor [31]:

$$F_V(t) = \frac{M_V^2}{M_V^2 - t} \exp\left\{-\frac{t}{96\pi^2 f^2} A^{(\pi)}(t)\right\}, \tag{34}$$

where $[\sigma_\pi \equiv \sqrt{1 - 4M_\pi^2/t}]$

$$A^{(\pi)}(t) \equiv \sigma_\pi^3 \log\left(\frac{\sigma_\pi + 1}{\sigma_\pi - 1}\right) + \log\left(\frac{M_\pi^2}{\mu^2}\right) + 8\frac{M_\pi^2}{t} - \frac{5}{3} - 128\pi^2 \delta L_9^r(\mu), \tag{35}$$

is the regularized one-loop function describing two intermediate pions (the small $K\bar{K}$ loop contribution has been neglected), which arises here from an integration over the $I = J = 1$ $\pi\pi$ phase shift at leading order in χPT,

$$F_V(t) = Q_n(t) \exp\left\{\frac{s^n}{\pi} \int_{4M_\pi^2}^{\infty} \frac{dz}{z^n} \frac{\delta_1^1(z)}{z - t - i\varepsilon}\right\}. \tag{36}$$

This expression is valid in the elastic region and has a polynomic ambiguity which is compensated by the subtraction function $Q_n(t)$. Only the logarithmic corrections are unambiguous. The ambiguity has been solved by matching the Omnès solution both to the χPT and large–N_C (SRA) results. There remains a local indetermination at higher orders, made explicit through the constant $\delta L_9^r(\mu) \equiv L_9^r(\mu) - L_9^{N_c \to \infty}$, which is next-to-leading in $1/N_C$ and does not contain any large infrared logarithm when $\mu \sim M_V$.

Equation (34) has obvious shortcomings. We have used an $O(p^2)$ approximation to the $\pi\pi$ phase shift, $\delta_1^1(t) = t \sigma_\pi^3/(96\pi f_\pi^2)$, which is a very poor (and even wrong) description at the higher end of the dispersive integration region. Nevertheless, one can always take a sufficient number of subtractions to emphasize numerically the low-energy region. Since our matching has fixed an infinite number of subtractions, this

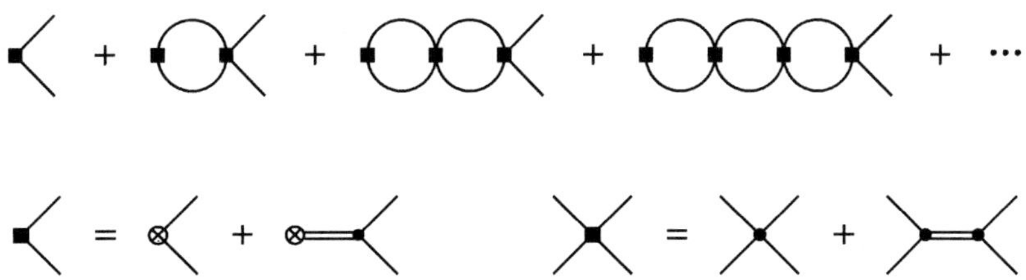

FIGURE 2. Dyson-Schwinger resummation of $F_V(t)$ with effective vertices. The cross indicates a vector-current insertion and the double line is a resonance propagator

result should give a good approximation for values of s not too large. Moreover, this can be phenomenologically improved with the use of the measured phase shifts [32].

A more important question concerns the ρ meson pole, which needs a proper treatment if one aims to describe physics around or above the resonance peak. The pole is regulated by the ρ width, which vanishes at $N_C \to \infty$. The dressed propagator can be calculated through a Dyson-Schwinger resummation constructed from effective Goldstone vertices containing both the local χPT interaction and the resonance-exchange contributions [31, 32, 33]:

$$F_V(t) = \frac{M_V^2}{M_V^2 - t + \xi_\rho(t) - i M_V \Gamma_\rho(t)}, \tag{37}$$

where

$$\xi_\rho(t) - i M_V \Gamma_\rho(t) = \frac{t M_V^2}{96 \pi^2 f^2} A^{(\pi)}(t). \tag{38}$$

Thus,

$$\Gamma_\rho(t) = \theta(t - 4M_\pi^2) \frac{t M_V}{96 \pi f^2} \sigma_\pi^3, \tag{39}$$

which at $t = M_\rho^2$ gives $\Gamma_\rho(M_\rho^2) = 144$ MeV, in reasonable agreement with the measured ρ width. The intermediate $K\bar{K}$ contributions can be included through a coupled-channel resummation [33]; the only modification is the change $A^{(\pi)}(t) \to A^{(\pi)}(t) + \frac{1}{2} A^{(K)}(t)$.

Equations (34) and (37) represent different resummations of higher-order corrections. They agree, by construction, at $O(p^4)$ in χPT and at the leading order in $1/N_C$. The result can be further improved by inserting into the Omnès exponential (36) the phase shift predicted in (37),

$$\delta_1^1(t) = \arctan\left\{ \frac{M_V \Gamma_\rho(t)}{M_V^2 - t + \xi_\rho(t)} \right\} = \frac{t \sigma_\pi^3}{96 \pi f_\pi^2} + \cdots, \tag{40}$$

and imposing the appropriate matching conditions [32].

Similar unitarization procedures have been applied to amplitudes with $I = J = 0$, which get large corrections from infrared chiral logarithms [25, 34, 35].

13

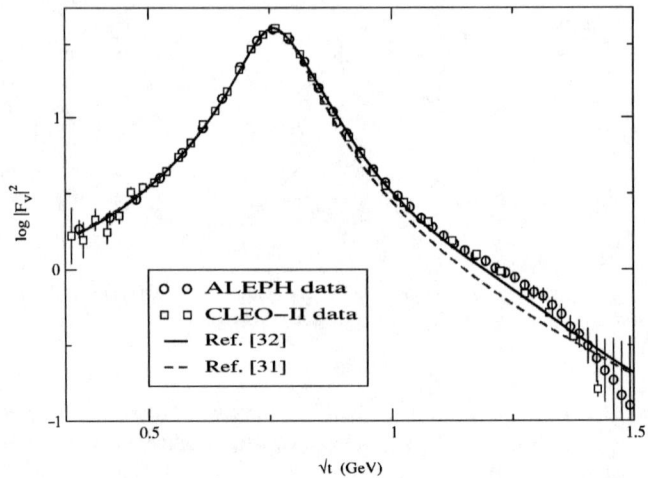

FIGURE 3. Comparison of $\tau \to \nu_\tau \pi \pi$ data with Omnès predictions for $F_V(t)$. The dashed line [31] corresponds to Eq. (34), with the term $iM_V\Gamma_\rho(t)$ shifted to the ρ propagator to regulate the pole. The continuous line is the 3-subtracted result [32], using the full phase shift (40) for $M_\rho \leq \sqrt{t}$ and $\delta_1^1(t)$ data at higher values of t

SUMMARY

The large–N_C limit provides a sensible approximation to the $N_C = 3$ hadronic world. Assuming confinement, the strong dynamics at $N_C \to \infty$ is given by tree diagrams with infinite sums of hadron exchanges, which correspond to the tree approximation to some local effective Lagrangian. Hadronic loops generate corrections suppressed by factors of $1/N_C$.

At very low energies the hadronic EFT describing the lightest pseudoscalar nonet is χPT, while resonance chiral theory provides the correct framework to incorporate the massive mesonic states. The short-distance properties of QCD at large N_C imply strong constraints on the chiral couplings, which result in a significative reduction on the number of free parameters. A very succesful prediction of the $O(p^4)$ χPT couplings is obtained, under the very reasonnable assumption that the lightest resonance multiplets give the dominant effects at low energies.

The expansion in powers of $1/N_C$ turns out to be a very useful tool for quantitative non-perturbative analyses. While there is a very successful leading-order phenomenology, some important physical effects only appear at subleading topologies: the $U(1)_A$ anomaly, the infrared χPT logarithms, the resonance widths, etc. Those effects can be rigorously analyzed with appropriate quantum field theory tools.

In recent years, the large–N_C techniques have provided a deeper understanding of strangeness–changing weak transitions [1]. Since weak currents factorize at large–N_C,

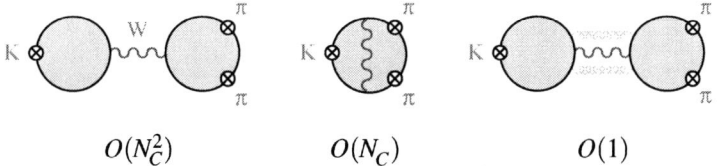

$$O(N_C^2) \qquad\qquad O(N_C) \qquad\qquad O(1)$$

FIGURE 4. Diagrammatic topologies contributing to $K \to \pi\pi$

a naive $1/N_C$ description of $K \to \pi\pi$ would imply $A(K^0 \to \pi^0\pi^0) = 0$. In terms of isospin amplitudes, $A_0 = \sqrt{2}A_2$; *i.e.*, there is no $\Delta I = 1/2$ enhancement at leading order in $1/N_C$. Owing to the presence of very different mass scales ($M_\pi < M_K < \Lambda_\chi \ll M_W$), the $1/N_C$ corrections induced by the strong interactions are amplified by large logarithms. These logarithmic corrections can be correctly implemented with the help of the renormalization-group at short distances and χPT techniques, supplemented with unitarity, at long distances. One gets in this way the right phenomenological pattern of kaon decays [35, 36, 37, 38].

A detailed $1/N_C$ analysis of the CP-violating ratio ε'/ε, taking into account all large logarithmic corrections, has been performed recently. The Standard Model prediction has been found to be [38]

$$\mathrm{Re}\left(\varepsilon'/\varepsilon\right) = \left(1.7 \pm 0.2^{+0.8}_{-0.5} \pm 0.5\right) \cdot 10^{-3} = (1.7 \pm 0.9) \cdot 10^{-3}, \qquad (41)$$

in excellent agreement with the experimental value $\mathrm{Re}\left(\varepsilon'/\varepsilon\right) = (1.66 \pm 0.16) \cdot 10^{-3}$ [39, 40, 41, 42]. The largest theoretical uncertainties originate in the numerical value of the strange quark mass and subleading $1/N_C$ corrections which are not enhanced by any logarithms. While a better determination of m_s can be expected soon, the control of these non-logarithmic corrections at the next-to-leading order in $1/N_C$ remains a challenge for future investigations [43].

ACKNOWLEDGMENTS

I would like to thank the organizers for their kind hospitality and for hosting an enjoyable workshop. This work has been partly supported by MCYT, Spain (Grant No. FPA-2001-3031), by GVA, Spain (Grant No. CTIDIB/2002/24), by EU funds for regional development and by the EU RTN *EURIDICE* (HPRN-CT-2002-00311).

REFERENCES

1. Pich, A., "Colourless Mesons in a Polychromatic World", in *The Phenomenology of Large–N_C QCD* (Tempe, Arizona, 9-11 January 2002), edited by R. Lebed, World Scientific, Singapore, 2002 [hep-ph/0205030].
2. 't Hooft, G., *Nucl. Phys. B* **72**, 461 (1974); **75**, 461 (1974).
3. Witten, E., *Nucl. Phys. B* **160**, 57 (1979).

4. Manohar, A.V., "Large N_C QCD", in *Probing the Standard Model of Particle Interactions*, Les Houches LXVIII (1997), edited by R. Gupta et al., Elsevier, Amsterdam, 1999, Vol. II, p. 1091-1169 [hep-ph/9802419].
5. Coleman, S., and Witten, E., *Phys. Rev. Lett.* **45**, 100 (1980).
6. Goldstone, J., *Nuovo Cimento* **19**, 154 (1961).
7. Weinberg, S., *Physica A* **96**, 327 (1979).
8. Pich, A., "Effective Field Theory", in *Probing the Standard Model of Particle Interactions*, Les Houches LXVIII (1997), edited by R. Gupta et al., Elsevier, Amsterdam, 1999, Vol. II, p. 949-1049 [hep-ph/9806303].
9. Gasser, J., and Leutwyler, H., *Nucl. Phys. B* **250**, 465, 517, 539 (1985).
10. Ecker, G., *Prog. Part. Nucl. Phys.* **35**, 1 (1995).
11. Pich, A., *Rep. Prog. Phys.* **58**, 563 (1995).
12. Adler, S.L., *Phys. Rev.* **177**, 2426 (1969); Bardeen, W.A., *Phys. Rev.* **184**, 1848 (1969); Bell, J.S., and Jackiw, R., *Nuovo Cimento A* **60**, 47 (1969).
13. Wess, J., and Zumino, B., *Phys. Lett. B* **37**, 95 (1971).
14. Witten, E., *Nucl. Phys. B* **223**, 422 (1983).
15. Kaiser, R., and Leutwyler, H., *Eur. Phys. J.* C **17**, 623 (2000).
16. Witten, E., *Nucl. Phys. B* **156**, 269 (1979).
17. Rosenzweig, C., Schechter, J., and Trahern, G., *Phys. Rev. D* **21**, 3388 (1980); Witten, E., *Ann. Phys., NY* **128**, 363 (1980); Di Vecchia, P., and Veneziano, G., *Nucl. Phys. B* **171**, 253 (1980).
18. H. Leutwyler, *Phys. Lett. B* **374**, 163 (1996).
19. Herrera-Siklódy, P., Latorre, J.I., Pascual, P., and Taron, J., *Nucl. Phys. B* **497**, 345 (1997); *Phys. Lett. B* **419**, 326 (1998).
20. Ecker, G., Gasser, J., Pich, A., and de Rafael, E., *Nucl. Phys. B* **321**, 311 (1989).
21. Peris, S., and de Rafael, E., *Phys. Lett. B* **348**, 539 (1995).
22. Ecker, G., Gasser, J., Leutwyler, H., Pich, A., and de Rafael, E., *Phys. Lett. B* **223**, 425 (1989).
23. Floratos, E.G., Narison, S., and de Rafael, E., *Nucl. Phys. B* **155**, 115 (1979); Pascual, P., and de Rafael, E., *Z. Phys. C* **12**, 127 (1982).
24. Weinberg, S., *Phys. Rev. Lett.* **18**, 507 (1967).
25. Jamin, M., Oller, J.A., and Pich, A., *Nucl. Phys. B* **622**, 279 (2002); **587**, 331 (2000).
26. Shifman, M.A., Vainstein, A.I., and Zakharov, V.I., *Nucl. Phys. B* **147**, 385, 448 (1979); Jamin, M., and Münz, M., *Z. Phys. C* **66**, 633 (1995).
27. Golterman, M., and Peris, S., *Phys. Rev. D* **61**, 034018 (2000).
28. Knecht, M., and de Rafael, E., *Phys. Lett. B* **424**, 335 (1998); Knecht, M., Peris, S., and de Rafael, E., *Phys. Lett. B* **443**, 255 (1998); Golterman, M., et al., *JHEP* **0201**, 024 (2002).
29. Knecht, M., and Nyffeler, A., *Eur. Phys. J.* C **21**, 659 (2001).
30. Omnès, R., Nuovo Cimento **8**, 316 (1958).
31. Guerrero, F., and Pich, A., *Phys. Lett. B* **412**, 382 (1997).
32. Gómez Dumm, D., Pich, A., and Portolés, J., *Phys. Rev. D* **62**, 054014 (2000); Pich, A., and Portolés, J., *Phys. Rev. D* **63**, 093005 (2001).
33. Sanz-Cillero, J.J., and Pich, A., hep-ph/0208199.
34. Oller, J.A. et al., *Phys. Rev. D* **63**, 114009 (2001); **60**, 074023 (1999); **59**, 074001 (1999); *Phys. Rev. Lett.* **80**, 3452 (1998).
35. Pallante, E., and Pich, A., *Nucl. Phys. B* **592**, 294 (2000); *Phys. Rev. Lett.* **84**, 2568 (2000).
36. Bardeen, W.A., Buras, A.J., and Gerard, J.M., *Nucl. Phys. B* **293**, 787 (1987); **264**, 371 (1986); *Phys. Lett. B* **211**, 343 (1988); **192**, 138 (1987); **180**, 133 (1986); W.A., Buras, A.J., and Gerard, J.M., *Phys. Lett. B* **192**, 156 (1987); Hambye, T., et al., *Eur. Phys. J.* C **10**, 271 (1999).
37. Pich, A., and de Rafael, E., *Nucl. Phys. B* **358**, 311 (1991); *Phys. Lett. B* **374**, 186 (1996); Jamin, M., and Pich, A., *Nucl. Phys. B* **425**, 15 (1994); Bertolini, S., et al., *Rev. Mod. Phys.* **72**, 65 (2000).
38. Pallante, E., Pich, A., and Scimemi, I., *Nucl. Phys. B* **617**, 441 (2001).
39. NA48 Collaboration, *Phys. Lett. B* **544**, 97 (2002); **465**, 335 (1999); *Eur. Phys. J.* C **22**, 231 (2001).
40. NA31 Collaboration, *Phys. Lett. B* **317**, 233 (1993).
41. KTeV Collaboration, hep-ex/0208007; *Phys. Rev. Lett.* **83**, 22 (1999).
42. E731 Collaboration, *Phys. Rev. Lett.* **70**, 1203 (1993).
43. de Rafael, E., "Analytic Approaches to Kaon Physics", hep-ph/0210317.

QCD string and chiral symmetry

A. A. Andrianov* and D. Espriu†

*St.Petersburg State University, St. Petersburg, Russia and INFN, Sez. di Bologna, Via Irnerio, 46, Bologna, Italy
†Departament d'Estructura i Constituents de la Matèria, Universitat de Barcelona, Diagonal, 647, Barcelona, Spain

Abstract. We assume that QCD can be described in a certain kinematical regime by an effective string theory. This hadronic string must couple to background chiral fields in a chirally invariant manner, taking into account the true chirally non-invariant QCD vacuum. By requiring conformal symmetry of the string and the unitarity constraint on chiral fields we reconstruct the equations of motion for the latter ones. These provide a consistent background for the propagation of the string. By further requiring locality of the effective action we recover the Lagrangian of non-linear sigma model of pion interactions. The prediction is totally unambiguous and parameter-free. The estimated chiral structural constants of Gasser and Leutwyler fit very well the phenomenological values.

INTRODUCTION

The history of attempts to describe the hadrons in the framework of a string theory within QCD or beyond encompasses already more than 30 years (see,[1]-[8] as well as the reviews [9]-[11]). Commonly cited arguments to justify the string description of QCD are the dominance of planar gluon diagrams in the large N limit[12], being interpreted as the world-sheet of a string, the expansion in terms of surfaces built out of plaquettes in strong-coupling lattice QCD[13], and the undeniable success of Regge phenomenology[10, 14].

There is a motivated agreement that in a certain kinematic regime the Nambu-Goto or, equivalently, the Polyakov string must provide a satisfactory effective description of the flux-tube linking one quark and an antiquark. Unfortunately, it has been known for a long time that the hadronic amplitudes derived from such type of strings, if taken at face value, are not physically consistent. To illuminate their flaws we recall the original Veneziano amplitude[1], which can be derived from Nambu-Goto string and supposedly describes the scattering amplitude of four pions. One can show that in this amplitude the scalar resonance is a tachyon and the vector state (which we should identify with the rho particle) is massless. Finally, such an amplitude does not have the appropriate Adler zero, i.e. the property that at $s = t = 0$ the pion scattering amplitude vanishes.

It seems very reasonable to assume that the main reason for the presence of a tachyon in the spectrum and the wrong chiral properties lies in a wrong choice of the vacuum[15]. Finding the 'correct' vacuum of string theory in four dimensions seems however a hopelessly difficult task, so we have to resort to some indirect method. Even more difficult it seems to find the 'correct' vertex operators implementing the different excitations, assuming that they exist at all. A possible way to take into account all the non-perturbative

CP660, *Hadron Physics: Effective Theories of Low Energy QCD*, edited by A. H. Blin et al.
© 2003 American Institute of Physics 0-7354-0120-9/03/$20.00

properties of the QCD vacuum and excitations was suggested in [16] and developed in [17]. Namely, one can assume that in QCD chiral symmetry breaking takes place and the massless (in the chiral limit) pseudoscalar mesons form the background of the QCD vacuum in which the string propagates. The string itself is assumed to contain all the other (massive) excitations of QCD. The massless pion fields can be collected in a unitary matrix $U(x)$ belonging to $SU(2)$ group (here we consider non-strange Goldstone mesons only). It describes excitations around the non-perturbative vacuum induced by chiral symmetry breaking. From the string point of view $U(x)$ is nothing but a collection of couplings involving the string variable $x_\mu(\tau, \sigma)$. It has to be coupled to the boundary of the string where flavor is attached. Our goal is to find a consistent string propagation in this non-perturbative background.

An essential property of string theory is conformal invariance. Since it must hold when perturbing the string around any vacuum we demand the new coupling to chiral fields, living on the boundary, to preserve it.

Thus our proposal is to introduce the general reparameterization-invariant boundary interaction to chiral fields and derive all the divergences induced by this interaction. We shall need additional dimensional operators in the boundary action to renormalize divergences. From the condition of vanishing β functions for $U(x)$ the equations of motion for chiral fields are obtained in the low-momentum (derivative) expansion. We consistently implement the unitarity constraint on the chiral fields and locality of the chiral Lagrangian and finally calculate the $O(p^4)$ terms of the Gasser and Leutwyler[18] effective Lagrangian without any additional assumptions or free parameters to adjust. A strikingly good correspondence with the phenomenological values is found.

ATTACHING PIONS TO THE QCD STRING AND DIAGRAMMAR

The hadronic string in the conformal gauge is described by the following conformal field theory action which has four dimensional Euclidean space-time as target space

$$\mathscr{W}_{str} = \frac{1}{4\pi\alpha'} \int d^{2+\varepsilon}\sigma \left(\frac{\varphi}{\mu}\right)^{-\varepsilon} \partial_i x_\mu \partial_i x_\mu, \tag{1}$$

where for $\varepsilon = 0$ one takes $x_\mu = x_\mu(\tau, \sigma)$, $-\infty < \tau < \infty, 0 < \sigma < \infty$, $i = \tau, \sigma, \mu = 1,...,4$. The conformal factor $\varphi(\tau, \sigma)$ is introduced to restore the conformal invariance in $2 + \varepsilon$ dimensions. The Regge trajectory slope (related to the inverse string tension) is known to be universal $\alpha' \simeq 0.9 \text{ GeV}^{-2}$ [19].

We would like to couple in a chiral invariant manner the matrix in flavor space $U(x)$ containing the meson fields to the string degrees of freedom while preserving general covariance in the two dimensional coordinates and conformal invariance under local scale transformations of the two-dimensional metric tensor.

Since the string variable x does not contain any flavor dependence, we introduce two dimensionless Grassmann variables ('quarks') living on the boundary of the string sheet: $\psi_L(\tau), \psi_R(\tau)$. They transform in the fundamental representation of the light flavor group ($SU(2)$ in the present paper). A local hermitean action $S_b = \int d\tau L_f$ is then introduced on the boundary $\sigma = 0$ to describe the interaction with background chiral fields $U(x(\tau)) = \exp(i\pi(x)/f_\pi)$, where the normalization scale is set to $f_\pi \simeq 93$ MeV, the weak pion decay constant.

The boundary Lagrangian is chosen to be reparameterization invariant and in its minimal form reads

$$
\begin{aligned}
L_f &= \frac{1}{2} i \left(\bar{\psi}_L U (1-z) \dot{\psi}_R - \dot{\bar{\psi}}_L U (1+z) \psi_R \right. \\
&\quad \left. + \bar{\psi}_R U^+ (1+z^*) \dot{\psi}_L - \dot{\bar{\psi}}_R U^+ (1-z^*) \psi_L \right),
\end{aligned} \tag{2}
$$

herein and further on a dot implies a τ derivative: $\dot{\psi} \equiv d\psi/d\tau$. It can be seen that this is the most general form of the boundary lagrangian compatible with all the symmetries.

A further restriction is obtained by requiring CP invariance,

$$
U \leftrightarrow U^+, \quad \psi_L \leftrightarrow \psi_R. \tag{3}
$$

The above Lagrangian is CP symmetric for $z = -z^* = ia$. The fulfillment of this symmetry happens to be crucial to preserve conformal symmetry in the presence of the added boundary interaction.

Now we expand the function $U(x)$ in powers of the string coordinate field $x_\mu(\tau) = x_{0\mu} + \tilde{x}_\mu(\tau)$ around a constant x_0,

$$
U(x) = U(x_0) + \tilde{x}_\mu(\tau) \partial_\mu U(x_0) + \frac{1}{2} \tilde{x}_\mu(\tau) \tilde{x}_\nu(\tau) \partial_\mu \partial_\nu U(x_0) + \dots. \tag{4}
$$

and look for the potentially divergent one particle irreducible diagrams. The two-fermion, N-boson vertex operators are generated by the expansion (4), from the generating functional $Z_b = \langle \exp(iS_b) \rangle$ and eq.(2). Each additional loop comes with a power of α'. One can find a resemblance to the familiar derivative expansion of chiral perturbation theory [18].

The free fermion propagator is

$$
\langle \psi_R(\tau) \bar{\psi}_L(\tau') \rangle = \langle \psi_L(\tau) \bar{\psi}_R(\tau') \rangle^+ = U^{-1}(x_0) \theta(\tau - \tau'), \tag{5}
$$

if we impose CP symmetry for unitary chiral fields $U(x)$.

The free boson propagator projected on the boundary is

$$
\langle x_\mu(\tau) x_\nu(\tau') \rangle = \delta_{\mu\nu} \Delta(\tau - \tau') = -2\delta_{\mu\nu} \alpha' \ln(|\tau - \tau'|\mu). \tag{6}
$$

The normalization of the string propagator is inferred [17] from the definition of the kernel of the N-point tachyon amplitude for the open string [9]. In dimensional regularization one uses $\Delta(0) \sim \alpha'/\varepsilon$ and $\Delta'(0) = 0$.

To implement the renormalization process we perform a loop (equivalent to a derivative) expansion, we then proceed to determine the counterterms required to make the theory finite and, finally, we impose the condition of a vanishing beta functional for the coupling $U(x)$ to ensure the absence of conformal anomaly.

RENORMALIZATION AT ONE AND TWO LOOPS

Using the above set of Feynman rules one arrives at the one-loop divergent part of the propagator

$$-\theta(A-B)U^{-1}\delta U U^{-1}, \quad \delta U \equiv \Delta(0)\left[\frac{1}{2}\partial_\mu^2 U - \frac{3+z^2}{4}\partial_\mu U U^{-1}\partial_\mu U\right]. \tag{7}$$

This divergence is eliminated by introducing an appropriate counterterm $U \to U + \delta U$. Conformal symmetry is restored (the beta-function is zero) if the above contribution vanishes, $\delta U = 0$.

Let us find out for which value of z this variation of U is compatible with its unitarity.

$$\delta(UU^+) = U \cdot \delta U^+ + \delta U \cdot U^+ = 0. \tag{8}$$

A simple calculation shows that this takes place for $z = \pm i$. The local classical action which has $\delta U = 0$ as equation of motion is

$$W^{(2)} = \frac{f_\pi^2}{4}\int d^4x \mathrm{tr}\left[\partial_\mu U \partial_\mu U^+\right], \tag{9}$$

i.e. the well known non-linear sigma model of pion interactions.

We have thus found the chiral action induced by the QCD string. It has all the required properties of locality, chiral symmetry and proper low momentum behavior (Adler zero) and describes massless pions. However f_π, the overall normalization scale, cannot be predicted from these arguments.

Before proceeding to a full two loop calculation we have to check whether the minimal Lagrangian (2) is sufficient to renormalize also the vertices containing the boson legs. It turns out that it is not.

To obtain the divergences for vertices with external boson lines we introduce an external background boson field $\bar{x}_\mu$ and split $x_\mu = \bar{x}_\mu + \eta_\mu$. The free propagator for the fluctuating field η_μ coincides with the one for x_μ.

The total one-loop divergence in the vertex with two fermions and one boson line can be represented by the following vertex operator in the Lagrangian

$$\frac{i}{2}\left(\bar{\psi}_L\Phi^{(1)}\psi_R - \bar{\psi}_L\Phi^{(2)}\psi_R\right) + \mathrm{h.c.}, \quad \Phi^{(1,2)} \equiv \bar{x}_\mu(\tau)(1\mp z)\left[\partial_\mu(\delta U)\mp \phi_\mu\right]. \tag{10}$$

The terms proportional to derivatives of δU are automatically eliminated by the renormalization of the one-loop propagator. But the part proportional to ϕ_μ remains and to absorb these divergences new counterterms are required. The latter ones can be parameterized with three bare constants g_1, g_2 and g_3, which are real if the CP symmetry for $z = -z^*$ holds

$$\begin{aligned}
\Delta L_{bare} &= \frac{i}{8}(1-z^2)\bar{\psi}_L\left((g_1 - zg_2)\partial_\nu \dot{U} U^{-1}\partial_\nu U - (g_1 + zg_2)\partial_\nu U U^{-1}\partial_\nu \dot{U}\right. \\
&\quad \left. + 2zg_3\partial_\nu U U^{-1}\dot{U}U^{-1}\partial_\nu U\right)\psi_R + \mathrm{h.c.}
\end{aligned} \tag{11}$$

Renormalization is accomplished by the subtraction

$$g_i = g_{i,r} - \Delta(0). \tag{12}$$

The constants $g_{i,r}$ are finite, but in principle scheme dependent. The counterterms are of higher dimensionality than the original Lagrangian (2) and the couplings g_i are of dimension α'. Since (2) was the most general coupling permitted by the symmetries of the model, one concludes that conformal symmetry seems to be broken by these boundary couplings already at tree level. However in spite of the fact that the new couplings are dimensional, it turns out [17] that their contribution into the trace of the energy-momentum tensor vanishes once the requirements of unitarity of U and CP invariance are taken into account. Therefore conformal invariance at the world-sheet level is not broken at the order we are working.

However the appearance of new vertices changes the fermion propagator. One obtains from such terms the following contribution to the propagator

$$
\begin{aligned}
&\theta(A-B)\frac{1}{16}\Delta(0)(1-z^2)U^{-1}\Big\{2(g_{1,r}-z^2 g_{2,r})\partial_\rho UU^{-1}\partial_\mu\partial_\rho UU^{-1}\partial_\mu U \\
&-(1+z)(g_{1,r}+zg_{2,r})\partial_\rho UU^{-1}\partial_\mu UU^{-1}\partial_\rho\partial_\mu U \\
&-(1-z)(g_{1,r}-zg_{2,r})\partial_\rho\partial_\mu UU^{-1}\partial_\rho UU^{-1}\partial_\mu U \\
&+4z^2 g_{3,r}\partial_\rho UU^{-1}\partial_\mu UU^{-1}\partial_\rho UU^{-1}\partial_\mu U\Big\}U^{-1} \\
&\equiv -\theta(A-B)\Delta(0)U^{-1}\delta^{(4)}UU^{-1},
\end{aligned} \tag{13}
$$

One should add this divergence to the one-loop result, thereby modifying the U field renormalization and equations of motion

$$\bar\delta U = \Delta(0)\left[\frac{1}{2}\partial_\mu^2 U - \frac{3+z^2}{4}\partial_\mu UU^{-1}\partial_\mu U + \delta^{(4)}U\right] = 0. \tag{14}$$

This is one source of $O(p^4)$ terms and we shall see that there is yet another contribution at the two loop order.

As to the other vertices it can be proven [17] that any diagram with an arbitrary number of external boson lines and two fermion lines, i.e. any vertex of those generated by the perturbative expansion of (2) is rendered finite by the previous counterterms. This completes the renormalization program at one loop.

There are 10 two-loop one-particle irreducible diagrams which were analytically calculated in [17]. The divergences in the propagator consist of the double divergent part, $\sim \Delta^2(0)$ and of the single divergent contributions, $\sim \Delta(0)$. A substantial part of these divergences are eliminated by performing the one-loop renormalization. The structure of the double pole divergences is compatible with renormalization group arguments.

Some single-pole divergences remain however. Namely, there are divergences linear in $\Delta(0)$ which come from irreducible two-loop diagrams with maximal number of vertices. These divergences are

$$
\begin{aligned}
-\Delta(0)U^{-1}\delta^{(4)}_{2-l}U^{-1} &\equiv c\Delta(0)\big[U^{-1}\partial_\rho UU^{-1}\partial_\mu UU^{-1}\partial_\mu UU^{-1}\partial_\rho UU^{-1}- \\
&-U^{-1}\partial_\rho UU^{-1}\partial_\mu UU^{-1}\partial_\rho UU^{-1}\partial_\mu UU^{-1}\big].
\end{aligned} \tag{15}
$$

with $c = \alpha'(1 - z^2)^2/8 = \alpha'/2$ for $z = \pm i$. This term survives after adding all the counterterms. It must therefore modify the equation of motion (14) at the next order in the α' expansion, $\delta^{(4)}U \to \delta^{(4)}U + \delta^{(4)}_{2-l}$. Its presence allows for non zero solutions for the coupling constants g_i and therefore for nonzero values for the Gasser-Leutwyler $O(p^4)$ coefficients.

LOCAL INTEGRABILITY OF THE EQUATIONS OF MOTION

The equation of motion, $\delta U = 0$, can be derived from the dimension-two local action (9), involving a unitary matrix $U(x)$, only for $z = \pm i$. If the four-derivative part of the equations of motion can be derived from dimension-four operators in a local effective Lagrangian then certain constraints are to be imposed on constants $g_{i,r}$.

Such a Lagrangian has only two terms compatible with the chiral symmetry,

$$\mathscr{L}^{(4)} = f_\pi^2 \mathrm{tr}\left(K_1 \partial_\mu U \partial_\rho U^+ \partial_\mu U \partial_\rho U^+ + K_2 \partial_\mu U \partial_\mu U^+ \partial_\rho U \partial_\rho U^+\right). \tag{16}$$

Other terms containing $\partial_\mu^2 U$ are reduced to the set (16) with the help of the lowest-order equations of motion.

The variation of the previous Lagrangian must saturate the dimension-four component of the equations of motion. From this we identify the low-energy constants with the coupling constants arising from the equations of motion (14) supplemented with (15) and after applying the $O(p^2)$ equations of motion. Then one obtains the following set of coefficients for the various chiral field structures

$$-2(2K_1 + K_2) = \frac{1}{16}(1 - z^2)(1 \pm z)(g_{1,r} \pm z g_{2,r}),$$

$$-4K_2 = \frac{1}{8}(1 - z^2)(-g_{1,r} + z^2 g_{2,r}), \quad 2[(1 - z^2)K_1 + K_2] = -c;$$

$$-2z^2 K_2 = 0, \quad 4[K_1 + K_2] = -\frac{1}{4}(1 - z^2)z^2 g_{3,r} + c. \tag{17}$$

For $z^2 = -1$ only one solution is possible,

$$K_2 = 0, \quad K_1 = -\frac{1}{4}c = -\frac{\alpha'}{8}, \quad g_{1,r} = -g_{2,r} = -g_{3,r} = 4c. \tag{18}$$

Thus, comparing eq.(16) with the usual parameterization of the Gasser and Leutwyler Lagrangian[18],

$$L_1 = \frac{1}{2}L_2 = -\frac{1}{4}L_3 = -\frac{1}{2}K_1 f_\pi^2 = \frac{f_\pi^2 \alpha'}{16}. \tag{19}$$

For $\alpha' = 0.9$ GeV^{-2} and $f_\pi \simeq 93$ MeV it yields $L_2 \simeq 0.9 \cdot 10^{-3}$ which is quite a satisfactory result[20].

The relation $L_1 = 1/2L_2 = -1/4L_3$ was established earlier in bosonization models [21] and in the chiral quark model[22] by means of a derivative expansion of quark determinant. However at that time its possible connection with a string description of

QCD was not recognized. The first attempt to derive the chiral coefficients from the Veneziano-type dual amplitude was undertaken in [23] where a similar relation was found but with different numerical values for the L_i. However the specific choice of dual amplitude in [23] cannot be derived from any known string theory.

Another check comes from the compatibility of the unitarity of U and the equations of motion at the two-loop level. It turns out that if one accepts arbitrary real coefficients in the set of dimension-four operators then the only solution compatible with the unitarity is given by the parameterization with constants K_1 and K_2.

CONCLUSIONS

In our talk we have reported on a simplified, but hopefully not unrealistic, model of the QCD string. Requiring its conformal invariance around a chirally non-invariant vacuum leads to the Gasser and Leutwyler Lagrangian. However the bosonic string action used here is of course not totally satisfactory. For instance, it does not prevent large Euclidean world sheets from crumpling [24], something that looks very unphysical. It does not also describe correctly the high-temperature behavior of large N QCD [25] either. To correct some of these shortcomings a proper QCD-induced string must be modified [24, 26] suitably by including operators breaking manifestly conformal symmetry on the world-sheet for large strings. Nevertheless we are concerned here with the low-energy string properties and therefore do not expect that the strategy and technique to derive the chiral field action needs any significant changes to be adjusted to a modified QCD string action.

We have restricted ourselves here to the $SU(2)$ global flavor group. In this case only parity-even terms in the equations of motion can be revealed from the simple fermion Lagrangian (2). In order to obtain the parity-odd WZW Lagrangian relevant for the case of three flavors one possibility would be to extend the boundary fermion action supplementing one-dimensional fermions with true spinor degrees of freedom. Another possibility is to include some topological terms in the world-sheet action (such as the self-intersection number, that involves the ε-symbol). This issue is under active investigation now.

We are also considering the coupling to external left and right gauge fields. So far we have been able to reproduce the covariant equations of motion at the leading order and we are actively working in the determination of L_9 and L_{10}.

Many other open questions can be formulated in connection with this work. For instance, we are obtaining the pion scattering amplitude in a momentum expansion. Is there any way of getting a closed expression similar to the Veneziano amplitude? What is the role of crossing in this approach? Can the conformal factor (φ) dependence be related to Λ_{QCD}? What is the connection to the parton model? Clearly a lot of work remains to be done.

ACKNOWLEDGMENTS

We express our gratitude to the organizers of the 2nd International Workshop on Hadron Physics - Effective Theories of Low Energy QCD for their hospitality and enthusiasm and, of course, for allowing us to present our work. The research contained in this talk is supported by projects MCyT FPA2001-3598 and CIRIT 2001SGR-00065 and the European Networks EURODAPHNE and EUROGRID, as well as by Grant RFFI 01-02-17152, Russian Ministry of Education Grant E00-33-208 and by The Program *Universities of Russia: Fundamental Investigations* (Grant 992612). By this work A.A. and D.E. contribute to the fulfillment of Project INTAS 2000-587.

REFERENCES

1. G. Veneziano, Nuovo Cim. 57A (1968) 190.
2. C. Lovelace, Phys. Lett. 28B (1968) 264;
   J. Shapiro, Phys. Rev. 179 (1969) 1345.
3. Y. Nambu, in *"Symmetries and Quark Models"*, R. Chand, ed., Gordon and Breach, 1970;
   L. Susskind, Nuovo Cim. 69A (1970) 457.
4. A.M. Polyakov, Phys. Lett. 103B (1981) 207.
5. G.P. Pronko and A.V. Razumov, Theor. Math. Phys. 56 (1984) 760.
6. H. Kleinert, Phys. Lett. 174B (1986) 335.
7. D. Antonov, D. Ebert, and Y.A. Simonov, Mod. Phys. Lett. A11 (1996) 1905.
8. L.D. Solovev, Phys. Rev. D58 (1998) 035005.
9. C.Rebbi, Phys.Rep.C12 (1974) 1.
10. P. Frampton, *"Dual Resonance Models"*, Benjamin, 1974.
11. A.M.Polyakov, *"Gauge Fields and Strings"*, Harwood, Chur, Switzerland, 1987.
12. G. 'tHooft, Nucl. Phys. B72 (1974) 461;
    G. Veneziano, Nucl. Phys. B117 (1976) 519.
13. K.G. Wilson, Phys. Rev. D10 (1974) 2445.
14. R. Kirschner, L.N. Lipatov and L. Szymanowski, Nucl. Phys. B425 (1994) 579.
15. E. Cremmer and J. Scherk, Nucl. Phys. B72 (1974) 117.
16. J. Alfaro, A. Dobado and D. Espriu, Phys. Lett. B460 (1999) 447.
17. J.Alfaro, A.A.Andrianov, L.Balart and D.Espriu, *"Hadronic string, conformal invariance and chiral symmetry"*, hep-th/0203215 (2002) 35p.
18. J. Gasser and H. Leutwyler, Nucl. Phys. B250 (1985) 465.
19. S. Filipponi, G. Pancheri and Y. Srivastava, Phys. Rev. Lett. 80 (1998) 1838.
20. G. Amoros, J. Bijnens, P. Talavera, Nucl. Phys. B602 (2001) 87.
21. A.A. Andrianov and L.Bonora, Nucl. Phys. B233 (1984) 232; 247
    A.A. Andrianov, Phys.Lett.B157 (1985) 425.
22. D. Espriu, E. de Rafael and J. Taron, Nucl. Phys. B345 (1990) 22: Erratum-*ibid.* B355 (1991) 278.
23. M. Polyakov and V. Vereshagin, Phys. Rev. D54 (1996) 1112.
24. A.M. Polyakov, Physica Scripta T15 (1987) 191.
25. J. Polchinski and Zhu Yang, Phys. Rev. D46 (1992) 3667.
26. M.C. Diamantini, H. Kleinert and C.A. Trugenberger, Phys. Rev. Lett. 82 (1999) 267.

Dispersive treatment of $K \to \pi\pi$

Gilberto Colangelo

Institut für Theoretische Physik der Universität Bern, Sidlerstr. 5, CH–3012 Bern

Abstract. We write a set of dispersion relations for the $K \to \pi\pi$ amplitude in which the weak Hamiltonian carries momentum. A soft pion theorem relates this amplitude to the $K \to \pi$ amplitude, and can be used to determine one of the two subtraction constants – the second constant is at present known only to leading order in chiral perturbation theory. We solve the dispersion relations numerically and express the result in terms of the unknown higher order corrections to this subtraction constant. We present preliminary results of an analysis of the effect of inelastic channels.

INTRODUCTION

The calculation of the $K \to \pi\pi$ amplitude in the Standard Model still remains one of the most difficult and yet unsolved problems of today's particle physics, despite many years of efforts and progress in our understanding of various related physics aspects. Indeed we lack yet a satisfactory explanation of the $\Delta I = 1/2$ rule, and do not yet have a calculation of ε'/ε in the Standard Model which would make a comparison to the measured value somewhat useful – the typical size of the theoretical uncertainties attached to any of the calculations available in the literature is around 50% (see, e.g. Ref. [1] and references therein), much larger than the experimental uncertainty [2]. A recurring discussion in the literature on calculations of $K \to \pi\pi$ concerns the role of final state interactions (FSI) – in some approaches (like $1/N_C$ or lattice QCD, in their simplest implementation[3]) these have been neglected as higher order corrections, whereas various authors [4] have stressed their importance and the necessity to take them into account, in order to reach a useful level of accuracy.

We have recently proposed [7] a new method for properly evaluating the effect of FSI. The method is based on a set of dispersion relations for the $K \to \pi\pi$ amplitude in which the weak Hamiltonian carries momentum. Kinematically, this amplitude is analogous to a scattering amplitude – indeed the dispersion relations are very similar to those valid for the $K\pi$ scattering amplitude. A chiral Ward identity relates this amplitude in an unphysical point to the amplitude with one pion less in the final state: such an identity is known with the name of soft–pion theorem, and allows to fix one subtraction constant in the dispersion relation in terms of the $K \to \pi$ amplitude. Lattice calculations typically give the latter [5], because it is relatively easier to calculate. As we will show, the dispersion relations require as input a second subtraction constant. Given the two subtraction constants one can solve numerically the dispersion relation and obtain the physical $K \to \pi\pi$ amplitude. The main advantage of this method is a theoretically clean separation between quantities (the subtraction constants) which do not suffer from strong FSI corrections, and the effect of FSI, which is calculated via the solution of the

CP660, *Hadron Physics: Effective Theories of Low Energy QCD*, edited by A. H. Blin et al.
© 2003 American Institute of Physics 0-7354-0120-9/03/$20.00

dispersion relations.

The method described here does *not* rely on off-shell extrapolations of the $K \to \pi\pi$ amplitude, which has been proposed and discussed by various authors [6]. As is well known, such off-shell extrapolations are arbitrary, and such an arbitrariness is unavoidably reflected in the final result. In our opinion any application of dispersion relations to these off–shell extrapolated amplitudes can hardly lead to any improvement in the calculation of the on-shell amplitude. The reader interested in this specific issue is referred to Ref. [8, 9] for a detailed discussion of the problems related to the off-shell extrapolation. Here we concentrate on the dispersion relation for the $K \to \pi\pi$ amplitude with momentum carried by the Hamiltonian.

DISPERSION RELATIONS AND SOFT PION THEOREMS

We consider the amplitude

$$\langle \pi(p_1)\pi(p_2)(I=0)|\mathscr{H}_W^{1/2}(0)|K(q_1)\rangle =: T^+(s,t,u) \tag{1}$$

with the Mandelstam variables: $s = (p_1+p_2)^2$, $t = (q_1-p_1)^2$, $u = (q_1-p_2)^2$, related by $s+t+u = 2M_\pi^2+M_K^2+q_2^2$, where q_2 is the momentum carried by the weak Hamiltonian. From now on we set $q_2^2 = 0$ (but $q_2^\mu \neq 0$ in general). The physical decay amplitude is obtained by setting $q_2^\mu = 0$ ($s = M_K^2$, $t = u = M_\pi^2$).

By neglecting the imaginary parts of D and higher waves the analytic structure of the amplitude simplifies and it can be decomposed into a combination of functions of a single variable:

$$
\begin{aligned}
T^+(s,t,u) &= M_0(s) + \frac{1}{3}\left[N_0(t) + N_0(u)\right] + \frac{2}{3}\left[R_0(t) + R_0(u)\right] \\
&+ \frac{1}{2}\left[\left(s-u-\frac{M_\pi^2\Delta}{t}\right)N_1(t) + \left(s-t-\frac{M_\pi^2\Delta}{u}\right)N_1(u)\right] ,
\end{aligned}
\tag{2}
$$

where $\Delta = M_K^2 - M_\pi^2$. The terms proportional to N_1 drop out from the physical decay amplitude:

$$\mathscr{A}(K \to \pi\pi) = T^+(M_K^2, M_\pi^2, M_\pi^2) = M_0(M_K^2) + \frac{2}{3}[N_0(M_\pi^2) + 2R_0(M_\pi^2)]. \tag{3}$$

Each of the single variable functions appearing in Eq. (2) is analytic in the complex plane except for a cut starting at $4M_\pi^2$ for M_0 and at $(M_K+M_\pi)^2$ for the remaining ones. These functions are *defined* to have the discontinuity on the positive real axis identical to that of a specific partial wave: M_0 to the $I = 0$ S-wave in the s channel, whereas in the t channel, N_0 and N_1 to the $I = 1/2$ S- and P- wave respectively, and R_0 to the $I = 3/2$ S-wave. Below the inelastic threshold, the elastic unitarity condition for these functions reads

$$\mathrm{disc}M_0(s) = \sin\delta_0^0(s)e^{-i\delta_0^0}\left[M_0(s) + \hat{M}_0(s)\right] \tag{4}$$

where δ_0^0 is the $\pi\pi$ phase shift. In the analogous formulae for the N's and R functions the πK phase shifts appear. The hat functions denote contributions from the other channels coming in via angular averages (see Ref. [7] for more details).

If one is only interested in the low–energy region, one can neglect the inelastic channels in first approximation: the solution of the dispersion relation for each of the functions is well approximated by the Omnès function [10] times a polynomial. It is therefore convenient to write the dispersion relation for the functions divided by the corresponding Omnès function:

$$M_0(s) = \Omega_0^0(s) \left\{ a + b(s-s_0) + \frac{(s-s_0)^2}{\pi} \int_{4M_\pi^2}^{\Lambda_1^2} \frac{\sin \delta_0^0(s') \hat{M}_0(s') ds'}{|\Omega_0^0(s')|(s'-s)(s'-s_0)^2} \right\} . \tag{5}$$

$\Omega_0^0(s)$ is the Omnès function subtracted at $s = s_0$: $\Omega_0^0(s_0) = 1$. The dispersion relation for the N's and R function is analogous to that for M_0 (5). An important point concerns the overall number of subtraction constants which is two for the amplitude T^+. It can be shown that these two constants can be shuffled into M_0 such that the dispersion relations for the other functions do not involve any other subtraction constants.

In order to use the dispersion relations to calculate the $K \to \pi\pi$ amplitude, one has to provide as input the two subtraction constants. One of the two can be given in terms of the $K \to \pi$ amplitude – a soft–pion theorem relates the amplitude at the soft pion point to the $K \to \pi$ amplitude up to terms of order M_π^2:

$$-\frac{1}{2F_\pi} \mathscr{A}(K \to \pi) = a + \frac{1}{3} \left[N_0(M_K^2) + 2R_0(M_K^2) \right] + \mathscr{O}(M_\pi^2) . \tag{6}$$

Notice that although the process involves a kaon, the relation is based on the use of the $SU(2)$ symmetry, and therefore suffers from corrections of order M_π^2 only. The key of the problem is how to calculate b. This constant is related to the derivative in s of the amplitude T^+ at the soft pion point. However there is a Ward identity that relates this derivative to another amplitude:

$$\frac{\partial}{\partial s} T^+ (s, \Sigma - s, M_\pi^2)_{|s=M_\pi^2} = \frac{1}{2} C(M_\pi^2, M_K^2, M_\pi^2) + \mathscr{O}(M_\pi^2) , \tag{7}$$

where $C(s,t,u)$ is an amplitude defined as:

$$\frac{i}{F_\pi} \int dx e^{ip_1 x} \langle \pi(p_2) | T \mathscr{H}_W^{1/2}(0) A^\mu(x) | K(q_1) \rangle = ip_1^\mu B + iq_1^\mu C + iq_2^\mu D , \tag{8}$$

where $A^\mu(x)$ is the axial current that couples to the pion removed from the outgoing state.

Although b is directly related to the derivative of the amplitude at the soft–pion point, one could also determine it in different ways. To fix the two subtraction constants one only needs two conditions on the amplitude. Evaluating the $K \to \pi\pi$ amplitude with momentum carried by the weak Hamiltonian in any convenient kinematical point different from the soft–pion point is enough to determine b. One kinematical configuration that

27

could be implemented on the lattice [11] is the following

$$s = 4M_\pi^2, \quad t = u = \frac{M_K^2}{2} - M_\pi^2 \tag{9}$$

with four-momenta $q_1 = q_2 + p_1 + p_2$ given by

$$p_1 = p_2 = (M_\pi, \mathbf{0}), \quad q_1 = (\sqrt{M_K^2 + q^2}, q, 0, 0), \quad q_2 = (q, q, 0, 0), \tag{10}$$

where

$$q = \frac{M_K^2}{4M_\pi} - M_\pi . \tag{11}$$

Lacking any numerical results on b we proceed by fixing b at a certain value and then varying it within a fairly wide range. To fix the central value and the range we use CHPT as a guide. At leading order, CHPT dictates the following relation between a and b:

$$b = \frac{3a}{M_K^2 - M_\pi^2} \left(1 + X + \mathcal{O}(M_K^4)\right) . \tag{12}$$

The size of the correction is at the moment unknown, but nothing protects it from being of order M_K^2. The explicit calculation in CHPT can be found in [7].

NUMERICAL TREATMENT

In our numerical study we have used $X = \pm 30\%$. The normalization of the amplitude is irrelevant here, and we have fixed it to $T^+(M_\pi^2, M_K^2, M_\pi^2) = 1$. The $\pi\pi$ phase shifts are taken from [12], with the scattering lengths determined in [13], and the πK phase shifts from [14]. Our results are shown in Fig. 1, where we have plotted $|T^+(s, \Sigma - s, M_\pi^2)|$ versus s, comparing our numerical solution of the dispersion relations to the CHPT leading order formula. The latter is what has been used so far whenever a number for the $K \to \pi\pi$ matrix element extracted from the lattice has been given. Our treatment shows that large corrections with respect to leading–order CHPT are to be expected. One source of large corrections is the Omnès factor due to $\pi\pi$ rescattering in the final state. The other potentially dangerous source is represented by X, the next-to-leading order correction to the relation (12) between a and b. The latter could (depending on the sign) in principle double, or to a large extent reabsorb the correction due to final state interactions.

The numerical solution shown in Fig. 1 is obtained in the elastic approximation. The uncertainties related to this approximation can be estimated by varying the cutoff Λ_1 appearing in the Omnès function (the dependence of the numerical solution on the cutoff in the Omnès functions in the t and u channels, and other details in the evaluation of the corresponding dispersive integrals is negligible). The dependence of the solution on this parameter is shown in Fig. 2. The figure clearly shows the necessity to bring the dependence on this cutoff under good control if one wants to obtain a high level of accuracy through this dispersion relation. If the contributions to the dispersive

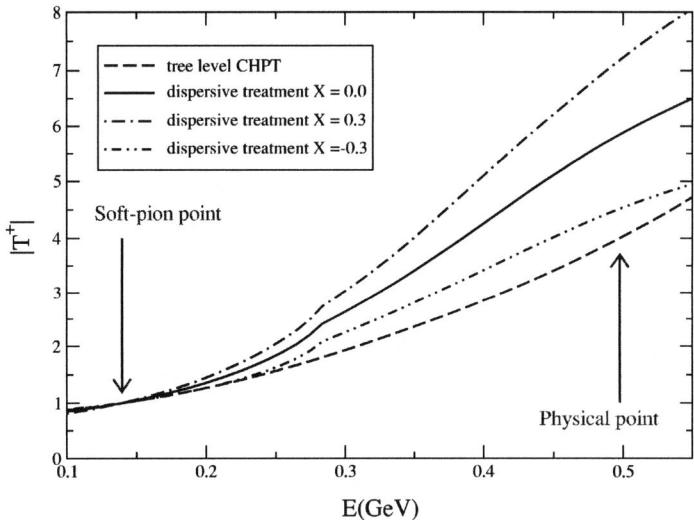

FIGURE 1. The function $|T^+(s,t,u)|$ plotted *vs.* $E = \sqrt{s}$ along the line of constant $u = M_\pi^2$: the result of our numerical study for different values of X are compared to tree level CHPT.

integrals coming from energy regions above the inelastic threshold (which, to a good approximation is at $\sqrt{s} = 2M_K$ in the s channel) are relevant, one must extend the Omnès method and explicitly take into account the inelastic channels.

We have recently started such an analysis which takes into account only the contribution from $\bar{K}K$ intermediate states in the s channel, which represent the most important contribution. In a schematic description our analysis goes along the following steps[1]:

- We solve the unitarity condition for the $K \to \pi\pi$ and the $K \to K\bar{K}$ at the same time:

$$
\begin{aligned}
\mathrm{Im}\,M_0^\pi &= \theta(s - 4M_\pi^2)(M_0^\pi(s) + \hat{M}_0^\pi(s))t_0^0(s)_{\pi\pi}^* \\
&+ \theta(s - 4M_K^2)(M_0^K(s) + \hat{M}_0^K(s))t_0^0(s)_{\pi K}^* \\
\mathrm{Im}\,M_0^K &= \theta(s - 4M_K^2)(M_0^K(s) + \hat{M}_0^K(s))t_0^0(s)_{KK}^* \\
&+ \theta(s - 4M_\pi^2)(M_0^\pi(s) + \hat{M}_0^\pi(s))t_0^0(s)_{\pi K}^*
\end{aligned}
\tag{13}
$$

- We ignore the t and u dependence of both the $K \to \pi\pi$ and the $K \to \bar{K}K$ amplitudes:

$$
T_{\pi,K}^+(s,t,u) \simeq M_0^{\pi,K}(s) \quad \Rightarrow \quad \hat{M}_0^{\pi,K}(s) = 0 \ .
$$

In this approximation the problem is completely analogous to the scalar form factor problem, which has been numerically treated in Ref. [15].

---

[1] In a self–explanatory notation, all the quantities denoted by π (K) in the superscript refer to the $K \to \pi\pi$ ($K \to \bar{K}K$) amplitude.

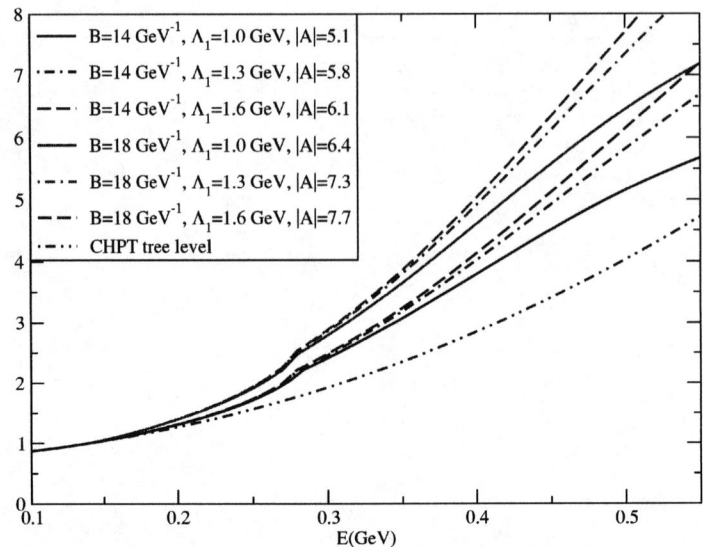

FIGURE 2. Dependence of the solution of the dispersion relation on the cutoff Λ_1. The value of the second subtraction constant is specified by $B = b/a = 3/(M_K^2 - M_\pi^2)(1+X)$. Its value in CHPT to leading order is $B \sim 13$ GeV^{-1}. In order to ease the comparison between the various curves, the legend gives also the numerical value of the amplitude at $E = M_K$ – the blue curve, corresponding to T^+ in CHPT to leading order gives $|A| = 4$.

- The numerical solution depends on four initial conditions:

$$a^\pi = M_0^\pi(M_\pi^2) \ , \qquad B^\pi = \frac{1}{a^\pi}\frac{\partial M_0^\pi}{\partial s}\Big|_{s=M_\pi^2} \qquad (14)$$

$$a^K = M_0^K(M_\pi^2) \ , \qquad B^K = \frac{1}{a^K}\frac{\partial M_0^K}{\partial s}\Big|_{s=M_\pi^2} \qquad (15)$$

The initial conditions for M_0^π are directly related to the subtraction constants a and b discussed so far. In this coupled channel treatment we need two more subtraction constants for the unphysical amplitude $K \to K\bar{K}$. In our numerical analysis we take these from LO CHPT and attach to them a very large uncertainty.

The numerical results are shown in Fig. 3 – they have been obtained using the standard solutions given in Ref. [15] [2]. The dependence on the parameters that determine the initial conditions for the $K \to \bar{K}K$ amplitude appears to be very mild. Even an estimate of these parameters on the basis of LO CHPT appears sufficient: a 50% uncertainty on these estimates has a barely visible effect on the final result. This shows that once this analysis will be completed we will have a method that allows a calculation of the on–shell amplitude in terms of the two subtraction constants a and b with very small

---

[2] We thank J. Gasser for providing us with the corresponding fortran code.

30

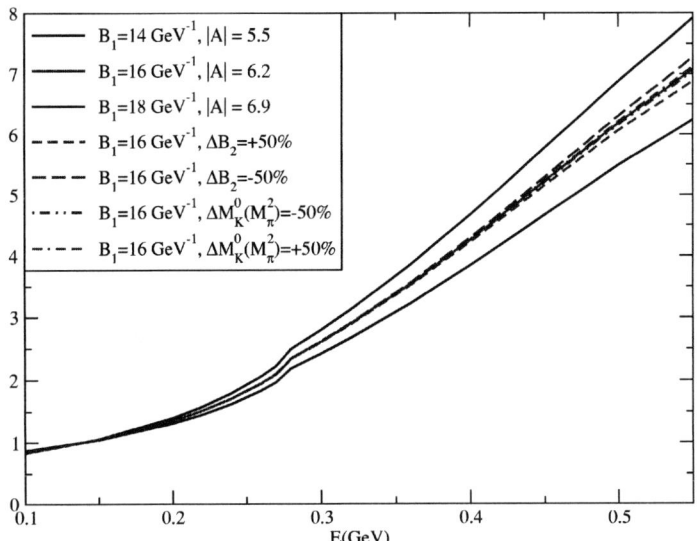

FIGURE 3. Results of a coupled channel analysis for the $K \to \pi\pi$ amplitude. In the legend we have used the notation $B_1 = B^\pi$ and $B_2 = B^K$. The red curves show the dependence of the final result on the input for the $K \to \bar{K}K$ amplitude.

uncertainties.

DISCUSSION AND OUTLOOK

We have set up a dispersive framework for the $K \to \pi\pi$ amplitude that allows one to evolve the amplitude from the soft pion point (where it is given by the $K \to \pi$ amplitude) to the physical point, taking into account all the main physical effects. The dispersion relation requires a second subtraction constant, the derivative of the amplitude at the soft pion point which, to the best of our knowledge, has not been calculated so far. Our numerical work, however, shows that the amplitude at the physical point depends strongly on the value of the slope at the soft pion point. A nonperturbative calculation of the second subtraction constant b is necessary in order to obtain an accurate result with this method.

Our method, in contrast with all the recent literature on the subject of final state interactions [4] does not rely neither on off–shell extrapolations of the $K \to \pi\pi$ amplitude, nor on the application of CHPT. It is rigorous and even more economical: if one wants to go beyond the simple recipe of Truong [6] of multiplying the leading order CHPT amplitude by the Omnès function, one has to estimate several low energy constants that appear in the $O(p^4)$ CHPT Lagrangian. Unfortunately, our knowledge of the $O(p^4)$ CHPT weak Lagrangian is rather meagre and would not allow any prediction beyond an order of magnitude estimate. With the method proposed here one does not have to refer to these several constants in any way, but needs as input only two amplitudes in QCD.

Recently, direct calculations of the $K \to \pi\pi$ matrix element on the lattice have been proposed [16]. Such a method does not rely on anything but the lattice formulation of QCD. Its practical implementation, however, is still way beyond what today's computers and algorithms would allow. As for the future, it is difficult to predict how long it will take before such a method will yield the first results, and even more the first ones that will have an accuracy comparable to the experimental ones. In view of this, I believe that it is wise to try to perform the same calculation with different methods. The one proposed here offers the advantage of separating in a well defined and rigorous manner the effect of FSI interactions from the rest. The calculation of the effect of FSI through dispersion relations relies mostly on our knowledge of the $\pi\pi$ phase shifts, which is extremely accurate thanks to recent work on Roy equations and the chiral expansion [12, 13]. In view of this high level of accuracy, even when direct calculations of the $K \to \pi\pi$ amplitude will be available, a comparison between the latter and the results obtained with the method proposed here will provide a powerful check.

ACKNOWLEDGMENTS

I thank M. Büchler, J. Kambor and F. Orellana for a pleasant collaboration on the subject discussed here, and the organizers for their invitation to a very interesting conference in an exceptionally beautiful surroundings.

REFERENCES

1. S. Bertolini, arXiv:hep-ph/0206095.
2. J. R. Batley *et al.* [NA48 Collaboration], Phys. Lett. B **544** (2002) 97 [arXiv:hep-ex/0208009].
   A. Alavi-Harati *et al.* [KTeV Collaboration], arXiv:hep-ex/0208007.
3. W. A. Bardeen, A. J. Buras and J. M. Gerard, Phys. Lett. B **180** (1986) 133; Nucl. Phys. B **293** (1987) 787; Phys. Lett. B **192** (1987) 138.
   C. Bernard et al., Phys. Rev. D **32** (1985) 2343.
4. T. N. Truong, Phys. Lett. **B207** (1988) 495. J. Kambor, J. Missimer and D. Wyler, Phys. Lett. B **261** (1991) 496. S. Bertolini et al., Nucl. Phys. B **514** (1998) 63 [hep-ph/9705244]; Nucl. Phys. B **514** (1998) 93 [hep-ph/9706260]. J. Bijnens and J. Prades, JHEP**9901** (1999) 023 [hep-ph/9811472]; JHEP**0006** (2000) 035 [hep-ph/0005189]. T. Hambye, G. O. Kohler and P. H. Soldan, Eur. Phys. J. C **10** (1999) 271 [hep-ph/9902334]. T. Hambye et al. Nucl. Phys. B **564** (2000) 391 [hep-ph/9906434]. E. Pallante and A. Pich, Phys. Rev. Lett. **84** (2000) 2568 [hep-ph/9911233]; Nucl. Phys. **B592** (2000) 294 [hep-ph/0007208]. E. Pallante, A. Pich and I. Scimemi, Nucl. Phys. B **617** (2001) 441 [arXiv:hep-ph/0105011].
5. T. Blum *et al.* [RBC Collaboration], arXiv:hep-lat/0110075.
   J. I. Noaki *et al.* [CP-PACS Collaboration], arXiv:hep-lat/0108013.
   For direct $K \to \pi\pi$ matrix element calculations see:
   D. Becirevic *et al.* [SPQCDR Collaboration], arXiv:hep-lat/0209136.
6. T. N. Truong in [4], E. Pallante and A. Pich in [4].
   M. Suzuki, Int. J. Mod. Phys. A **16** (2001) 4637 [arXiv:hep-ph/0102028].
   E. A. Paschos, hep-ph/9912230.
7. M. Büchler et al., Phys. Lett. B **521** (2001) 22 [arXiv:hep-ph/0102287].
8. M. Büchler et al., Phys. Lett. B **521** (2001) 29 [arXiv:hep-ph/0102289].
9. G. Colangelo Nucl. Phys. Proc. Suppl. **106** (2002) 53 [arXiv:hep-lat/0111003].
10. R. Omnès, Nuovo Cim. **8** (1958) 316.

11. C. Dawson et al., Nucl. Phys. B **514** (1998) 313 [hep-lat/9707009].
12. B. Ananthanarayan, G. Colangelo, J. Gasser and H. Leutwyler, Phys. Rept. **353** (2001) 207 [arXiv:hep-ph/0005297].
13. G. Colangelo, J. Gasser and H. Leutwyler, Phys. Lett. B **488** (2000) 261 [hep-ph/0007112]. Nucl. Phys. B **603** (2001) 125 [arXiv:hep-ph/0103088].
14. M. Jamin, J. A. Oller and A. Pich, Nucl. Phys. B **587** (2000) 331 [hep-ph/0006045].
15. J. F. Donoghue, J. Gasser and H. Leutwyler, Nucl. Phys. B **343** (1990) 341.
16. L. Lellouch, and M. Lüscher, Commun. Math. Phys. **219** (2001) 31 [arXiv:hep-lat/0003023]. C. J. Lin, G. Martinelli, C. T. Sachrajda and M. Testa, Nucl. Phys. B **619** (2001) 467 [arXiv:hep-lat/0104006].

Meson-Baryon Interactions in Unitarized Chiral Perturbation Theory [1]

G. García Recio*, J. Nieves*, E. Ruiz Arriola* and M. Vicente Vacas†

*Departamento de Física Moderna, Universidad de Granada. 18071-Granada (Spain)
†Departamento de Física Teórica and IFIC, Centro Mixto Universidad de Valencia-CSIC, Ap.
Correos 22085, E-46071 Valencia (Spain)

Abstract. Meson-Baryon Interactions can be successfully described using both Chiral Symmetry and Unitarity. The $s-$wave meson-baryon scattering amplitude is analyzed in a Bethe-Salpeter coupled channel formalism incorporating Chiral Symmetry in the potential. Two body coupled channel unitarity is exactly preserved. The needed two particle irreducible matrix amplitude is taken from lowest order Chiral Perturbation Theory in a relativistic formalism. Off-shell behavior is parameterized in terms of low energy constants. The relation to the heavy baryon limit is discussed. The position of the complex poles in the second Riemann sheet of the scattering amplitude determine masses and widths baryonic resonances of the $N(1535)$, $N(1670)$, $\Lambda(1405)$ and $\Lambda(1670)$ resonances which compare well with accepted numbers.

INTRODUCTION

The existence of baryon resonances in meson-baryon reactions is a non-perturbative feature of QCD at intermediate energies. Although Relativistic Invariance, Crossing Symmetry, Unitarity and Chiral Symmetry undoubtedly provide powerful constraints, the usual high and low energy simplifications do not directly apply here just because there is no obvious expansion parameter and some special methods and approximations have to be developed.

The nature of baryonic resonances is not fully unambiguous since their lifetime, τ_R, is very short and hence they cannot be observed decaying into final products. The standard and accepted definition is that resonances correspond to poles in the complex CM energy plane in unphysical sheets, $-2\mathrm{Im}\sqrt{s_R} = \Gamma_R = 1/\tau_R$ complying with causality requirements. It is remarkable that although this definition has long been known, it has only been very recently incorporated explicitly in the Particle Data [1]. Other definitions (zeros of the K-matrix, phase shift passing through 90°, maximum of cross section, etc.), although simpler to work with, coincide with the former one only in the limit $\Gamma_R \to 0$. For non-relativistic potential scattering this limit produces the exponential law and corresponds to the picture that a stable particle tunnels through the barrier to the continuum. Nevertheless, one should pay attention to the fact that going to the complex $s-$plane in terms of data on the real axis may be a model dependent operation,

---

[1] Presented by E. Ruiz Arriola at the 2nd International Workshop On Hadron Physics: Effective Theories Of Low-Energy QCD Coimbra 2002

particularly for not very sharp resonances. Conversely, going from the pole in the unphysical sheet to the real scattering line is also not uniquely defined, and the energy dependence may be model dependent as well. This ambiguity is enhanced in the case where there are several open channels, because there are many ways of parameterizing a multichannel scattering amplitude with a given analytical structure [2]. The situation is aggravated by the fact that very often the analysis of data in terms of partial wave amplitudes are not provided with error estimates.

In the quark model baryon resonances are naturally interpreted as bound state composites of three valence quarks, and their widths are computed as matrix elements of hadronic transition operators between the bound and continuum states [3]. This approach corresponds to the picture of unstable particles weakly coupled to a continuum. To improve on this, the scattering problem has to be solved. The information entering such a problem is baryon masses and baryon-meson form factors; the only specific reminder of the underlying quark degrees of freedom has to do with these form factors. Although quark models are very fruitful and provide a lot insight into the problem they suffer at least from two deficiencies regarding the description of meson-baryon scattering. Firstly, assuming that masses of hadrons come out properly as a result of a quark model calculation, there is still the possibility that it fails when describing the scattering process. This could only be interpreted as a failure of the particular set of interactions or approximations used to solve the quark model, but does not prevent from finding another quark model providing a better description. In the second place, at the hadronic level symmetries, such as relativistic invariance and chiral symmetry, which are known to work well, are difficult to impose at the hadronic level if hadrons are described as composites.

An alternative and not necessarily incompatible (although perhaps more economical) point of view to the quark model calculations is to formulate the whole problem directly in terms of hadronic degrees of freedom. This allows to impose all known information, symmetries in particular, from the beginning, and offers the possibility to falsify all unknown information. It is of course impossible to do this at all energies, but it may be achieved at low energies, in terms of a finite and countable number of unknown parameters, the so-called low energy constants. This is the point of view of Chiral Perturbation Theory [4, 5]. At the level of a relativistic Lagrangean this corresponds to write down the most general infinite set of tree level operators compatible with the known symmetries. Since a tree level Lagrangean only produces real amplitudes and hence violates unitarity it is necessary to incorporate quantum (loop) corrections.

In the meson-baryon system there is a problem in Chiral Perturbation Theory already found long ago [6] because in the standard dimensional regularization, heavy particles do not decouple. This result is counter-intuitive because it means that particles with a very small Compton wavelength propagate. Two approaches have been suggested to overcome this difficulty. In Heavy Baryon Chiral Perturbation Theory (HBChPT) [7, 8] one takes the non-relativistic limit first and then proceeds in dimensional regularization. The heavy particle decoupling is explicitly built in, but relativistic invariance is not manifest at any step of the calculation. Within this framework elastic πN scattering has been studied to third [9, 10] and fourth order [11] in the chiral expansion and also $\bar{K}N$ elastic scattering [14]. A more recent proposal keeps relativistic invariance explicitly at any step of the calculation but introduces a new so-called infrared regularization which complies with decoupling in the heavy particle baryon [12] and allows a satisfactory description of πN elastic scattering [13]. In either case, although crossing is exact at any order of the expansion unitarity is only built in perturbatively. The need for unitarization in the $S = -1$ channel becomes obvious after the work of Ref. [14] where it is shown that HBChPT to one loop fails completely in the $\bar{K}N$ channel already at threshold due to nearby subthreshold $\Lambda(1405)$-resonance.

The Bethe-Salpeter equation (BSE) provides the framework beyond perturbation theory to treat the relativistic two body problem from a Quantum Field Theory point of view. This approach allows to treat the scattering problem preserving exact unitarity. In practical applications, however, the number of particles is kept fixed and other approximations are done, violating crossing symmetry. At the level of partial waves unitarity implies a right cut discontinuity while crossing generates a left cut for the scattering amplitude, but in the scattering region one expects the energy dependence to by mainly determined by the right cut.

The $s-$wave meson-baryon scattering incorporating chiral symmetry and unitarization for several open channels has been studied in previous works [15, 16, 17, 18, 19, 20, 21]. The purpose of the present contribution is to give a brief overview of *a possible* approximation scheme for meson-baryon scattering based on the BSE. This is the natural extension of work previously done for meson-meson scattering [22] to the meson-baryon system [23] for heavy baryons and in a relativistic formulation [24, 25, 26].

S-WAVE MESON-BARYON SCATTERING

The coupled channel scattering amplitude for the baryon-meson process in given isospin channel I is given by

$$(T_P)_{BA} = \bar{u}_B(P - k', s_B) t_P(k, k') u_A(P - k, s_A) \tag{1}$$

Here, $u_A(P - k, s_A)$ and $u_B(P - k', s_B)$ are baryon Dirac spinors normalized as $\bar{u}u = 2M$, P is the conserved total CM four momentum, $P^2 = s$, and $t_P(k, k')$ is a matrix in the Dirac and coupled channel spaces. Details on normalizations and definitions of the amplitudes can be seen in Ref. [24].

On the mass shell and using the equations of motion for the free Dirac spinors $(\not{P} - \not{k} - M_A)u_A(P - k) = 0$ and its transposed $\bar{u}_A(P - k)(\not{P} - \not{k} - M_A) = 0$ the parity and Lorentz invariant amplitude t_P relevant $s-$wave scattering can be simply written as a matrix function in coupled channel space of $\not{P}$ with P the total CM momentum

$$t_P(k, k')|_{\text{on-shell}} = t(\not{P}) \tag{2}$$

In terms of the matrix $t(\not{P})$ defined in Eq. (2), the $s-$wave coupled-channel matrix, $f_0^{\frac{1}{2}}(s)$, is simply given by:

$$\left[f_0^{\frac{1}{2}}(s) \right]_{B \leftarrow A} = -\frac{1}{8\pi\sqrt{s}} \sqrt{\frac{|\vec{k}_B|}{|\vec{k}_A|}} \sqrt{E_B + M_B} \sqrt{E_A + M_A} \left[t(\sqrt{s}) \right]_{BA} \tag{3}$$

36

where the CM three–momentum moduli read

$$|\vec{k}_i| = \frac{\lambda^{\frac{1}{2}}(s,M_i,m_i)}{2\sqrt{s}} \qquad i = A,B \tag{4}$$

with $\lambda(x,y,z) = x^2 + y^2 + z^2 - 2xy - 2xz - 2yz$ and $E_{A,B}$ the baryon CM energies. The phase of the matrix T_P is such that the relation between the diagonal elements ($A = B$) in the coupled channel space of $f_0^{\frac{1}{2}}(s)$ and the in-elasticities (η) and phase-shifts (δ) is the usual one,

$$\left[f_0^{\frac{1}{2}}(s)\right]_{AA} = \frac{1}{2i|\vec{k}_A|}\left(\eta_A(s)e^{2i\delta_A(s)} - 1\right) \tag{5}$$

Hence, the optical theorem reads, for $s \geq (M_A + m_A)^2$,

$$\frac{4\pi}{|\vec{k}_A|}\mathrm{Im}\left[f_0^{\frac{1}{2}}(s)\right]_{AA} = \sum_B \sigma_{B \leftarrow A} = 4\pi\sum_B\left|\left[f_0^{\frac{1}{2}}(s)\right]_{BA}\right|^2 = \sigma_{AA} + \frac{\pi}{|\vec{k}_A|^2}\left(1 - \eta_A^2\right) \tag{6}$$

where in the right hand side only open channels contribute.

CHIRAL BARYON-MESON LAGRANGIAN

At lowest order in the chiral expansion the chiral baryon meson Lagrangian contains kinetic and mass baryon pieces and meson-baryon interaction terms and is given by [5]

$$\mathscr{L}_1 = \mathrm{Tr}\{\bar{B}(i\slashed{\nabla} - M_B)B\} + \frac{1}{2}\mathscr{D}\,\mathrm{Tr}\{\bar{B}\gamma^\mu\gamma_5\{u_\mu,B\}\} + \frac{1}{2}\mathscr{F}\,\mathrm{Tr}\{\bar{B}\gamma^\mu\gamma_5[u_\mu,B]\}, \tag{7}$$

The meson kinetic and mass pieces and the baryon mass chiral corrections are second order and read

$$\begin{aligned}\mathscr{L}_2 &= \frac{f^2}{4}\mathrm{Tr}\left\{u_\mu^\dagger u^\mu + (U^\dagger\chi + \chi^\dagger U)\right\} \\ &- b_0\mathrm{Tr}(\chi_+)\mathrm{Tr}(\bar{B}B) - b_1\mathrm{Tr}(\bar{B}\chi_+ B) - b_2\mathrm{Tr}(\bar{B}B\chi_+)\end{aligned} \tag{8}$$

where "Tr" stands for the trace in $SU(3)$. In addition,

$$\begin{aligned}\nabla_\mu B &= \partial_\mu B + \frac{1}{2}[u^\dagger\partial_\mu u + u\partial_\mu u^\dagger, B], \\ U = u^2 &= e^{i\sqrt{2}\Phi/f}, \qquad u_\mu = iu^\dagger\partial_\mu U u^\dagger \\ \chi_+ &= u^\dagger\chi u^\dagger + u\chi^\dagger u, \qquad \chi = 2B_0\mathcal{M}\end{aligned} \tag{9}$$

M_B is the common mass of the baryon octet, due to spontaneous chiral symmetry breaking for massless quarks. The $SU(3)$ coupling constants which are determined by semileptonic decays of hyperons are $\mathscr{F} \sim 0.46$, $\mathscr{D} \sim 0.79$ ($\mathscr{F} + \mathscr{D} = g_A = 1.25$). The constants B_0 and f (pion weak decay constant in the chiral limit) are not determined by the symmetry. The current quark mass matrix is $\mathscr{M} = \text{Diag}(m_u, m_d, m_s)$. The parameters b_0, b_1 and b_2 are coupling constants with dimension of an inverse mass. The values of b_1 and b_2 can be determined from baryon mass splittings, whereas b_0 gives an overall contribution to the octet baryon mass M_B. Neglecting octet-singlet mixing, the $SU(3)$ matrices for the meson and the baryon octet are written in terms of the meson and baryon spinor fields respectively and are given by

$$\Phi = \begin{pmatrix} \frac{1}{\sqrt{2}}\pi^0 + \frac{1}{\sqrt{6}}\eta & \pi^+ & K^+ \\ \pi^- & -\frac{1}{\sqrt{2}}\pi^0 + \frac{1}{\sqrt{6}}\eta & K^0 \\ K^- & \bar{K}^0 & -\frac{2}{\sqrt{6}}\eta \end{pmatrix}, \tag{10}$$

and

$$B = \begin{pmatrix} \frac{1}{\sqrt{2}}\Sigma^0 + \frac{1}{\sqrt{6}}\Lambda & \Sigma^+ & p \\ \Sigma^- & -\frac{1}{\sqrt{2}}\Sigma^0 + \frac{1}{\sqrt{6}}\Lambda & n \\ \Xi^- & \Xi^0 & -\frac{2}{\sqrt{6}}\Lambda \end{pmatrix}. \tag{11}$$

respectively. The $MB \to MB$ vertex obtained from the former Lagrangian reads

$$\mathscr{L}_{MB \to MB} = \frac{i}{4f^2} \text{Tr}\left\{ \bar{B}\gamma^\mu \left[[\Phi, \partial_\mu \Phi], B \right] \right\}. \tag{12}$$

Assuming isospin conservation, the scattering amplitude (convention $-iT_{MB \to MB} = +i\mathscr{L}_{MB \to MB}$) in the Dirac spinor basis, at lowest order is given by

$$t_P^{(1)}(k, k') = \frac{D}{f^2}(\slashed{k} + \slashed{k}') \tag{13}$$

where k and k' are incoming and outgoing meson momenta and D a coupled-channel matrix. On the mass shell one can use Dirac's equation and gets

$$t^{(1)}(\slashed{P}) \equiv t_P^{(1)}(k, k')|_{\text{on-shell}} = \frac{1}{f^2}\left\{ \slashed{P} - \hat{M}, D \right\} \tag{14}$$

which depends only on the CM momentum $\slashed{P}$. Obviously, the s-wave scattering amplitude $f_{BA}(s)$, defined through Eq. (3) is real and hence cannot satisfy the optical theorem, Eq. (6). Thus, there is need for unitarization.

To take into account the some chiral symmetry breaking effects physical mass splittings and different decay constants must be accounted for. This can be easily accomplished through the prescription

$$D/f^2 \to \hat{f}^{-1} D \hat{f}^{-1} \tag{15}$$

where $\hat{f}$ is a diagonal matrix in the coupled channel space. We will use the D/f^2 notation throughout, meaning Eq. (15) in practice.

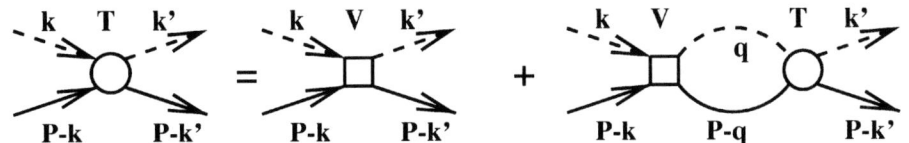

FIGURE 1. Diagrammatic representation of the Bethe-Salpeter Equation for Meson-Baryon Scattering. k is the meson momentum and $P - k$ the Baryon momentum. $s = P^2$.

BETHE-SALPETER EQUATION

Two Particle Unitarity

To evaluate the meson-baryon scattering amplitude t_P we use the BSE (see Fig. (1)) in a given isospin and strangeness channel,

$$t_P(k,k') = v_P(k,k') + i \int \frac{d^4 q}{(2\pi)^4} t_P(q,k') \Delta(q) S(P-q) v_P(k,q) \tag{16}$$

where $t_P(k,k')$ is the scattering amplitude defined in Eq. (1), $v_P(k,k')$ the two particle irreducible Green's function (or *potential*), and $S(P-q)$ and $\Delta(q)$ the baryon and meson exact propagators respectively. The above equation turns out to be a matrix one, both in the coupled channel and Dirac spaces. For any choice of the *potential* $v_P(k,k')$, the resulting scattering amplitude $t_P(k,k')$ fulfills the coupled channel unitarity condition,

$$t_P(k,k') - \bar{t}_P(k',k) = i \int \frac{d^4 q}{(2\pi)^4} t_P(q,k') \operatorname{Disc}\left[\Delta(q) S(P-q)\right] \bar{t}_P(q,k) \tag{17}$$

where $\bar{t}_P(k,p) = \gamma_0 t_P^\dagger(k,p) \gamma_0$ and $t_P^\dagger(k,p)$ is the total adjoint in the Dirac and coupled channel spaces, including also the discontinuity change $s + i\varepsilon \to s - i\varepsilon$). The two particle discontinuity is given by Cutkosky's rules,

$$\operatorname{Disc}\left[\Delta(q) S(P-q)\right] = (-2\pi i)^2 \delta^+\left[q^2 - \hat{m}^2\right]\left(\slashed{P} - \slashed{q} + \hat{M}\right) \delta^+\left[(P-q)^2 - \hat{M}^2\right] \tag{18}$$

$\hat{m}$ and $\hat{M}$ are meson and baryon (diagonal) mass matrices respectively and $\delta^+(p^2 - m^2) = \Theta(p^0)\delta(p^2 - m^2)$ is the on-shell condition. If the on-shell amplitude depends on $\slashed{P}$ only, one gets the very simple relation discontinuity equation for the inverse amplitude,

$$\operatorname{Disc} t(\slashed{P})^{-1} = -\operatorname{Disc} J(\slashed{P}) \tag{19}$$

where the quadratic and logarithmically divergent one-loop integral

$$J(\slashed{P}) = i \int \frac{d^4 q}{(2\pi)^4} \frac{1}{q^2 - \hat{m}^2} \frac{1}{\slashed{P} - \slashed{q} - \hat{M}} \tag{20}$$

has been introduced. Direct calculation leads to

$$J(\slashed{P}) = \slashed{P}\left[\left(\frac{s - \hat{m}^2 + \hat{M}^2}{2s}\right) J_0(s) + \frac{\Delta_{\hat{m}} - \Delta_{\hat{M}}}{2s}\right] + \hat{M} J_0(s) \tag{21}$$

where the quadratic divergences

$$\Delta_{\hat{m}} = i \int \frac{d^4q}{(2\pi)^4} \frac{1}{q^2 - \hat{m}^2}, \qquad \Delta_{\hat{M}} = i \int \frac{d^4q}{(2\pi)^4} \frac{1}{q^2 - \hat{M}^2} \tag{22}$$

and the standard logarithmically divergent one loop function

$$J_0(s) = i \int \frac{d^4q}{(2\pi)^4} \frac{1}{q^2 - \hat{m}^2} \frac{1}{(P-q)^2 - \hat{M}^2} = \bar{J}_0(s) + \hat{J}_{mM} \tag{23}$$

have been introduced. Here $\bar{J}_0(s)$ is normalized to vanish at threshold, $s = (m+M)^2$, and $\hat{J}_{mM}$ is a divergent subtraction constant.

Lowest Order Solution

The BSE requires some input potential and baryon and meson propagators to be solved. At lowest order of the BSE-based chiral expansion, we approximate the iterated *potential* by the chiral expansion lowest order meson-baryon amplitudes in the desired strangeness and isospin channels, and the intermediate particle propagators by the free ones (which are diagonal in the coupled channel space). From the meson-baryon chiral Lagrangian, one gets at lowest order for the *potential*

$$v_P(k,k') = t_P^{(1)}(k,k') = \frac{D}{f^2}(\not{k} + \not{k}') \tag{24}$$

The s−wave BSE can be solved up to a numerical matrix inversion in the coupled channel space [24]. The result for the inverse on-shell amplitude reads

$$t(\not{P})^{-1} + = -J(\not{P}) + \frac{\Delta_{\hat{m}}}{\not{P} - \hat{M}} + \left\{ v(\not{P})^{-1} + \frac{1}{f^4}(\not{P} - \hat{M})D\frac{\Delta_{\hat{m}}}{\not{P} - \hat{M}}D(\not{P} - \hat{M}) \right\}^{-1} \tag{25}$$

This solution manifestly fulfills the on-shell unitarity condition, Eq. (19). Notice that if $\Delta_{\hat{m}} = \Delta_{\hat{M}} = 0$ the tree level *on-shell* potential $v(\not{P}) = t^{(1)}(\not{P})$ determines the amplitude up to a constant J_{mM}. Thus, the difference to do with off-shell effects which we parameterize in terms of divergent constants. Assuming a specific cut-off method for the divergent integrals J_{mM}, Δ_m and Δ_M embodies specific correlations among them, but quite generally one expects them to be uncorrelated.

Renormalization

To renormalize the amplitude given in Eq. (25), we note that in the spirit of an Effective Field Theory (EFT) all possible counter-terms should be considered. This can be achieved in our case in a perturbative manner, making use of the formal expansion of the bare amplitude $T = V + VG_0V + VG_0VG_0V + \cdots$, where G_0 is the two particle propagator. Thus, a counter-term series should be added to the bare amplitude such that the sum of both becomes finite. At each order in the perturbative expansion, the divergent part of the counter-term series is completely determined. However, the finite piece remains arbitrary as long as the used *potential V* and the meson and baryon propagators are approximated rather than being the exact ones. Our renormalization scheme is such that the renormalized amplitude can be cast, again, as in Eq. (25). This amounts in practice, to interpret the previously divergent quantities $\hat{J}_{mM}$, $\Delta_{\hat{m}}$ and $\Delta_{\hat{M}}$ as 12 renormalized free parameters for the s−wave lowest order amplitude in a given isospin-strangeness channel. These parameters and therefore the renormalized amplitude can be expressed in terms of physical (measurable) magnitudes. In principle, they encode the unknown short distance behavior, in particular the composite nature of hadrons which becomes relevant when they start overlapping. In practice it seems convenient to fit them to the available data, although they might be computed within models.

Number of parameters

Within the spirit of an effective field theory at the hadronic level, the number of adjustable parameters should not be smaller than those allowed by the symmetry; this is the only way both to falsify all possible theories embodying the same symmetry principles and to make wider the energy window which is being described. The opposite situation, i.e. a redundancy of parameters is also not desirable, but less problematic because it may be detected. The precise number of unknown parameters is mainly controlled to any order of the calculation by crossing symmetry. In a unitarized approach, the best way to avoid this parameter redundancy is to match the unitarized amplitude to one obtained from a Lagrangian formalism as was explicitly done for meson-meson scattering [22]. Unfortunately, there is no standard one loop ChPT calculation for the meson-baryon reaction with open channels to compare with. We comment on this matching below. At present the only practical, but indirect, way to detect such a parameter redundancy is through a fit to experimental data if the correlations in some parameters turn out to be very strong.

Numerical Results

To finish the presentation, we show some selected numerical results for the $I = 1/2$, $S = 0$ and $I = 0$, $S = -1$ channels. Details of the fitting procedure and the relevant experimental data as well as a thorough discussion of errors can be found in Refs. [24] and [26] respectively. In Fig. 2 the πN phase shift and inelastity in the S_{11} channel, as well as transition cross sections are depicted. Finally, in Fig. 3 we show our results for several amplitudes and cross sections in the $I = 0$, $S = -1$ channel, compared to the work of Ref. [20] where $\Delta_m = \Delta_M = 0$. As we see, the description is quite satisfactory.

Scattering lengths for the are given in Tab. 1, compared experiment and other determinations for the $I = 1/2$, $S = 0$ and $I = 0$, $S = -1$ channels. Errors are assigned by propagating best fit parameter uncertainties including possible correlations [24, 26].

41

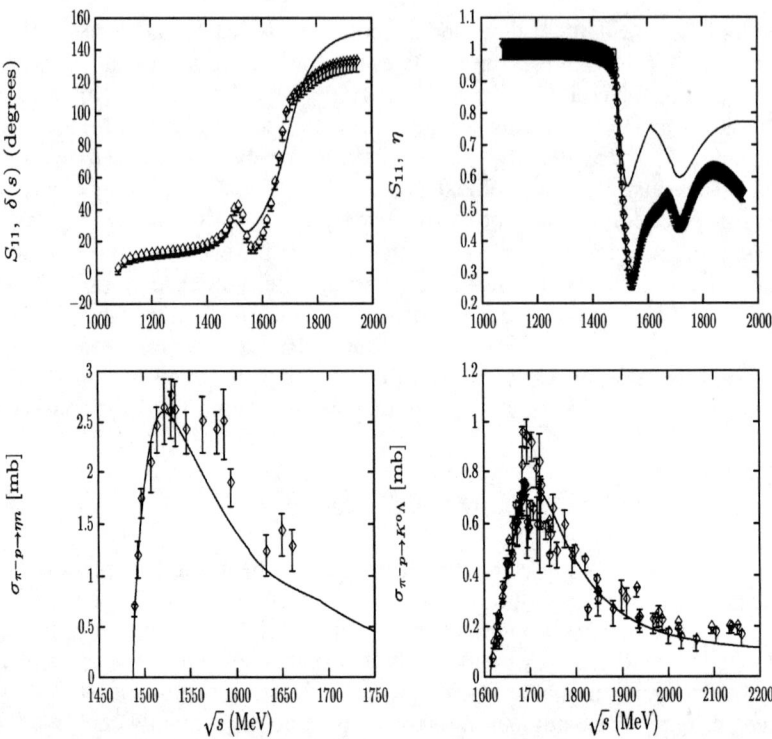

FIGURE 2. πN scattering BSE results as a function of C.M. energy $\sqrt{s}$. Upper left figure: S_{11} phase shifts. Upper right figure: inelasticity inqy the πN channel. Lower left figure: $\pi N \to \eta N$ cross section. Lower right figure: $\pi N \to K\Lambda$ cross section. Data from Ref. [27]. (See Ref. [24] for further details.)

The static limit

In the static limit, baryons behave like fixed sources, and consequently the two particle problem should reduce to a potential scattering problem (in our case of meson-baryon scattering it would correspond to a Klein-Gordon equation with a spin-dependent potential). It is a known fact that the BSE has difficulties in reproducing this heavy-light limit in certain situations (ladder approximation to one boson exchange [36]). The remedy to this situation is to make use of the Gross Equation, which essentially

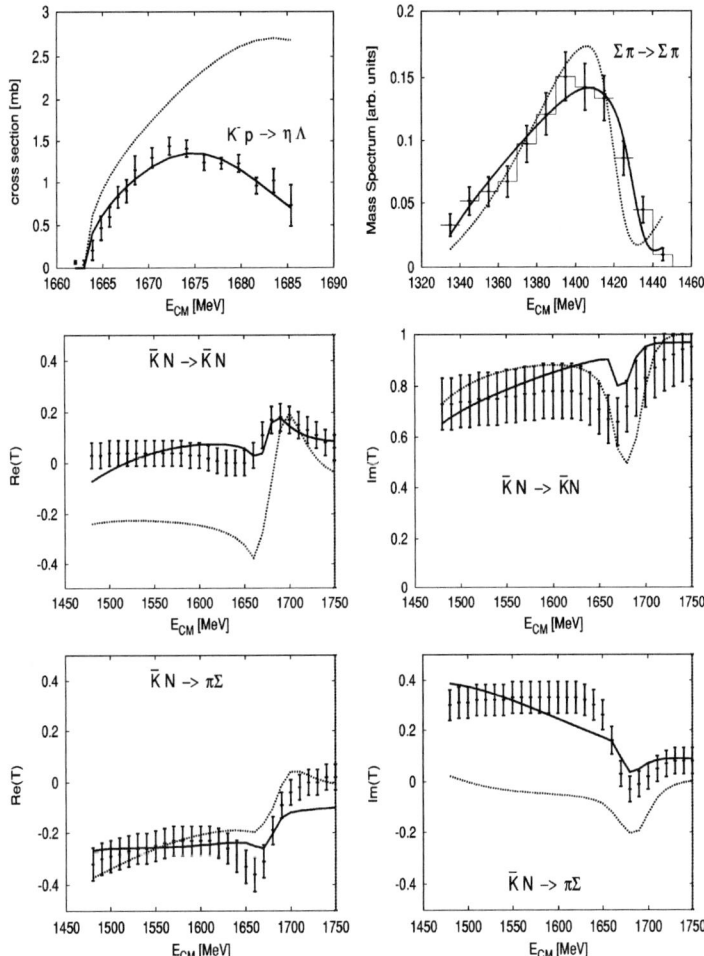

FIGURE 3. Best fit results for the Bethe-Salpeter equation in the $S = -1, I = 0$ channel (solid lines) compared to Ref. [20] (doted lines) where $\Delta_m = \Delta_M = 0$. **Upper panel:** Experimental data for $\pi\Sigma \to \pi\Sigma$ and $K^- p \to \eta\Lambda$ are from Refs. [29] and [31], respectively. **Middle panel:** The real (left panel) and imaginary (right panel) parts of the s-wave T-matrix, with normalization specified in the main text, for elastic $\bar{K}N \to \bar{K}N$ process in the $I = 0$ isospin channel as functions of the CM energy. Experimental data are taken from the analysis of Ref. [28] with the errors stated in Ref. [26]. **Lower Pannel:** Same as middle pannel for the inelastic channel $\bar{K}N \to \pi\Sigma$.

TABLE 1. Real and imaginary parts of scattering lengths (in fm) for best fit BSE results in several isospin and strangeness elastic channels.

	I	S	Re a (fm)	Im a (fm)	Re a (fm)	Im a (fm)
πN	1/2	0	0.179(4)	0	0.252(6)*	0
					0.19(5)†	0
ηN	1/2	0	0.772(5)	0.217(3)	0.68**	0.24
$K\Lambda$	1/2	0	0.0547(5)	0.032(4)		
$\pi\Sigma$	0	-1	1.10(6)	0		
$\bar{K}N$	0	-1	-1.20(9)	1.29(9)	-1.71‡	0.68
$\eta\Lambda$	0	-1	0.50(5)	0.27(1)		

* Experiment Ref.[33]
† HBChPT Ref.[38]
** Potential model Ref.[27]
‡ Experiment, Ref. [35]

means putting the would-be heavy particle on the mass shell from the very beginning. In Chiral Perturbation Theory this is also a tricky point in the relativistic formulation, since dimensional regularization does not lead to decoupling of heavy particles, which has only been solved after a clever choice of renormalization scheme, the so-called infrared regularization [12].

The heavy baryon expansion may be taken by making the baryon masses $\hat{M} \to \infty$ but keeping the meson masses, $\hat{m}$, and the meson momentum, q, finite and considering baryon mass splittings as higher order effects, see e.g. [5],

$$\hat{M} = M_B + \Delta\hat{M} \tag{26}$$

with $M_B \to \infty$ the common mass of the baryon octet which is proportional to the identity matrix. Accordingly, in the πN elastic channel we take

$$\sqrt{s} = E + \omega = M_N + \omega + \frac{\omega^2 - m_\pi^2}{2M_N} + \cdots \tag{27}$$

where $M_N = M_B + \Delta M_N$. Following Ref. [24] in the static limit one obtains from Eqs. (3) and (25) $(f(\omega) \to -t(s)/(4\pi))$

$$f(\omega)^{-1} = 8\pi \left[\bar{K}_{\hat{m}}(\omega) + \frac{1}{16\pi^2} \ln \frac{\hat{M}^2}{\hat{m}^2}(\hat{m} - \omega) + \hat{M}J^0_{\hat{m},\hat{M}} + \frac{\Delta^0_{\hat{m}\hat{M}}}{4\hat{M}} \right]$$
$$- \frac{4\pi}{\omega} \left\{ \Delta^0_{\hat{m}} - \left[\frac{2}{f^2}D + \frac{1}{f^4}D\Delta^0_{\hat{m}}D \right]^{-1} \right\} \tag{28}$$

with the heavy baryon one-loop integral

$$K_{\hat{m}}(\omega) = \frac{1}{i} \int \frac{d^4q}{(2\pi)^4} \frac{1}{q^2 - \hat{m}^2} \frac{1}{\omega - v \cdot q} \tag{29}$$

and $\bar{K}_{\hat{m}}(\omega) = K_{\hat{m}}(\omega) - K_{\hat{m}}(m) - (\omega - m)K'_{\hat{m}}(m)$ and the heavy baryon approximation of the subtraction constants is denoted by the superscript 0. Eq. (28) corresponds, as it should, to a one particle scattering problem, fulfilling the coupled channel unitarity condition

$$\text{Im} f(\omega)^{-1} = -\sqrt{\omega^2 - \hat{m}^2}\, \theta(\omega - \hat{m}) \tag{30}$$

The pole in Eq. (28) for the inverse amplitude is a static limit reminiscent from the baryonic Adler zero, $\sqrt{s} - \hat{M} = 0$, of the lowest order potential. The constant combination appearing in the inverse amplitude, Eq. (28), $\hat{M} J^0_{\hat{m}\hat{M}} + \Delta^0_{\hat{m}\hat{M}}/4\hat{M}$ should go to some definite value in the static limit, $M \to \infty$. In case it would diverge, the scattering amplitude would become trivial. Numerical estimates in Ref. [24] suggest that the dangerous combination is not unnaturally large since

$$\hat{M} J^0_{\hat{m}\hat{M}} + \Delta^0_{\hat{m}\hat{M}}/4\hat{M} \sim \hat{m} \tag{31}$$

Chiral and heavy baryon expansions at threshold

If, in addition to a heavy baryon expansion, a chiral expansion in powers of $1/f^2$ is carried out, one should recover in this double expansion some form of the results found in Ref. [9, 10, 38] within HBChPT for the elastic πN scattering amplitude. The results of Ref. [24] show that the matching to HBChPT is indeed possible, although there is an ambiguity related to the absence of a left hand cut in our BSE amplitude, Eq. (25), which is approximated with a polynomial in the scattering region. The net result is that there is a large degree of redundancy in the BSE parameters, which hardly impose any practical constraints on them[2].

Moreover, a simple expansion in $1/f^2$ of the BSE amplitude keeping the baryon masses does not converge well as one expects from the fact that in the Born approximation the matrix elements $\pi N \to \eta N$, $\eta N \to \eta N$ vanish, although in the full amplitude the $N(1535)$ resonance is developed indicating strong rescattering effects (see also below). For instance, taking πN scattering length for $I = 1/2'$ obtained from a best fit to the data as an example one gets [24]

$$a^{I=1/2}_{\pi N} = \underbrace{0.22}_{1/f^2} + \underbrace{\overbrace{0.22}^{\pi N} - \overbrace{1.06}^{K\Lambda} + \overbrace{0.18}^{K\Sigma}}_{1/f^4} + \ldots = 0.179 \text{ fm} \tag{32}$$

Note that the first two contributions leads to a value -0.43fm.

---

[2] In this regard one should mention that matching of unitarized amplitudes to HBChPT does not always work due to a lack of convergence in HBChPT. See the discussion in Refs. [23, 37, 38].

TABLE 2. Resonance Masses and Widths in MeV found for the best fit BSE calculation, compared with PDG data [1].

I	S	M_R(MeV)	Γ_R(MeV)	PDG	M_R(MeV)	Γ_R(MeV)
1/2	0	1497(1)	83(1)	N(1535)	1505(10)	170(80)
1/2	0	1684(1)	194(1)	N(1660)	1660(20)	160(10)
0	-1	1368(12)	250(23)	Unknown		
0	-1	1443(3)	50(7)	Λ (1405)	1407(4)	50(2)
0	-1	1678(1)	29(2)	Λ (1670)	1670(10)	35(12)
0	-1				1673(2)*	23(6)

* Experiment. Ref. [32]

Resonances as poles in the second Riemann-sheet

The unitarity condition for the inverse amplitude, expressed as a discontinuity equation reads [24]

$$\text{Disc}\,[t^{-1}(s)] \equiv t_I^{-1}(s+i\varepsilon) - t_I^{-1}(s-i\varepsilon) = 2i\rho(s) \qquad s > (m+M)^2 \tag{33}$$

with real s, where the phase space function

$$\rho(s) = \frac{\lambda^{1/2}(s,m_\pi,M_N)}{16\pi s} \times \frac{(\sqrt{s}+M_N)^2 - m_\pi^2}{2\sqrt{s}} \tag{34}$$

has been introduced, understanding that $\rho(s)$ is a function of the real variable s. Then, analytically continuing the phase space function to all complex plane, one finds the amplitude in the second Riemann sheet,

$$t_{II}^{-1}(z) = t_I^{-1}(z) - 2i\rho(z) \tag{35}$$

Masses and widths computed as poles in the second Riemann sheet $z_R = M_R^2 - iM_R\Gamma_R$, can be looked up at Tab. 2. The quoted theoretical errors are obtained propagating the uncertainties in the best fit parametrs.

It is striking that the unobserved resonance has also been obtained by the authors of Ref. [39] although with $M = 1390$ MeV and $\Gamma = 66$ MeV. Finally, one finds [26] that the resonances *are not* of the Breit-Wigner form, so that the simple relation between residues and branching ratios does not hold[3]. Instead, an extrapolation to the real axis is required to extract branching ratios as provided in the PDG [1] for which often Breit-Wigner or other particular forms are assumed. In this regard, it would be desirable that future editions of the PDG would also incorporate the residues.

---

[3] In the one channel case the Breit-Wigner form implies a direct relation between the residue at the pole and the imaginary part of the pole. This relation only holds in the limit of sharp resonances.

ACKNOWLEDGMENTS

E.R.A. thanks the organizers of the workshop for the invitation and the kind atmosphere provided in Coimbra. We warmly thank E. Oset and A. Ramos for useful discussions. This research was supported by DGES under contracts BFM2000-1326 and PB98-1367 and by the Junta de Andalucia.

REFERENCES

1. K. Hagiwara *et al.*, Phys. Rev. **D 66** (2002) 010001.
2. For a review see, e.g. A. M. Badalyan, L. P. Kok, M. L. Polykarpov and Yu. A. Simonov, Phys. Rep. **82** (1982) 31 and references therein.
3. For a review see e.g. M. Arima, K. Masutani and T. Sato, Prog. Theor. Phys. Suppl. **137** (2000) 169 and references therein.
4. J. Gasser and H. Leutwyler, Annals Phys. **158**, 142 (1984).
5. For a review see e.g., A. Pich, Rep. Prog. Phys. **58** (1995) 563 and references therein.
6. J. Gasser, M.E. Sainio and A. Svarc, Nucl. Phys. **B307** (1988) 779.
7. E. Jenkins and A. V. Manohar, Phys. Lett. **B 255** (1991) 558.
8. V. Bernard, N. Kaiser, J. Kambor and Ulf G. Meissner, Nucl. Phys. **B 388** (1992) 315.
9. M. Mojzis, Eur. Phys. J. **C 2** (1998) 181.
10. N. Fettes, Ulf-G. Meissner and S. Steininger; Nucl. Phys.**A 640** (1988) 199.
11. N. Fettes and Ulf. G. Meißner, Nucl.Phys. **A 676** (2000) 311.
12. T. Becher and H. Leutwyler, Eur. Phys. Jour. **C9** (1999) 643.
13. T. Becher and H. Leutwyler, JHEP **0106**, 017 (2001)
14. N. Kaiser, Phys. Rev. **C 64** (2001) 045204.
15. N. Kaiser, P.B. Siegel and W. Weise, Nucl. Phys. **A594** (1995) 325; *ibid*. Phys. Lett. **B362** (1995) 23.
16. E. Oset and A. Ramos, Nucl. Phys. **A635** (1998) 99.
17. B. Krippa, Phys. Rev. **C 58** (1998) 1333.
18. J. A. Oller and Ulf. G. Meißner, Physl. Lett. **500** (2001) 263.
19. M.Th. Keil, G. Penner and U. Mosel, Phys. Rev. **C 63** (2001) 045202.
20. E. Oset, A. Ramos and C. Bennhold, Phys. Lett. **B 527** 99 (2002), Erratum *ibid*. **B 530** (2002) 260.
21. M. F. M. Lutz and E. E. Kolomeitsev, Nucl. Phys.**A 700** (2002) 193.
22. J. Nieves and E. Ruiz Arriola, Phys. Lett. **B 455** (1999) 30; *ibidem* Nucl. Phys. **A679** (2000) 57.
23. J. Nieves and E. Ruiz Arriola, Phys. Rev. **D63**, (2001) 076001 .
24. J. Nieves and E. Ruiz Arriola, Phys. Rev. **D64** (2001) 116008.
25. C. García Recio, J. Nieves, E. Ruiz Arriola and M. J. Vicente-Vacas, contributed talk at QNP2002, 9-14 June 2002, KFZ Jülich (Germany) (to appear in the proceedings), nucl-th/0209053.
26. C. García Recio, J. Nieves, E. Ruiz Arriola and M. J. Vicente-Vacas, hep-ph/0210311.
27. R.A. Arndt, I.I. Strakovsky, R.L. Workman, and M.M. Pavan, Phys. Rev. **C52**, 2120 (1995).
28. G.P. Gopal *et al.,* Nucl. Phys. **B119** (1977) 362.
29. R.J. Hemingway, Nucl Phys. **B253** (1984) 742.
30. R.H. Dalitz and A. Deloff, J. Phys. G: Nucl. Part. Phys. **17** (1991) 289.
31. A. Starostin *et al.*, Phys. Rev. **C 64** (2001) 055205.
32. D.M. Manley *et al.*, Phys. Rev. Let. **88** (2002) 012002-1.
33. H.-Ch. Schröder *et al.*, Phys. Lett. **B 469** (1999) 25.
34. A. Baldini *et al.*, in Landolt-Börnstein, Vol. 12a (Springer, Berlin, 1988)
35. A.D. Martin, Nucl. Phys. **B 179** (1981) 33.
36. F. Gross, Phys. Rev. **C26** (1982) 2203.
37. J.R. Peláez and A. Gómez Nicola, Phys. Rev. **D62** (2000) 017502.
38. J. Nieves and E. Ruiz Arriola, *hep-ph/0001013*; A. Gómez Nicola, J. Nieves, J.R. Peláez and E. Ruiz Arriola, Phys. Lett. **B486** (2000) 77.
39. D. Jido, A. Hosaka, J. C. Nacher, E. Oset and A. Ramos, Phys. Rev. **C 66**, 025203 (2002)

Excited Baryon Spectroscopy in the $1/N_c$ Expansion

Norberto N. Scoccola

Physics Dept., Comisión Nac. Energía Atómica, Libertador 8250, (1429) Buenos Aires, Argentina.
Universidad Favaloro, Solís 453, (1078) Buenos Aires, Argentina.
Consejo Nacional de Investigaciones Científicas y Técnicas, Argentina.

Abstract. We analyze the masses of the negative parity $SU(6)$ **70**-plet baryons using the $1/N_c$ expansion to order $1/N_c$ and to first order in $SU(3)$ breaking. At this level of precision there are twenty predictions which include the well known Gell-Mann Okubo and equal spacing relations together with four new relations involving $SU(3)$ breaking splittings in different $SU(3)$ multiplets. Although the breaking of $SU(6)$ symmetry occurs at zeroth order in $1/N_c$, it turns out to be small. The dominant source of the breaking is the hyperfine interaction which is of order $1/N_c$. The spin-orbit interaction, of zeroth order in $1/N_c$, is entirely fixed by the splitting between the singlet states $\Lambda(1405)$ and $\Lambda(1520)$, and the spin-orbit puzzle is solved by the presence of other zeroth order operators involving flavor exchange.

INTRODUCTION

The understanding of baryons and their excitations from QCD still represents a wide open chapter in strong interaction physics. With some exceptions, such understanding is largely based on data of different sorts and its analyses by means of models, most prominently the constituent quark model in its different versions [1]. One exception are the ground state baryons, namely the spin 1/2 octet and spin 3/2 decouplet, where definite progress in relating observables to QCD has been achieved by means of effective theories. At low energies Chiral Perturbation Theory (ChPT) is such an effective theory[2]. Beyond the low energy domain the availability of effective theories has been limited. In particular, in the resonance domain (excited baryons) there has not been a well established model independent analysis scheme. In order to formulate an effective theory, it is necessary to identify the small expansion parameters available. In the resonance domain, besides the light quark masses, one such expansion parameter is provided by $1/N_c$. Based on general arguments, it is believed that an expansion in $1/N_c$ as first proposed by 't Hooft [3] should hold in QCD in all regimes. The application of the $1/N_c$ expansion to baryons starting with the pioneering work of Witten [4] has served to understand numerous issues. In the past several years the $1/N_c$ expansion in the ground state baryons has been extensively investigated as summarized in several reviews [5]. The expansion in $1/N_c$ for excited baryons has been proposed and applied in several works [6, 7, 8, 9, 10] with encouraging success.

The importance of having a model independent approach to the physics of excited baryons should be emphasized. On one hand it is still mysterious how and why constituent quark models are a good qualitative, and at times also quantitative, description

CP660, *Hadron Physics: Effective Theories of Low Energy QCD*, edited by A. H. Blin et al.
© 2003 American Institute of Physics 0-7354-0120-9/03/$20.00

of that sector. It is not quite clear either what the specific deficiencies of the different versions of the quark model are. On the other hand, there is currently important experimental improvement being achieved in particular thanks to the N^* program at Jefferson Lab [11], where the study of non-strange resonances is being carried out with unprecedented precision, and also there is important progress towards studying the baryon spectrum by means of lattice QCD simulations [12]. Sorting out and understanding the physics emerging from the old and new experimental data as well as from the lattice would be greatly optimized if an effective theory is available.

Although baryons in the large N_c limit belong to increasingly large representations of the spin and flavor groups, thus giving the appearance that they are increasingly complicated, there is instead a great degree of simplification emerging in that limit. In the ground state sector it was discovered [13, 14] that the requirement that unitarity be fulfilled in π-N elastic scattering in the large N_c limit requires a dynamical spin-flavor symmetry that leads to the mentioned simplification. This dynamical symmetry is a contracted $SU(2F)$ spin-flavor symmetry, where F is the number of flavors. Up to corrections of order $1/N_c$ it is possible to replace the contracted symmetry group by the ordinary $SU(2F)$ group [15]. The possibility of building an effective theory based on this dynamical symmetry rests on the fact that for ground state baryons the symmetry is broken at order $1/N_c$. It is therefore possible to implement the $1/N_c$ expansion around the spin-flavor symmetry limit. Once the spin-flavor multiplet has been identified, which in the case of ground state baryons is the totally symmetric $SU(2F)$ representation with N_c fundamental indexes, the $1/N_c$ expansion of different observables can be carried out in terms of an expansion in composite operators which are sorted according to their order in $1/N_c$. For different static observables, such as masses, magnetic moments, axial couplings, etc., using the Wigner-Eckart theorem a basis of operators can be built in terms of products of the generators of the spin-flavor group. This method was applied in the sixties [16], and when combined with the $1/N_c$ expansion it has lead to very successful analyses of various ground state baryon observables (see Refs.[5] and references therein).

While the $1/N_c$ expansion is implemented rigorously along those lines for the ground state baryons, for excited baryons there is one difficulty of principle. This has to do with the observation [7], to be made more explicit later, that in general spin-flavor symmetry is not exact for excited baryons even in the $N_c \to \infty$ limit. The breaking of spin-flavor symmetry at zeroth order in $1/N_c$ is identified in the constituent quark picture by the coupling of the orbital angular momentum. Such a breaking can give rise to spin-flavor configuration mixing, *i.e.* mixing of different spin-flavor representations, at zeroth order. This would suggest that spin-flavor symmetry cannot be used to formulate the $1/N_c$ expansion. However, it is well established from phenomenology that the orbital angular momentum couples very weakly. This is shown in analyses in the quark model [17, 18] as well as analyses in the $1/N_c$ expansion along the lines followed in this work. The zeroth order spin-flavor symmetry breaking turns out to be in the real world with $N_c = 3$ similar in magnitude to that of order $1/N_c$ breaking effects. This is illustrated by comparing the spin-orbit splitting of 115 MeV between the negative parity singlet Λs, namely the $\Lambda(1405)$ and the $\Lambda(1520)$, to the hyperfine splittings that are of order $1/N_c$ and about 150 MeV in the corresponding negative parity states. This suggests that configuration mixing will as well be small. With this latter assumption, it is appropriate in practice to neglect configuration mixing in studying the spectrum to a level of precision of order $1/N_c$ when $N_c = 3$. In the present analysis of the negative parity baryons that belong primarily to the **70**-plet of $SU(6)$, a systematic error of order $\delta_{mix}^2/\Delta M$ where δ_{mix} is the mixing component of the mass Hamiltonian and ΔM is the splitting between the **70**-plet and the **56**-plet with which it mixes. Unfortunately no solid information exists about negative parity baryons that could be assigned primarily to a **56**-plet, and therefore no convincing estimate can at present be made about that systematic error. Thus, with the hypothesis that configuration mixing is small, the implementation of the $1/N_c$ expansion for excited baryons masses by working within a spin-flavor multiplet can be carried out along similar lines as in the ground state baryons as it was shown in [7, 9, 19, 20].

It should be noticed that excited baryons are not narrow in the large N_c limit [6, 8, 21] as the coupling of pions and kaons mediating the transitions to ground state baryons are of order N_c^0. For this reason, the possibility that excited baryons are built as meson-baryon resonances in large N_c is quite open. At this point it is important to emphasize that the $1/N_c$ analysis does not imply a specific picture of the excited baryons, as it relies entirely on group theoretical arguments and in ordering effective operators in powers of $1/N_c$, and will in particular include the possibility that excited baryons are to a large extent resonances.

In the early study of negative parity baryon masses the non-strange baryons were considered, where the expansion was carried out up to order $1/N_c^2$ [9]. In this contribution we review the work reported in Refs.[19, 20] where the $SU(6)$ **70**-plet masses were analyzed to order $1/N_c$ and to order ε where this latter parameter is a measure of the magnitude of $SU(3)$ breaking by the strange quark mass.

NEGATIVE PARITY BARYON STATES

The excited baryon states that correspond in the constituent quark model to the first radial and orbital excitations fit quite well into respectively a positive parity **56**$^+$ and a negative parity **70**$^-$ irreducible representation of the spin-flavor group $SU(6)$. Although in both cases not all the states have been experimentally determined, it seems that with those that are well established, namely those assigned a status of at least three stars by the Particle Data Group [22], it is safe to establish that observation. This conclusion is reinforced by the success of the analyses of the masses in both groups of states [9, 19, 21].

In a constituent quark picture and in the large N_c limit, the lowest baryonic excitations consist of a core of $N_c - 1$ quarks in the ground state of the Hartree effective potential, the core being therefore in the totally symmetric spin-flavor representation, and an excited quark, which for the negative parity baryons discussed in this work is in an $\ell = 1$ state [7, 6]. The states therefore fill the $(3, \mathbf{70})$ multiplet of the group $O(3) \otimes SU(6)$. Phenomenology strongly indicates that even when the current quark masses are small and the constituent quark picture is most likely not accurate, the states can still be identified as belonging to the $(3, \mathbf{70})$. Since for the group theoretical aspects of the analysis in this work the use of the constituent quark picture can be made with no loss of generality, throughout the constituent quark terminology will be often used.

In order to explicitly build the states, it is first convenient to give a brief review of the $SU(6)$ group. It has thirty five generators, namely $\{S_i, T_a, G_{ia}\}$, with $i = 1, 2, 3$ and $a = 1, \cdots, 8$, where the first three are the generators of the spin $SU(2)$, the second eight are the generators of flavor $SU(3)$, and the last twenty four can be identified as an octet of axial-vector currents in the limit of zero momentum transfer. The algebra of $SU(6)$ has the following commutation relations that fix the normalizations of the generators:

$$
\begin{aligned}
\left[S_i, S_j\right] &= i\varepsilon_{ijk}S_k \\
\left[T_a, T_b\right] &= if_{abc}T_c \\
\left[G_{ia}, G_{jb}\right] &= \frac{i}{4}\delta_{ij}f_{abc}T_c + \frac{i}{2}\varepsilon_{ijk}\left(\frac{1}{3}\delta_{ab}S_k + d_{abc}G_{kc}\right) \\
\left[S_i, G_{ja}\right] &= i\varepsilon_{ijk}G_{ka} \\
\left[T_a, G_{ib}\right] &= if_{abc}G_{ic},
\end{aligned}
\tag{1}
$$

where d_{abc} and f_{abc} are the usual $SU(3)$ symmetric and antisymmetric tensors, respectively. In the non-relativistic quark picture these generators can be expressed in terms of the quark fields:

$$
S_i = q^\dagger \frac{\sigma_i}{2} q \quad , \quad T_a = q^\dagger \frac{\lambda_a}{2} q \quad , \quad G_{ia} = q^\dagger \frac{\sigma_i \lambda_a}{4} q,
\tag{2}
$$

where the Gell-Mann matrices are normalized as $\mathrm{Tr}\lambda_a\lambda_b = 2\delta_{ab}$.

The states in the totally symmetric irreducible representation S are given by a Young tableau that consists of a single row of N_c boxes, and the mixed symmetric irreducible representation MS relevant to this work consists of a row with $N_c - 1$ boxes and a second row with one box.

In order to build the states belonging to the mixed symmetric $SU(6)$ irreducible representation it is convenient to start by considering the states

$$
|SS_z; (p,q), Y, II_z; S^c > = \sum \begin{pmatrix} S^c & \frac{1}{2} & S \\ S^c_z & s_z & S_z \end{pmatrix} \begin{pmatrix} (p^c, q^c) & (1,0) & (p,q) \\ (Y^c, I^c I^c_z) & (y, \frac{1}{2} i_z) & (Y, II_z) \end{pmatrix}
$$

$$
\times \ |S^c S^c_z; (p^c, q^c), Y^c, I^c I^c_z\rangle \, |\frac{1}{2} s_z; (1,0), y, \frac{1}{2} i_z\rangle,
\tag{3}
$$

where S is the the total spin of the baryon associated with the spin group $SU(2)$, S^c is the core spin (the core is in the S representation of SU(6)), Y and I are the hypercharge and isospin respectively, and (p,q) indicates the $SU(3)$ irreducible representation. For $SU(3)$ a Young tableau denoted by the pair (p,q) consists of $p+q$ boxes in the first row and q boxes in the second. From the decomposition of the S representation of $SU(6)$ as a sum of direct products of irreducible representations of $SU(2) \otimes SU(3)$ it results that $p^c + 2q^c = N_c - 1$ and $p^c = 2S^c$. This latter relation is a consequence of the fact that for the S representation the two factors in the direct products involved in the decomposition have the same Young tableau (p,q). The rule then results from the fact that $p = 2S$ in $SU(2)$. The $p^c = 2S^c$ relation is a generalization of the so called $I = J$ rule for two flavors.

Not all the states displayed in Eqn. (3) are in the MS irreducible representation of $SU(6)$. While states with $p \neq 2S$ belong automatically in the MS representation, those with $p = 2S$ are a linear combination of states in the S and MS representations. The corresponding S and MS states are easily obtained by considering in each representation the quadratic Casimir invariant of SU(6), namely $C^{(2)}_{SU(6)} = 2\,G_{ia}G_{ia} + \frac{1}{2}C^{(2)}_{SU(3)} + \frac{1}{3}C^{(2)}_{SU(2)}$. For the S representation $C^{(2)}_{SU(6)} = 5N_c(N_c + 6)/12$, and for the MS representation $C^{(2)}_{SU(6)} = N_c(5N_c + 18)/12$. Making use of these relations the $p = 2S$ states in the MS representation turn out to be given by

$$
\begin{aligned}
| SS_z; (p = 2S, q), Y, II_z >_{\text{MS}} &= \sqrt{\frac{S(N_c + 2(S+1))}{N_c(2S+1)}} \; | SS_z; (p,q), Y, II_z; \; S^c = S + \frac{1}{2} > \\
&- \sqrt{\frac{(S+1)(N_c - 2S)}{N_c(2S+1)}} \; | SS_z; (p,q), Y, II_z; \; S^c = S - \frac{1}{2} > . \quad (4)
\end{aligned}
$$

The states belonging to the $(3,\mathbf{70})$ of $O(3) \otimes SU(6)$ are now expressed by including the orbital angular momentum:

$$
|JJ_z; S; (p,q), Y, II_z >_{\text{MS}} = \sum \begin{pmatrix} S & \ell & J \\ S_z & m & J_z \end{pmatrix} |SS_z; (p,q), Y, II_z >_{\text{MS}} | \ell\, m \rangle, \quad (5)
$$

where J is the total angular momentum of the baryon. For $N_c = 3$ the negative parity states span the $(3,\mathbf{70})$ irreducible representation. Expressing them in the obvious notation $^{2S+1}d_J$, they are as follows: five $SU(3)$ octets ($^2 8_{\frac{1}{2}}$, $^2 8_{\frac{3}{2}}$, $^4 8_{\frac{1}{2}}$, $^4 8_{\frac{3}{2}}$, $^4 8_{\frac{5}{2}}$), two decouplets ($^2 10_{\frac{1}{2}}$, $^2 10_{\frac{3}{2}}$), and two singlets ($^2 1_{\frac{1}{2}}$, $^2 1_{\frac{3}{2}}$). The corresponding states in these multiplets are shown in the first column of Table 2.

The octets ($p = 1$) of spin 1/2 satisfy $p = 2S$ and therefore involve a linear combination of core states as specified in Eqn. (4). All other states have $p \neq 2S$ and have therefore core states with well defined spin: the spin 3/2 octets as well as the decouplets ($p = 3$) have $S^c = 1$, and the two singlet Λ states ($p = 0$) have $S^c = 0$.

In the limit of exact $SU(3)$ symmetry there are two possible mixings induced by interactions that break spin-flavor symmetry. These mixings are between the pairs of states that are in the two octets with same J. The mixing angles are defined according to:

$$
\begin{pmatrix} 8_J \\ 8_J' \end{pmatrix} = \begin{pmatrix} \cos\theta_{2J} & \sin\theta_{2J} \\ -\sin\theta_{2J} & \cos\theta_{2J} \end{pmatrix} \begin{pmatrix} ^2 8_J \\ ^4 8_J \end{pmatrix}, \quad (6)
$$

where $J = \frac{1}{2}$ and $\frac{3}{2}$, $8_J^{(\prime)}$ are mass eigenstates, and the mixing angles are constrained to be in the interval $[0, \pi)$. At this point it is easy to check that in the $SU(3)$ limit there are nine masses and two mixing angles, i.e., eleven observables.

MASS OPERATORS

In the subspace of the MS states, the mass operator can be expressed in terms of a linear combination of composite operators sorted according to their order in $1/N_c$. A basis of such composite operators can be constructed using the $O(3) \otimes SU(6)$ generators, distinguishing two sets according to whether the generator acts on the core or on the excited quark. Generators acting on the core will be denoted by $\{S_i^c, T_a^c, G_{ia}^c\}$ and those acting on the excited quark by $\{s_i, t_a, g_{ia}\}$. Operators can be classified according to their n-body character. Thus, operators containing a product of n $SU(6)$ generators acting only on the core, and operators containing a product of $n-1$ $SU(6)$ generators acting on the core and at least one generator of $O(3) \otimes SU(6)$ acting on the excited quark, are said to be n-body operators. The $1/N_c$ counting can then be determined by the following two criteria:

i) An n-body operator requires that at least $(n-1)$ gluons be exchanged between the n quarks, and thus, the coefficient that multiplies the operator in the effective theory is proportional to $1/N_c^{(n-1)}$. Henceforth, this factor will be absorbed in the definition of the operator.

ii) In the large N_c limit the $SU(6)$ generators G_{ia}^c with $a = 1, 2, 3$, and T_8^c have matrix elements of order N_c between states with spin and strangeness of order N_c^0, and they are therefore called coherent generators. The presence of coherent factors in a composite operator leads to an enhancement of its matrix elements between such states given by a power of N_c. In order to determine that power it is necessary to first reduce the products of generators by means of the commutation relations.

At this point it should be noted that there is an ambiguity in the identification of the physical states for $N_c = 3$ with the states in the large N_c spin-flavor multiplet. As in previous works, we resolve this ambiguity by identifying the physical states with those of spin and strangeness of order N_c^0 so that (ii) can be applied.

In the construction of composite operators identities for certain products of generators valid in a given irreducible representation of $SU(6)$ should be used. These identities or reduction rules have been given in Ref. [23] for the case of the S representation of $SU(6)$.

The mass operators in the basis must be rotationally invariant, parity and time reversal even and isospin symmetric. At order N_c the basis of $SU(3)$ singlet operators consist of one operator, namely the identity operator that essentially counts the number of valence quarks in the baryon and, therefore, preserves spin-flavor symmetry. At order N_c^0 only operators involving factors of the orbital angular momentum appear. It is not difficult to understand why there are operators of order N_c^0: by looking at the constituent quark picture, the excited quark moves in the effective potential of order N_c^0 generated by the core giving rise to a spin-orbit interaction (e.g. the $\ell \cdot s$ operator) with a strength of order N_c^0. The matrix elements of the excited quark spin in the MS representation are of order N_c^0, and so the spin-orbit interaction is of that order as well. There are three linearly independent operators of order N_c^0, the operators O_2, O_3 and O_4 listed in Table 1. The 1-body operator O_2 is the ordinary spin-orbit operator, while the remaining operators are 2-body and involve factors carrying flavor. The dynamics giving rise to composite operators involving flavor exchange is not understood but it is likely that long distance effects due to the pion and kaon clouds give an important part of the strength to these

TABLE 1. List of operators and the coefficients resulting from the best fit to the known **70**-plet masses and mixings.

Order	Operator		Fitted coef. [MeV]		
N_c^1	$O_1 = N_c\,1$	$c_1 =$	449	$\pm$	2
N_c^0	$O_2 = l_i\,s_i$	$c_2 =$	52	$\pm$	15
	$O_3 = \frac{3}{N_c}\,l_{ij}^{(2)}\,g_{ia}\,G_{ja}^c$	$c_3 =$	116	$\pm$	44
	$O_4 = \frac{4}{N_c+1}\,l_i\,t_a\,G_{ia}^c$	$c_4 =$	110	$\pm$	16
N_c^{-1}	$O_5 = \frac{1}{N_c}\,l_i\,S_i^c$	$c_5 =$	74	$\pm$	30
	$O_6 = \frac{1}{N_c}\,S_i^c\,S_i^c$	$c_6 =$	480	$\pm$	15
	$O_7 = \frac{1}{N_c}\,s_i\,S_i^c$	$c_7 =$	-159	$\pm$	50
	$O_8 = \frac{2}{N_c}\,l_{ij}^{(2)}\,s_i\,S_j^c$	$c_8 =$	3	$\pm$	55
	$O_9 = \frac{3}{N_c^2}\,l_i\,g_{ja}\{S_j^c,G_{ia}^c\}$	$c_9 =$	71	$\pm$	51
	$O_{10} = \frac{2}{N_c^2}\,t_a\{S_i^c,G_{ia}^c\}$	$c_{10} =$	-84	$\pm$	28
	$O_{11} = \frac{3}{N_c^2}\,l_i\,g_{ia}\{S_j^c,G_{ja}^c\}$	$c_{11} =$	-44	$\pm$	43
$\varepsilon\,N_c^0$	$\bar{B}_1 = t_8 - \frac{1}{2\sqrt{3}N_c}O_1$	$d_1 =$	-81	$\pm$	36
	$\bar{B}_2 = T_8^c - \frac{N_c-1}{2\sqrt{3}N_c}O_1$	$d_2 =$	-194	$\pm$	17
	$\bar{B}_3 = \frac{10}{N_c}\,d_{8ab}\,g_{ia}\,G_{ib}^c + \frac{5(N_c^2-9)}{8\sqrt{3}N_c^2(N_c-1)}O_1 +$				
	$\quad + \frac{5}{2\sqrt{3}(N_c-1)}O_6 + \frac{5}{6\sqrt{3}}O_7$	$d_3 =$	-15	$\pm$	30
	$\bar{B}_4 = 3\,l_i\,g_{i8} - \frac{\sqrt{3}}{2}O_2$	$d_4 =$	-27	$\pm$	19

operators. In particular pion-exchange quark models [24, 25] lead naturally to important flavor exchange contributions. It is a straightforward exercise to check that any other operators of order N_c^0 are linearly dependent with the identity and the three operators of order N_c^0 up to terms of order $1/N_c$. There are eleven operators of order $1/N_c$ in the basis. Since for $N_c = 3$ there are eleven $SU(3)$ singlet mass observables as indicated earlier, and four basis operators have already been identified at orders N_c and N_c^0, it results that for $N_c = 3$ only seven operators of order $1/N_c$ are independent. We consider those listed in Table 1. Among them is the very important hyperfine operator O_6, known to play a crucial role in baryon spectroscopy. Note that there are also three 3-body operators as well.

The breaking of $SU(3)$ symmetry is driven by the mass difference $m_s - \hat{m}$, where $\hat{m}$ is the average of the u- and d-quark masses. A measure of the breaking is given by the ratio $\varepsilon = (M_K^2 - M_\pi^2)/\Lambda^2$ where Λ is a light hadronic scale, for instance a vector meson mass. Here $SU(3)$ breaking will be included to order ε. Note that for $N_c = 3$, ε and $1/N_c$ are of similar size, and therefore corrections of order ε/N_c are neglected. At order ε the effective operators are obviously octets. Explicit construction gives four basis operators, two 1-body and two 2-body operators. Listed in Table 1 are improved operators obtained by combining the octet pieces with singlet operators in such a way that the resulting operators, $\bar{B}_{1,\dots,4}$, have vanishing matrix elements between non-strange baryons. Note that the operator $\bar{B}_2$ consists of two pieces of order N_c, namely T_8^c and the operator O_1. However, the order N_c pieces cancel and $\bar{B}_2$ is actually of order N_c^0. It should be mentioned that, for reasons explained below, several singlet and octet operators differ from those in [9] and [19] by a scaling factor.

The calculation of the matrix elements of these operators in the basis described in the previous section is a rather lengthy task. For this purpose the Wigner-Eckart theorem has been repeatedly used to express any given matrix element in terms of combinations of $SU(2)$ and $SU(3)$ Clebsch-Gordan coefficients and the reduced matrix elements of the elementary operators acting on the core and excited quark, $\{S_i^c, T_a^c, G_{ia}^c\}$ and $\{s_i, t_a, g_{ia}\}$, respectively. As usual such reduced matrix elements were expressed in terms of Clebsch-Gordan coefficients and some particular matrix elements which are simple to evaluate explicitly. Fortunately, for the cases of interest there exist analytic formulas for all the $SU(2)$ coefficients and $SU(3)$ isoscalar factors involved in the calculations. Consequently, it has been possible to derive explicit expressions, in terms of N_c, for all the matrix elements of the singlet and octet operators included in the present analysis[20]. It also is possible to check that certain combinations of operators are demoted to become of higher order in $1/N_c$ in the sector of non-singlet baryons. As it was observed in [9] the combination of zeroth order operators $O_2 + O_4$ is of order $1/N_c$ and the combination of zeroth and first order operators $O_2 + O_4 + \frac{2}{3}O_5 + \frac{8}{3}O_9$ is of order $1/N_c^2$ in that sector.

FITS AND DISCUSSIONS

In terms of the basis of operators introduced in the previous section the **70**-plet mass operator up to order $1/N_c$ and order εN_c^0 has the most general form:

$$M_{70} = \sum_{i=1}^{11} c_i O_i + \sum_{i=1}^{4} d_i \bar{B}_i \quad , \tag{7}$$

where c_i and d_i are the unknown real coefficients to be determined by fitting to the known masses and mixing angles. These coefficients encode the non-perturbative QCD dynamics that cannot be constrained by symmetries. Calculating these coefficients would be equivalent to solving QCD in this baryon sector. Fortunately, the experimental data available in the case of the **70**-plet is enough to obtain them by performing a fit[19]. The inputs to the fit consist of seventeen masses of negative parity baryons which have been assigned a status of three or more stars by the Particle Data Group [22], and the two leading order mixing angles $\theta_1 = 0.61 \pm 0.09$ and $\theta_3 = 3.04 \pm 0.15$ on which there is a rather good consensus about their values as obtained from strong decays of the non-strange members of the multiplet [26, 17, 6]. Note that θ_3 is consistent with zero mod π. Thus, the fifteen coefficients are fitted to these nineteen observables. Of course, a larger number of inputs would have been desirable in order to increase the confidence level of the results. It is however expected that the chief features of the results that are found here are sufficiently well established with these inputs.

Before proceeding to the numerical analysis of the masses, it is important to establish relations among observables that must hold to the order of this analysis. Including $SU(3)$ breaking to all orders there are fifty observables, namely thirty masses and twenty mixing angles. Since there are fifteen operators in the basis being considered, there must be thirty five relations. On the other hand, if $SU(3)$ breaking is considered only to order ε, there are thirty five observables where twenty one of them are masses and fourteen are mixing angles. This then implies that up to order ε and order $1/N_c$ there are twenty relations. Among these relations there are those independent of the leading order mixing angles resulting among traces of mass matrices for states with the same quantum numbers I and J. There are thirteen such relations, which are independent of the coefficients c_i and b_i, all of them involving mass splittings. Five of them are Gell-Mann Okubo relations (one per octet), four equal spacing rules (two per decouplet), and four novel relations that involve mass splittings across $SU(3)$ multiplets. These relations are given by:

$$14(s_{\Lambda_{3/2}} + s_{\Lambda'_{3/2}}) + 63s_{\Lambda_{5/2}} + 36(s_{\Sigma_{1/2}} + s_{\Sigma'_{1/2}}) = 68(s_{\Lambda_{1/2}} + s_{\Lambda'_{1/2}}) + 27s_{\Sigma_{5/2}}$$

$$14(s_{\Sigma_{3/2}} + s_{\Sigma'_{3/2}}) + 21s_{\Lambda_{5/2}} - 9s_{\Sigma_{5/2}} = 18(s_{\Lambda_{1/2}} + s_{\Lambda'_{1/2}}) + 2(s_{\Sigma_{1/2}} + s_{\Sigma'_{1/2}})$$

$$14s_{\Sigma''_{1/2}} + 49s_{\Lambda_{5/2}} + 23(s_{\Sigma_{1/2}} + s_{\Sigma'_{1/2}}) = 45(s_{\Lambda_{1/2}} + s_{\Lambda'_{1/2}}) + 19s_{\Sigma_{5/2}}$$

$$14s_{\Sigma''_{3/2}} + 28s_{\Lambda_{5/2}} + 11(s_{\Sigma_{1/2}} + s_{\Sigma'_{1/2}}) = 27(s_{\Lambda_{1/2}} + s_{\Lambda'_{1/2}}) + 10s_{\Sigma_{5/2}}, \qquad (8)$$

where s_{B_i} is the mass splitting between the state B_i and the non-strange states in the $SU(3)$ multiplet to which it belongs. For the purpose of identifying the states within an $SU(3)$ multiplet the order ε mixing is disregarded. It is important to stress here that these relations are independent of the leading order as well as the order ε mixings.

As already mentioned, several of the singlet and octet operators defined in this article differ from those in [9] and [19] by a scaling factor. The reason for this is that here the operators have been defined in such a way that their matrix elements are of the same order in $1/N_c$ at which the operator contributes. With this, the natural size of the coefficients of singlet operators is about 500 MeV as the analysis below shows, and that of the coefficients of $SU(3)$ breaking is about $\varepsilon \times 500$ MeV $\sim 150 - 200$ MeV.

The fit has been performed by treating the singlet pieces of the mass operator exactly and the $SU(3)$ breaking to first order of perturbation theory in ε. This approach is justified in practice by the fact that the hyperfine interaction turns out to be the dominant spin-flavor breaking piece. The results of the fit are given in Table 1, where the natural size of the coefficients associated with the singlet operators is seen to be set by the coefficient of O_1. In fitting the data the experimental errors given by the Particle Data Group [22] are taken whenever they are larger than the expected magnitude of higher order corrections in the analysis. These corrections are of order ε^2 or ε/N_c, and their magnitude is taken to be about 15 MeV. Although this is not crucial for the outcome of the fit, the resulting χ^2 is more realistic. For instance, the singlet Λ masses are known experimentally within 5 MeV, and taking this error, which in magnitude would correspond to a higher order of precision in the expansion in ε and $1/N_c$, would be unrealistic.

Table 2 displays the empirical masses together with the masses, whose χ^2 per degree of freedom turns out to be 1.29. Also given are the masses provided by the Isgur-Karl model[17].

In order to better understand the outcome of the fit, it is convenient to first emphasize the hierarchy that emerges from the analysis. As it was already found in the analysis of

TABLE 2. Masses (in MeV) as predicted by the $1/N_c$ expansion as compared with empirical values of all states with a status of three or more stars in [22] and those of the quark model calculation of Isgur and Karl [17].

State	Expt.	Large N_c	QM
$N_{1/2}$	1538 ± 18	1541	1490
$\Lambda_{1/2}$	1670 ± 10	1667	1650
$\Sigma_{1/2}$	(1620)	1637	1650
$\Xi_{1/2}$		1779	1780
$N_{3/2}$	1523 ± 8	1532	1535
$\Lambda_{3/2}$	1690 ± 5	1676	1690
$\Sigma_{3/2}$	1675 ± 10	1667	1675
$\Xi_{3/2}$	1823 ± 5	1815	1800
$N'_{1/2}$	1660 ± 20	1660	1655
$\Lambda'_{1/2}$	1785 ± 65	1806	1800
$\Sigma'_{1/2}$	1765 ± 35	1755	1750
$\Xi'_{1/2}$		1927	1900
$N'_{3/2}$	1700 ± 50	1699	1745
$\Lambda'_{3/2}$		1864	1880
$\Sigma'_{3/2}$		1769	1815
$\Xi'_{3/2}$		1980	1985
$N_{5/2}$	1678 ± 8	1671	1670
$\Lambda_{5/2}$	1820 ± 10	1836	1815
$\Sigma_{5/2}$	1775 ± 5	1784	1760
$\Xi_{5/2}$		1974	1930
$\Delta_{1/2}$	1645 ± 30	1645	1685
$\Sigma''_{1/2}$		1784	1810
$\Xi''_{1/2}$		1922	1930
$\Omega_{1/2}$		2061	2020
$\Delta_{3/2}$	1720 ± 50	1720	1685
$\Sigma''_{3/2}$		1847	1805
$\Xi''_{3/2}$		1973	1920
$\Omega_{3/2}$		2100	2020
$\Lambda''_{1/2}$	1407 ± 4	1407	1490
$\Lambda''_{3/2}$	1520 ± 1	1520	1490

the non-strange baryons [9], the singlet operators of order N_c^0 give contributions to the masses that are much smaller than the natural size expected at this order. Indeed, they turn out to be of similar or even smaller magnitude than the natural size expected at order $1/N_c$. Among the operators of order $1/N_c$, the dominant operator is the hyperfine operator O_6 that gives the chief spin-flavor breaking, with all other operators giving contributions that are suppressed with respect to the natural size. This observed hierarchy that goes beyond the simple ordering in powers of $1/N_c$ reflects the dynamics of QCD, and it is interesting to note that in its general aspects it agrees with the hierarchy that results in constituent quark models.

In what follows we present a more detailed description of the role of the different singlet and SU(3) breaking operators.

The hyperfine operator O_6 gives the gross spin-flavor breaking features in the **70**-plet. This operator does not affect the singlet Λ states which involve core states with $S^c = 0$ only, while it increases the masses of the rest of the states according to the $S^c = 1$ content of the cores. The typical mass shifts produced by this operator are 160 MeV, which is the natural size expected from $1/N_c$ counting alone. In particular, this makes clear why the singlet Λ's are the lightest states in the **70**-plet, as it is well known from the early works in the constituent quark model based on QCD [27]. In the large N_c limit the hyperfine interaction in the core should be the same as in the ground state baryons. The $\Delta - N$ splitting gives for the ground state baryons a strength of 100 MeV for the hyperfine interaction defined by $\sum_{i \neq j}^{N_c} s_i \cdot s_j$, while the corresponding strength implied by the result obtained for the coefficient c_6 is equal to 150 MeV. This disagreement is a manifestation of higher order corrections in $1/N_c$ and is in line with the expected magnitude for $N_c = 3$. It should be mentioned here that, as it occurs in the ground state baryons, the hyperfine operator has actually coherent matrix elements between states with spin of order N_c, for which the splittings between states are of zeroth order.

The spin-orbit operator O_2 has the peculiarity of being one of the two operators that affect the singlet Λ masses (note that improved $SU(3)$ breaking operators also do, but only through their singlet pieces proportional to the unity and spin-orbit operators, since the octet piece has vanishing matrix elements for these states). The splitting between the singlet Λ's is therefore a direct measure of the spin-orbit interaction. The fact that the splitting is only 113 MeV indicates the weakness of the spin-orbit effect. This is perhaps the best indication that the formal problem originating in the fact that spin-flavor symmetry is broken at order N_c^0 is rather harmless in practice. Understanding the smallness of the spin-orbit interaction from QCD is an important open dynamical problem.

One interesting observation can be made concerning the splittings between spin-orbit partners in the **56**-plet. The positive parity **56**-plet with $\ell = 2$ containing the spin-orbit partner N^* states $N_{3/2}(1720)$ and $N_{5/2}(1680)$, and the Δ^* states $\Delta_{1/2}(1910)$, $\Delta_{3/2}(1920)$, $\Delta_{5/2}(1905)$ and $\Delta_{7/2}(1950)$ [22] shows very small splittings. Indeed, these splittings are suppressed by a factor one third or smaller with respect to the spin-orbit splitting between the singlet Λ states in the **70**-plet. Since the coupling of orbital angular momentum within the **56**-plet is of order $1/N_c$, it is possible that this suppression is just a manifestation of the $1/N_c$ expansion.

The remaining two operators of order N_c^0 involve flavor exchange and give also contributions that are rather small, but important in the understanding of two issues.

i) The first issue is the so called spin-orbit puzzle in the quark model that can be summarized by the incompatibility between such splittings in the sector of non-strange baryons and in the singlet Λ's. The operator O_4 gives contributions that compensate those of O_2 to the splitting between the spin-orbit partner states $\Delta_{1/2}$ and $\Delta_{3/2}$, where the manifestation of the spin-orbit puzzle was most dramatic. The operators O_3 and O_5 do also give some relevant contributions to such splittings for other states, but they are smaller than those of O_4. A conclusion that can be drawn here is that the spin-orbit puzzle in quark models is resolved by the flavor-exchange effective interactions not included in that model and that appear naturally in the $1/N_c$ analysis.

ii) The second issue is the leading order mixings that are due to the off diagonal matrix elements in the two octet mass matrices. In particular, the mixing angle θ_1, which is the only significant one of the two leading order angles, is almost entirely determined by the operator O_3. The angle θ_3 receives contributions from several operators $(O_{2,\cdots,5,9,11})$ that are of similar magnitude and tend to cancel.

The hyperfine operator O_7 gives very small contributions. This operator involves the spin-spin interactions between the excited quark and the core, which according to the constituent quark picture is suppressed by the centrifugal barrier. The current analysis shows that this operator gives splittings much smaller than those by O_6 and of the order of 25 MeV, in qualitative agreement with the quark picture.

The operator O_5 and the three-body operators give contributions whose magnitude is in the few tens of MeV, i.e. much smaller than the natural size of $1/N_c$ contributions, and have no clearly definite effect with which they could be associated. They do however contribute to the ultimate quality of the fit. Finally, the operator O_8 is clearly irrelevant.

Concerning $SU(3)$ breaking, only three out of the four $SU(3)$ breaking operators are significant giving natural size contributions. $\bar{B}_3$ is weak and can to some extent be disregarded. The dominance of the O_6 operator may indicate that associated octet operators such as $\frac{1}{N_c}S_i^c G_{i8}^c$, which is of order ε/N_c, would be important, even when it appears at higher order than the ones considered in this paper. Although this may be so, the available data for splittings does not allow to pin down the relevance of such an operator. The fact is that with the four improved leading order operators already included the fit is very good, and the inclusion of such operator does not lead to a significant improvement. More data on $SU(3)$ splittings would be required to clarify this issue. The inclusion of such an operator would spoil the splitting relations of Eq. (8). The main observations on $SU(3)$ breaking are the following:

- Only one relation can be tested with available data, namely the Gell-Mann Okubo relation for the $J = 3/2$ octet that is predominantly $S = 1/2$. Determining the masses of the $\Xi'_{1/2}$ and $\Xi_{5/2}$ would complete two more octets and test the corresponding relations. A further octet can be completed that has one two-star state, the $\Sigma_{1/2}(1620)$ state, by finding the corresponding Ξ state predicted by the analysis to have a mass of 1779 MeV.

- The test of equal spacing relations is not possible. Only the Δ states in the two decouplets are known. It is clearly very important at some point in the future to have further decouplet states experimentally pinned down for that purpose. The results obtained here indicate that the splittings in both decouplets are similar and in the range $125 - 135$ MeV, which is the typical splitting produced by one unit of strangeness.

- The four new splitting relation of Eq.(8) cannot be tested at this point because the masses of $\Lambda'_{3/2}$, $\Sigma'_{3/2}$, $\Sigma''_{1/2}$, and $\Sigma''_{1/2}$ that enter respectively in the four relations need to be known. After replacing the known experimental values, each of the splitting relations gives a prediction, namely: $s_{\Lambda'_{3/2}} = 149$ MeV, $s_{\Sigma'_{3/2}} = 55$ MeV,

$s_{\Sigma''_{1/2}} = 154$ MeV, and $s_{\Sigma''_{3/2}} = 134$ MeV. If the operator $\bar{B}_3$ is ignored, five relations result that were given in [19]. In that case one relation could be tested, namely the relation $9(s_{\Sigma_{1/2}} + s_{\Sigma'_{1/2}}) + 21s_{\Lambda_{5/2}} = 17(s_{\Lambda_{1/2}} + s_{\Lambda'_{1/2}}) + 5s_{\Sigma_{5/2}}$; by including the two-star state $\Sigma_{1/2}(1620)$ as input that relation is satisfied to a few percent.

CONCLUSIONS

The $1/N_c$ expansion for excited baryons has been implemented under the assumption that there is an approximate spin-flavor symmetry in the large N_c limit. This assumption relies on the observation that zeroth order violations of this symmetry are very small in practice. Consequently, the only effects that have been left out in the analysis carried out for the $SU(6)$ **70**−plet are related to spin-flavor configuration mixing. Since the analysis shows that the zeroth order spin-flavor breaking in the **70**−plet has a magnitude smaller than the natural size of first order contributions, the scheme is phenomenologically sound.

The analysis also shows that the $1/N_c$ expansion can be consistently applied because there are no corrections that are unnaturally large. On the other hand, a hierarchy emerges in the form of effective coefficients being unnaturally small. In its gross features the picture that emerges is similar to the quark model one, but at a finer level the suppressed dynamics manifests itself in particular through flavor exchange effective interactions, largely absent in most quark models, which are important in describing two effects, namely, the zeroth order mixings and the resolution of the spin-orbit puzzle.

At the level of $SU(3)$ singlet spin-flavor symmetry breaking the level of predictivity is quite limited, the reason being that the number of observables is equal to the number of operators in the singlet basis up to order $1/N_c$. There is however predictivity at the level of $SU(3)$ breaking to order $\varepsilon \times N_c^0$: besides Gell-Mann Okubo and equal spacing relations there are four new relations across different $SU(3)$ multiplets. Unfortunately, with the available data only one Gell-Mann Okubo relation can be tested. This should be a motivation to experimentally establish a few more key states in the **70**−plet.

The present analysis provides a useful framework to sort out and understand results from lattice QCD simulations of excited baryons. The $1/N_c$ expansion allows to separate the contributions that follow from the dynamical $SU(2F)$ symmetry and its breaking, from the non-perturbative reduced matrix elements of the QCD operators. In particular, the $\Lambda(1405)$ appears naturally as the lightest state and a spin-orbit partner of the $\Lambda(1520)$. The spin-spin and spin-orbit interactions that give the gross structure of the **70**-plet are especially interesting and lattice simulations together with the $1/N_c$ analysis could help to further understand their nature.

ACKNOWLEDGMENTS

The results reported is this contribution have been obtained in collaboration with J.L. Goity and C.L. Schat. I would like to thank the organizers for their warm hospitally during my stay in Coimbra. This work was partially supported by the ANPCYT (Argentina) through grant # 03-08580 and by Fundación Antorchas (Argentina).

REFERENCES

1. S. Capstick and W. Roberts, Prog. Part. Nucl. Phys. **45**, S241 (2000).
2. U.-G. Meißner, in "Encyclopedia of Analytic QCD", Vol. 1, 417 (2000), M. Shifman editor; World Scientific (2000).
3. G. 't Hooft, Nucl. Phys. **B72**, 461 (1974).
4. E. Witten, Nucl. Phys **B160**, 57 (1979).
5. A. V. Manohar, "Large N QCD", hep-ph/9802419. Published in Proceedings of "Probing the Standard Model of Particle Interactions". F. David and R. Gupta editors; E. Jenkins, Ann. Rev. Nucl. and Part. Sci. **48**, 81 (1998); R. F. Lebed, Czech. J. Phys. **49**, 1273 (1999).
6. C. D. Carone, H. Georgi, L. Kaplan and D. Morin, Phys. Rev. **D50**, 5793 (1994).
7. J. L. Goity, Phys. Lett. **B414**, 140 (1997).
8. D. Pirjol and T.-M. Yan, Phys. Rev. **D57**, 1449 (1998); **D57**, 5434(1998).
9. C. E. Carlson, C. D. Carone, J. L. Goity, and R. F. Lebed, Phys. Lett. **B438**, 327 (1998); Phys. Rev. **D59**, 114008 (1999).
10. Z. A. Baccouche, C. K. Chow, T. D. Cohen and B. A. Gelman, Nucl. Phys. **A696**, 638 (2001).
11. V. Burkert, "The N^* program at Jefferson Lab: status and prospects", hep-ph/0207149 and references therein.
12. D. G. Richards et al., (QCDSF/UKQCD/LHPC Coll.), Nucl. Phys. Proc. Suppl. **109**, 89 (2002).
    M. Göckeler et al., (QCDSF/UKQCD/LHPC Collaboration), Phys. Lett. **B532**, 63 (2002).
    W. Melnitchouk et al., hep-lat/0202022, and Nucl. Phys. Proc. Suppl. **109**, 96 (2002).
    J. M. Zanotti et al. (CSSM Lattice Collaboration), Phys. Rev. **D65**, 074507 (2002).
    S. Sasaki, T. Blum and S. Ohta, Phys. Rev. **D65**, 074503 (2002).
13. J. L. Gervais and B. Sakita, Phys. Rev. Lett. **52**, 87 (1984); Phys. Rev. **D30**, 1795 (1984).
    K. Bardakci, Nucl. Phys. **B243** 197 (1984).
14. R. Dashen and A. V. Manohar, Phys. Lett. **B315**, 425 (1993); Phys. Lett. **B315**, 438 (1993).
15. C. D. Carone, H. Georgi and S. Osofsky, Phys. Lett. **B322**, 227 (1994).
    M. A. Luty and J. March-Russell, Nucl. Phys. **B426**, 71 (1994).
    M. A. Luty, J. March-Russell, M. White, Phys. Rev. **D51**, 2332 (1995).
    P. Bedaque and M. Luty, Phys. Rev. **D54**, 2317 (1996).
16. O. W. Greenberg and M. Resnikoff, Phys. Rev. **163**, 1844 (1967).
17. N. Isgur and G. Karl, Phys. Rev. **D18**, 4187 (1978).
18. S. Capstick and N. Isgur, Phys. Rev. **D34**, 2809 (1986).
19. C. L. Schat, J. L. Goity and N. N. Scoccola, Phys. Rev. Lett. **88**, 102002 (2002).
20. J. L. Goity, C. L. Schat and N. N. Scoccola, Phys. Rev. **D**, in press (hep-ph/0209174).
21. C. E. Carlson and C. D. Carone, Phys. Lett. **B441**, 363 (1998); Phys. Rev. **D58**, 053005 (1998); Phys. Lett. **B484**, 260 (2000).
22. Particle Data Group (D. E. Groom et al.), Eur. Phys. J. **C15**, 1 (2000).
23. R. Dashen, E. Jenkins, and A. V. Manohar, Phys. Rev. **D51**, 3697 (1995).
24. L. Ya. Glozman and D. O. Riska, Phys. Rept. **268**, 263 (1996).
25. H. Collins and H. Georgi, Phys. Rev. **D59**, 094010 (1999).
26. A. J. Hey, P. J. Litchfield and R. J. Cashmore, Nucl. Phys. **B95**, 516 (1975).
    D. Faiman and D. E. Plane, Nucl. Phys. **B50**, 379 (1972).
27. A. De Rújula, H. Georgi and S. L. Glashow, Phys. Rev. **D12**, 147 (1975).

Generalized Heat Kernel Coefficients for a New Asymptotic Expansion

Alexander A. Osipov[*][†] and Brigitte Hiller[*]

[*]Centro de Física Teórica, Departamento de Física, Universidade, P3004-516 Coimbra, Portugal
[†]Laboratory of Nuclear Problems, JINR, 141980 Dubna, Russia

Abstract. The method which allows for asymptotic expansion of the one-loop effective action $W = \ln \det A$ is formulated. The positively defined elliptic operator $A = U + \mathcal{M}^2$ depends on the external classical fields taking values in the Lie algebra of the internal symmetry group G. Unlike the standard method of Schwinger – DeWitt, the more general case with the nongenerate mass matrix $\mathcal{M} = \mathrm{diag}(m_1, m_2, \ldots)$ is considered. The first coefficients of the new asymptotic series are calculated and their relationship with the Seeley – DeWitt coefficients is clarified.

INTRODUCTION

The method of Schwinger – DeWitt [1, 2] allows for calculations of radiative corrections in the coordinate space. It can be used in any field theory but reveals to be an extremely powerful tool when explicit covariance of calculations is needed at all intermediate steps. Gauge theory [2], quantum gravity [3], chiral field theory [4] belong to those cases. In these and in many other cases it is necessary to evaluate the determinant of the positively defined elliptic operator which describes quadratic fluctuations of quantum fields in the presence of some background field. It contains in compact form the whole information about the one-loop contribution of quantum fields. The proper time formalism of Schwinger solves this problem. The result is an asymptotic expansion of the effective action in powers of proper time with Seeley – DeWitt coefficients a_n [2, 5], which accumulate the whole dependence on background fields. It is remarkable that every term of the expansion is invariant with respect to transformations of the internal symmetry group. This is a consequence of the general covariance inherent to the formalism. At the present time the asymptotic coefficients a_n are well known up to and including $n = 5$ for a general operator of Laplace type. The details can be found, for instance, in [4, 6].

In the case of massive quantum fields with a degenerate mass matrix $\mathcal{M} = \mathrm{diag}(m, m, \ldots)$, it is not difficult to derive from the proper time expansion an expansion in inverse powers of m^2, since the mass dependence is easily factorized and a subsequent integration over proper time leads to the desired result. The resulting asymptotic coefficients remain the same. If the mass matrix is not degenerate, i.e. $\mathcal{M} = \mathrm{diag}(m_1, m_2, \ldots)$, than its total factorization is impossible because of the non-commutativity of $\mathcal{M}$ with the rest of the elliptic operator. At the same time a naive factorization by parts breaks the covariant character of the asymptotic series. The natural question arises: is there any simple way to follow which leads to factorization,

CP660, *Hadron Physics: Effective Theories of Low Energy QCD*, edited by A. H. Blin et al.

conserving at the same time the explicit covariance of the expansion? Fortunately, there is such a method [7]. It leads to the generalization of the Seeley – DeWitt coefficients.

INTEGRAL REPRESENTATION FOR THE ONE-LOOP CONTRIBUTION

The logarithm of a formal determinant describes the low order radiative corrections to the classical theory. Let fermion fields play the role of virtual quanta producing those corrections and the scalar and pseudoscalar mesons be external background fields. In this case the real part of the corresponding effective action can be represented as a proper time integral

$$W[Y] = -\ln|\det D| = \frac{1}{2}\int_0^\infty \frac{dt}{t}\rho(t,\Lambda^2)\mathrm{Tr}\left(e^{-tD^\dagger D}\right). \tag{1}$$

Here the Dirac operator D depends on the background fields, which are collected in the hermitian second order differential elliptic operator $D^\dagger D = \mathcal{M}^2 + B = -\partial^2 + Y + \mathcal{M}^2$ in the term Y. We use the euclidian metric to define the effective action $W[Y]$. To make the integral over t convergent at the lower limit, the regulator $\rho(t,\Lambda^2)$ has been added in (1). All our conclusions are independent of the form of this function which includes the ultraviolet cut off Λ.

Let us assume that we are dealing with chiral field theory, which possesses the global $U(N_f)_L \times U(N_f)_R$ symmetry if the fermion fields are massless. The typical example is quantum chromodynamics. It is known that the vacuum state of low energy QCD is noninvariant with respect to the action of the chiral group and the whole system makes a phase transition to the state with massive quarks. If one takes into account that the explicit chiral symmetry breaking takes place in QCD through the mass terms of current quarks one can conclude that not equal current quark masses will lead also to not equal constituent quark masses. In order to study this system at large distances we will need the expansion of the effective action in inverse powers of the non degenerate mass matrix

$$\mathcal{M} = \sum_{i=1}^{N_f} \mathcal{M}_i E_i, \quad (E_i)_{jk} = \delta_{ij}\delta_{ik}, \quad E_i E_j = \delta_{ij} E_j. \tag{2}$$

The orthonormal basis E_i belongs to the flavor space where the chiral group acts on quarks and background fields in accordance with their transformation properties.

The heat kernel $\mathrm{Tr}[\exp(-tD^\dagger D)]$ can be represented [1] as a matrix element of an operator acting in the abstract (unphysical) Hilbert space:

$$\mathrm{Tr}\left(e^{-tD^\dagger D}\right) = \int d^4x\, \mathrm{tr} < x|e^{-tD^\dagger D}|x> . \tag{3}$$

The plane wave basis $|p>$ significantly simplifies our calculations and leads to the integral representation

$$W[Y] = \frac{1}{2}\int d^4x \int \frac{d^4p}{(2\pi)^4} \int_0^\infty \frac{dt}{t^3}\rho(t,\Lambda^2)e^{-p^2}\mathrm{tr}\left(e^{-t(\mathcal{M}^2+A)}\right)\cdot 1. \tag{4}$$

Here $A = B - 2ip\partial/\sqrt{t}$, and the trace is calculated in flavor space.

ASYMPTOTIC SCHWINGER – DEWITT EXPANSION

Before proceeding with our calculations this is the right place to say several words about the standard asymptotic Schwinger – DeWitt expansion. If the mass matrix $\mathcal{M}$ has the degenerate form, than $[\mathcal{M}, A] = 0$ and we find

$$\text{tr}\left(e^{-t(\mathcal{M}^2 + A)}\right) = e^{-tm^2}\text{tr}\left(e^{-tA}\right) = e^{-tm^2}\text{tr}\left(\sum_{n=0}^{\infty} t^n a_n\right). \tag{5}$$

Here a_n are the Seeley – DeWitt coefficients which depend on background fields and their derivatives. The integration over momentum and proper time in (4) can be readily done and we obtain the well known result

$$W[Y] = \int \frac{d^4x}{32\pi^2} \sum_{n=0}^{\infty} J_{n-1}(m^2)\text{tr}(a_n), \tag{6}$$

where integrals $J_n(m^2)$ are given by

$$J_n(m^2) = \int_0^{\infty} \frac{dt}{t^{2-n}} e^{-tm^2} \rho(t, \Lambda^2). \tag{7}$$

Independently on the type of regularization the following property is fulfilled

$$J_n(m^2) = \left(-\frac{\partial}{\partial m^2}\right)^n J_0(m^2). \tag{8}$$

Choosing $\rho(t, \Lambda^2) = 1 - (1 + t\Lambda^2)e^{-t\Lambda^2}$, which corresponds to two subtractions, one finds

$$J_0(m^2) = \Lambda^2 - m^2 \ln\left(1 + \frac{\Lambda^2}{m^2}\right). \tag{9}$$

We see that functions $J_n(m^2)$, starting from $n > 1$, have the asymptotic behavior $J_n(m^2) \sim m^{-2(n-1)}$ at large values of m^2, i.e. we obtain the inverse mass expansion. To warrant the convergence of the series it is necessary not only that the mass m be different from zero, but also that the background fields change slowly at distances of the order of the Compton wavelength $(1/m)$ of the fermion field. If these criteria are not fulfiled then the creation of real pairs gets essential and this expansion is not useful for calculations.

The remarkable property of the considered expansion is the gauge covariance of the Seeley – DeWitt coefficients.

NON-DEGENERATE MASS CASE

Let us return back to the formula (4) and show the way how to develop the abovementioned tool for the case $[\mathcal{M}, A] \neq 0$. As a first step let us use the formula

$$\text{tr}\left(e^{-t(\mathcal{M}^2+A)}\right) = \text{tr}\left(e^{-t\mathcal{M}^2}\left[1 + \sum_{n=1}^{\infty}(-1)^n f_n(t,A)\right]\right), \tag{10}$$

to factorize an exponent with the non commuting mass matrix $\mathcal{M}$. Here

$$f_n(t,A) = \int_0^t ds_1 \int_0^{s_1} ds_2 \ldots \int_0^{s_{n-1}} ds_n A(s_1)A(s_2)\ldots A(s_n), \tag{11}$$

where $A(s) = e^{s\mathcal{M}^2}Ae^{-s\mathcal{M}^2}$. It is true, that

$$\text{tr}\left[e^{-t\mathcal{M}^2}f_n(t,A)\right] = \frac{t^n}{n!}\sum_{i_1,i_2,\ldots i_n}^{N_f} c_{i_1 i_2 \ldots i_n}(t)\frac{1}{n}\sum_{perm}\text{tr}(A_{i_1}A_{i_2}\ldots A_{i_n}), \tag{12}$$

where $A_i = E_i A$ by definition, a second summation means that n possible cyclic permutations of operators inside the trace must be done. The coefficients $c_{i_1 i_2 \ldots i_n}(t)$ are totally symmetric with respect to permutations of indices and are calculated easily [7], for instance

$$c_i(t) = e^{-tm_i^2}, \quad c_{ij}(t) = \frac{e^{-tm_i^2} - e^{-tm_j^2}}{t\Delta_{ji}},$$

$$c_{ijk}(t) - \frac{2}{t^2}\left(\frac{e^{-tm_i^2}}{\Delta_{ji}\Delta_{ki}} + \frac{e^{-tm_j^2}}{\Delta_{kj}\Delta_{ij}} + \frac{e^{-tm_k^2}}{\Delta_{ik}\Delta_{jk}}\right),$$

$$c_{ijkl} = \frac{3!}{t^3}\left(\frac{e^{-tm_i^2}}{\Delta_{li}\Delta_{ki}\Delta_{ji}} + \frac{e^{-tm_j^2}}{\Delta_{ij}\Delta_{lj}\Delta_{kj}} + \frac{e^{-tm_k^2}}{\Delta_{jk}\Delta_{ik}\Delta_{lk}} + \frac{e^{-tm_l^2}}{\Delta_{kl}\Delta_{jl}\Delta_{il}}\right). \tag{13}$$

Here $\Delta_{ij} \equiv m_i^2 - m_j^2$. In the case of coincidence of indices one can get that $c_i = c_{ii} = c_{ii\ldots i}$.

Therefore the heat kernel is represented as a sum, every term of which contains the coefficient $c_{ij\ldots}(t)$, multiplyed by the corresponding trace from the product of operators A_i. The following integration over t replaces the t-dependent part in these terms by the integrals $J_l(m_i^2)$ and we finally will face the following problem: every term of this expansion will not be invariant with respect to the transformations of the chiral group, although the total effective action $W[Y]$ will possess this property. One needs the algorithm which automatically groups the terms in chiral invariant blocks. This problem has been solved in the papers [7] on the basis of recurrence relations

$$J_l(m_j^2) - J_l(m_i^2) = \sum_{n=1}^{\infty}\frac{\Delta_{ij}^n}{2^n n!}\left[J_{l+n}(m_i^2) - (-1)^n J_{l+n}(m_j^2)\right]. \tag{14}$$

There it has been shown that the essence of the problem is confined to the correct choosing of the factorized combination, built from the functions $J_i(m_j^2)$, namely, the effective action $W[Y]$ must be represented in the form

$$W[Y] = \int \frac{d^4x}{32\pi^2} \sum_{i=0}^{\infty} I_{i-1} \text{tr}(b_i), \quad I_i \equiv \frac{1}{N_f} \sum_{j=1}^{N_f} J_i(m_j^2). \tag{15}$$

In this case the background fields are *automatically* combined in the covariant coefficients b_i. For example, for $N_f = 3$ the first four coefficients are

$$
\begin{aligned}
b_0 &= 1, \quad b_1 = -Y, \quad b_2 = \frac{Y^2}{2} + \frac{\Delta_{12}}{2} \lambda_3 Y + \frac{1}{2\sqrt{3}} (\Delta_{13} + \Delta_{23}) \lambda_8 Y, \\
b_3 &= -\frac{Y^3}{3!} - \frac{1}{12}(\partial Y)^2 - \frac{1}{12} \Delta_{12} (\Delta_{31} + \Delta_{32}) \lambda_3 Y \\
&+ \frac{1}{12\sqrt{3}} \left[\Delta_{13}(\Delta_{21} + \Delta_{23}) + \Delta_{23}(\Delta_{12} + \Delta_{13}) \right] \lambda_8 Y \\
&+ \frac{1}{4\sqrt{3}} (\Delta_{31} + \Delta_{32}) \lambda_8 Y^2 + \frac{1}{4} \Delta_{21} \lambda_3 Y^2.
\end{aligned}
\tag{16}
$$

Several comments are in order here. First of all it is obvious that in the limiting case of equal masses $m_1 = m_2 = \ldots = m_{N_f}$ our result coincides with the standard Schwinger – DeWitt expansion. If the masses are not equal, the series (15) is a generalization of the well-known result (6). Instead of the asymptotic Seeley – DeWitt coefficients a_n come the coefficients b_n. Indeed one can check that if the operator $D^\dagger D$ transforms in the adjoint representation $\delta(D^\dagger D) = i[\omega, D^\dagger D]$, then b_n are also covariant, i.e. $\delta b_n = i[\omega, b_n]$, where $\omega = \alpha + \gamma_5 \beta$ are the parameters of the global infinitesimal chiral transformations. It also should be mentioned that, rigorously speaking, the obtained expansion is not an exact inverse mass expansion. Although the integrals I_l for $l \geq 1$ possess the necessary form for asymptotic behaviour $I_{l+1}(m_i^2) \sim m_i^{-2l}$, the coefficients b_l depend on the differences of masses, which may influence the character of the expansion. However general symmetry requirements are a more serious and stringent argument in favour of the obtained series as compared to the result of the thorough study [8] based only on the idea of a $1/m^2$ expansion.

RELATION BETWEEN COEFFICIENTS B_N AND A_N

The problem of the calculation of the generalized heat kernel coefficients is a more complicated mathematical problem than the calculation of the standard Seeley – DeWitt coefficients. However one can significantly simplify this problem [9], if from the very beginning one uses the transformation properties of the coefficients b_n. Indeed, let us return back to the formula (5) and extend it to the case $[\mathcal{M}, Y] \neq 0$, omitting for simplicity all terms with derivatives in A. It is remarkable that already at this stage one can take into account the two main conclusions which we have found in the previous section: the form of the factorized part depending on the mass and gauge covariance of coefficients b_n. The first aim can be reached through the definition

$$< A >\equiv \frac{\mathrm{tr}(A)}{\mathrm{tr}(1)}, \quad \bar{\mathscr{M}}^2 = \mathscr{M}^2 - < \mathscr{M}^2 >, \tag{17}$$

the second one through putting

$$\left(\bar{\mathscr{M}}^2 + Y\right)^n \equiv \sum_{k=0}^n (-1)^k C_n^k < \left(\bar{\mathscr{M}}^2\right)^{n-k} > b_k. \tag{18}$$

Since the left side of Eq.(18) transforms covariantly, i.e. $(\bar{\mathscr{M}}^2 + Y)^\Omega = \Omega^{-1}(\bar{\mathscr{M}}^2 + Y)\Omega$, and on the right side the term $< \bar{\mathscr{M}}^2 >$ is invariant with respect to the chiral transformations Ω, it is obvious, that $b_k(Y^\Omega, \bar{\mathscr{M}}^{2\Omega}) = \Omega^{-1} b_k(Y, \bar{\mathscr{M}}^2)\Omega$. To see how these definitions work let us consider the exponent in the heat kernel (5)

$$e^{-t(\mathscr{M}^2 + Y)} = e^{-t < \mathscr{M}^2 >} \sum_{n=0}^\infty \frac{(-t)^n}{n!} \left(\bar{\mathscr{M}}^2 + Y\right)^n$$

$$= e^{-t < \mathscr{M}^2 >} < e^{-t\bar{\mathscr{M}}^2} > \sum_{n=0}^\infty \frac{t^n}{n!} b_n = < e^{-t\mathscr{M}^2} > \sum_{n=0}^\infty \frac{t^n}{n!} b_n. \tag{19}$$

Since

$$< e^{-t\mathscr{M}^2} >= \sum_{i=1}^{N_f} e^{-tm_i^2} < E_i >= \frac{1}{N_f} \sum_{i=1}^{N_f} e^{-tm_i^2}, \tag{20}$$

the following integration over t leads us to the known result (15), and the coefficients b_n can be found using the definition (18). The same definition allows us to relate b_n with the linear combination of the standard Seeley – DeWitt coefficients $a_i(Y)$, where the replacement $Y \to Y + \bar{\mathscr{M}}^2$ should be done [9]

$$b_n = \sum_{i=0}^n \frac{\beta_{n-i}}{(n-i)!} a_i(Y \to Y + \bar{\mathscr{M}}^2). \tag{21}$$

The parameters β_i depend only on the masses of fermion fields. To establish the form of this dependence it is sufficient to calculate b_n in the simplest case, when in the elliptic operator under consideration all terms with derivatives are omitted. This problem is much simpler to solve than the calculation of the coefficients b_n from the very beginning.

ACKNOWLEDGMENTS

This work is supported by grants provided by Fundação para a Ciência e a Tecnologia, POCTI/2000/FIS/35304 and NATO "Outreach" Cooperation Program.

REFERENCES

1. Schwinger, J., *Phys. Rev.* **82**, 664 (1951).
2. DeWitt, B.S., *Dynamical Theory of Groups and Fields,* Gordon and Breach, New York, 1965.
3. DeWitt, B.S., *Phys. Rep.* **19**, 295 (1975).
4. Ball, R.D., *Phys. Rep.* **182**, 1 (1989) (and references therein).
5. Seeley, R., *Am. Math. Soc. Proc. Symp. Pure Math.* **10**, 288 (1967); CIME, 167 (1968).
6. Gilkey, P., *Invariance Theory, the Heat Equation, and the Atiyah-Singer Index Theorem,* 2nd ed., CRC, Boca Raton, FL, 1994; Branson, T.P, Gilkey, P.B., and Vassilevich, D.V., *Jour. Math. Phys.* **39**, 1040 (1998); Van de Ven, A.E.M., *Class. Quant. Grav.* **15**, 2311 (1998).
7. Osipov A.A., and Hiller, B., *Phys. Rev. D* **63**, 094009 (2001); *Phys. Letters B* **515**, 458 (2001); *Phys. Rev. D* **64**, 087701 (2001).
8. Lee, C., Min H., and Pac, P.Y., *Nucl. Phys.* **B202**, 336 (1982); Lee, C., Lee, T., and Min, H., *Phys. Rev. D* **39**, 1681 (1989); Lee, C., Lee, T., and Min, H., *ibid.* **39**, 1701 (1989).
9. Salcedo, L.L., *Eur. Phys. Journal C* **14**, 1 (2001).

't Hooft Determinant: fluctuations and the structure of the vacuum

Brigitte Hiller* and Alexander A. Osipov*†

*Centro de Física Teórica, Departamento de Física, Universidade, P3004-516 Coimbra, Portugal
†Laboratory of Nuclear Problems, JINR, 141980 Dubna, Russia

Abstract. The 't Hooft six quark flavor mixing interaction ($N_f = 3$) is bosonized by the path integral method. The considered complete Lagrangian is constructed on the basis of the combined 't Hooft and $U(3) \times U(3)$ extended chiral four fermion Nambu – Jona-Lasinio interactions. The method of the steepest descents is used to derive the effective mesonic Lagrangian. Additionally to the known lowest order stationary phase (SP) result of Reinhardt and Alkofer we obtain the contribution from the small quantum fluctuations of bosonic configurations around their stationary phase trajectories. Fluctuations give rise to multivalued solutions of the gap equations. Using the gap equations we construct the effective potential, from which the structure of the vacuum can be settled. We obtain that from the several extremal solutions, only one is a minimum. The others are either maxima or a saddle point. The effective potential reveals furthermore the existence of logartithmically divergent attractive wells at certain field expectation values, signaling caustic regions.

INTRODUCTION

The global $U_L(3) \times U_R(3)$ chiral symmetry of the QCD Lagrangian (for massless quarks) is broken by the $U_A(1)$ Adler-Bell-Jackiw anomaly of the $SU(3)$ singlet axial current $\bar{q}\gamma_\mu \gamma_5 q$. Through the study of instantons [1, 2], it became clear that effective $2N_f$ quark interactions, known as 't Hooft interactions, violate $U_A(1)$, but are still invariant under chiral $SU(N_f) \times SU(N_f)$. In the case of two flavors they are four-fermion interactions, and the resulting low-energy theory resembles the Nambu Jona-Lasinio model [3]. In the case of three flavors they are six-fermion interactions which are responsible for the correct description of η and η' physics, and additionally lead to the OZI-violating effects [4, 5],

$$\mathscr{L}_{\text{det}} = \kappa(\det\bar{q}P_R q + \det\bar{q}P_L q) \tag{1}$$

where the matrices $P_{R,L} = (1 \pm \gamma_5)/2$ are projectors and the determinant is over flavor indices.

The physical degrees of freedom of QCD at low-energies are mesons. The bosonization by the path integral approach of the effective quark interaction (1) together with the four-quark interactions has been considered in [6], where the lowest order stationary phase approximation (SPA) has been used to estimate the leading contribution from the 't Hooft determinant. In this approximation the functional integral is dominated by the stationary trajectories $r_{\text{st}}(x)$, determined by the extremum condition $\delta S(r) = 0$ of the action $S(r)$. The lowest order SPA corresponds to the case in which the integrals associated with $\delta^2 S(r)$, for the path $r_{\text{st}}(x)$ are neglected and only $S(r_{\text{st}})$ contributes to the

CP660, *Hadron Physics: Effective Theories of Low Energy QCD*, edited by A. H. Blin et al.
© 2003 American Institute of Physics 0-7354-0120-9/03/$20.00

generating functional. The subject of the present work is to include the contribution associated with $\delta^2 S(r)$. Our motivation is the following: let us consider the one-dimensional integrals of the form

$$F(\lambda) = \int_{-\infty}^{\infty} dt \exp[i\lambda f(t)] \tag{2}$$

What is sought by the method of SP is the dominant contribution to F as $\lambda \to \infty$. The dominant contribution to the integral comes from regions of t where f' vanishes. Supposing that f' vanishes at only one point t_0 and neglecting contributions to the integral from regions far from t_0, one can obtain the result

$$F(\lambda) = \sqrt{\frac{2\pi i}{\lambda f''(t_0)}} \, e^{i\lambda f(t_0)}. \tag{3}$$

Hence $F(\lambda)$ goes to zero like $1/\sqrt{\lambda}$. The lowest order SPA would correspond to the result $F(\lambda) = const \cdot \exp[i\lambda f(t_0)]$, which has the incorrect asymptotic behavior $F(\lambda) = \mathcal{O}(1)$. If results related to finite-dimensional integrals, such as $F(\lambda)$, mean anything with regard to corresponding functional integrals, one can conclude that the leading SPA should include the contribution from the integrals associated with the second functional derivative $\delta^2 S(r_{st})$.

PATH INTEGRAL BOSONIZATION

To be definite, let us consider the theory of the quark fields in four dimensional Minkowski space, with dynamics defined by the Lagrangian density

$$\mathcal{L} = \mathcal{L}_{NJL} + \mathcal{L}_{det}. \tag{4}$$

The first term here is the extended version of the Nambu – Jona-Lasinio (NJL) Lagrangian $\mathcal{L}_{NJL} = \mathcal{L}_0 + \mathcal{L}_{int}$, consisting of the free field part

$$\mathcal{L}_0 = \bar{q}(i\gamma^\mu \partial_\mu - \hat{m})q, \tag{5}$$

and the $U(3)_L \times U(3)_R$ chiral symmetric four-quark interaction

$$\mathcal{L}_{int} = \frac{G}{2}[(\bar{q}\lambda_a q)^2 + (\bar{q}i\gamma_5\lambda_a q)^2]. \tag{6}$$

We assume that quark fields have color and flavor indices running through the set $i = 1, 2, 3$; λ_a are the standard $U(3)$ Gell-Mann matrices with $a = 0, 1, \ldots, 8$. The current quark mass, $\hat{m}$, is a nondegenerate diagonal matrix with elements $diag(\hat{m}_u, \hat{m}_d, \hat{m}_s)$, it explicitly breaks the global chiral $U(3)_L \times U(3)_R$ symmetry of the $\mathcal{L}_{NJL}$ Lagrangian. The second term in (4) is given by (1). Letting

$$s_a = -\bar{q}\lambda_a q, \quad p_a = \bar{q}i\gamma_5\lambda_a q, \quad s = s_a\lambda_a, \quad p = p_a\lambda_a \tag{7}$$

70

yields

$$\mathcal{L}_{\text{det}} = -\frac{\kappa}{64}\left[\det(s+ip)+\det(s-ip)\right]. \tag{8}$$

The dynamics of the system is described by the vacuum transition amplitude in the form of the path integral

$$Z = \int \mathscr{D}q\mathscr{D}\bar{q}\exp\left(i\int d^4x\mathscr{L}\right). \tag{9}$$

which can be written in the equivalent form [6]

$$Z = \int \mathscr{D}\sigma_a\mathscr{D}\phi_a\mathscr{D}q\mathscr{D}\bar{q}\exp\left(i\int d^4x\mathscr{L}_q\right)\int \mathscr{D}r_{1a}\mathscr{D}r_{2a}\exp\left(i\int d^4x\mathscr{L}_r\right) \tag{10}$$

where

$$\mathscr{L}_q(\bar{q},q,\sigma,\phi) = \bar{q}(i\gamma^\mu\partial_\mu - \hat{m} - \sigma + i\gamma_5\phi)q, \tag{11}$$

$$\begin{aligned}\mathscr{L}_r(\sigma,\phi,r_1,r_2) &= 2G\left[(r_{1a})^2 + (r_{2a})^2\right] - 2(r_{1a}\sigma_a + r_{2a}\phi_a)\\ &\quad - \frac{\kappa}{8}\left[\det(r_1+ir_2)+\det(r_1-ir_2)\right]. \end{aligned}\tag{12}$$

The Fermi fields enter the action bilinearly, we can always integrate over them. At this stage one should also shift the scalar fields $\sigma_a \to \sigma_a + \Delta_a$ by demanding that the vacuum expectation values of the shifted fields vanish $< 0|\sigma_a|0 >= 0$, yielding gap equations to fix parameters $\Delta_a = m_a - \hat{m}_a$, where m_a denotes the constituent quark masses [1]. To evaluate path integrals over $r_{1,2}$ one has to use the method of stationary phase, or, after the formal analytic continuation in the time coordinate $x_4 = ix_0$, the method of steepest descents. The analytical continuation of the Euclidean version of the path integral under consideration is (see [8] for details),

$$J(\sigma,\phi) = \int_{-i\infty+r_{\text{st}}}^{+i\infty+r_{\text{st}}} \mathscr{D}r_{1a}\mathscr{D}r_{2a}\exp\left(\int d^4x\mathscr{L}_r(\sigma,\phi,r_1,r_2)\right). \tag{13}$$

Near the saddle point r_{st}^a,

$$\mathscr{L}_r \approx \mathscr{L}_r(r_{\text{st}}) + \frac{1}{2}\sum_{\alpha,\beta}\tilde{r}_\alpha\mathscr{L}_{\alpha\beta}''(r_{\text{st}})\tilde{r}_\beta \tag{14}$$

where r_{st}^a is a solution of the equations $\mathscr{L}_r'(r_1,r_2) = 0$ determining a flat spot of the surface $\mathscr{L}_r(r_1,r_2)$,

$$\begin{cases} 2Gr_1^a - (\sigma+\Delta)_a - \frac{3\kappa}{8}A_{abc}(r_1^br_1^c - r_2^br_2^c) &= 0 \\ 2Gr_2^a - \phi_a + \frac{3\kappa}{4}A_{abc}r_1^br_2^c &= 0. \end{cases} \tag{15}$$

---

[1] The shift by the current quark mass is needed to hit the correct vacuum state, see e.g. [7].

with totally symmetric constants, A_{abc}, closely related with the $U(3)$ constants d_{abc}. We use in (14) symbols $\tilde{r}^a$ for the differences $(r^a - r^a_{st})$. To deal with the multitude of integrals in (13) we define a column vector $\tilde{r}$ with eighteen components $\tilde{r}_\alpha = (\tilde{r}^a_1, \tilde{r}^a_2)$ and with the matrix $\mathscr{L}''_{\alpha\beta}(r_{st})$ being equal to

$$\mathscr{L}''_{\alpha\beta}(r_{st}) = 4GQ_{\alpha\beta}, \quad Q_{\alpha\beta} = \begin{pmatrix} \delta_{ab} - \frac{3\kappa}{8G}A_{abc}r^c_{1st} & \frac{3\kappa}{8G}A_{abc}r^c_{2st} \\ \frac{3\kappa}{8G}A_{abc}r^c_{2st} & \delta_{ab} + \frac{3\kappa}{8G}A_{abc}r^c_{1st} \end{pmatrix}. \quad (16)$$

Eq.(13) can now be concisely written as

$$J(\sigma, \phi) = \exp\left(\int d^4x \mathscr{L}_r(r_{st})\right) \int_{-i\infty}^{+i\infty} \mathscr{D}\tilde{r}_\alpha \exp\left(2G\int d^4x \tilde{r}^t Q(r_{st})\tilde{r}\right)[1 + \mathscr{O}(\hbar)]. \quad (17)$$

The solutions of Eqs.(15) are the following even and odd parity combinations r^a_{1st} and r^a_{2st} expressed in the form of increasing powers in σ_a, ϕ_a

$$r^a_{1st} = h_a + h^{(1)}_{ab}\sigma_b + h^{(1)}_{abc}\sigma_b\sigma_c + h^{(2)}_{abc}\phi_b\phi_c + \ldots \quad (18)$$

$$r^a_{2st} = h^{(2)}_{ab}\phi_b + h^{(3)}_{abc}\phi_b\sigma_c + \ldots \quad (19)$$

Putting these expansions in Eqs.(15) one obtains a series of selfconsistent equations to determine the constants $h_a, h^{(1)}_{ab}, h^{(2)}_{ab}$,

$$2Gh_a - \Delta_a - \frac{3\kappa}{8}A_{abc}h_bh_c = 0, \quad (20)$$

$$2G\left(\delta_{ac} - \frac{3\kappa}{8G}A_{acb}h_b\right)h^{(1)}_{ce} = \delta_{ae}, \quad (21)$$

$$2G\left(\delta_{ac} + \frac{3\kappa}{8G}A_{acb}h_b\right)h^{(2)}_{ce} = \delta_{ae}. \quad (22)$$

As a result we get

$$\mathscr{L}_r(r_{st}) = -2h_a\sigma_a - h^{(1)}_{ab}\sigma_a\sigma_b - h^{(2)}_{ab}\phi_a\phi_b + \mathscr{O}(\text{field}^3). \quad (23)$$

We now turn to the evaluation of the path integral in Eq.(17). In order to define the measure $\mathscr{D}\tilde{r}_\alpha$ more accurately we expand $\tilde{r}_\alpha$ in a Fourier series

$$\tilde{r}_\alpha(x) = \sum_{n=1}^{\infty} c_{n,\alpha}\varphi_n(x), \quad (24)$$

assuming that suitable boundary conditions are imposed. The set of the real functions $\{\varphi_n(x)\}$ form an orthonormal and complete sequence, therefore

$$\int \mathscr{D}\tilde{r}_\alpha \exp\left(2G\int d^4x \tilde{r}^t Q(r_{st})\tilde{r}\right) = \frac{C}{\sqrt{\det(2G\lambda^{\alpha\beta}_{nm})}}. \quad (25)$$

72

with

$$2G\lambda_{nm}^{\alpha\beta} = \begin{pmatrix} h_{ac}^{(1)-1} & 0 \\ 0 & h_{ac}^{(2)-1} \end{pmatrix}_{\alpha\sigma} \left(\delta_{\sigma\beta}\delta_{nm} + \int d^4x \varphi_n(x) F_{\sigma\beta}(x)\varphi_m(x) \right) \tag{26}$$

and

$$F_{\sigma\beta} = \frac{3\kappa}{4} A_{eba} \begin{pmatrix} -h_{ce}^{(1)}(r_{1st}^a - h_a) & h_{ce}^{(1)} r_{2st}^a \\ h_{ce}^{(2)} r_{2st}^a & h_{ce}^{(2)}(r_{1st}^a - h_a) \end{pmatrix}_{\sigma\beta}. \tag{27}$$

Only the matrix $F_{\sigma\beta}$ depends here on fields σ, ϕ. By absorbing in C the irrelevant field independent part of $2G\lambda_{nm}^{\alpha\beta}$, and expanding the logarithm in the representation $\det(1+F) = \exp\operatorname{tr}\ln(1+F)$, one can obtain finally for the complete action

$$S_r = \int d^4x \left\{ \mathscr{L}_r(r_{st}) + \frac{a}{2G^2} \sum_{n=1}^{\infty} \frac{(-1)^n}{n} \operatorname{tr}\left[F_{\alpha\beta}^n(r_{st}) \right] \right\} \tag{28}$$

proposing that the undetermined dimensionless constant a will be fixed by confronting the model with experiment afterwards [9]

THE GROUND STATE

Let us study the ground state of the model under consideration, then properties of the excitations will follow naturally. In Eq.(12) the field coefficients h_i obey

$$2Gh_i - \Delta_i = \frac{\kappa}{8} t_{ijk} h_j h_k \tag{29}$$

with the totally symmetric coefficients t_{ijk} equal to zero except for $t_{uds} = 1$ and with order parameters $\Delta_i \neq 0$ $(i = u, d, s)$. The t_{ijk} are related to coefficients A_{abc} by the embedding formula $3\omega_{ia} A_{abc} e_{bj} e_{ck} = t_{ijk}$ where matrices ω_{ia}, and e_{ai} are defined as follows

$$e_{ai} = \frac{1}{2\sqrt{3}} \begin{pmatrix} \sqrt{2} & \sqrt{2} & \sqrt{2} \\ \sqrt{3} & -\sqrt{3} & 0 \\ 1 & 1 & -2 \end{pmatrix}, \qquad \omega_{ia} = \frac{1}{\sqrt{3}} \begin{pmatrix} \sqrt{2} & \sqrt{3} & 1 \\ \sqrt{2} & -\sqrt{3} & 1 \\ \sqrt{2} & 0 & -2 \end{pmatrix}. \tag{30}$$

Here the index a runs $a = 0, 3, 8$ (for the other values of a the corresponding matrix elements are equal to zero). We have also $h_a = e_{ai} h_i$, and $h_i = \omega_{ia} h_a$. Similar relations can be obtained for Δ_i and Δ_a. In accordance with these notations we will use, for instance, that $h_{ci}^{(1)} = \omega_{ia} h_{ca}^{(1)}$.

A tadpole graphs calculation gives for the gap equations the following result

$$2h_i + \frac{3a\kappa}{8G^2}\left(h_{ab}^{(2)} - h_{ab}^{(1)} \right) A_{abc} h_{ci}^{(1)} = \frac{N_c}{2\pi^2} m_i J_0(m_i^2) \tag{31}$$

73

where the left hand side is the contribution from (28) and the right hand side is the contribution of the quark loop from (10), [8].

The second term on the left hand side of Eq.(31) is the correction resulting from the Gaussian integrals of the steepest descent method, comprising the effects of small fluctuations around the stationary path. If one puts for a moment $a = 0$ in Eq.(31), and combines the result with Eqs.(29), one finds gap equations which are very similar to the ones obtained in [4]. For this case one obtains values of (m_u, m_s) which are uniquely related to values of (G, κ).

The general case, which we have when $a > 0$ in Eq.(31), yields in turn a region for m_u, m_s, where three values of couplings (G, κ) are possible.

Conversely, one can study the solutions: $m_u = m_u(G, \kappa)$, $m_s = m_s(G, \kappa)$ at fixed values of input parameters: $\Lambda, \hat{m}_u = \hat{m}_d, \hat{m}_s$. Again we find several extremal solutions. We refer to [8] for graphical displays. Multivalued solutions of the gap equations have been obtained in a different context in [10].

To be able to pin down which of these solutions are minima, we will next derive the effective potential. We show how to obtain it for the special case $m_u = m_d = m_s$, which displays already the main characteristics of the vacuum. The more general case $m_u = m_d \neq m_s$ will be derived and discussed elsewhere [11], but we present below the final result. The effective potential can be generated from the gap equations [12]. Next we rewrite the gap equations in a form which can be readily integrated over m_u analytically. In Eqs. (20) we obtain that

$$A_{acb}h_b = -\frac{1}{3}h_u \begin{pmatrix} -2 & 0 \\ 0 & \vec{1} \end{pmatrix}, \qquad \delta_{ae} = 2G \begin{pmatrix} 1 \mp \frac{\kappa h_u}{4G} & 0 \\ 0 & \vec{1} \pm \frac{\kappa h_u}{8G}\vec{1} \end{pmatrix}_{ac} h_{ce}^{(1,2)}. \qquad (32)$$

Here the first element of the matrix in the first equation is the (0,0) component and the bold unit element in the diagonal represents a 8×8 matrix, with lines/columns ordered as $(3, 8, 1, 2, 4, 5, 6, 7)$. In the second equation the matrix notation is the same, but now the upper (lower) sign corresponds to the two possible types of marices $h_{ce}^{(1)}$ and $h_{ce}^{(2)}$ respectively. >From here we get immediatly that the difference $h_{ab}^{(2)} - h_{ab}^{(1)} = h_{ab}^{(d)}$ has elements

$$h_{00}^{(d)} = \frac{-\kappa h_u}{4G^2(1 - \kappa^2 h_u^2/(4G)^2)},$$

$$h_{ii}^{(d)} = \frac{\kappa h_u}{8G^2(1 - \kappa^2 h_u^2/(8G)^2)}, \qquad (i = 1,...8). \qquad (33)$$

The final analytical result for the gap equation is then

$$\frac{16Gy}{\kappa} - \frac{\kappa y}{2G^4} \frac{(1 - 3y^2)}{(1 - 4y^2)(1 - y^2)(1 - 2y)} - \frac{N_c}{2\pi^2}m_u J_0(m_u^2) = 0 = \frac{dU(m_u)}{dm_u}, \qquad (34)$$

where

$$y = \frac{\kappa h_u}{8G} = \frac{1}{2}\left(1 \pm \sqrt{1 - m_u\kappa/(4G^2)}\right). \qquad (35)$$

and a has been chosen to be equal to 1. The vanishing of y for $\kappa \to 0$ implies that the minus solution is the correct one. Integration over m_u yields the effective potential U

$$
U = ch_u^2 - \frac{4}{3}bh_u^3 - \frac{N_c}{8\pi^2}\left[\Lambda^2 m_u^2 + \Lambda^4 \ln\left(1 + \frac{m_u^2}{\Lambda^2}\right) - m_u^4 \ln\left(1 + \frac{\Lambda^2}{m_u^2}\right)\right]
$$
$$
+ \frac{4}{3c^2}\ln\left[\left(1 - 4y^2\right)\left(1 - y^2\right)^8\right] \tag{36}
$$

where $c = 2G$, $b = \kappa/4$. It is now easy to identify logarithmic divergencies at the values $\Delta_u = \frac{3c^2}{4|b|}$ and $\Delta_u = \frac{2c^2}{|b|}$, which are related to the vanishing of one or more eigenvalues of the second derivative of the action at these points. These divergencies are known as caustics and can be treated by formulating the path integral near these points as Airy integrals, see e.g. [13]. This will be done in a forthcoming work [11]. Between these two points there is a maximum of the gap equation and before the first logarithmic well we find two extrema, a minimum and a maximum. This picture of the vacuum is quite different from the one obtained previously in [6] and [4], where fluctuations have not been considered. In the latter approach the minimum is the only stable solution, contrary to the present case, where an extra domain of caustic points must be considered. For completeness we give, without derivation, the expression for the effective potential in the case $m_u = m_d \neq m_s$, $\kappa < 0$,

$$
U(m_u, m_s) = \phi(m_u) + \frac{1}{2}\phi(m_s)
$$
$$
+ 3cQ\left[1 + \frac{3b^2}{2c^2}Q\left(1 + \frac{8}{3}\sin^2\frac{\theta}{3}\cos\frac{2\theta}{3}\right)\right]
$$
$$
+ \frac{2}{c^2}\ln\left|\left[(1 - 2y_u^2)^2 - (y_s)^2\right](1 - y_s^2)^3(1 - y_u^2)^4\right| \tag{37}
$$

where

$$
\phi(m_i) = -\frac{N_c}{8\pi^2}\left[\Lambda^2 m_i^2 + \Lambda^4 \ln\left(1 + \frac{m_i^2}{\Lambda^2}\right) - m_i^4 \ln\left(1 + \frac{\Lambda^2}{m_i^2}\right)\right], \quad i = u, s \tag{38}
$$

$$
Q = \frac{b\Delta_s - c^2}{3b^2}, \tag{39}
$$

$$
\sin\theta = \frac{c\Delta_u}{2b^2\sqrt{-Q^3}} \tag{40}
$$

$$
y_u = \frac{\kappa h_u}{8G}, \quad h_u = 2\sqrt{-Q}\sin(\theta/3)
$$
$$
y_s = \frac{\kappa h_s}{8G}, \quad h_s = \frac{1}{c}(\Delta_s + bh_u^2). \tag{41}
$$

Here the extremum solutions of the gap equations unfold as one minimum, one saddle point and two maxima. Again, for certain configurations, one obtains caustics.

CONCLUDING REMARKS

The purpose of this work has been twofold. Firstly we have developed the technique which is necessary to go beyound the lowest order SPA in the problem of the path integral bosonization of the 't Hooft six quark interaction. This technique is rather general and can be readily used in other applications. Second, we have explored with considerable detail the implications of taking the quantum fluctuations in account in the description of the hadronic vacuum. We encountered several extrema as solutions to the gap equations. We have derived from the gap equations the effective potential, from which we can identify only one of the solutions to be a minimum. However the effective potential reveals also the existence of logarithmic divergent wells, regions of caustics. The physical and numerical consequences of this vacuum will be investigated in a forthcoming work.

Acknowledgements: We are very grateful to G. Ripka and A. Pich for many interesting discussions.

This work is supported by Fundação para a Ciência e a Tecnologia, POCTI/35304/-FIS/2000 and NATO "Outreach" Cooperation Program.

REFERENCES

1. A. M. Polyakov, Phys. Lett. B 59 (1975) 82; Nucl. Phys. B 121 (1977) 429. A. A. Belavin, A. M. Polyakov, A. Schwartz and Y. Tyupkin, Phys. Lett. B 59 (1975) 85. G. 't Hooft, Phys. Rev. Lett. 37 (1976) 8; Phys. Rev. D 14 (1976) 3432. C. Callan, R. Dashen and D. J. Gross, Phys. Lett. B 63 (1976) 334. R. Jackiw and C. Rebbi, Phys. Rev. Lett. 37 (1976) 172. S. Coleman, "The uses of instantons" Erice Lectures, 1977.
2. D. Diakonov, "Chiral symmetry breaking by instantons", Lectures at the Enrico Fermi School in Physics, Varenna, June 27 - July 7 (1995); hep-ph/9602375.
3. V. G. Vaks, A. I. Larkin ZhETF 40 (1961) 282; Y. Nambu, G. Jona-Lasinio, Phys. Rev. 122 (1961) 345; 124 (1961) 246.
4. V. Bernard, R. L. Jaffe, U.-G. Meißner, Nucl. Phys. B 308 (1988) 753.
5. T. Kunihiro and T. Hatsuda, Phys. Lett. B 206 (1988) 385. T. Hatsuda, Phys. Lett. B 213 (1988) 361. Y. Kohyama, K. Kubodera and M. Takizawa, Phys. Lett. B 208 (1988) 165. M. Takizawa, Y. Kohyama and K. Kubodera, Prog. Theor. Phys. 82 (1989) 481.
6. H. Reinhardt and R. Alkofer, Phys. Lett. B 207 (1988) 482.
7. A.A. Osipov and B. Hiller, Phys. Rev. D 62 (2000) 114013; idem Phys. Rev. D 63 (2001) 094009.
8. A.A. Osipov and B. Hiller, Phys. Lett. B 539 (2002) 76.
9. R. Jackiw, Int. J. Mod. Phys. B 14 (2000) 2011; hep-th/9903044.
10. P.J. Bicudo, J.E. Ribeiro and A.V. Nefediev, Phys. Rev. D 65 (2002) 085026.
11. A.A. Osipov and B. Hiller, in preparation;
12. M. Sher, Phys. Reports 179 (1989) 273;
13. L.S. Schulman, *Techniques and Applications of Path Integration* (Wiley-Interscience, New York, 1981)

Models of color confinement based on dual superconductors

Georges Ripka*† and Jiří Hošek**

*Service de Physique Théorique, Centre d'Etudes de Saclay, F-91191 Gif-sur-Yvette, France
†ECT*, Villa Tambosi, Strada delle Tabarelle 286, I-38050 Villazzano (Trento), Italy
**Department of Theoretical Physics, Nuclear Physics Institute
25068 Řež near Prague, Czech Republic

Abstract. Recently, the relatively old speculation that the physical QCD vacuum might be a kind of dual superconductor, in which color-magnetic monopoles have condensed, seems to have received some "experimental" confirmation in lattice calculations. The lattice calculations do not dictate, however, the form of the effective low-energy theory. And indeed, a rather wide panoply of possible effective theories has been proposed. The purpose of this talk is to review them in order to contrast their properties.

INTRODUCTION

Models of the confinement of color are in fact the models of the QCD ground state. Most of them view the QCD vacuum as a medium [1] with prescribed properties. Perhaps the simplest realization of color confinement is provided by the bag model[2]: colored quarks and gluons, being themselves the sources of the (chromo)electric field, exist in spacially finite bubbles with a (chromo)dielectric function $\varepsilon = 1$. The medium outside hadrons is impermeable to the (chromo)electric field i.e., $\varepsilon = 0$. Due to the Lorentz invariance, the responses of the QCD vacuum to the electric and magnetic fields are related: the magnetic permeability μ equals to $1/\varepsilon$. A perfect confining diaelectric ($\varepsilon \to 0$) implies a perfectly paramagnetic behavior ($\mu \to \infty$). Inside ordinary superconductors the experimentally observed perfect diamagnetism corresponds to $\mu = 0$. In this sense any confining QCD vacuum medium could be called a dual (color) superconductor. In the literature the name is reserved, however, to a restricted class of models in which $\varepsilon = 0$ is due to the condensate of the (chromo)magnetic monopoles, by analogy to the perfect diamagnetism of ordinary superconductors, which is caused by the condensate of the electric monopoles carried by the electron Cooper pairs. There exist also models of color confinement which are not related to the condensate of (chromo)magnetic monopoles. Only experiment, be it a numerical lattice calculation, can justify or invalidate them. In this talk we discuss simple properties of some lagrangian models which ascribe Ginzburg-Landau-like order parameters to the color-confining vacuum.

CP660, *Hadron Physics: Effective Theories of Low Energy QCD*, edited by A. H. Blin et al.
© 2003 American Institute of Physics 0-7354-0120-9/03/$20.00

THE VACUUM AS A PERFECT DIA-ELECTRIC MEDIUM

The idea is quite simple. Assume that the inverse propagator of the gauge field (either a photon or an Abelian gluon) acquires the form:

$$\Delta_{\mu\nu}^{-1} = \left(g_{\mu\nu}k^2 - k_\mu k_\nu\right)\frac{k^2}{k^2 - S^2} \tag{1}$$

which means that the dielectric function $\varepsilon(k) = \frac{k^2}{k^2-S^2}$ vanishes at small momenta. The action of the gauge field A_μ has the form:

$$I_j(A) = \frac{1}{(2\pi)^4}\int d^4k\left(-\frac{1}{2}A^\mu(k)\left(g_{\mu\nu}k^2 - k_\mu k_\nu\right)\frac{k^2}{k^2 - S^2}\right)A^\nu(-k) - j_\mu A^\mu \tag{2}$$

In the presence of a static (time-independent) source of charge $j^\mu = \delta_0^\mu \rho(\vec{r})$, the photon field reduces to its Coulomb part: $A^\mu = (\phi, 0)$, with a Fourier component $\phi_{\vec{k}} = -\frac{\vec{k}^2+S^2}{\vec{k}^4}\rho_{\vec{k}}$. This equation is usually expressed in terms of a potential $v(\vec{r})$ defined thus:

$$\phi(\vec{r}) = \int d^3r\, v(\vec{r} - \vec{r}')\rho(\vec{r}') \tag{3}$$

In the present case, we obtain a confining potential:

$$v(\vec{r}) = -\frac{1}{(2\pi)^3}\int d^3k\, e^{-i\vec{k}\cdot\vec{r}}\frac{\vec{k}^2+S^2}{\vec{k}^4} = v(r_0) - \frac{1}{4\pi r} + \frac{S^2}{8\pi}r \tag{4}$$

The infra-red divergence of the integral is absorbed into the value $v(r_0)$ of the potential at an arbitrary point. Note that S serves as a continuous order parameter: in the normal "perturbative" phase we have $S = 0$ and a Coulomb potential develops, whereas in the confined phase $S \neq 0$ and an additional linear potential develops. In a 'normal' Lorentz-invariant superconductor described by the Abelian Higgs model it is the magnetic permeability $\mu(k) = \frac{1}{\varepsilon(k)}$ which vanishes and not the dielectric function $\varepsilon(k)$. We shall discuss confinement in terms of models which are akin to *dual superconductors*, that is, superconductors in which the role of the electric and magnetic fields are exchanged. Dual superconductors exhibit a Meissner effect in which the electric field (and not the magnetic field) is expelled. This leads to the formation of an Abrikosov string and of a corresponding asymptotically linear potential between electric or color charges [3, 4].

COUPLING OF THE GAUGE FIELD TO A MASSIVE GHOST VECTOR FIELD

There are several ways to obtain a propagator such as (1). One way is to couple the gauge field A^μ to a massive vector ghost field B^μ [5, 6, 7]. The model is described by the action:

$$I_{j,S}(A,B) = \int d^4x \left(-\frac{1}{2}(\partial \wedge A)^2 + \frac{1}{2}(\partial \wedge B)^2 - \frac{S^2}{2}(A-B)^2 - j \cdot A \right) \tag{5}$$

with the obvious notation $-\frac{1}{2}(\partial \wedge A)^2 = -\frac{1}{4}(\partial_\mu A_\nu - \partial_\nu A_\mu)(\partial^\mu A^\nu - \partial^\nu A^\nu)$ and $A \cdot B = A_\mu B^\mu$. The action is a simple quadratic form of the fields and it is invariant under the joint gauge transformation:

$$A_\mu \to A_\mu + \partial_\mu \chi \qquad B_\mu \to B_\mu - \partial_\mu \chi \tag{6}$$

which means that we can fix the gauge of A or B but not of both independently. In the Landau gauge $\partial_\mu B^\mu = 0$, the equation of motion for the field B reads $(\partial^2 + S^2) B^\mu = S^2 A^\mu$. We can thus eliminate the field B from the action which becomes:

$$I_{j,S}(A) = \int d^4x \left(\frac{1}{2} A_\mu \left(g^{\mu\nu} - \frac{\partial^\mu \partial^\nu}{\partial^2} \right) \frac{\partial^4}{\partial^2 + S^2} A_\nu + \frac{1}{2} A_\mu \frac{\partial^\mu \partial^\nu}{\partial^2} A_\nu - j_\mu A^\mu \right) \tag{7}$$

The longitudinal part of the gauge field is thus described by an action identical to (2) and, in the presence of a static source of charge, a linear potential of the form (4) is formed. Two comments are called for.

1. The vector field B^μ is called a "ghost field" because it has a kinetic energy $+\frac{1}{2}(\partial \wedge B)^2$ with the wrong sign. This entails that the energy of the system is not bounded from below. It is not clear whether this constitutes a serious objection to the model, as long as the ghost field can be integrated out and no sources exist for the ghost field.

2. The constant S serves as an order parameter. In the confining phase, $S \neq 0$. In the normal "perturbative" phase, $S = 0$ and the system consists of massless gauge particles A^μ and a decoupled ghost B^μ. The order parameter may be promoted to a dynamical field by adding to the Lagrangian a term of the form:

$$L(S) = \frac{1}{2}(\partial S)^2 - \frac{b}{2}(S^2 - v^2)^2 \tag{8}$$

The frequently quoted "London limit" consists in keeping the field S constant and in omitting the term (8).

3. A Lagrangian of the form above would probably yield a Van der Waals type potential $1/r^n$ between color singlets, and not an exponetially decaying potential as observed.

COUPLING OF THE GAUGE FIELD TO A MASSIVE GHOST ANTISYMMETRIC TENSOR FIELD

Another way to obtain a propagator of the form (1) consists in coupling the gauge field to an antisymmetric tensor field $\Phi_{\mu\nu}(x) = -\Phi_{\nu\mu}(x)$. An action for such a model is [8]:

$$I_j(A,\Phi,S) = \int d^4x \left(\frac{1}{2}(\partial \cdot \Phi)^2 - \frac{1}{2}(F - \kappa S\Phi)^2 - j \cdot A + L(S) \right) \tag{9}$$

using a notation in which $F\Phi = \frac{1}{2}F_{\mu\nu}\Phi^{\mu\nu}$ and $(\partial \cdot \Phi)^\mu = \partial_\nu \Phi^{\nu\mu}$. In the action (9), the field tensor $F^{\mu\nu} = \partial^\mu A^\nu - \partial^\nu A^\mu$ is Abelian and κ is a dimensionless coupling constant. The action (9) is invariant with respect to the Abelian gauge transformation $A_\mu \to A_\mu + \partial_\mu \chi$. The equation of motion for the field Φ is:

$$\left(K + \kappa^2 S^2\right)\Phi = \kappa SF, \qquad \Phi = \frac{1}{K + \kappa^2 S^2}\kappa SF \tag{10}$$

where K is the differential operator:

$$K_{\mu\nu,\alpha\beta} = g_{\mu\alpha}\partial_\nu\partial_\beta - g_{\mu\beta}\partial_\nu\partial_\alpha + g_{\nu\beta}\partial_\mu\partial_\alpha - g_{\nu\alpha}\partial_\mu\partial_\beta \tag{11}$$

which has the properties $KK = \partial^2 K$ and $KF = \partial^2 F$. The field Φ can be eliminated from the action, which reduces to:

$$I_j(A,S) = \int d^4x \left(-\frac{1}{2}F \left(1 - \kappa S\frac{1}{K + \kappa^2 S^2}\kappa S \right) F - j \cdot A + L(S) \right) \tag{12}$$

When the field S is constant, as in the ground state where $S = v$, it commutes with the operator K and the action reduces to:

$$I_j(A) = \int d^4x \left(-\frac{1}{2}F \left(\frac{\partial^2}{\partial^2 + \kappa^2 S^2} \right) F - j \cdot A \right) \tag{13}$$

which is identical to the action (2). In the presence of a static source of charge, a confining potential develops, equal to $v(\vec{r}) = v(r_0) - \frac{1}{4\pi r} + \frac{\kappa^2 S^2}{8\pi}r$. This model is subject to the same criticism as the preceding one. It involves a ghost field Φ which can be integrated out and which decouples from the gauge field in the perturbative phase where $S = 0$.

AN ALTERNATIVE COUPLING TO AN ANTISYMMETRIC TENSOR FIELD

The gauge field may be coupled to an antisymmetric tensor field, which is *not* a ghost field. The model action is:

$$I_j(A,\Phi,S) = \int d^4x \left(-\frac{1}{2}(\partial \cdot \overline{\Phi})^2 - \frac{1}{2}(F - S\Phi)^2 - j \cdot A - J \cdot \Phi \right) \tag{14}$$

where j^μ and $J^{\mu\nu}$ are source terms discussed below. The main difference with the model action (9) resides in the kinetic term, which involves the *dual* tensor field $\bar{\Phi}_{\mu\nu} = \frac{1}{2}\varepsilon_{\mu\nu\alpha\beta}\Phi^{\alpha\beta}$, and which has an opposite sign, so that $\bar{\Phi}$ is *not* a ghost field and the energy of the system is bounded from below. The model action (14) has a double gauge invariance. Since it is expressed in terms of $F_{\mu\nu}$, it is invariant with respect to the Abelian gauge transformation $A_\mu \to A_\mu + \partial_\mu\chi$. However, it has the additional gauge invariance:

$$A^\mu \to A^\mu + S\Lambda^\mu \qquad \Phi_{\mu\nu} \to \Phi_{\mu\nu} + \partial_\mu\Lambda_\nu - \partial_\nu\Lambda_\mu \qquad (15)$$

provided that the sources j and J are related by the equation $\partial_\nu J^{\nu\mu} = S j^\mu$. Effectively, one can work in a gauge where the gluon field A^μ is set to zero. The Lagrangian (14) was studied by Kalb and Ramond in another context [9]. It is the London limit of a model proposed by Hosek and Ripka [10]. More recently, Ellwanger and Wschebor [11, 12] as well as Deguchi and Okubo [13] have shown that this model confines.

LANDAU-GINZBURG MODEL OF A DUAL SUPERCONDUCTOR

A widely used model of color confinement is the Landau-Ginzburg model of a dual superconductor. A dual superconductor is a superconductor in which the roles of the electric and magnetic fields are interchanged. It is then convenient to express the electric and magnetic fields in terms of the dual field tensor $\bar{F}_{\mu\nu} = \frac{1}{2}\varepsilon_{\mu\nu\alpha\beta}F^{\alpha\beta}$:

$$F^{\mu\nu} = \begin{pmatrix} 0 & -E_x & -E_y & -E_z \\ E_x & 0 & -H_z & H_y \\ E_y & H_z & 0 & -H_x \\ E_z & -H_y & H_x & 0 \end{pmatrix} \qquad \bar{F}^{\mu\nu} = \begin{pmatrix} 0 & -H_x & -H_y & -H_z \\ H_x & 0 & E_z & -E_y \\ H_y & -E_z & 0 & E_x \\ H_z & E_y & -E_x & 0 \end{pmatrix} \qquad (16)$$

The duality transformation $F \to \bar{F}$ is simply an exchange of the electric and magnetic fields: $\vec{E} \to \vec{H}$ and $\vec{H} \to -\vec{E}$. One is then tempted to define the dual field tensor $\bar{F}$ in terms of a gauge potential B^μ, by writing $\bar{F}_{\mu\nu} = \partial_\mu B_\nu - \partial_\nu B_\mu$ The gauge potential B^μ must not be confused with the gauge potential A^μ associated to the field tensor $F = \partial \wedge A$ and this is a source of complications which will be discussed below. The electric and magnetic fields are expressed in terms of the gauge potential B^μ as follows:

$$\vec{H} = -\partial_t\vec{B} - \vec{\nabla}\chi \qquad \vec{E} = -\vec{\nabla} \times \vec{B} \qquad (17)$$

A Landau-Ginzburg model of a dual superconductor can be obtained by writing the Landau-Ginzburg Lorentz-invariant action of a normal superconductor [14] in terms of the gauge field B^μ. It is simply a minimal coupling of the gauge field B^μ to a complex scalar field ψ:

$$I(\psi, \psi^*, B) =$$

$$\int d^4x \left(-\frac{1}{4}\bar{F}_{\mu\nu}\bar{F}^{\mu\nu} + \frac{1}{2}\left(\partial_\mu\psi^* - iqB_\mu\psi^*\right)\left(\partial^\mu\psi + iqB^\mu\psi\right) - \frac{1}{2}b\left(\psi^*\psi - v^2\right)^2 \right) \quad (18)$$

where q is a dimensionless *magnetic* charge. The physics becomes more transparent if the complex scalar field is expressed in polar form $\psi(x) = S(x) e^{iq\varphi(x)}$. The action (18) then reads, in our previous notation:

$$I(S, \varphi, B) = \int d^4x \left(-\frac{1}{2}F^2 + \frac{q^2 S^2}{2}(B + \partial\varphi)^2 + L(S) \right) \tag{19}$$

Let us distinguish the time and space components by writing $B^\mu = \left(\chi, \vec{B} \right)$. When the fields are time-independent, the energy of the system can be reduced to the form:

$$\mathcal{E} = \int d^3r \left[\frac{1}{2}\left(\vec{\nabla}S\right)^2 + \frac{1}{2}b\left(S^2 - v^2\right)^2 + \frac{q^2 S^2}{2}\left(\vec{B} - \vec{\nabla}\varphi\right)^2 + \frac{1}{2}\left(\vec{\nabla} \times \vec{B}\right)^2 \right] \tag{20}$$

When $S \neq 0$, the term $\frac{q^2 S^2}{2}\left(\vec{B} - \vec{\nabla}\varphi\right)^2$ is minimal at $\vec{B} = \vec{\nabla}\varphi$, and the *electric* field $\vec{E} = -\vec{\nabla} \times \vec{B}$ vanishes. This is the celebrated Meissner effect: in a dual superconductor, the electric field is expelled. This property was exploited in 1973 by Nielsen and Olesen [3], who displayed a stationary solution of the equations of motion with cylindrical symmetry, the perfect analogue of the Abrikosov string of a normal superconductor. The fields which make the energy (20) stationary satisfy the equations:

$$\left[-\vec{\nabla}^2 + 2b\left(S^2 - v^2\right) + q^2\left(\vec{B} - \vec{\nabla}\varphi\right)^2 \right] S = 0$$

$$\vec{\nabla} \times \left(\vec{\nabla} \times \vec{B}\right) + q^2 S^2\left(\vec{B} - \vec{\nabla}\varphi\right) = 0, \qquad \vec{\nabla} \cdot \left(\vec{B} - \vec{\nabla}\varphi\right) = 0 \tag{21}$$

A solution with cylindrical symmetry ($\vec{r} = (\rho, \theta, z)$) exists :

$$\vec{B} - \vec{\nabla}\varphi = \vec{n}_\theta C(\rho) \qquad S(\vec{r}) = S(\rho) \tag{22}$$

where the radial functions satisfy the equations:

$$-\frac{1}{\rho}\frac{\partial}{\partial\rho}\left(\rho\frac{\partial S}{\partial\rho}\right) + 2b\left(S^2 - v^2\right)S + q^2 C^2 S = 0$$

$$-\frac{1}{\rho}\frac{\partial}{\partial\rho}\left(\rho\frac{\partial C}{\partial\rho}\right) + q^2 S^2 C = 0 \tag{23}$$

together with the boundary conditions $S(\rho) \underset{\rho\to\infty}{\to} v$ and $C(\rho) \underset{\rho\to\infty}{\to} 0$. The solution is localized close to the z-axis. The energy density is independent of z and the expression (20) yields directly the *string tension*, that is, the energy per unit length parallel to the z-axis:

$$\frac{\partial\mathcal{E}}{\partial z} = 2\pi \int_0^\infty \rho d\rho \left[\frac{1}{2}\left(\frac{\partial S}{\partial\rho}\right)^2 + \frac{1}{2}b\left(S^2 - v^2\right)^2 + \frac{q^2}{2}S^2 C^2 + \frac{1}{2\rho^2}\left(\frac{\partial}{\partial\rho}(\rho C)\right)^2 \right] \tag{24}$$

Such a solution is called an "Abrikosov-Nielsen-Olesen string". It is created by well separated equal and opposite *electric* (or color-electric) charges. The geometry of the string ensures that an asymptotically linear confining potential develops between the charges. In the neighborhood of an Abrikosov-Nielsen-Olesen string, the electric field $\vec{E}$ is parallel to the z-axis and a magnetic current $\vec{j}_{mag}$ circulates around the z-axis:

$$\vec{E} = -\vec{\nabla} \times \vec{B} = -\vec{n}_z \frac{1}{\rho}\frac{\partial}{\partial\rho}\left(\rho C_\theta\right) \quad \vec{j}_{mag} = \vec{\nabla} \times \left(\vec{\nabla} \times \vec{B}\right) = -\vec{n}_\theta \frac{1}{\rho}\frac{\partial}{\partial\rho}\left(\rho \frac{\partial C}{\partial\rho}\right) \quad (25)$$

A vortex is thus created in the dual superconductor. At first sight, the scalar field S does not appear to be a valid order parapeter. Although the superconducting phase occurs when $S \neq 0$, if we set $S = 0$ the Lagrangian reduces to $-\frac{1}{4}\overline{F}_{\mu\nu}\overline{F}^{\mu\nu}$, which does not describe free gauge particles (as $-\frac{1}{4}F_{\mu\nu}F^{\mu\nu}$ would). However, this problem can be avoided if the model is expressed in the Zwanziger formalism [15], which involves the two gauge potentials A^μ and B^μ. In that case,, when $S = 0$, the gauge field B^μ can be integrated out and only the term $-\frac{1}{4}F_{\mu\nu}F^{\mu\nu}$ remains.

EVIDENCE FOR THE FORMATION OF THE ABRIKOSOV-NIELSEN-OLESEN STRING ON THE LATTICE

Figure 1 shows the profiles of the electric field and the magnetic current (25) observed in a lattice calculation and compared to the corresponding profiles of an Abrikosov-Nielsen-Olesen string. They are encouragingly similar. The fit was obtained with the model parameters:

$$m_H = 2v\sqrt{b} = 1.3123 \pm 0.0771\,GeV \quad m_V = qv = 1.3614 \pm 0.0143\,GeV \quad (26)$$

THE OCCURRENCE OF MONOPOLES IN AN ABELIAN GAUGE

The lattice calculations, depicted on Figure 1 show the electric field and magnetic currents of gluon fields in an Abelian gauge. The Abelian gauge is usually defined in terms of a scalar field V which transforms as a vector in color space: $V(x) = V_a(x)T_a$, where T_a are the $SU(N)$ generators of rotations in color space. The field $V(x)$ is a matrix in color space which can be diagonalized. In $SU(2)$ and $SU(3)$ respectively, the diagonal form is obtained by a rotation $\Omega(x)$ in color space:

$$V \to \Omega V \Omega^\dagger = V'T_3 \quad V \to \Omega V \Omega^\dagger = V'_3 T_3 + V'_8 T_8 \quad (27)$$

The rotation $\Omega(x)$ is a gauge transformation, under which the gluon field is transformed to $A_\mu \to A_\mu^\Omega = \Omega\left(A_\mu + \frac{1}{ie}\partial_\mu\right)\Omega^\dagger$, where e is the QCD coupling constant, which plays the role of an electric charge. In 1981 't Hooft pointed out that, if at any point $\vec{r}_0$, two

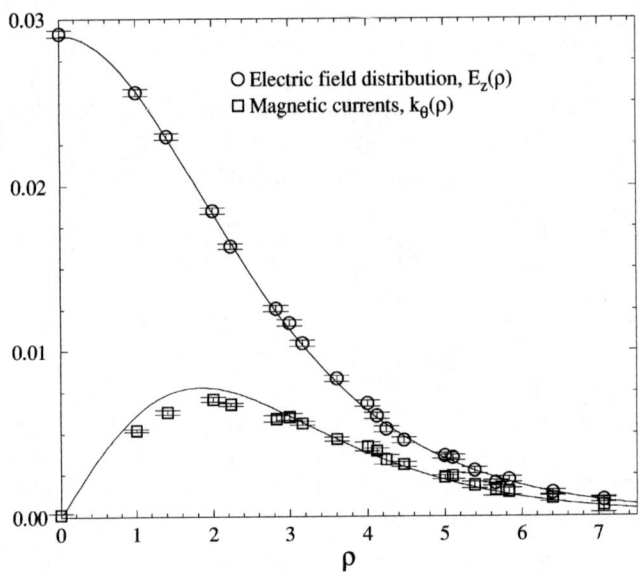

FIGURE 1. The electric field and the magnetic current (25) of an Abrikosov-Nielsen-Olesen string are compared to lattice data in the maximal Abelian gauge. Figure taken from Ref. [17].

eigenvalues of the matrix $V\left(\vec{r}_0\right)$ coincide, then the Abelian part of transformed gluon field A_μ^Ω develops a singularity and, in the vicinity of this point, it acquires the same shape as it would have in the vicinity of a magnetic monopole [4]. To see this, consider, for simplicity, the case of $SU(2)$. The argument can easily be extended to the $SU(N)$ case. The diagonal form of the field V is $V'\left(\vec{r}\right)T_3$ and a degeneracy of the eigenvalues can only occur at points $\vec{r}_0$ in space where the field $V\left(\vec{r}_0\right)$ vanishes. In the vicinity of this point, the field $V\left(\vec{r}\right)$ can be expressed in terms of a Taylor expansion:

$$V\left(\vec{r}\right)=V_a\left(\vec{r}\right)T_a=T_a C_{ab}\left(x_b-x_{0b}\right) \qquad C_{ab}=\left.\frac{\partial V_a}{\partial x_b}\right|_{\vec{r}=\vec{r}_0} \qquad (28)$$

The matrix C_{ab} defines a coordinate system in which the field $V\left(\vec{r}'\right)$ has the hedgehog form $V\left(\vec{r}'\right)=x_a'T_a$ with $x_a'=C_{ab}\left(x_b-x_{0b}\right)$. Dropping the primes on x', we can express the field V in spherical coordinates:

$$V\left(\vec{r}\right)=x_a T_a=T_1 r\sin\theta\cos\varphi+T_2 r\sin\theta\sin\varphi+T_3 r\cos\theta \qquad (29)$$

The matrix $\Omega\left(\vec{r}\right)$, which diagonalizes V, is:

$$\Omega(\theta,\varphi)=\begin{pmatrix} e^{i\varphi}\cos\frac{\theta}{2} & \sin\frac{\theta}{2} \\ -\sin\frac{\theta}{2} & e^{-i\varphi}\cos\frac{\theta}{2} \end{pmatrix} \qquad (30)$$

84

The transformed gluon field may be expressed in spherical coordinates. We find:

$$\vec{A}^{\Omega} = \Omega \left(\vec{A} + \frac{1}{ie} \vec{\nabla} \right) \Omega^{\dagger} = \Omega \vec{A} \Omega^{\dagger} + \tag{31}$$

$$+ \frac{1}{er} \left(-\vec{n}_{\theta} T_2 e^{i\varphi} + \vec{n}_{\varphi} \frac{1}{\sin\theta} \left(-(1 + \cos\theta) T_3 + \frac{1}{2} T_+ \sin\theta e^{i\varphi} + \frac{1}{2} T_- \sin\theta e^{-i\varphi} \right) \right)$$

The terms are all regular, except for the term $\frac{1}{ie} \left(\Omega \vec{\nabla} \Omega^{\dagger} \right)_{sg} = -\frac{1}{e} \vec{n}_{\varphi} \frac{1+\cos\theta}{r\sin\theta} T_3$, which becomes singular when $\theta \to 0$, that is, on the positive z-axis. We see that, in the vicinity of the points where $V(\vec{r}) = 0$, the *diagonal* (Abelian) part $\vec{A}_3(\vec{r})$ of the gluon field acquires a singularity, and we can recognize that the expression (31) is exactly the form which a vector field acquires in the vicinity of a magnetic monopole with a magnetic charge equal to $q = -\frac{4\pi}{e}$, a value which satisfies the Dirac quantization condition.

EVIDENCE FOR THE CONDENSATION OF MONOPOLES IN THE QCD GROUND STATE

In the last few years, the condensation of monopoles in the QCD ground state has been observed in several lattice simulations. The Pisa group has made an extensive study of monopole condensation [18, 19, 20] in both $SU(2)$ and $SU(3)$ simulations. They construct an operator $\mu_{op}(\vec{r}_0)$ which creates a gluon field, in an Abelian gauge, with the shape which it acquires in the vicinity of a magnetic monopole [21]. The vacuum expectation value of this operator is non vanishing in the QCD ground state at zero temperature, and it becomes zero at the critical temperature at which deconfinement occurs. The operator may thus serve as an order parameter which distinguishes the confined and deconfined phases. A most interesting and troubling result found by the Pisa group is the apparent irrelevance of the choice of the Abelian gauge. Monopole condensation appears to be independent of the choice of the field $V(x)$ chosen to fix the gauge. This may be a salutary feature because, in any given gauge, one gluon at least carries no color charge and is therefore not confined. Several other groups have observed monopole condensation in the QCD ground state. See, for example, references [17, 22] and [23, 24, 25]. However, in the maximal Abelian gauge, the gauge in which the field $V(x)$ is chosen so as to minimize the effect of the off-diagonal (non-Abelian) gluons, the string tension generated by monopole condensation is found to be close to the full string tension [26, 27].

CONCLUSION

How can we incorporate the properties observed on a lattice in models of color confinement? At first sight, the Landau-Ginzburg model of a dual superconductor appears to be the obvious choice. The lattice observations seem to point to the action (19) in which the field tensor refers to the Abelian parts B_3^{μ} and B_8^{μ} of the dual gauge field. However, we need to specify how the quark current couples to the gluon system. The standard quark current $j_{a\mu} = \bar{q}\frac{1}{2}\lambda_a\gamma_\mu q$ couples to the gluon field A_a^μ, but not to the dual field B^μ. The Maxwell equation $\partial_\nu F^{\nu\mu} = j^\mu$ cannot however be satisfied when the dual field tensor $\bar{F}^{\mu\nu} = \partial^\mu B^\nu - \partial^\nu B^\mu$ is expressed in terms of a dual potential B^μ, because, in that case, we have identically $\partial_\nu F^{\nu\mu} = 0$. We encounter here exactly the same problem which confronted Dirac in 1948 when he attempted to incorporate magnetic monopoles into electrodynamics [28]. We can adopt his solution by replacing the equation $\bar{F} = \partial \wedge B$ by the modified equation $\bar{F} = \partial \wedge B - \frac{1}{n \cdot \partial}\overline{n \wedge j}$ which satisfies the equation $\partial \cdot \bar{F} = j$. The second term is the well known Dirac string term which involves an arbitrary vector n^μ. It introduces a singularity which occurs along a line (the Dirac string) in 3-dimensional space the purpose of which is to propagate the flux of the magnetic charge. In 1971, Zwanziger developed a more powerful method of dealing with this problem by formulating a *local* Lagrangian which involves both potentials A^μ and B^μ [15]. Extensive applications of the Landau-Ginzburg action (19), suitably modified either by the introduction of Dirac strings, or by the use of the Zwanziger formalism, have been made, notably by several groups in Japan [29, 30]. The references [31] and [32] provide useful reviews of papers which use this model to display color confinement and even spontaneously broken chiral symmetry. We need to specify the color index of the gluon field A^μ. In some formulations [5, 8, 10], the action is made a color singlet by summing over the color indices. For example, the action (14) becomes:

$$I_j(A,\Phi,S) = \int d^4x \left(-\frac{1}{2}\left(\partial \cdot \overline{\Phi_a}\right)^2 - \frac{1}{2}\left(F_a - m\Phi_a\right)^2 - j_a \cdot A_a - J_a \cdot \Phi_a \right) \qquad (32)$$

where $a = 1,...,N_c^2 - 1$ is a color index. In other formulations [11, 12], it is assumed that the model is expressed in the "Abelian projection", which means that only the Abelian gluons (A_3^μ in the case of $SU(2)$ and A_3^μ, A_8^μ in the case of $SU(3)$) appear in the action. But in all cases, only the Abelian part $F^{\mu\nu} = \partial^\mu A^\nu - \partial^\nu A^\mu$ of the field tensor is retained. In all fairness, we should mention a frequently voiced and serious objection to all the models akin to dual superconductors: in the confined phase they involve colored massive vector fields which are not observed experimentally. This may, however, reflect the repeatedly verified fact that, at a given time, known realizations of a beautiful principle (the dual color superconductivity) are less perfect than the principle itself.

The authors wish to thank Nicolas Wschebor for illuminating discussions.

REFERENCES

1. T.D.Lee. *Particle Physics and Introduction to Field Theory*. Harwood Academic Press (New-York), 1981.
2. K.Johnson C.B.Thorn A.Chodos, R.L.Jaffe and V.F.Weisskopf. *Phys.Rev. D9*, page 3471, 1974.
3. H. B. Nielsen and P. Olesen. *Nucl. Phys. B61*, page 45, 1973.
4. G. 't Hooft. *Nucl. Phys. B190*, page 455, 1981.
5. Jiří Hošek. *Phys. Lett. B226*, page 377, 1989.
6. H. Narnhofer and W. Thirring. In Vol 119 Springer Tracts in Modern Physics, editor, *Rigourous Methods in Particle Physics*, page 1. Springer-Verlag, Berlin, 1990.

7. Jiří Hošek. *Czechoslovak Journal of Physics 43*, page 309, 1993.
8. Jiří Hošek. *Phys. Rev. D46*, page 3645, 1992.
9. M.Kalb and P.Ramond. *Phys. Rev. D9*, page 2273, 1974.
10. Jiří Hošek and Georges Ripka. *Z. Phys. A354*, page 177, 1996. hep-ph/9412285.
11. Ulrich Ellwanger and Nicolas Wschebor. *Phys. Lett. B517*, page 462, 2001. hep-th/0107093.
12. Ulrich Ellwanger and Nicolas Wschebor. Confinement with Kalb-Ramon fields. hep-th/0107196, 2001.
13. Shinichi Deguchi and Yousuke Kokubo. Quantization of massive abelian antisymmetric tensor field and linear potential. hep-th/0204087.
14. V.L.Guinzburg and L.D.Landau. *Zh.Eksp.Teor.Fiz. 20*, page 1064, 1950. An English translation may be found in "Men of Physics: L.D.Landau", vol.I,p.138.
15. Daniel Zwanziger. *Phys. Rev. D3*, page 880, 1970.
16. G.S. Bali. The mecanism of quark confinement. *Newport News 1998, Quark confinement and the hadron spectrum*, 1998. hep-ph/9809351.
17. G.S. Bali K. Schilling and C. Schlichter. *Prog. Theor. Phys. Suppl. 131*, page 645, 1998. hep-lat/9808039.
18. L. Montesi A. Di Giacomo, B. Lucini and G. Paffuti. *Phys. Rev. D61*, page 034503, 2000. hep-lat/9906024.
19. L. Montesi A. DiGiacomo, B. Lucini and G. Paffuti. *Phys. Rev. D61*, page 034504, 2000. hep-lat/9906025.
20. A. DiGiacomo B. Lucini Anf G. Paffuti J.M. Carmona, M. D'Elia. *Phys. Rev. D64*, page 114507, 2001. hep-lat/0103005.
21. A. Di Giacomo and G. Pafutti. *Phys. Rev. D56*, page 6816, 1997. hep-lat/9707003.
22. G.S. Bali C. Schlichter and K. Schilling. *Nucl. Phys. Proc. Suppl. 63*, page 519, 1998. hep-lat/9709114.
23. M.I. Polikarpov F.G. Gubarev, E.-M. Ilgenfritz and T. Suzuki. *Phys. Lett. B468*, page 134, 1999. hep-lat/9909099.
24. M.I. Polikarpov M.N. Chernodub, F.V. Gubarev and V.I. Zakharov. *Nucl. Phys. B592*, page 107, 2001. hep-th/0003138.
25. M.I. Polikarpov M.N. Chernodub, F.V. Gubarev and V.I. Zakharov. *Nucl. Phys. B600*, page 163, 2001. hep-th/0010265.
26. T.Suzuki E.-M.Ilgenfritz Y.Koma, M.Koma and M.I.Polikarpov. A fresh look on the flux-tube in abelian-projected SU(2) gluodynamics. hep-lat/020014, 2002.
27. Richard W.Haymaker Srinath Cheluvaraja and Takayuki Matsuki. Dual abrikosov vortex between confined charges. hep-lat/0210016, 2002.
28. P.A.M. Dirac. *Phys. Rev. 74*, page 817, 1948.
29. S. Sasaki H. Suganuma and H. Toki. *Nucl. Phys. B435*, page 207, 1995. hep-ph/9409407.
30. H. Suganuma S. Umisedo and H. Toki. *Phys. Rev. D57*, page 1605, 1998. hep-ph/9710231.
31. Hiroko Ichie and Hideo Suganuma. *Nucl. Phys. B574*, page 70, 2000. hep-lat/9808054.
32. H. Toki and H. Suganuma. *Progr. Part. Nucl. Phys. 45*, page S397, 2000.

Casimir Energies in the Light of Renormalizable Quantum Field Theories

H. Weigel

Institute for Theoretical Physics, Tübingen University
Auf der Morgenstelle 14, D–72076 Tübingen, Germany

Abstract. Effective hadron models commonly require the computation of functional determinants. In the static case these are one–loop vacuum polarization energies, known as Casimir energies. In this talk I will present general methods to efficiently compute renormalized one–loop vacuum polarization energies and energy densities and apply these methods to construct soliton solutions within a variational approach. This calculational method is particularly useful to study singular limits that emerge in the discussion of the *classical* Casimir problem which is usually posed as the response of a fluctuating quantum field to externally imposed boundary conditions.

INTRODUCTION

Many models in hadron physics originate from integrating out the more fundamental degrees of freedom like, for example, quarks. This change of field variables requires efficient tools to compute functional determinants like

$$\text{Det}\left(i\slashed{\partial}+\Gamma_i\Phi_i-m\right),\tag{1}$$

in the Nambu–Jona–Lasinio model [1] for quark flavor dynamics [2]. Here Φ_i denote background fields that couple to the internal symmetries via the generators Γ_i. Functional determinants are highly ultraviolet divergent and thus need to be regularized, and eventually renormalized by perturbative counterterms. This is particularly elaborate when the perturbative expansions become invalid, as for example for soliton configurations [3]. These configurations are usually static, $\Phi_i(x) = \Phi_i(\vec{x})$, such that the determinant becomes proportional to the vacuum polarization energy

$$\tfrac{1}{2}\sum_n\left(\omega_n - \omega_n^{(0)}\right).\tag{2}$$

Here ω_n are the eigenvalues of the single particle Hamiltonian in the presence of $\Phi_i(\vec{x})$ while $\omega_n^{(0)}$ denote the eigenvalues in the case that the Φ_i assume their vacuum values.

In this talk I will describe tools to unambiguously regularize and renormalize such vacuum polarization energies and energy densities starting from the energy density operator in quantum field theory.

This presentation is based on work with E. Farhi, N. Graham, R. L. Jaffe, V. Khemani, M. Quandt, and M. Scandurra. The publications [4, 5, 6, 7, 8] of these collaborations should be consulted for further details.

CP660, *Hadron Physics: Effective Theories of Low Energy QCD*, edited by A. H. Blin et al.

METHOD

In quantum field theories energies and energy densities are computed as renormalized matrix elements of the energy density operator $\hat{T}_{00}$. Here I will present the method to unambiguously compute such a matrix element when the fluctuating quantum field is coupled to a classical background. The method is based on expressing this matrix element in terms of a Green's function with appropriate boundary conditions. Then the energy density is given by a sum over bound states plus an integral over the continuum scattering states. Ample use will be made of analytic properties of scattering data especially to deform momentum integrals along a cut on the positive imaginary axis. To regulate the ultraviolet divergences of the theory, which corresponds to eliminating the contribution associated with the semi–circle at infinite complex momenta, the leading Born approximations to the Green's function are subtracted and later exactly added back in as Feynman diagrams. These diagrams are then regularized and renormalized in ordinary Feynman perturbation theory.

Formalism

Consider a static, spherically symmetric background potential $\sigma = \sigma(r)$ with $r = |\vec{x}|$ in n spatial dimensions. The symmetric energy density operator for a real scalar field coupled to σ is

$$\hat{T}_{00}(x) = \frac{1}{2}\left[\dot{\phi}^2 + \phi\left(-\vec{\nabla}^2 + m^2 + \sigma(r)\right)\phi\right] + \frac{1}{4}\vec{\nabla}^2\left(\phi^2\right) \tag{3}$$

with the spatial derivative term rearranged for later use of the Schrödinger equation to evaluate the expression in brackets. The "vacuum" is the state $|\Omega\rangle$ of lowest energy in the background σ. The "trivial vacuum" is the state $|0\rangle$ of lowest energy when $\sigma \equiv 0$. The vacuum energy density is the renormalized expectation value of $\hat{T}_{00}$ with respect to the vacuum $|\Omega\rangle$, $\langle\Omega|\hat{T}_{00}(x)|\Omega\rangle_{\text{ren}}$, which includes the matrix elements of the counterterms. The energy density only depends on the radial coordinate r since σ spherically symmetric,

$$\varepsilon(r) = \frac{2\pi^{n/2}}{\Gamma(\frac{n}{2})}r^{n-1}\langle\Omega|\hat{T}_{00}(x)|\Omega\rangle_{\text{ren}}. \tag{4}$$

The wavefunctions factorize in radial functions $\phi_\ell(t,r)$ and angle dependent spherical harmonics. The Fock decomposition for the radial functions reads

$$\begin{aligned}
\phi_\ell(t,r) &= \frac{1}{r^{\frac{n-1}{2}}}\int_0^\infty \frac{dk}{\sqrt{\pi\omega}}\left[\psi_\ell(k,r)e^{-i\omega t}a_\ell(k) + \psi_\ell^*(k,r)e^{i\omega t}a_\ell^\dagger(k)\right] \\
&\quad + \frac{1}{r^{\frac{n-1}{2}}}\sum_j \frac{1}{\sqrt{2\omega_{\ell j}}}\left[\psi_{\ell j}(r)e^{-i\omega_{\ell j}t}a_{\ell j} + \psi_{\ell j}(r)e^{i\omega_{\ell j}t}a_{\ell j}^\dagger\right].
\end{aligned} \tag{5}$$

This decomposition contains scattering states with $\omega = \sqrt{k^2 + m^2}$ and bound states with $\omega_{\ell j} = \sqrt{m^2 - \kappa_{\ell j}^2}$. The total angular momentum assumes integer values $\ell = 0,1,2,\ldots$ in all dimensions except for $n = 1$, where $\ell = 0$ and 1 only, corresponding to the symmetric and antisymmetric channels respectively. The radial wavefunctions ψ are solutions to the Schrödinger–like equation

$$-\psi'' + \frac{1}{r^2} \left(v - \tfrac{1}{2}\right)\left(v + \tfrac{1}{2}\right) \psi + \sigma(r)\psi - k^2\psi = 0 \tag{6}$$

where $v = \ell - 1 + \frac{n}{2}$. In each angular momentum channel, the wavefunctions are normalized to satisfy the completeness relation

$$\frac{2}{\pi} \int_0^\infty dk\, \psi_\ell^*(k,r)\psi_\ell(k,r') + \sum_j \psi_{\ell j}(r)\psi_{\ell j}(r') = \delta(r - r'). \tag{7}$$

Then the standard equal time commutation relations for the quantum field ϕ yield canonical commutation relations for the creation and annihilation operators, $[a_\ell(k), a_{\ell'}^\dagger(k')] = \delta(k - k')\delta_{\ell\ell'}$ and $[a_{\ell j}, a_{\ell' j'}^\dagger] = \delta_{jj'}\delta_{\ell\ell'}$. All other commutators vanish. The vacuum $|\Omega\rangle$ is annihilated by all of the $a_\ell(k)$ and $a_{\ell j}$. The matrix element (4) can now be computed by inserting eq. (5) into eq. (3):

$$\varepsilon(r) = \sum_\ell N_\ell \left[\int_0^\infty \frac{dk}{\pi} \omega \psi_\ell^*(k,r)\psi_\ell(k,r) + \sum_j \frac{\omega_j}{2}\psi_{\ell j}(r)^2 \right] \tag{8}$$

$$+ \tfrac{1}{4} D_r \sum_\ell N_\ell \left[\int_0^\infty \frac{dk}{\pi\omega} \psi_\ell^*(k,r)\psi_\ell(k,r) + \sum_j \frac{1}{2\omega_{\ell j}}\psi_{\ell j}(r)^2 \right] - \varepsilon^{(0)}(r) + \varepsilon_{\mathrm{CT}}(r).$$

Here N_ℓ is the degeneracy factor, $D_r = \frac{\partial}{\partial r}\left(\frac{\partial}{\partial r} - \frac{n-1}{r}\right)$, $\varepsilon_{\mathrm{CT}}(r)$ is the counterterm contribution, and $\varepsilon^{(0)}(r)$ indicates the subtraction of the energy density in the trivial vacuum.

The scattering state contribution is identified with the Green's function by defining the local spectral density

$$\rho_\ell(k,r) \equiv \frac{k}{i} G_\ell(r,r,k), \tag{9}$$

where

$$G_\ell(r,r',k) = -\frac{2}{\pi} \int_0^\infty dq\, \frac{\psi_\ell^*(q,r)\psi_\ell(q,r')}{(k+i\varepsilon)^2 - q^2} - \sum_j \frac{\psi_{\ell j}(r)\psi_{\ell j}(r')}{k^2 + \kappa_{\ell j}^2}, \tag{10}$$

so that for real k

$$\psi_\ell^*(k,r)\psi_\ell(k,r) = \mathrm{Im}\left\{ k G_\ell(r,r,k) \right\} = \mathrm{Re}\left\{ \rho_\ell(k,r) \right\}. \tag{11}$$

The $i\varepsilon$ prescription has been chosen so that this Green's function is meromorphic in the upper half–plane, with simple poles at the imaginary momenta $k = i\kappa_{\ell j}$ corresponding to bound states. For real k, the imaginary part of the Green's function at $r = r'$ is an odd function of k, while the real part is even. Hence eq. (8) can be expressed as a contour integral in the upper half–plane. The contribution from the semi–circular contour at large $|k|$ with $\mathrm{Im}(k) \geq 0$ must be eliminated. Subtracting sufficiently many terms in the Born series from the Green's function yields a convergent integral. Then I add back exactly what I subtracted in the form of Feynman diagrams. The integrand (8) has a branch cut along the imaginary axis, $k \in [im, +i\infty]$, and simple poles at the bound state momenta $k = i\kappa_j$. The corresponding residues cancel the explicit bound state contributions. The discontinuity along the cut is hence all what is left to be considered

$$
\begin{aligned}
\varepsilon(r) &= -\sum_\ell N_\ell \int_m^\infty \frac{dt}{\pi} \sqrt{t^2 - m^2} \left[1 - \frac{1}{4(t^2 - m^2)} D_r \right] [\rho_\ell(it, r)]_N + \sum_{i=1}^N \varepsilon_{FD}^{(i)}(r) + \varepsilon_{CT}(r) \\
&\equiv \bar{\varepsilon}(r) + \sum_{i=1}^N \varepsilon_{FD}^{(i)}(r) + \varepsilon_{CT}(r),
\end{aligned} \tag{12}
$$

where

$$
\begin{aligned}
[\rho_\ell(k, r)]_N &\equiv \rho_\ell(k, r) - \rho_\ell^{(0)}(k, r) - \rho_\ell^{(1)}(k, r) \ldots - \rho_\ell^{(N)}(k, r) \\
&= \frac{k}{i} \left[G_\ell(r, r, k) - G_\ell^{(0)}(r, r, k) - G_\ell^{(1)}(r, r, k) - \ldots - G_\ell^{(N)}(r, r, k) \right]. \tag{13}
\end{aligned}
$$

The superscript (j) indicates the term of order j in the Born expansion. Subtraction of the free Green's function $G_\ell^{(0)}(r, r, k)$ corresponds to subtracting $\varepsilon^{(0)}(r)$ above. The potentially divergent pieces are precisely identified [9, 10] with Feynman diagrams $\varepsilon_{FD}^{(i)}(r)$, which are regularized and renormalized using standard methods. When combined with the contribution from the counterterms $\varepsilon_{CT}(r)$ they yield finite contributions to the energy density (for smooth backgrounds).

The Radial Green's Function

A variety of solutions to eq. (6) is distinguished by different boundary conditions:

free Jost solution $\quad\quad w_\ell(kr)$: $\quad (-1)^\nu \sqrt{\frac{\pi}{2} kr} \left[J_\nu(kr) + i Y_\nu(kr) \right]$

Jost solution $\quad\quad f_\ell(k, r)$: $\quad \lim_{r \to \infty} \frac{f_\ell(k, r)}{w_\ell(kr)} = 1$

regular solution $\quad\quad \phi_\ell(k, r)$: $\quad \lim_{r \to 0} \frac{\Gamma(\nu+1)}{\sqrt{\pi}} \left(\frac{r}{2} \right)^{-(\nu+\frac{1}{2})} \phi_\ell(k, r) = 1$

physical scattering solution $\quad \psi_\ell(k, r)$: $\quad \psi_\ell(k, r) = \frac{k^{\nu+\frac{1}{2}}}{F_\ell(k)} \phi_\ell(k, r)$

The physical scattering solution is normalized with respect to the Jost function, $F_\ell(k)$ that is obtained as the ratio of the interacting and free Jost solutions at $r = 0$,

$$
F_\ell(k) = \lim_{r \to 0} \frac{f_\ell(k, r)}{w_\ell(kr)}. \tag{14}
$$

In particular, two regular solutions emerge: ϕ_ℓ has a simple boundary condition at $r = 0$, so that it is analytic in the upper half k–plane; ψ_ℓ has a physical boundary condition at $r \to \infty$, corresponding to incoming and outgoing spherical waves.

The Green's function has the simple representation

$$
G_\ell(r, r', k) = \frac{\phi_\ell(k, r_<) f_\ell(k, r_>)}{F_\ell(k)} (-k)^{\nu - \frac{1}{2}}, \tag{15}
$$

where $r_>$ $(r_<)$ denotes the larger (smaller) of the two arguments r and r'. The poles of $G_\ell(r,r',k)$ occur at the zeros of the Jost function, which are the imaginary bound state momenta. These are the only poles of eq. (10) in the upper half–plane and, since the two functions in eq. (10) and eq. (15) obey the same inhomogeneous differential equation, they are indeed identical.

Although G_ℓ is analytic in the upper half–plane, f_ℓ and ϕ_ℓ contain pieces that oscillate for real k and exponentially increase or decrease when k has an imaginary part. Actually, only the case $r = r'$ is interesting, cf. eq. (13). Then the product $f_\ell\phi_\ell$ is well–behaved. This motivates to factorize the dangerous exponential components[1],

$$f_\ell(k,r) \equiv w_\ell(kr)g_\ell(k,r) \quad \text{and} \quad \phi_\ell(k,r) \equiv \frac{(-k)^{-\nu+\frac{1}{2}}}{2\nu}\frac{h_\ell(k,r)}{w_\ell(kr)}, \tag{16}$$

where w_ℓ is the free Jost solution introduced above. With these definitions,

$$G_\ell(r,r,k) = \frac{h_\ell(k,r)g_\ell(k,r)}{2\nu g_\ell(k,0)}. \tag{17}$$

The definition of h_ℓ does *not* just remove the free part. Instead, it enforces the cancellation of w_ℓ in the Green's function. After analytically continuing to $k = it$, $g_\ell(it,r)$ obeys

$$g_\ell''(it,r) = 2t\xi_\ell(tr)g_\ell'(it,r) + \sigma(r)g_\ell(it,r) \tag{18}$$

with the boundary conditions $\lim_{r\to\infty} g_\ell(it,r) = 1$ and $\lim_{r\to\infty} g_\ell'(it,r) = 0$. A prime indicates a derivative with respect to the radial coordinate r. Using these boundary conditions, one integrates the differential equation numerically for $g_\ell(it,r)$, starting at $r = \infty$ and proceeding to $r = 0$. Similarly, $h_\ell(it,r)$ obeys

$$h_\ell''(it,r) = -2t\xi_\ell(tr)h_\ell'(it,r) + \left[\sigma(r) - 2t^2\frac{d\xi_\ell(\tau)}{d\tau}\bigg|_{\tau=tr}\right]h_\ell(it,r). \tag{19}$$

The factors in eq. (16) were chosen to yield simple boundary conditions: $h_\ell(it,0) = 0$ and $h_\ell'(it,0) = 1$. The numerical integration for h starts at $r = 0$ and runs to $r = \infty$. For real τ,

$$\xi_\ell(\tau) \equiv -\frac{d}{d\tau}\ln\left[w_\ell(i\tau)\right] \tag{20}$$

is real with $\lim_{\tau\to\infty}\xi_\ell(\tau) = 1$, so the two functions $h_\ell(it,r)$ and $g_\ell(it,r)$ are manifestly real. They are also holomorphic in the upper half k–plane and, most importantly, they are bounded according to $|g_\ell(k,r)| \leq$ const. and $|h_\ell(k,r)| \leq$ const.$[\nu r/(1+|k|r)]$. Thus the representation of the partial wave Green's function in terms of g_ℓ and h_ℓ is smooth and numerically tractable on the positive imaginary axis.

The computation of the Born series, eq. (13), is also straightforward in this formalism. The solutions to the differential equations eq. (18) and eq. (19) are expanded about the

---

[1] For $n = 1$ and $n = 2$, the case of $\ell = 0$ is somewhat different [5].

free solutions,

$$
\begin{aligned}
g_\ell(it,r) &= 1 + g_\ell^{(1)}(it,r) + g_\ell^{(2)}(it,r) + \dots \\
h_\ell(it,r) &= 2vrI_v(tr)K_v(tr) + h_\ell^{(1)}(it,r) + h_\ell^{(2)}(it,r) + \dots ,
\end{aligned}
\tag{21}
$$

where the superscript labels the order of the background potential σ. The higher order components obey inhomogeneous linear differential equations such that σ is the source term for $g^{(1)}$, $\sigma g^{(1)}$ is the source term for $g^{(2)}$, and so on. Substituting these solutions in the expansion of eq. (17) with respect to the order of the background potential finally yields the Born series for the local spectral density

$$
[\rho_\ell(it,r)]_N = \left[t \frac{h_\ell(it,r)g_\ell(it,r)}{2vg_\ell(it,0)} \right]_N .
\tag{22}
$$

Thus I have available a computationally robust representation for the Born subtracted energy density $\bar{\varepsilon}(r)$, cf. eq. (12).

Feynman Diagram Contribution

To one–loop order, the Feynman diagrams of interest are generated by expanding

$$
\langle 0 | \hat{T}_{00}(x) | 0 \rangle \sim \tfrac{i}{2} \mathrm{Tr} \left[\hat{T}_x \left(-\partial^2 - m^2 - \sigma \right)^{-1} \right]
\tag{23}
$$

to order N in the background σ. Here $\hat{T}_x$ is the coordinate space operator corresponding to the insertion of the energy density defined by eq. (3) at the spacetime point x, and the trace includes space–time integration. The Feynman diagrams are obtained in ordinary perturbation theory, thus the matrix element in eq. (23) is evaluated between the trivial vacuum state, which is annihilated by the plane wave annihilation operators. The energy density operator has pieces of order σ^0 and σ^1: $\hat{T}_x = \hat{T}_x^{(0)} + \hat{T}_x^{(1)}$. The computation of its vacuum matrix element is most conveniently performed in momentum space. The relevant matrix elements are

$$
\langle k' | \hat{T}_x^{(0)} | k \rangle = e^{i(k'-k)x} \left[k^{0\prime}k^0 + \vec{k}' \cdot \vec{k} + m^2 \right] \quad \text{and} \quad \langle k' | \hat{T}_x^{(1)} | k \rangle = \sigma(x) e^{i(k'-k)x}.
\tag{24}
$$

Here I will explicitly consider the contributions to $\langle 0 | \hat{T}_{00}(x) | 0 \rangle$ that are linear in σ. The first contribution of this order comes directly from $\hat{T}_x^{(1)}$

$$
\tfrac{i}{2} \mathrm{Tr} \left(\frac{1}{-\partial^2 - m^2} \hat{T}_x^{(1)} \right) = \tfrac{i}{2}\sigma(x) \int \frac{d^d k}{(2\pi)^d} \frac{1}{k^2 - m^2} .
\tag{25}
$$

This local contribution is ultraviolet divergent for $d = n + 1 \geq 2$. It is canceled identically by the counterterm in the no–tadpole renormalization scheme. An additional contribution at order σ originates from $\hat{T}_x^{(0)}$ and the first–order expansion of the propagator,

$$
\tfrac{i}{2} \mathrm{Tr} \left(\frac{1}{-\partial^2 - m^2} \hat{T}_x^{(0)} \frac{1}{-\partial^2 - m^2} \sigma \right) .
\tag{26}
$$

At $\mathcal{O}(\sigma)$ the renormalized Feynman diagram contribution to the energy density becomes

$$\varepsilon_{\text{FD}}^{(1)}(r) + \varepsilon_{\text{CT}}(r) = C_d r^{n-1} \int \frac{d^{d-1}q}{(2\pi)^{d-1}} \tilde{\sigma}(\vec{q}) e^{i\vec{q}\cdot\vec{x}} \int_0^1 d\zeta \frac{\zeta(1-\zeta)\vec{q}^2}{\left[m^2 + \zeta(1-\zeta)\vec{q}^2\right]^{2-d/2}}, \quad (27)$$

with $C_d = 2\pi^{\frac{d-1}{2}}\Gamma(2-\frac{d}{2})/[\Gamma(\frac{d-1}{2})(4\pi)^{d/2}]$ and $\tilde{\sigma}(q) = 2\pi\delta(q^0)\tilde{\sigma}(\vec{q})$ is the Fourier transform of the (time independent) background field. This piece is finite for $d = n < 4$ and does not contribute to the total energy because it vanishes when integrated over space. The extension to higher order Feynman diagrams is straightforward.

Total Energy

The total energy is simply the integrated energy density of eq. (12),

$$E[\sigma] = \int_0^\infty \varepsilon(r)\,dr. \quad (28)$$

Both, the t integral and the sum over channels in eq. (12), are absolutely convergent. This is a consequence of deforming the momentum integral in the upper–half plane[2]. Thus the order of integration can be interchanged

$$E[\sigma] = -\sum_\ell N_\ell \int_m^\infty \frac{dt}{\pi} \sqrt{t^2 - m^2} \int_0^\infty dr \, [\rho_\ell(it,r)]_N + \sum_{i=1}^N E_{\text{FD}}^{(i)} + E_{\text{CT}}, \quad (29)$$

where the total derivative term has integrated to zero. As already explained, a sufficient number, N, of Born approximations to the local spectral density $\rho_\ell(it,r)$ must be subtracted to render the t integral convergent. These subtractions are then added back in as the contribution to the total energy from the Feynman diagrams. Combined with the contribution from the counterterms this gives a finite result,

$$\sum_{i=1}^N E_{\text{FD}}^{(i)} + E_{\text{CT}} = \int_0^\infty dr \left[\sum_{i=1}^N \varepsilon_{\text{FD}}^{(i)}(r) + \varepsilon_{\text{CT}}(r) \right]. \quad (30)$$

In practice, $E[\sigma]$ is more efficiently computed directly from the perturbation series of the total energy. First, I'd like to recall that

$$2\int_0^\infty dr \, [\rho(it,r)]_N = \frac{d}{dt} \left[\ln F_\ell(it)\right]_N = \frac{d}{dt} \left[\ln g_\ell(it,0)\right]_N \quad (31)$$

is valid[3] for $\text{Re}(t) > 0$, cf. Appendix A of Ref. [5]. This allows me to write

$$E[\sigma] = \sum_\ell N_\ell \int_m^\infty \frac{dt}{2\pi} \frac{t}{\sqrt{t^2 - m^2}} \left[\beta_\ell(t,0)\right]_N + \sum_{i=1}^N E_{\text{FD}}^{(i)} + E_{\text{CT}}. \quad (32)$$

---

[2] For real momenta, $k = it$, this amounts to performing the momentum integral before the radial integral.
[3] In general, the case $\text{Re}(t) = \text{Im}(k) = 0$ causes uncontrollable oscillations at large r: for real k the integral $\int_0^\infty dr \, [\rho(k,r)]_N$ does not exist and the integrated local spectral density cannot be related to the phase shift.

The real function $\beta_\ell(t,r) = \ln g_\ell(it,r)$ is determined by the differential equation

$$-\beta_\ell''(t,r) - \left[\beta_\ell'(t,r)\right]^2 + 2t\xi_\ell(tr)\beta_\ell'(k,r) + \sigma(r) = 0 \tag{33}$$

with the boundary conditions $\lim_{r\to\infty} \beta(t,r) = \lim_{r\to\infty} \beta'(t,r) = 0$. In $\left[\beta_\ell(t,0)\right]_N$, the first N Born terms must again be subtracted. They are obtained by iterating the differential equation (33) according to the expansion of $\beta_\ell(t,r)$ in powers of σ.

To make contact with previous work [4, 6, 7, 10],

$$E[\sigma] = \sum_\ell N_\ell \left[\int_0^\infty \frac{dk}{2\pi} \sqrt{k^2 + m^2} \frac{d}{dk}[\delta_\ell(k)]_N + \frac{1}{2}\sum_j \omega_{\ell j} \right] + \sum_{i=1}^N E_{FD}^{(i)} + E_{CT} \tag{34}$$

observe that along the real axis the phase of the Jost function is the scattering phase shift,

$$i\ln F_\ell(k) = i\ln|F_\ell(k)| + \delta_\ell(k). \tag{35}$$

Equations (32) and (34) are proven identical by first noticing that for real k $|F_\ell(k)|$ and $\delta_\ell(k)$ are respectively even and odd functions and then computing the momentum integrals along the branch cut $k \in [im, +i\infty]$.

SOLITON FORMATION IN A D=1+1 CHIRAL MODEL

The described method is well suited to efficiently and unambiguously compute vacuum polarization energies in renormalizable quantum field theories. Then the total energy, *i.e.* the sum of the classical and vacuum polarization energies, is a functional of the background field. Varying this background field maps an energy surface. The existence of a local minimum on that surface indicates the existence of an energetically stable solution to the equation of motion, a soliton[4]. Of particular interest are models that do not contain soliton solutions at the classical level such that solitons get stabilized by quantum corrections.

Now I would like to consider this idea in the framework of a simple chiral model in $D = 1+1$ [6]. The realistic $D = 3+1$ case is more difficult and a discussion is presented in Ref. [7]. In this two–dimensional model a two–component boson field $\vec{\phi} = (\phi_1, \phi_2)$ couples chirally to a fermion Ψ that come in N_f (equivalent) modes:

$$\mathscr{L} = \frac{1}{2}\partial_\mu\vec{\phi}\cdot\partial^\mu\vec{\phi} + \sum_{n=1}^{N_f} \bar{\Psi}_i\left\{i\partial\!\!\!/ - G\left(\phi_1 + i\gamma_5\phi_2\right)\right\}\Psi_i. \tag{36}$$

where the potential for the boson field

$$V(\vec{\phi}) = \frac{\lambda}{8}\left[\vec{\phi}\cdot\vec{\phi} - v^2 + \frac{2\alpha v^2}{\lambda}\right]^2 - \alpha v^3\left(\phi_1 - v\right) + \text{const.} \tag{37}$$

---

[4] The minimum itself is not necessarily a soliton because the space of variational parameters is limited and the exact soliton might have even lower energy.

contains a term (proportional to α) that breaks the chiral symmetry explicitly in order to avoid problems stemming from (unphysical) infra–red singularities that occur when the vacuum configuration would be determined via the naïve treatment of spontaneous symmetry breaking[11]. In this manner it is guaranteed that the VEV is given by $\langle \vec{\phi} \rangle = (v, 0)$. Here the counterterm Lagrangian is not presented explicitly. It is determined such that the quantum corrections lead to a vanishing tadpole diagram for the boson field. Considering only the classical contribution does <u>not</u> support a stable soliton soliton.

In the limit that the number of fermion modes becomes large with $v^2/N_f \sim \mathcal{O}(1)$ only the classical and one fermion loop pieces contribute. In the following I will only consider that limit, *i.e.* $E_{\text{tot}} = E_{\text{cl}} + E_{\text{F}}$. The fermion contribution can be split into two pieces $E_{\text{F}} = E_{\text{vac}} + E_{\text{val}}$. The valence part E_{val} is given in terms of the bound state energies such as to saturate the total fermion number that is fixed to be N_F. The vacuum piece is computed according to the formalism described in the preceding section:

$$E_{\text{vac}}[\vec{\phi}] = -\frac{1}{2} \sum_i^{\text{b.s.}} \left(|\omega_i| - Gv \right) - \int_0^\infty \frac{dk}{2\pi} \left(\omega_k - Gv \right) \frac{d}{dk} \left(\delta_{\text{F}}(k) - \delta^{(1)}(k) \right), \qquad (38)$$

which is obtained from Eq (34) by employing Levinson's theorem. Note the overall "-" sign for fermions and recall that the single particle spectrum is not charge conjugation invariant. Furthermore, δ_{F} denotes the sum of the eigenphase shifts[5]. The subtraction

$$\delta^{(1)}(k) = \frac{2G^2}{k} \int_0^\infty dx \left(v^2 - \vec{\phi}^2(x) \right) \qquad (39)$$

that renders E_{vac} finite contains both first and second order Born approximants in the fluctuations of $\vec{\phi}$ about $\langle \vec{\phi} \rangle$. The first order is unambiguously fixed by the no–tadpole renormalization condition and the second order by the chiral symmetry.

Having established the energy functional I now consider variational *Ansätze* for the background field that turn this functional in a function of the variational parameters. As an example I assume

$$\phi_1 + i\phi_2 = v \{ 1 - R + R \exp [i\pi (1 + \tanh(Gvx/w))] \} \qquad (40)$$

that introduces width (w) and amplitude (R) parameters. For prescribed model parameters (G,v,etc.) the energy must be minimized with respect to w and R. The resulting binding energy $\mathcal{B} = E_{\text{tot}} - Gv$ is shown in Fig. 1. Even though the *Ansatz* (40) may not be the final answer to the minimalization problem, $\mathcal{B}$ is definitely negative. Thus a solitonic configuration is energetically favored showing that indeed quantum fluctuations can create a soliton that is not stable at the classical level.

CASIMIR ENERGIES

As noted earlier, the presented method to compute vacuum polarization energies is not limited to the case of smooth background fields. It is particularly interesting to employ

---

[5] The eigenchannels are labeled by parity and the sign of the single particle eigenenergies.

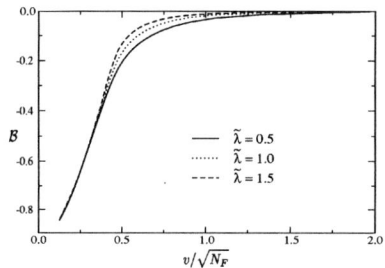 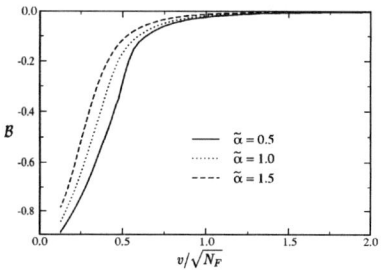

FIGURE 1. The maximal binding energy as a function of the model parameters as obtained from the *Ansatz* (40) in units of Gv; $\tilde{\alpha} = \alpha/G^2$ and $\tilde{\lambda} = \lambda/G^2$.

singular background fields to imitate boundary conditions of the fluctuating quantum field. It is known for some time [12] that the vacuum polarization energy diverges when the fluctuating field is constrained by boundary conditions. *Ad hoc* schemes for their removal have been proposed [13]. However, the method of renormalization in continuum quantum field theory represents the *only* physical way to treat these divergences.

Dirichlet Points in One Space Dimension

The simplest example to be considered is that of a massive scalar field $\phi(t,x)$ in one dimension, constrained to vanish at $x = -a$ and a. The standard approach, in which the boundary conditions are imposed *a priori*, gives an energy [14]

$$\widetilde{E}_2(a) = -\frac{m}{2} - \frac{2a}{\pi} \int_m^\infty dt \, \frac{\sqrt{t^2 - m^2}}{e^{4at} - 1} .$$ (41)

The tilde denotes the imposition of the Dirichlet boundary condition at the outset. This expression yields an attractive force between the two Dirichlet points,

$$\widetilde{F}(a) = -\frac{d\widetilde{E}_2}{d(2a)} = -\int_m^\infty \frac{dt}{\pi} \frac{t^2}{\sqrt{t^2 - m^2}(e^{4at} - 1)} .$$ (42)

In the massless limit it simplifies considerably: $\widetilde{E}_2(a) = -\pi/48a$ and $\widetilde{F}(a) = -\pi/96a^2$. But this result is not internally consistent: as $a \to \infty$, $\widetilde{E}_2(a) \to 0$, indicating that the energy of an isolated "Dirichlet point" is zero. The limit $a \to 0$ also describes a single Dirichlet point, but $\widetilde{E}_2(a) \to \infty$ as $a \to 0$. Also note that $\widetilde{E}_2(a)$ is well defined as $m \to 0$, even though scalar field theories in one space dimension become infrared divergent when $m \to 0$.

The Dirichlet point problem can nicely be studied with the presented method. A single delta–function background

$$\sigma_1 = \lambda \delta(x - a)$$ (43)

has the Green's function

$$G_\lambda(x,y) = G_0(x,y) - \frac{\lambda G_0(x,a)G_0(a,y)}{1 + \lambda G_0(a,a)} \tag{44}$$

where $G_0(x,y)$ is the Green's function in the non–interacting case. The momentum argument has been omitted. Obviously the limit $\lambda \to \infty$ gives Dirichlet boundary conditions, $G_\infty(x,a) = G_\infty(a,y) = 0$. The vacuum polarization energy associated with eq. (43) is [5]

$$E_1(\lambda) = \int_m^\infty \frac{dt}{2\pi} \frac{t\ln\left[1 + \frac{\lambda}{2t}\right] - \frac{\lambda}{2}}{\sqrt{t^2 - m^2}} \tag{45}$$

To study the problem of two Dirichlet points it is obvious to consider

$$\sigma_2(x) = \lambda\left[\delta(x+a) + \delta(x-a)\right]. \tag{46}$$

The renormalized Casimir energy for this potential has also been computed in Ref. [5],

$$E_2(a,\lambda) = \int_m^\infty \frac{dt}{2\pi}\frac{1}{\sqrt{t^2 - m^2}}\left\{t\ln\left[1 + \frac{\lambda}{t} + \frac{\lambda^2}{4t^2}(1 - e^{-4at})\right] - \lambda\right\} \tag{47}$$

For any finite coupling λ, the inconsistencies noted in $\widetilde{E}_2(a)$ do not afflict $E_2(a,\lambda)$: as $a \to \infty$, $E_2(\lambda) \to 2E_1(\lambda)$, and as $a \to 0$, $E_2(a,\lambda) \to E_1(2\lambda)$. Also $E_2(a,\lambda)$ diverges logarithmically in the limit $m \to 0$ as it should. The *force*, obtained by differentiating eq. (47) with respect to $2a$, agrees with eq. (42) in the limit $\lambda \to \infty$. However $E_2(a,\lambda)$ *diverges* like $\lambda\log\lambda$ as $\lambda \to \infty$. Thus the renormalized Casimir *energy* diverges as the Dirichlet boundary condition is imposed, a physical effect which is missed in the pure boundary condition calculation.

The counterterms to the energy density vanish away from $x = \pm a$ because they are local functions of $\sigma(x)$. Therefore the Casimir *energy density* for $x \neq \pm a$ can be calculated assuming Dirichlet boundary conditions from the start simply by subtracting the density in the absence of boundaries without encountering any further divergences [14],

$$\widetilde{\varepsilon}_2(x,a) = -\frac{m}{8a} - \int_m^\infty \frac{dt}{\pi}\frac{\sqrt{t^2 - m^2}}{e^{4at} - 1} - \frac{m^2}{4a}\sum_{n=1}^\infty \frac{\cos\left[\frac{n\pi}{a}(x-a)\right]}{\sqrt{(\frac{n\pi}{2a})^2 + m^2}} \quad \text{for } |x| < a$$

$$\widetilde{\varepsilon}_2(x,a) = -\frac{m^2}{2\pi}K_0(2m|x-a|) \quad \text{for } |x| > a. \tag{48}$$

This result excludes the points $x = \pm a$. For finite λ the Casimir energy density, $\varepsilon_2(x,a,\lambda)$, was also computed in Ref. [5] and is displayed in Fig. 2. The energy density between the isolated points is negative and approaches the boundary condition limit (48) in a non–uniform manner. In the limit $\lambda \to \infty$ it agrees with eq. (48) except at $x = \pm a$ where it contains extra, singular contributions. If one integrates eq. (48) over all x, ignoring the singularities at $x = \pm a$, one obtains eq. (41). Including the singular contributions at $\pm a$ by integrating $\varepsilon_2(x,a,\lambda)$ gives eq. (47).

This simple example illustrates the principal situation: In the Dirichlet limit the renormalized Casimir energy diverges because the energy density on the "surface", $x = \pm a$ diverges. However the Casimir force and the Casimir energy density for all $x \neq \pm a$ remain finite and equal to the results obtained by imposing the boundary conditions *a priori*, eqs. (42) and (48).

Two Space Dimensions

A scalar field in two space dimensions constrained to vanish on a circle of radius a presents a more complex problem. For smooth backgrounds only the local *tadpole* diagram diverges and thus the no–tadpole renormalization condition is still sufficient to render the theory finite. For $\sigma(\vec{x}) = \lambda \delta(r - a)$ and $r \neq a$ the subtracted local spectral density $[\rho_\ell(it,r)]_0$ vanishes *exponentially* as the momentum along the cut in eq. (12) increases. For finite λ, both the t-integral and the ℓ-sum are uniformly convergent so $\lambda \to \infty$ can be taken under the sum and integral. The resulting energy density, $\varepsilon(r,\lambda)$, agrees with $\tilde{\varepsilon}(r)$, obtained when the Dirichlet boundary condition, $\phi(a) = 0$, is assumed from the start. As in one dimension, nothing can be said about the total energy because $\tilde{\varepsilon}(r)$ is not defined at $r = a$, but unlike the one dimensional case, the integral $\int dr\, \varepsilon(r,\lambda)$ now diverges even in the sharp limit for finite λ.

To understand the situation better, let's consider $\sigma(\vec{x})$ to be a narrow Gaußian of width w centered at $r = a$ and explore the sharp limit where $w \to 0$ and $\sigma(\vec{x}) \to \lambda \delta(r - a)$. For $w \neq 0$, σ does not vanish at any value of r, so $[\rho_\ell(it,r)]_0$ no longer falls exponentially at large t (and $r \neq a$), and subtraction of the first Born approximation to $\rho_\ell(it,r)$ is necessary, *i.e.* $N = 1$ in eq. (12). As noted above, the compensating tadpole graph can be canceled against the counterterm, $c_1 \lambda \sigma(\vec{x})$. The result is a renormalized Casimir energy density, $\varepsilon(r,w,\lambda)$, and Casimir energy, $E(w,\lambda) = \int_0^\infty dr\, \varepsilon(r,w,\lambda)$, both of which are finite. However as $w \to 0$ both $\varepsilon(a,w,\lambda)$ and $E(w,\lambda)$ diverge.

The divergence can be traced to the $\mathcal{O}(\lambda^2)$ Feynman diagram. This diagram is separated by subtracting the *second* Born approximation to $\rho_\ell(it,r)$, *i.e.* $N = 2$ in eq. (12).

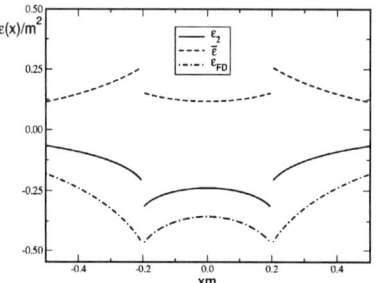
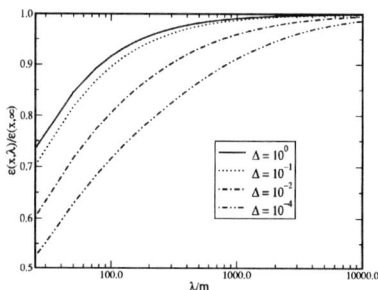

FIGURE 2. The energy density for the two delta function background at $a = 0.2/m$ computed from eqs (12) and (27). Left panel: $\lambda = 3m$. The distinct contributions associated with eqns. (12) and (27) are disentangled but the singular contributions at $x = \pm a$ are omitted. Right panel: Dirichlet limit $\lambda \to \infty$ in comparison to the the boundary condition result (48); $\Delta = (a - x)/a$.

Then the ℓ-sum and t-integral no longer diverge in the sharp limit, $w \to 0$ but the equivalent diagram must be added back explicitly. In the limit $w \to 0$ it contributes

$$-\frac{\lambda^2 a^2}{8} \int_0^M dp\, J_0^2(ap) \arctan\frac{p}{2m} \tag{49}$$

to the total energy. This diverges logarithmically as $M \to \infty$. The divergence originates from the high momentum components in the Fourier transform of $\sigma(\vec{x}) = \lambda \delta(r-a)$ rather than the high energy behavior of the loop integral. This divergence gives an infinite contribution to the stress because it varies with the radius of the circle. This divergence only gets worse in higher dimensions (in contrast to the claim of Ref. [15]). For example, for $\sigma(r) = \lambda \delta(r-a)$ in three space dimensions the *renormalized* two point function is proportional to $\lambda^2 a^4 \int_0^M dp\, f(p)$ with $f(p) = p^2 j_0^2(pa) \ln p$ for large p. This integral diverges like $M \ln(M)$.

The imposed upper limit, M, in eq. (49) plays the role of a *physical* cutoff that regulates divergences localized on the surface. It is not related to the regulator of the ultraviolet divergences in loop integrals. Hence divergences like in eq. (49) are not renormalized by standard counterterms whose (divergent) coefficients are independent of the considered background because they are fixed by renormalization conditions on Green's functions at some prescribed finite external momenta. Divergences that emerge as $M \to \infty$ indicate that even the sharp limit $w \to 0$ does not exist. The strong coupling limit $\lambda \to \infty$ makes the divergence even worse. If the divergent terms depend on the quantity conjugate to the force (tension) under consideration[6] the force (tension) cannot be defined independently of the structure of the material. This is the case for the shell or the sphere, but not for rigid bodies.

CONCLUSION

In this talk I have presented an efficient method to compute vacuum polarization energies in renormalizable quantum field theories for static background fields. Starting point for this method is the energy density operator. Its matrix element in the ground state is expressed in terms of the Green's function which subsequently is parameterized by data from scattering off the background field. To compute the momentum integrals as contour integrals in the upper–half plane, this approach makes ample use of the identity of Feynman diagrams and Born approximants to Casimir energies. More importantly, this identity allows one to implement standard (perturbative) renormalization conditions on the divergent, low order Green's functions. In this way the removal of the ultraviolet loop divergences is independent of the considered background.

Utilizing a variational approach to the so–obtained total energy, solitons can be constructed. As an application I have shown that in a $1+1$ dimensional chiral model quantum corrections create a soliton that is classically unstable. In similar $3+1$ dimensional

---

[6] For example, the distance between plates or the radius of the shell.

models the situation is more complex [7], as issues like Landau poles [16] and sphaleron barriers [17] complicate matters.

The divergences that arise when a quantum field is forced to vanish on a surface can be nicely studied in this approach by implementing a boundary condition as the limit of a less singular background. Energy densities away from the surfaces or quantities like the force between rigid bodies, for which the surfaces can be held fixed, are finite and independent of the material cutoffs. Observables that require deformation or change in area of the surface cannot be defined independently of the other material properties.

ACKNOWLEDGMENTS

I would like to thank the organizers for this interesting and worthwhile workshop. Furthermore I appreciate helpful remarks on the manuscript by M. Quandt and acknowledge support by the Deutsche Forschungsgemeinschaft (DFG) in form of a Heisenberg–fellowship under contract We 1254/3-2.

REFERENCES

1. Y. Nambu and G. Jona–Lasinio, Phys. Rev. **122** (1961) 345, **124** (1961) 246.
2. D. Ebert and H. Reinhardt, Nucl. Phys. B **271** (1986) 188.
3. R. Alkofer, H. Reinhardt, and H. Weigel, Phys. Rept. **265** (1996) 139.
4. For a comprehensive review see: N. Graham, R. L. Jaffe, and H. Weigel [arXiv:hep-th/0201148] in M. Bordag, ed., *Proceedings of the Fifth Workshop on Quantum Field Theory Under the Influence of External Conditions*, Intl. J. Mod. Phys. A **17** (2002) No. 6 & 7.
5. N. Graham, R. L. Jaffe, V. Khemani, M. Quandt, M. Scandurra, and H. Weigel, Nucl. Phys. B **645**, 49 (2002) [arXiv:hep-th/0207120].
6. E. Farhi, N. Graham, R. L. Jaffe, and H. Weigel, Phys. Lett. B **475**, 335 (2000) [arXiv:hep-th/9912283], Nucl. Phys. B **585**, 443 (2000) [arXiv:hep-th/0003144], Nucl. Phys. B **595** (2001) 536 [arXiv:hep-th/0007189], H. Weigel, arXiv:hep-th/0108180.
7. E. Farhi, N. Graham, R. L. Jaffe, and H. Weigel, Nucl. Phys. B **630**, 241 (2002) [arXiv:hep-th/0112217], E. Farhi, N. Graham, R. L. Jaffe, V. Khemani, and H. Weigel, *in preparation*.
8. N. Graham, R. L. Jaffe, V. Khemani, M. Quandt, M. Scandurra, and H. Weigel, arXiv:hep-th/0207205.
9. J. Schwinger, Phys. Rev. **94** (1954) 1362, J. Baacke, Z. Phys. C **53** (1992) 402.
10. E. Farhi, N. Graham, P. Haagensen, and R. L. Jaffe, Phys. Lett. B **427**, 334 (1998) [arXiv:hep-th/9802015], N. Graham and R. L. Jaffe, Nucl. Phys. B **544** (1999) 432 [arXiv:hep-th/9808140].
11. S. Coleman, Commun. Math. Phys. **31** (1973) 259.
12. D. Deutsch and P. Candelas, Phys. Rev. D **20** (1979) 3063, P. Candelas, Ann. Phys. **143** (1982) 241.
13. K. Symanzik, Nucl. Phys. B **190** (1981) 1, A. A. Actor, Ann. Phys. **230** (1994) 303, Fort. Phys. **43** (1995) 141.
14. V. M. Mostepanenko and N. N. Trunov, *The Casimir Effect and its Application*, Clarendon Press, Oxford (1997), K. A. Milton, *The Casimir Effect: Physical Manifestations Of Zero-Point Energy*, River Edge, USA: World Scientific (2001).
15. K. A. Milton, arXiv:hep-th/0210081.
16. G. Ripka and S. Kahana, Phys. Rev. D **36** (1987) 1233, J. Hartmann, F. Beck, and W. Benz, Phys. Rev. C **50** (1994) 3088. [arXiv:hep-ph/9410263].
17. G. 't Hooft, Phys. Rev. Lett. **37** (1976) 8, N. S. Manton, Phys. Rev. D **28** (1983) 2019, F. R. Klinkhamer and N. S. Manton, Phys. Rev. D **30** (1984) 2212.

Light meson resonances from unitarized Chiral Perturbation Theory

J.R.Peláez*† and A.Gómez Nicola†

*Dip. di Fisica. Universita' degli Studi and INFN, Firenze, Italy
†Departamento de Física Teórica II
Universidad Complutense
28040 Madrid, Spain

Abstract. We report on our recent progress in the generation of resonant behavior in unitarized meson-meson scattering amplitudes obtained from Chiral Perturbation Theory. These amplitudes provide simultaneously a remarkable description of the resonance region up to 1.2 GeV as well as the low energy region, since they respect the chiral symmetry expansion. By studying the position of the poles in these amplitudes it is possible to determine the mass and width of the associated resonances, as well as to get a hint on possible classification schemes, that could be of interest for the spectroscopy of the scalar sector.

THE LIGHT MESON PUZZLE

In this work we review our recent progress in determining the position of the poles [1] that appear associated to resonant behavior in meson-meson scattering amplitudes, obtained from unitarized one-loop Chiral Perturbation Theory [2]. This apparently formal interest is motivated by the spectroscopy of light mesons, whose present status is somewhat controversial. Poles in the second Riemann sheet of partial wave scattering amplitudes are of relevance because when they are close to the real, physical values of the center of mass energy $\sqrt{s}$, we can neglect all other terms in the partial wave and simply write

$$t(s) = \frac{R_R}{s - s_{pole}} = \frac{R_R}{s - (\mathrm{Re}\sqrt{s_{pole}})^2 - (\mathrm{Im}\sqrt{s_{pole}})^2 - i\,2\,\mathrm{Re}\sqrt{s_{pole}}\,\mathrm{Im}\sqrt{s_{pole}}} \tag{1}$$

where R_R would be some real residue that can be calculated but is irrelevant for us here. Furthermore, if by "close to the real axis" we mean that $\mathrm{Im}\sqrt{s_{pole}} \ll \mathrm{Re}\sqrt{s_{pole}}$, then, we can approximate:

$$t(s) \simeq \frac{R_R}{s - (\mathrm{Re}\sqrt{s_{pole}})^2 - i\,2\,\mathrm{Re}\sqrt{s_{pole}}\,\mathrm{Im}\sqrt{s_{pole}}} \equiv \frac{R_R}{s - M_R^2 + i M_R \Gamma_R} \tag{2}$$

where in order to write our equation in the familiar Breit-Wigner form, in the last step we have identified $\sqrt{s_{pole}} \simeq M_R - i\Gamma_R/2$. Breit-Wigner (BW) resonances yield the familiar and experimentally distinct resonant shape in the cross section and its associated fast phase movement, which increases by π in a very small energy range. The quantum

CP660, *Hadron Physics: Effective Theories of Low Energy QCD*, edited by A. H. Blin et al.
© 2003 American Institute of Physics 0-7354-0120-9/03/$20.00

numbers of the resonances correspond to those of the partial wave where the pole is sitting.

However, the farther away from the real axis the poles are, the lousier becomes the connection with resonance parameters. Let us remark that in order to have a BW shape, it is essential for the pole to be near the real axis, or more quantitatively $M_R \gg \Gamma_R$. This allows us to neglect all other terms in the amplitude as well as terms of order Γ_R^2/M_R^2. Intuitively, the familiar resonances that are clearly seen or detected are quasi bound states whose decay time is large (their width is small) compared with their rest energy (their mass). Of course, between a nice BW resonant shape and the continuum, one could think of all intermediate situations, which, naively correspond to changing the pole position from the vicinity of the real axis to have an infinite imaginary part. In other words, starting from narrow resonances and moving the pole to $-i\infty$, we get broader structures, and finally, the continuum.

In particular, broad resonant structures seem to occur in the scalar channels in meson-meson scattering, where in the last decade there has been a renewed interest [3, 4] on the longstanding controversy about the existence of a broad scalar-isoscalar resonance in the low energy region: the so called σ, or $f_0(600)$ in the latest version of the Particle Data Group (PDG) Review [5]. Its experimental evidence only from $\pi\pi$ scattering is rather confusing, since it definitely does not display a Breit-Wigner shape, although many groups have been able to identify an associated pole in the amplitude, but deep in the complex plane A similar or even more confusing situation occurs in πK scattering, where another pole, the κ, has been suggested by many groups [6, 7], but again there is no trace of a BW shape in the scattering. For a compilation of σ and κ poles see the nice overview in [8].

Let us remark that meson-meson scattering data [9] are hard to obtain. As a matter of fact the problem is that they have been extracted from reactions like meson-N →meson-meson-N, but with assumptions like a factorization of the four meson amplitude, or that only one meson is exchanged and that it is more or less on shell, etc... All these approximations introduce large systematic errors. There are, however, other sources of information on meson-meson interactions like, for instance, the very precise determination of a combination of $\pi\pi$ phase shifts from K_{l4} decays [10]. At higher energies the decays of even heavier particles can be also used to study the previously mentioned and other scalar resonances like the $f_0(980)$ or the $a_0(980)$. For instance, very recently, results from charm decays [11], seem to find both the σ and κ poles in reasonable agreement with the groups mentioned above, but the controversy about their existence still lingers on.

Meson spectroscopy aims at classifying the bound states of QCD and at identifying their nature, that is, what are they made of. Starting with the scalar-isoscalar sector, its relevance is twofold: First, one of the most interesting features of QCD is its non-abelian nature, which implies that the carriers of the strong force, the gluons, interact among themselves, contrary to what happens with photons in QED. A possible consequence of this fact is the existence of bound states of gluons, or glueballs, which will certainly be isoscalars. In particular, the lightest ones are expected to be also scalars. Naively, once all the members of quark multiplets are identified in the scalar-isoscalar sector, what remains, if any, are good candidates for glueballs. Of course, the whole picture is much more messy due to mixing phenomena, so that the resonances we actually see are a superposition of different kind of states. Second, it is also understood that QCD has an spontaneous breaking of the chiral symmetry since its vacuum is not invariant under chiral transformations. The study of the scalar-isoscalar sector is relevant to understand the QCD vacuum, which has precisely those quantum numbers.

Nevertheless, we should not forget the other channels, since we can find there the other members of the multiplets, since all the channels are related by the chiral SU(3) symmetry of QCD. We cannot simply add BW resonances to different channels without carefully taking into account this symmetry. Concerning vector channels, there are clear BW resonances like the $\rho(770)$ in $\pi\pi$ scattering or the $K^*(892)$ in πK scattering, that the meson spectroscopy community identify with $q\bar{q}$ states. These are so clearly resonant that "vector meson dominance" is basically enough to describe the bulk of meson interactions.

POLES FROM CHIRAL SYMMETRY AND UNITARITY

The interest of our work in the context of meson spectroscopy is that we have been able to *generate* the resonant behavior present in meson-meson scattering. Our amplitudes [2] have been obtained by unitarizing the one-loop amplitudes obtained from Chiral Perturbation Theory (ChPT [12]), which is the most general effective Lagrangian built of pions, kaons and etas, that respects the chiral symmetry constraints of QCD. However, since the ChPT amplitudes behave as polynomials at high energy, they violate partial wave unitarity, which is imposed with unitarization methods: in our case, the Inverse Amplitude Method (IAM) [13, 4]. Note that *the resonances are not included explicitly*.

Part of this program had been first been carried out for partial waves in the elastic region [13, 4], for which a simple single channel approach could be used, finding the ρ and σ poles in $\pi\pi$ scattering and that of K^* in $\pi K \rightarrow \pi K$. For coupled channel processes, an *approximate* form of this approach had already been shown [7] to yield a remarkable description of the whole meson-meson scattering data up to 1.2 GeV. When these partial waves were continued to the second Riemann sheet, several poles were found, corresponding to the $\rho, K^*, f_0, a_0, \sigma$ and κ resonances (note that the κ pole could have also been obtained in the elastic single channel formalism). The approximations were needed because at that time not all the ChPT meson-meson amplitudes were known to one-loop. Hence, in [7] only the leading order and the dominant s-channel loops were considered in the calculation, neglecting crossed and tadpole loop diagrams. Of course, in this way the ChPT low energy expansion could only be recovered at leading order. Concerning the divergences, they were regularized with a cutoff, which violates chiral symmetry, making them finite, but not cutoff independent. Fortunately, the cutoff dependence was rather weak and the description of the data was remarkable for cutoffs of the size of the chiral scale. Nevertheless, due to this cutoff regularization, it was not possible to compare the eight parameters of the chiral Lagrangian, which are supposed to encode the underlying QCD dynamics, with those obtained from other low energy processes. That is, it was not possible to test the compatibility of the chiral parameters with the values already present in the literature.

Of course, due to the controversial nature, or even the doubts about the existence of the scalar states, it is very important to check that the poles are not just artifacts of the approximations, to estimate the uncertainties in their parameters, and to check their compatibility with other experimental information regarding ChPT. That was the reason why, in a first step, the $K\bar{K} \to K\bar{K}$ one-loop amplitudes were calculated in [14], also unitarizing them coupled to the $\pi\pi$ states, and reobtaining the σ, f_0 and ρ poles. The whole calculation of one-loop meson meson scattering has been recently completed with the totally new $K\eta \to K\eta, \eta\eta \to \eta\eta$ and $K\eta \to K\pi$ amplitudes [2]. In addition the other five existing independent amplitudes have also been recalculated. The reason for repeating those existing calculations is that, to one loop, one could choose to write all amplitudes in terms of just f_π, or use all f_π, f_K and f_η, or any other combination of them that is equivalent up to $O(p^4)$ etc... However, when one choice is made for one amplitude, the other ones have to be calculated consistently in order to keep the coupled channel perturbative unitarity, which is needed for the IAM. As commented before, with these unitarized amplitudes we obtained [2] a simultaneous description of meson meson scattering data in the resonant region up to 1.2 GeV, but also of the low energy region, with scattering lengths compatible with the most recent determinations. The fact that the calculation was complete to one loop and renormalized as in standard ChPT, also allowed us to show that the resulting set of chiral parameters was compatible with previous determinations in the literature.

The final step is therefore to extend analytically the amplitudes to the complex plane and search for poles in the second Riemann sheet. We will provide next a brief account of how we have built our amplitudes, how the data have been fitted, but also our first, preliminary, results for the poles, although a more detailed exposition and the final calculations will be presented somewhere else soon [1].

CHIRAL PERTURBATION THEORY AMPLITUDES

The QCD massless Lagrangian for the light u, d and s quarks is invariant under the $SU(3)_L \times SU(3)_R$ chiral symmetry, which rotates the Left (or Right) components of these quarks among them. There is also an small explicit breaking due to the small masses of those quarks, but at sufficiently high energies that effect should be rather small. Nevertheless the $SU(3)_L \times SU(3)_R$ symmetry is not seen in the physical spectrum, but only $SU(3)_{L+R}$ is realized approximately once the small explicit breaking is taken into account. The familiar isospin is nothing but the $SU(2)_{L+R}$ subgroup. The $SU(3)_{L-R}$ symmetry has to be spontaneously broken, and indeed, the pions, kaons and etas can be identified as the associated Goldstone bosons of this breaking. Once more, they are not massless, due to the small masses of those quarks, but they are much lighter (and much more stable) than other hadrons with their same quantum numbers, and than the generic hadronic scale of approximately 1 GeV.

These Goldstone bosons are expected to be the relevant degrees of freedom at low energies. Their low energy dynamics can then be described [15] by the most general Lagrangian made of pions, kaons and etas, that implements the symmetry breaking pattern described above, as well as other usual constraints like Lorentz invariance, locality, etc... This is called Chiral Perturbation Theory [12], and it corresponds to an expansion in external momenta, the energy or the mass of the mesons, generically p, over the chiral scale $\Lambda = 4\pi f_\pi \simeq 1.2\,\text{GeV}$. The leading term, $O(p^2)$ is nothing but the non-linear sigma model and only depends on the meson masses and the chiral scale $4\pi f$, where f is the meson decay constant at leading order. Since there are no more free parameters, it is universal, i.e., independent of the detailed mechanism of symmetry breaking. It is enough to reproduce the current algebra results of the 60's. At next to leading order $O(p^4)$, there are eight terms which now are multiplied by some arbitrary low energy constants $L_i(\mu)$, also called chiral parameters. These parameters contain information on the specific dynamics of the underlying theory, but are also needed for the renormalization of the divergences that appear at one-loop when one uses vertices from the lowest order Lagrangian. This renormalization procedure can be carried out to more loops by adding higher order terms in the Lagrangian. In this way it is possible to obtain finite calculations order by order, at the price of including an increasing number of parameters. However, these new terms will all be suppressed by additional powers of p^2/Λ^2 so that the lowest orders will be dominant at low energies. For our purposes it will be enough to work at one-loop, that is $O(p^4)$, so that we still have amplitudes with imaginary parts, as well as the eight L_i parameters that contain information on the specific QCD dynamics.

Therefore, the lowest order, $\mathscr{O}(p^2)$, meson-meson scattering amplitudes (called "low energy theorems" [15] because as we have just commented, they only depend on the symmetry breaking scale) are obtained just from the tree level diagrams of the lowest order Lagrangian. In contrast, the calculation of the $\mathscr{O}(p^4)$ contribution involves the evaluation of the following Feynman diagrams: First, the tree level graphs with the second order Lagrangian, which depend on the chiral parameters L_i. Second, the one-loop diagrams in Fig.1, whose divergences will be absorbed in the L_i through renormalization.

In particular, those graphs in Fig.1a provide an imaginary part to ensure perturbative unitarity, whereas those graphs in Fig.1e, provide the wave function, mass and *decay constant renormalizations*. As we will see the renormalization of the decay constant will play a subtle role in the determination of the $f_0(980)$ and $a_0(980)$ pole positions. Let us then explain this somewhat technical point: Note that the meson decay constants $f_\pi \simeq 94.4\,\text{MeV}$, $f_K = 1.22 f_\pi$ and $f_\eta = 1.3 f_\pi$ only differ at $O(p^4)$ [12, 2]. At leading order, all of them are equal to the only scale in the Lagrangian, f, which, after renormalization, is not directly the physical observable. As a consequence, if we want to write our amplitudes in terms of observable quantities, we could substitute f by f_π or f_K or f_η, or any combination of them. We could even make a different choice for each amplitude *as long as we do not couple the amplitudes among them*. However, if one wants to study a coupled channel process, once a choice is made for one amplitude, the choices for the coupled amplitudes have to be made consistently, if one wants to ensure perturbative unitarity. The same argument would follow for the masses, but they already differ at leading order, so that the numerical difference is irrelevant compared with the decay constant case.

The one-loop amplitudes of $\pi\pi \to \pi\pi$ [12], $\pi K \to \pi K$ [16] and that of $\pi\eta \to \pi\eta$ [16] were calculated more than a decade ago, because the thresholds of these reactions is low enough to apply the standard ChPT formalism. As explained in the introduction, the $K\bar{K} \to K\bar{K}$ one-loop amplitudes were calculated in [14], and those of $K\eta \to K\eta, \eta\eta \to \eta\eta$ and $K\eta \to K\pi$ in [2], much more recently since their thresholds are much higher and they only became interesting when the appropriate unitarization methods were developed. In [2], the other five one-loop amplitudes were recalculated in order to express all of them in terms of f_π only, and ensure exact perturbative partial wave unitarity, which we explain in the next section.

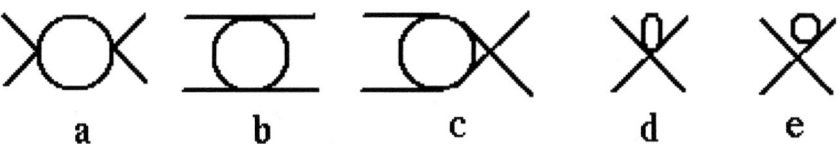

$$\begin{array}{ccccc} \text{a} & \text{b} & \text{c} & \text{d} & \text{e} \end{array}$$

FIGURE 1.   Generic one-loop Feynman diagrams that have to be evaluated in meson-meson scattering.

As we have already commented in the introduction, meson-meson scattering data is customarily presented using partial waves of definite isospin and angular momentum, t_{IJ}. In particular the data is given in terms of the complex phase of the amplitude, or phase shifts δ_{IJ} According to our previous discussion, the meson-meson partial waves within ChPT are thus obtained as series in the momenta, (some terms are also multiplied by chiral logarithms from the loops functions). Generically, in the chiral expansion we will then find, omitting the I,J subindices, $t \simeq t_2 + t_4 + ...$, where t_2 and t_4 the $\mathcal{O}(p^2)$ and $\mathcal{O}(p^4)$ contributions, respectively.

PARTIAL WAVE UNITARITY

The S matrix unitarity relation $SS^\dagger = 1$ translates into simple relations for the elements of the T matrix $t^{\alpha\beta}$ if they are projected into partial waves, where $\alpha, \beta, ...$ denote the different states physically available. For instance, if there is only one possible state, α, the partial wave $t^{\alpha\alpha}$ satisfies

$$\operatorname{Im} t^{\alpha\alpha} = \sigma_\alpha |t^{\alpha\alpha}|^2 \quad \Rightarrow \quad \operatorname{Im} \frac{1}{t^{\alpha\alpha}} = -\sigma_\alpha \quad \Rightarrow \quad t^{\alpha\alpha} = \frac{1}{\operatorname{Re} t^{\alpha\alpha} - i\sigma_\alpha} \qquad (3)$$

where $\sigma_\alpha = 2q_\alpha/\sqrt{s}$ and q_α is the C.M. momentum of the state α. Written in this way it can be readily noted that *we only need to know the real part of the Inverse Amplitude*. The imaginary part is fixed by unitarity. As a matter of fact, this relation *only holds above threshold* up to the energy where another state, β, is physically accessible. Above that point, the unitarity relation for the partial waves can be written as:

$$\begin{aligned} \operatorname{Im} t^{\alpha\alpha} &= \sigma_\alpha |t^{\alpha\alpha}|^2 + \sigma_\beta |t^{\alpha\beta}|^2, & (4) \\ \operatorname{Im} t^{\alpha\beta} &= \sigma_\alpha t^{\alpha\alpha} t^{12*} + \sigma_\beta t^{\alpha\beta} t^{22*}, \\ \operatorname{Im} t^{\beta\beta} &= \sigma_\alpha |t^{\alpha\beta}|^2 + \sigma_\beta |t^{\beta\beta}|^2. \end{aligned}$$

or, in matrix form (and only above the second threshold):

$$\operatorname{Im} T = T\Sigma T^* \quad \Rightarrow \quad \operatorname{Im} T^{-1} = -\Sigma \quad \Rightarrow \quad T = (\operatorname{Re} T - i\Sigma)^{-1} \qquad (5)$$

with

$$T = \begin{pmatrix} t^{\alpha\alpha} & t^{\alpha\beta} \\ t^{\alpha\beta} & t^{\beta\beta} \end{pmatrix} \quad , \quad \Sigma = \begin{pmatrix} \sigma_\alpha & 0 \\ 0 & \sigma_\beta \end{pmatrix}, \qquad (6)$$

which allows for a straightforward generalization to the case of n accessible states. Once more, unitarity means that we would only need to calculate the real part of the inverse amplitude matrix.

Coming back to ChPT, we can notice that the perturbative series of ChPT behave as polynomials with a higher order term $O(p^N/\Lambda^N)$. If we substitute them in the above unitarity relations for the imaginary parts of T, which are non-linear, we will have $O(p^N/\Lambda^N)$ on the left side, but $O(p^{2N}/\Lambda^{2N})$ on the right. Hence, ChPT amplitudes will never satisfy unitarity exactly. Nevertheless, ChPT partial waves satisfy unitarity perturbatively, that is, instead of eq.(3), they can satisfy:

$$\mathrm{Im}\, t_2^{\alpha\alpha} = 0, \quad \mathrm{Im}\, t_4^{\alpha\alpha} = \sigma_\alpha |t_2^{\alpha\alpha}|^2 \qquad (7)$$

for the single channel case, and instead of eq.(5), they can satisfy

$$\mathrm{Im}\, T_2 = 0, \quad \mathrm{Im}\, T_4 = T_2 \Sigma T_2^* \qquad (8)$$

for the coupled channel case. Note that , as we did for a single channel, we are using T_2 and T_4 for the $O(p^2)$ and $O(p^4)$ contributions to the scattering matrix. We say "can satisfy" because, generically, the above expressions for the one-loop contributions do not hold exactly, but only up to $O(p^6)$. However, when expressed in terms of physical decay constants, the above relations can even be satisfied exactly if the substitution of $1/f$ in terms of $1/f_\pi$ or $1/f_K$ or $1/f_\eta$ is made to match their corresponding powers on both sides of the above equations. In such case, the $O(p^6)$ can be made to vanish. (As we already commented, the masses also suffer the same subtlety and the same care has to be taken with them.)

Since in the literature the amplitudes had been calculated sometimes just in terms of $1/f_\pi$ but some other times using or $1/f_K$ or $1/f_\eta$ independently, we recalculated all of them in terms of just f_π in [2], the simplest choice. Nevertheless, we are also presenting here results with the much more natural choice of using the decay constants associated to each field in the process. From the formal point of view, the two choices are equivalent up to $O(p^4)$, but in the second one the resummation of the decay constants is implicitly carried out to higher orders. In addition, it has the advantage of using f_K when dealing with kaons or f_η when dealing with etas. Numerically, the differences could be sizable at high energies when using the unitarized amplitudes.

UNITARIZATION: THE INVERSE AMPLITUDE METHOD

Unitarity is a very important feature of scattering, and it is even more relevant when dealing with resonances, which generically saturate the unitarity bounds. This can be illustrated in the single channel case, where eq.(3) implies the following unitarity bound: $|t_{\alpha\alpha}| \leq 1/\sigma_\alpha$. Moreover, if we sit on top of a BW resonance, at $s = M_R^2$, we see from eq.(2), that the amplitude becomes purely imaginary, that is $\mathrm{Im}\, t_{\alpha\alpha} = |t_{\alpha\alpha}|$, and therefore, in this case eq.(3) implies $|t_{\alpha\alpha}| = 1/\sigma_\alpha$. The unitarity bound is saturated. Once more, the ChPT amplitudes if extrapolated to high enough energies, will violate also this bound, since they behave as polynomials in s.

In order to unitarize the ChPT amplitudes one of the simplest methods is to introduce the $\mathrm{Re}\,T$ in eq.(5), calculated as a ChPT expansion

$$T^{-1} \simeq T_2^{-1}(1 - T_4 T_2^{-1} + ...),\tag{9}$$

$$\mathrm{Re}\,T^{-1} \simeq T_2^{-1}(1 - (\mathrm{Re}\,T_4)T_2^{-1} + ...).\tag{10}$$

Taking into account the perturbative unitarity conditions, eq.(8), we thus find

$$T^{IAM} \simeq T_2(T_2 - T_4)^{-1}T_2,\tag{11}$$

which is the coupled channel Inverse Amplitude Method, which we have indeed used to unitarize *simultaneously* the whole set of one-loop ChPT meson-meson scattering amplitudes. Let us remark that if we reexpand eq.(11) at low energies, we recover the vary same chiral expansion, $T^{IAM} = T_2 + T_4 + ...$, which ensures that we are respecting the QCD chiral symmetry breaking pattern at low energies. In addition, it can be easily checked that T^{IAM} satisfies the partial wave unitarity conditions, eq.(5), *exactly*, above the thresholds of all the physically accessible channels. Let us also mention that the IAM can be also generalized to higher orders [13, 17], including the case when the leading order t_2 vanishes [18].

Let us finally remark that the IAM violates crossing symmetry, since obviously we are treating the right and the left cuts differently. The largest influence of the worse left cut approximation is on the closest point to the left cut, that is, the thresholds. We will see that the IAM threshold parameters are in good agreement both with data and with standard ChPT (which certainly respects crossing symmetry), therefore the crossing symmetry violation coming from the IAM itself seems to be small. However, as we have already explained, the meson-meson data is obtained using strong extrapolations. Hence, even the data carries its own amount of crossing violation if errors are not taken into account. When considering not only threshold data, but also experimental information in other regions, *including their uncertainties* it can be shown that the IAM yields indeed just an small crossing symmetry violation [17].

THE INVERSE AMPLITUDE METHOD FIT TO THE SCATTERING DATA

Once we had all the amplitudes calculated within the standard ChPT renormalization scheme (dimensional regularization in the $\overline{MS} - 1$ scheme), we first looked at the results using the IAM with previous determinations of the chiral parameters from other processes (see the ChPT column in Table 1). Due to their large error bars, the uncertainties thus obtained were rather large, but all the resonant behavior in meson-meson scattering was clearly recovered. For the detailed plots, we refer the reader to [2], but this already suggests that a description of the resonances is possible within the uncertainty limits of the chiral parameters.

TABLE 1. Different sets of chiral parameters ($\times 10^3$). The first column comes from recent analysis of K_{l4} decays [21] (L_4 and L_6 are set to zero). In the ChPT column L_1, L_2, L_3 come from [22] and the rest from [15]. The three last ones correspond to the values from the IAM including the uncertainty due to different systematic error used on different fits. Sets II and II are obtained using amplitudes expressed in terms of f_π, f_K and f_η, whereas the amplitudes in set I are expressed in terms of f_π only.

Parameter	K_{l4} decays	ChPT	IAM I	IAM II	IAM III
$L_1^r(M_\rho)$	0.46	0.4 ± 0.3	0.56 ± 0.10	0.59 ± 0.08	0.60 ± 0.09
$L_2^r(M_\rho)$	1.49	1.35 ± 0.3	1.21 ± 0.10	1.18 ± 0.10	1.22 ± 0.08
L_3	-3.18	-3.5 ± 1.1	-2.79 ± 0.14	-2.93 ± 0.10	-3.02 ± 0.06
$L_4^r(M_\rho)$	0 (fixed)	-0.3 ± 0.5	-0.36 ± 0.17	0.2 ± 0.004	0 (fixed)
$L_5^r(M_\rho)$	1.46	1.4 ± 0.5	1.4 ± 0.5	1.8 ± 0.08	1.9 ± 0.03
$L_6^r(M_\rho)$	0 (fixed)	-0.2 ± 0.3	0.07 ± 0.08	0 ± 0.5	-0.07 ± 0.20
L_7	-0.49	-0.4 ± 0.2	-0.44 ± 0.15	-0.12 ± 0.16	-0.25 ± 0.18
$L_8^r(M_\rho)$	1.00	0.9 ± 0.3	0.78 ± 0.18	0.78 ± 0.7	0.84 ± 0.23

Of course, a much better description could be obtained with a fit to the data. We therefore carried out a fit, using MINUIT [20], to the presently available data on meson-meson scattering. Due to the already commented problems with the systematic uncertainties in the data, which has not been quantified in the original articles, we performed fits adding a 1%, 3% or a 5% systematic error. The resulting curves are basically indistinguishable to the naked eye. The errors quoted in Table 1 for the IAM sets of fitted chiral parameters, correspond to those of MINUIT combined with a systematic error that covers the spread of values obtained when adding that 1%, 3% or 5% systematic error. Note that the values we obtain are compatible with previous determinations. In particular, we show in Table 2 the threshold parameters compared with existing data and plain ChPT determinations to one and two loops.

TABLE 2. Scattering lengths a_{IJ} and slope parameters b_{IJ} for different meson-meson scattering channels. For experimental references see [2]. Let us remark that our one-loop IAM results are very similar to those of two-loop ChPT.

Threshold parameter	Experiment	IAM fit I [2]	ChPT $\mathcal{O}(p^4)$ [4, 16]	ChPT $\mathcal{O}(p^6)$ [23]
a_{00}	0.26 ± 0.05	$0.231^{+0.003}_{-0.006}$	0.20	0.219 ± 0.005
b_{00}	0.25 ± 0.03	0.30 ± 0.01	0.26	0.279 ± 0.011
a_{20}	-0.028 ± 0.012	$-0.0411^{+0.0009}_{-0.001}$	-0.042	-0.042 ± 0.01
b_{20}	-0.082 ± 0.008	-0.074 ± 0.001	-0.070	-0.0756 ± 0.0021
a_{11}	0.038 ± 0.002	0.0377 ± 0.0007	0.037	0.0378 ± 0.0021
$a_{1/20}$	$0.13...0.24$	$0.11^{+0.06}_{-0.09}$	0.17	
$a_{3/20}$	$-0.13...-0.05$	$-0.049^{+0.002}_{-0.003}$	-0.5	
$a_{1/21}$	$0.017...0.018$	0.016 ± 0.002	0.014	
a_{10}		$0.15^{+0.07}_{-0.11}$	0.0072	

The IAM I fit was obtained expressing all the amplitudes in terms of just f_π, which, as we have already explained is somewhat unnatural when dealing with kaons or etas. The plots and the uncertainties of this fit were already given in [2], and therefore we have preferred to present here our first results using amplitudes written in terms of f_K and f_η when dealing with processes involving kaons or etas. In particular, we have rewritten our $O(p^2)$ amplitudes changing one factor of $1/f_\pi$ by $1/f_K$ for each two kaons present between the initial or final state, or by $1/f_\eta$ for each two etas appearing between the initial and final states. In the special case $K\eta \to K\pi$ we have changed $1/f_\pi^2$ by $1/(f_K f_\eta)$. Of course, these changes introduce some corrections at $O(p^4)$ which can be easily obtained using the relations between the decay constants and f provided in [12, 2]. The $1/f_\pi$ factor in each loop function at $O(p^4)$ (generically, the $J(s)$ given in the appendix of [2]) have to be changed according to eqs.(8). The amplitudes thus obtained are formally equivalent to the previous ones, up to $O(p^6)$ differences. However, at high energies there can be some small numerical differences when determining the poles. Obviously, the $\pi\pi \to \pi\pi$ amplitude remains unchanged.

The fit results using these more naturally normalized amplitudes are given in Fig.2, and the resulting new sets of parameters is also presented in Table 1 as the IAM set II. Note that the only parameters that suffer a sizable change are those related to the definition of decay constants: L_4 and L_5. As it happened in [2], the uncertainty bands are calculated from a MonteCarlo Gaussian sampling (1000 points) of the L_i sets within their error bars, assuming they are uncorrelated (and therefore they are conservative estimates).

We have even performed a third fit, the IAM III, by fixing L_4 to zero as in the most recent K_{l4} $O(p^4)$ determinations given also in Table 1.

Let us recall that in these proceedings we are still showing some preliminary results whose calculation is still in progress [1]. In a forthcoming work [1] we will provide the final numbers (mostly for the errors) and the threshold parameters for these other fits. Concerning the threshold parameters we do not expect relevant changes compared to data since the $\pi\pi \to \pi\pi$ amplitude has not changed and therefore the new numbers will remain almost identical to those of IAM I.

As we can see in Fig.2, we obtain again a nice description of meson-meson data up to 1.2 GeV, including once more all the resonant behaviors. One may wonder what would be the effect of applying the IAM to higher orders. Only the $\pi\pi \to \pi\pi$ amplitude has been calculated up to $O(p^6)$ and it has been unitarized in [17], using the higher order form of IAM. The results regarding poles and resonances in the single channel case are unchanged and the parameters are compatible with those of standard ChPT at $O(p^6)$.

Finally, let us remark that the IAM has also been applied to $\pi\pi$ elastic scattering in the $(I, J) = (0, 2)$ wave [18], whose leading order vanishes. The amplitude has to be considered up to $O(p^6)$ and add an approximation at $O(p^8)$, but the IAM is able to generate a pole associated to the $f_2(1200)$ BW resonance. The mass and widths are in fairly good agreement with data taking into account that that resonance has only an 80% decay into pions.

POLES IN MESON-MESON SCATTERING

In Table 3 we present the position of poles in the second Riemann sheet of meson-meson scattering calculated with the one-loop IAM. The names we provide refer to the most

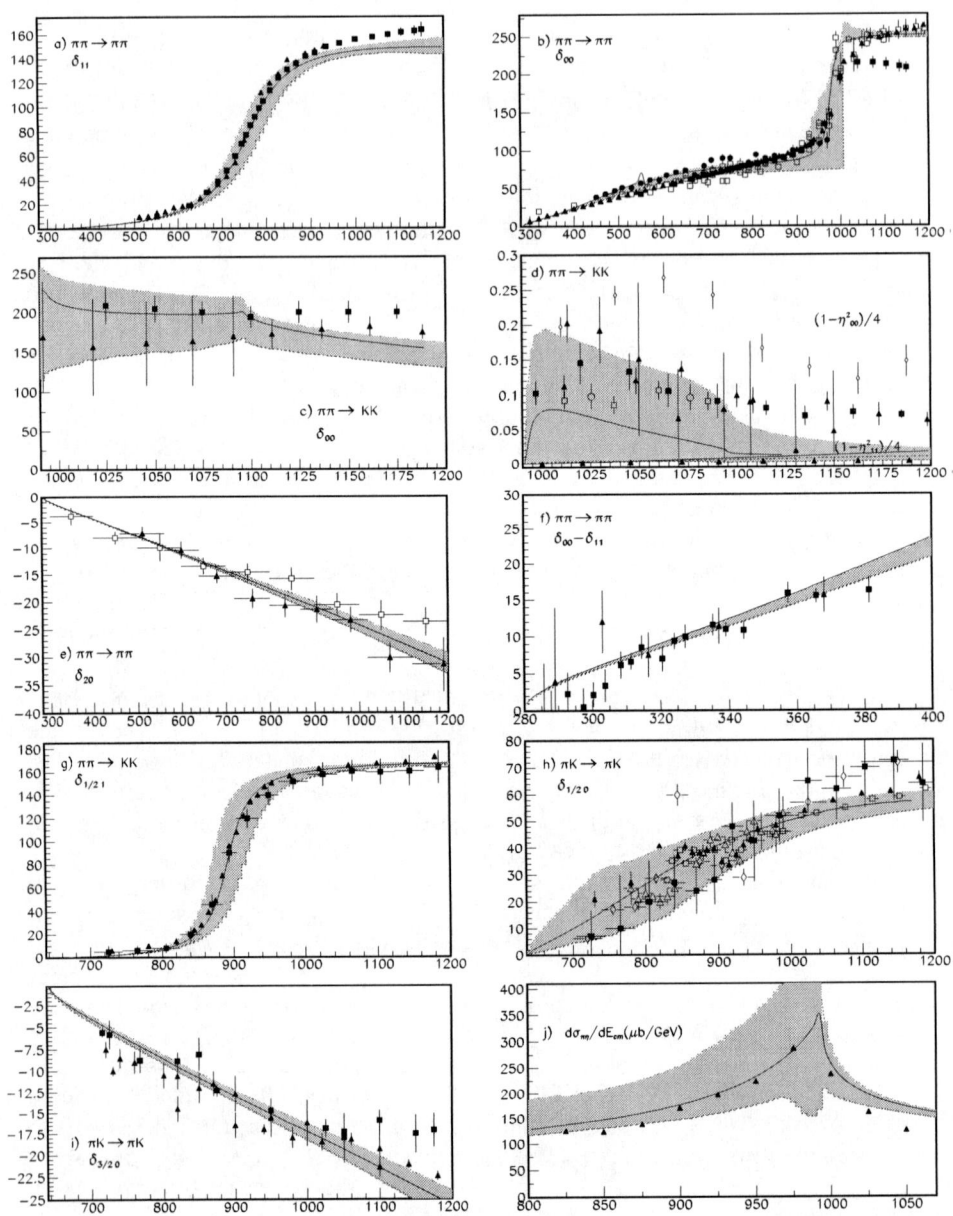

FIGURE 2. IAM fit to meson-meson scattering data, set II. The uncertainties cover also the estimated systematic errors. The statistical errors from the fit would be much smaller.

similar states that we have found in the literature, but that does not mean that from the present approach we could drag any conclusion on their nature. In Table 4 we provide either the mass and width of these resonances or their pole position as given in the PDG.

TABLE 3. Pole positions (with errors) in meson-meson scattering. When close to the real axis the mass and width of the associated resonance is $\sqrt{s_{pole}} \simeq M - i\Gamma/2$.

$\sqrt{s_{pole}}$(MeV)	ρ	K^*	σ	f_0	a_0	κ
IAM Approx (no errors)	759-i71	892-i21	442-i227	994-i14	1055-i21	770-i250
IAM I (errors)	760-i82 $\pm 52\pm i25$	886-i21 $\pm 50\pm i8$	443-i217 $\pm 17\pm i12$	988-i4 $\pm 19\pm i3$	cusp?	750-i226 $\pm18\pm i11$
IAM II (errors)	754-i74 $\pm 18\pm i10$	889-i24 $\pm 13\pm i4$	440-i212 $\pm 8\pm i15$	973-i11 $^{+39}_{-127}$ $^{+i189}_{-i11}$	1117-i12 $^{+24}_{-320}$ $^{+i43}_{-i12}$	753-i235 $\pm 52\pm i33$
IAM III (errors)	748-i68 $\pm 31\pm i29$	889-i23 $\pm 22\pm i8$	440-i216 $\pm 7\pm i18$	972-i8 $^{+21}_{-56}\pm i7$	1091-i52 $^{+19}_{-45}$ $^{+i21}_{-i40}$	754-i230 $\pm 22\pm i27$

TABLE 4. Mass and widths or pole positions of the light resonances quoted in the PDG. Recall that for narrow resonances $\sqrt{s_{pole}} \simeq M - i\Gamma/2$

PDG2002	$\rho(770)$	$K^*(892)^\pm$	σ or $f_0(600)$	$f_0(980)$	$a_0(980)$	κ
Mass (MeV)	771 ± 0.7	891.66 ± 0.26	(400-1200)-i(300-500)	980 ± 10	980 ± 10	not
Width (MeV)	149 ± 0.9	50.8 ± 0.9	(we list the pole)	40-100	50-100	listed

Let us briefly comment Table 3. In the first line we are giving the results already obtained in [7], with the approximated coupled channel IAM, using amplitudes with f_π, f_K and f_η. It can be noticed that there were nine scalar poles, the σ, the $f_0(980)$, the three states of the $a_0(980)$ as well as the four states of the κ. Since they were generated simultaneously, they could be a good candidate for a nonet, although clearly some mechanism should be producing the mass difference, very likely some kind of mixing with higher order states [24].

Concerning the results of the IAM, we see that there are always poles associated to the vector resonances ρ and K^*, in good agreement with the data and with the approximated method. The uncertainties in the pole positions have been obtained again using a MonteCarlo Gaussian sample (300 samples) of the L_i parameters, within the errors of each set. Let us note that the vector octet is complete, since we also obtain a pole in the $(I,J) = (0,1)$ below the $\bar{K}K$ threshold, but it is only a crude approximation to the Φ and ω states (it is the octet Φ indeed). The problem here is that the other relevant coupled channel that separates the Φ and the ω is a three pion state, that we cannot implement in the IAM. For details, we refer the reader to [19, 7, 2].

Concerning scalar states, from Table 3 we see that the results concerning the most controversial ones are consistent and in very good agreement between different IAM sets and also with the approximated IAM. In other words, the results for the σ and the κ poles are robust within this approach: *there are always "light" poles in the* $(I,J) = (0,0),(1/2,0)$ *channels, and their position is fairly well determined*, in round numbers, around $440 - i215$ MeV for the σ and $750 - i230$ MeV for the κ. The errors are comparatively small as it can be seen in Table 3.

The situation concerning f_0 is also rather stable for the mass, which is always around 975 MeV. In contrast, the uncertainty on the width is rather large. In particular, the central value is somewhat small when using set 1 (just one f_π) but in a fairly good agreement with data when considering sets 2 and 3 or the approximated IAM (all of them use f_π, f_K and f_η). As we argued before, it was natural to expect that the use of f_K and f_η when dealing with kaons or etas would provide better results.

Finally, the most sensible state seems to be the $a_0(980)$ resonance. It can be noticed that it is present as a pole in the second Riemann sheet in sets 2 and 3 as well as in the approximated IAM. However, it is not found as a pole with set 1, using just f_π. The fact that the $a_0(980)$ pole was absent if one uses only the tree level terms and the tadpoles of the complete amplitudes in [2] (again using just f_π) with the approximated IAM was first noted in [25] and has been interpreted as a possible cusp effect.

Given the uncertainty on the $a_0(980)$ it is hard to identify it conclusively as a pole or a cusp. However, we think that there is a somewhat stronger support for the pole interpretation, although with a strong threshold distortion: On the one hand, the width of the $f_0(980)$, which is closely related to the $a_0(980)$, is much better described by the IAM when using several decay constants, which then give a pole for the $a_0(980)$. On the other hand the existence of the $a_0(980)$ state seems much less controversial from other sources apart from meson-scattering data [5]. We remark, anyway, that the two possibilities can be accommodated within the IAM.

CONCLUSIONS

We have reported on our recent work where we have completed the meson-meson scattering amplitudes to one-loop within Chiral Perturbation Theory (ChPT). In order to extend the applicability of these amplitudes to the resonance region, we have unitarized them with the Inverse Amplitude Method (IAM). In this way, we have been able to describe the meson-meson scattering data up to 1.2 GeV, generating the resonant behaviors, but simultaneously respecting the chiral low energy expansion. These new amplitudes are unitarized in dimensional regularization in order to preserve chiral symmetry, avoiding the use of a cutoff. Thus we have been able to check that the chiral parameters obtained from the IAM description are compatible with previous determinations from other processes within standard ChPT.

In this workshop we have also shown our progress in determining the position of the poles that appear in the IAM amplitudes. When they are close to the real axis above threshold, the position of these poles is related to the mass and width of the associated narrow BW resonances.

In this way, we have been able to establish more robustly our results for the controversial σ and κ scalar states. They seem to be generated simultaneously with the $f_0(980)$ and the $a_0(980)$, and are therefore good candidates for a possible light scalar nonet. Nevertheless, the $a_0(980)$ is found to be very sensible to the choice on how to express the amplitudes in terms of the physical meson decay constants.

We hope these results could be of interest in the field of meson spectroscopy

ACKNOWLEDGMENTS

A.G.N and J.R.P wish to thank the organizers of the "2nd International Workshop on Hadron Physics" for their kind invitation and for their efforts to offer us such a pleasant and lively workshop in Coimbra. Work supported by the Spanish CICYT projects, FPA2000-0956, PB98-0782 and BFM2000-1326. J.R.P. acknowledges support from the CICYT-INFN collaboration grant 003P 640.15.

REFERENCES

1. A. Gómez Nicola and J. R. Peláez, in preparation.
2. A. Gómez Nicola and J. R. Peláez, Phys. Rev. D **65** (2002) 054009
3. R. Kaminski, L. Lesniak and J. P. Maillet, Phys. Rev. D **50** (1994) 3145. R. Delbourgo and M. D. Scadron, Mod. Phys. Lett. A **10** (1995) 251. S. Ishida *et al.*, Prog. Theor. Phys. **95** (1996) 745. M. Harada, F. Sannino and J. Schechter, Phys. Rev. D **54** (1996) 1991 N. A. Tornqvist and M. Roos, Phys. Rev. Lett. **76** (1996) 1575.
4. A. Dobado and J. R. Pelaez, Phys. Rev. D **47** (1993) 4883. Phys. Rev. D **56** (1997) 3057.
5. K. Hagiwara *et al.*, Phys. Rev. D **66**, 010001 (2002).
6. R.L. Jaffe, Phys. Rev. **D15** 267 (1977); Phys. Rev. **D15**, 281 (1977). E. van Beveren *et al.* Z. Phys. **C30**, 615 (1986). S. Ishida *et al*, Prog. Theor. Phys. **98**,621 (1997). D. Black, A. H. Fariborz, F. Sannino, J. Schechter. Phys. Rev. **D58**:054012,1998. E. van Beveren and G. Rupp, Eur. Phys. J. C **22** (2001) 493
7. J. A. Oller, E. Oset and J. R. Pelaez, Phys. Rev. Lett. **80** (1998) 3452; Phys. Rev. D **59** (1999) 074001 [Erratum-ibid. D **60** (1999) 099906].
8. E. van Beveren and G. Rupp, arXiv:hep-ph/0201006.
9. S. D. Protopopescu *et al.*, Phys. Rev. **D7**, (1973) 1279; P. Estabrooks and A.D.Martin, Nucl.Phys.**B79**, (1974) 301. G. Grayer *et al.*, Nucl. Phys. **B75**, (1974) 189. D. Cohen, Phys. Rev. **D22**, (1980) 2595. W. Hoogland *et al.*, Nucl. Phys. **B126** (1977) 109. M. J. Losty *et al.*, Nucl. Phys. **B69** (1974) 185. R. Mercer *et al.*, Nucl. Phys. **B32** (1971) 381. P. Estabrooks *et al.*, Nucl. Phys. **B133** (1978) 490. H. H. Bingham *et al.*, Nucl. Phys. **B41** (1972) 1. S. L. Baker *et al.*, Nucl. Phys. **B99** (1975) 211. D. Aston *et al.* Nucl. Phys. **B296** (1988) 493. D. Linglin *et al.*, Nucl. Phys. **B57** (1973) 64 .
10. S. Pislak *et al.* [BNL-E865 Collaboration], Phys. Rev. Lett. **87** (2001) 221801.
11. E791 Collaboration, Phys. Rev. Lett. **86**,(2001) 770. C. Gobel for the E791 Collab. hep-ex/0012009.
12. J. Gasser and H. Leutwyler, Annals Phys. **158** (1984) 142. Nucl. Phys. B **250** (1985) 465.
13. T. N. Truong, Phys. Rev. Lett. **61** (1988) 2526. Phys. Rev. Lett. **67**, (1991) 2260; A. Dobado, M.J.Herrero and T.N. Truong, Phys. Lett. **B235** (1990) 134.
14. F. Guerrero and J. A. Oller, Nucl. Phys. B **537** (1999) 459 [Erratum-ibid. B **602** (2001) 641].
15. S. Weinberg, Physica A**96** (1979) 327.
16. V. Bernard, N. Kaiser, U.G. Meissner, Phys. Rev. **D43** (1991) 2757; Nucl. Phys. **B357** (1991) 129; Phys. Rev. **D44** (1991) 3698.
17. J. Nieves, M. Pavon Valderrama and E. Ruiz Arriola, Phys. Rev. D **65** (2002) 036002.
18. A. Dobado and J. R. Pelaez, Phys. Rev. D **65** (2002) 077502.
19. J. A. Oller, E. Oset and J. R. Peláez, Phys. Rev. D **62** (2000) 114017.
20. F. James, Minuit Reference Manual D506 (1994).
21. G. Amorós, J. Bijnens and P. Talavera, Nucl. Phys. **B602** (2001) 87.
22. J. Bijnens, G. Colangelo and J. Gasser, Nucl. Phys. **B427** (1994) 427.
23. G. Amoros, J. Bijnens and P. Talavera, Nucl. Phys. B **585** (2000) 293 [Erratum-ibid. B **598** (2001) 665].
24. J. A. Oller and E. Oset, Phys. Rev. D **60** (1999) 074023.
25. M. Uehara, arXiv:hep-ph/0204020.

Nucleon-Nucleon interactions from effective field theory

José A. Oller [1]

Departamento de Física. Universidad de Murcia. E-30071, Murcia. Spain

Abstract. We have established a new convergent scheme to treat analytically nucleon-nucleon interactions from a chiral effective field theory. The Kaplan-Savage-Wise (KSW) amplitudes are resummed to fulfill the unitarity or right hand cut to all orders below pion production threshold. This is achieved by matching order by order in the KSW power counting the general expression of a partial wave amplitude with resummed unitarity cut, with the inverses of the KSW amplitudes. As a result, a new convergent and systematic KSW expansion is derived for an on-shell interacting kernel $\mathcal{R}$ in terms of which the partial waves amplitudes are computed. The agreement with data for the S-waves is fairly good up to laboratory energies around 350 MeV and clearly improves and reestablishes the phenomenological success of the KSW amplitudes when treated within this scheme.

INTRODUCTION

Effective field theories are the standard method to deal with strong interactions in the non-perturbative regime. As paradigm we have SU(2) Chiral Perturbation Theory (CHPT) for pion physics [1, 2, 3], where a convergent power counting is established in terms of derivatives, insertions of quark mass matrix and external sources. This has also been applied to pion-nucleon interactions with baryon number, B, equal to 1 [4]. Its extension to nucleon-nucleon physics [5, 6, 7, 8, 9] is not straightforward due to the appearance of two new scales: the large scattering lengths of the S-waves and the large nucleon mass, M. The latter can be easily handled for B=1 although this is no longer the case for B> 1 [5]. Large efforts have been devoted during the last years to end with a convergent effective field theory for nucleon-nucleon interactions including pions, for a review see e.g. [10]. On the one hand, we have the original Weinberg's proposal [5], with an undoubted phenomenological success [11]. In this scheme, the nucleon-nucleon potential is calculated in a chiral expansion up to some definite order and then is employed as a kernel in a Lippmann-Schwinger equation to calculate *numerically* the physical amplitudes. Nevertheless, this approach suffers from inconsistencies in the power counting due to the appearance of divergences that are enhanced by powers of the large nucleon mass. As a result, their coefficients are much larger than those expected from 'naive power counting'. Despite this, such divergences are finally reabsorbed by counterterms whose chiral orders are established by assuming

---

[1] oller@um.es

CP660, *Hadron Physics: Effective Theories of Low Energy QCD*, edited by A. H. Blin et al.

natural size for their coefficients. On the other hand, we have the Kaplan-Savage-Wise (KSW) effective field theory [12] with its consistent power counting that directly applies to the physical amplitude, like in standard CHPT [1, 2, 3], and one finally has *analytical* nucleon-nucleon amplitudes. Interestingly, in the KSW power counting all the ultraviolet divergences appearing in loops are canceled by contact operators appearing at either the same or lower order in the expansion. This is not the case in the Weinberg's approach where one has cut-off dependence which gives an estimate of the size of higher order corrections. However, the convergence and the phenomenological success of the KSW effective theory is just restricted to a very narrow region close to the nucleon-nucleon threshold despite having included explicitly the pion fields [13, 14]. This was clear from the analysis at NNLO in the KSW effective field theory performed in ref.[13] where the NNLO departs from data before the NLO and badly diverges for center-of-mass three-momentum above ~ 100 MeV. The reason for this bad convergence properties were identified with large contributions from the twice iterated pion exchange in the triplet channels. That is, pions cannot be treated perturbatively in all the nucleon-nucleon channels and in particular in the $S_1^3 - D_1^3$ one.

Hence, despite all the efforts made during the past years, the issue of deriving a consistent effective field theory for nucleon-nucleon interactions is still open. We perform in this paper a step forward by extending the range of convergence of the KSW amplitudes. In ref. [5] Weinberg proposed to calculate the nucleon-nucleon potential in a chiral expansion and then solve a Lippmann-Schwinger equation in order to treat properly the large enhancements, due to factors $2M/p$, from the reducible two nucleon diagrams which do not enter in the calculation of the potential, see also ref.[8]. From a S-matrix point of view the aim of the Weinberg's proposal can be recasted so that one should resum the unitarity or right-hand cut, which is the responsible for all the unitarity bubbles with their large $2M/p$ factors. Nevertheless, solving a Lippmann-Schwinger equation is not the only way to accomplish this [15]. We propose here a general scheme to resum the unitarity cut as given in ref.[16]. This scheme, originally motivated as an application of the N/D method [17, 18], was already employed in meson-meson [18, 19] and meson-baryon [20, 21] systems. This method should not be confused with Paddè resummations like those performed in the meson-meson or meson-baryon sectors [22]. Nevertheless, there are specific facts in the nucleon-nucleon scattering, associated with the largeness of the scattering lengths in the S-waves, that require a special treatment not present in any meson-meson or meson-nucleon system, as we explain below. It is also worth stressing that all the formalism presented here is analytic.

FORMALISM

It seems that pions cannot be treated perturbatively. This is indeed the main reason for the failure of convergence of the KSW scheme, which considers pions perturbatively, for center of mass three-momentum $p \gtrsim 100$ MeV. This was clearly established in ref.[13], as one can see when comparing the calculated mixing parameter ε_1 at NLO [12] and NNLO [13]. While for very low momentum the NNLO calculations agree better with data, unfortunately for momenta higher than 100 MeV the NNLO calculations badly diverge from experiment, much more than the NLO results, and no improvement is obtained despite pions are explicitly included. This should be compared with the phenomenological success of the Weinberg's scheme in [6] and particularly in [11], where the chiral expansion seems to be under control, at least at the level of the phenomenology. In the latter scheme there are still the issues of avoiding cut-off dependence and the associated inconsistencies in the Weinberg's power counting that were established in refs.[8, 12].

In this paper we make use of the KSW amplitudes which are calculated within an effective field theory and are properly renormalized. There is nonetheless a clear difference between the KSW effective field theory and CHPT. While the latter is convergent in the SU(2) sector in an energy window that could be expected from the scales that are involved in the problem, the same does not occur with the former [13].

Let us denote a partial wave by $T_{L_J^{2S+1}, \hat{L}_J^{2S+1}}$, where L and $\hat{L}$ refer to the orbital angular momentum of the initial and final state respectively, and S and J indicate the total spin and total angular momentum of the systems, in order. Then unitarity requires above the nucleon-nucleon threshold and below the $NN\pi$ one, which is around $p \simeq 280$ MeV, that:

$$\mathrm{Im}\,(T^{(2S+1)J})^{-1}_{ij} = -\frac{Mp}{4\pi}\delta_{ij}\,, \tag{1}$$

where M is the nucleon mass, $T^{(2S+1)J}$ is a matrix whose matrix elements are those triplet partial waves that mix each other, e.g. for $S_1^3 - D_1^3$ one has $T_{11}^{31} = T_{S_1^3, S_1^3}$, $T_{12}^{31} = T_{21}^{31} = T_{S_1^3, D_1^3}$ and $T_{22}^{31} = T_{D_1^3, D_1^3}$. In the previous equation we denote by $(T^{(2S+1)J})^{-1}$ the inverse of the $T^{(2S+1)J}$ matrix. If the partial waves do not mix then $T^{(2S+1)J}$ is just a number equal to $T_{L_J^{2S+1}, L_J^{2S+1}}$. The imaginary part in the previous equation is responsible for the unitarity cut. Thinking of a dispersion relation for the inverse of the amplitude this cut can be easily taken into account from eq.(1), as we previously did in the meson-meson [18] and meson-baryon [20, 21] sectors, and gives rise to the integral:

$$g_i(s) = \frac{1}{\pi}\int_{4M^2}^{\infty} \frac{Mp(s')}{4\pi}\frac{1}{s'-s+i0^+}\,ds'\,, \tag{2}$$

where s is the usual Mandelstam variable. This integral is divergent and requires a subtraction, $Mv_i/4\pi$:

$$g_i(s) = \frac{M}{4\pi}\left(v_i + ip + \frac{M\sigma(s)}{\pi}\log\frac{1-\sigma(s)}{1+\sigma(s)}\right)\,, \tag{3}$$

with $\sigma(s) = \sqrt{1 - \frac{4M^2}{s}}$. The logarithm is purely real in the physical region and just gives rise to relativistic corrections that are essentially negligible in the energy range that we consider. Once this is done, one can write for the full partial wave matrix or number $T^{(2S+1)J}$ the expression:

$$T^{(2S+1)J} = -\left[\mathscr{R}^{-1} + g\right]^{-1}\,. \tag{4}$$

118

In this expression all the other possible cuts of a nucleon-nucleon partial wave, either due to the exchange of other particles or because its helicity structure, are included in the input or kernel $\mathscr{R}$ that we still must fix. The unitarity requirements, resummation of the infinite set of reducible diagrams with two intermediate nucleons, are accomplished by $g(s)$, which as stated is a diagonal matrix in the case of mixed partial waves, or just one function for the unmixed ones. E.g., in the coupled partial waves S_1^3 and D_1^3, $g_{11}(s) = g_1(s)$ and corresponds to the S_1^3 channel and $g_{22}(s) = g_2(s)$ and refers to the D_1^3 channel. Of course, $g_{12}(s) = g_{21}(s) = 0$.

We now specify $\mathscr{R}$ in the key expression eq.(4). For that purpose we make use of the results of the KSW effective field theory for two nucleon systems and of its power counting, that we apply to $\mathscr{R}$ and $g(s)$ in eq.(4). An important point is to establish the chiral order of $g_i(s)$. This is a trivial task for the phase space and the logarithmic term in $g(s)$, eq.(3), since they can be expanded as a series in powers of p starting at order p. The only point we have to consider separately is the chiral order of v_i. For that purpose, we can easily see that considering v_i as a constant of order p^0 is a result when $g(s)$ is calculated with a finite cut-off. Let us perform this illustrative exercise of more than academic importance since the finite three-momentum cut-off is the regularization employed in the Weinberg's scheme. For a given nucleon-nucleon channel i, we can write:

$$g_i^c(s) = -i4M^2 \int \frac{d^4q}{(2\pi)^4} \frac{1}{(q^2 - M^2 + i0^+)((P-q)^2 - M^2 + i0^+)} , \qquad (5)$$

where P is the total four momentum of the two nucleon system, $P^2 = s$, and the superindex c in $g^c(s)$ indicates that is calculated with a cut-off. After performing the q^0 integrating we have:

$$g_i^c(s) = \frac{M^2}{2\pi^2} \int_M^\Lambda \frac{\sqrt{w^2 - M^2}}{w^2 - \frac{s}{4} - i0^+} dw , \qquad (6)$$

where $\Lambda = \sqrt{Q^2 + M^2}$ with Q the three-momentum cut-off. Finally we obtain the explicit result:

$$g_i^c(s) = \frac{M^2}{4\pi^2} \left(2\log\frac{\Lambda+Q}{M} + \sigma(s) \left[\log\frac{\sigma(s) - \frac{2M^2 + \Lambda\sqrt{s}}{Q\sqrt{s}}}{\sigma(s) + \frac{2M^2 - \Lambda\sqrt{s}}{Q\sqrt{s}}} + \log\frac{2\Lambda - \sqrt{s}}{2\Lambda + \sqrt{s}} \right] \right) . \qquad (7)$$

Performing a non-relativistic expansion in the previous equation we have:

$$g_i^c(s) = \frac{M^2}{4\pi^2} \left(2\log\frac{\Lambda+Q}{M} + i\frac{\pi p}{M} + \sigma(s)\log\frac{1 - \sigma(s)}{1 + \sigma(s)} + \mathcal{O}(\frac{p^2}{M^2}) \right) . \qquad (8)$$

Comparing with $g_i(s)$, eq.(3), we finally have:

$$v_i = \frac{2M}{\pi} \log\frac{\Lambda+Q}{M} , \qquad (9)$$

119

TABLE 1. Values for v_i from eq.(9) for different values of the three-momentum cut-off Q.

Q	v	Q	v	Q	v
100	64	500	310	900	510
200	130	600	360	1000	550
300	190	700	410	1100	600
400	250	800	460	1300	670

which is a quantity of order p^0 in the KSW power counting [12, 13]. In the KSW EFT Q is expected to be around 300 MeV, [12]. In table 1 we show the values of v_i in MeV from eq.(9) for different values of Q. We will obtain later, directly from fits to data, similar values of v_i.

Before applying the previous scheme to the KSW amplitudes, let us consider the expansion in powers of x of $f(x) = \cot(x) = \cos x / \sin x$. For that we write:

$$f(x) = -\frac{1}{\tau(x)^{-1} + \theta(x)} , \tag{10}$$

such that $\tau(x) = t_0 + t_1 x + t_2 x^2 + \mathcal{O}(x^3)$ and $\theta(x) = z_0 + z_1 x + z_2 x^2 + \mathcal{O}(x^3)$. We see here that although $f(x)$ starts at order x^{-1} we have considered the functions τ and θ to start at order x^0. In order to fix t_i in terms of the z_i (which are assumed to be known) and of the known expansion of $\cot x$, is simpler to expand the inverse of $f(x)$, then we obtain up to $\mathcal{O}(x^2)$:

$$\frac{\sin x}{\cos x} = x + \mathcal{O}(x^3) = -\frac{1}{t_0} + \frac{t_1}{t_0^2}x + \frac{t_2}{t_0^2}x^2 - \frac{t_1^2}{t_0^3}x^2 - z_0 - z_1 x - z_2 x^2 + \mathcal{O}(x^3) . \tag{11}$$

>From where one can fix the t_0, t_1 and t_2 coefficients. It is obvious how to proceed for higher orders. This simple example also illustrates that if we want to calculate $\tau(x)$ up to order x^i one needs to know $f(x)$ up to order x^{i-2} since $f(x)$ already starts at order x^{-1}.

- S_0^1 elastic partial wave.

  The KSW S_0^1 partial wave, $A_{S_0^1}^{KSW}$, was calculated at NLO (order p^0) in ref.[12] and then at NNLO (order p) in ref.[13]. Let us denote these partial waves by $A_{-1}(p)$, $A_0(p)$ and $A_1(p)$, where the subindex indicates the chiral order. As in the simple example of the $\cot(x)$, if we take as input the KSW amplitudes up to order p then we will be able to calculate $\mathscr{R}$ up to order p^3. This unambiguously fixes the order one has to calculate in the KSW EFT so that the $\mathscr{R}$ is obtained up to the required precision. Following the same notation as for the KSW amplitudes, let us write:

$$\mathscr{R} = R_0 + R_1 + R_2 + R_3 + \mathcal{O}(p^4) , \tag{12}$$

and $g(s) = Mv/4\pi + iMp/4\pi - p^2/2\pi^2 + \mathcal{O}(p^4)$. Then we we must match:

$$\frac{1}{A_{S_0^1}^{KSW}} = \left(\frac{1}{A_{-1}}\right) - \left(\frac{A_0}{A_{-1}^2}\right) + \left(\frac{A_0^2 - A_1 A_{-1}}{A_{-1}^3}\right) + \mathcal{O}(p^4) , \tag{13}$$

with

$$-\left(\frac{1}{\mathscr{R}}+g\right) = -\left(\frac{1}{R_0}+\frac{Mv}{4\pi}\right)+\left(\frac{R_1}{R_0^2}-i\frac{Mp}{4\pi}\right)+\left(\frac{p^2}{2\pi^2}+\frac{R_0R_2-R_1^2}{R_0^3}\right) \quad (14)$$

$$+\left(\frac{R_1^3-2R_0R_1R_2+R_0^2R_3}{R_0^4}\right)+\mathscr{O}(p^4) ,$$

where we have shown between brackets the different orders, from order p^0 up to
order p^3. As a result of the matching we can fix the different R_0, R_1, R_2 and R_3
in terms of v, $A_{-1}(p)$, $A_0(p)$ and $A_1(p)$. We follow ref.[13], so that any of the
previous amplitudes are scale independent at each order in the expansion.
Taking into account that

$$A_{-1} = -\frac{4\pi}{M}\frac{1}{\gamma+ip} , \quad (15)$$

following the notation of ref.[13], where γ is a quantity of order p, the expressions
for the R_i are explicitly given in ref.[16].
Hence working at NLO, $\mathscr{O}(p^0)$, in the KSW amplitudes [12] we will have $\mathscr{R}$ up to
$\mathscr{O}(p^2)$, as explained above,

$$\mathscr{R}^{NLO} = R_0+R_1+R_2 , \quad (16)$$

and matching with the ones at NNLO [13], $\mathscr{O}(p)$, we calculate $\mathscr{R}$ up to $\mathscr{O}(p^3)$:

$$\mathscr{R}^{NNLO} = R_0+R_1+R_2+R_3 . \quad (17)$$

The resulting $\mathscr{R}$ is substituted in eq.(4) and in this way we calculate the partial
waves at the different orders considered so forth. It is clear that the process above
can be done so as to match with a KSW amplitude calculated at any order. In this
way the precision of the resulting amplitude is increased order by order.
• S_1^3–D_1^3 coupled partial waves.
The resulting expressions to match between are the expansions of the matrix elements of the inverse of the KSW matrix of partial waves for the $S_1^3 - D_1^3$ sector,
$A_{S_1^3-D_1^3}^{KSW}$, and the ones that result from the expansion of eq.(4). In this case $\mathscr{R}$ is a
2×2 symmetric matrix:

$$\mathscr{R} = \left(\begin{array}{cc} R_{11} & R_{12} \\ R_{12} & R_{22} \end{array}\right) , \quad (18)$$

and $g(s) = diagonal(g_1(s),g_2(s))$ with its associated subtraction constants v_1 and
v_2. Like in the S_0^1 channel we will take $g_i = \mathscr{O}(p^0)$ and $R_{11} = R_{11,0}+R_{11,1}+$
$R_{11,2}+R_{11,3}+\mathscr{O}(p^4)$. In the KSW scheme at the leading order the S_1^3 and D_1^3
are uncoupled and the mixing starts at NLO, one order higher. Thus, we take
$R_{12} = R_{12,1}+R_{12,2}+\mathscr{O}(p^3)$, starting one order higher than R_{11}. Finally, since the

D_1^3 partial wave starts at order p^0, and is free of unnatural scattering lengths, we simply take $R_{22} = R_{22,0} + R_{22,1} + \mathcal{O}(p^2)$. The expansion of the inverse of eq.(4) can be done straightforwardly so that one can then easily solve for the $R_{ij,k}$ functions, see ref.[16].

Similarly as in the S_0^1 case, working at NNLO implies:

$$
\begin{aligned}
R_{11}^{NNLO} &= R_{11,0} + R_{11,1} + R_{11,2} + R_{11,3} , \\
R_{12}^{NNLO} &= R_{12,1} + R_{12,2} , \\
R_{22}^{NNLO} &= R_{22,0} + R_{22,1} .
\end{aligned}
\tag{19}
$$

For more details we refer to ref.[16].

- P, D, F_2^3 and G_3^3 partial waves except D_1^3.

For the P and D waves, the NLO KSW amplitudes [12] just contain one pion exchange and at NNLO [13, 24] they only include in addition the reducible part of the twice iterated one pion exchange. The physics behind this is then quite limited and this will show up in the phenomenology which, on the other hand, is of the same quality as that of the LO Weinberg's power counting scheme results, ref.[11], which contain a similar input for the potential. For these partial waves we follow the same kind of treatment as discussed above. For the elastic ones we have $\mathscr{R} = R_0 + R_1 + \mathcal{O}(p^2)$ and $g = \mathcal{O}(p^0)$ and for the coupled channel partial waves, namely $P_2^3 - F_2^3$ and $D_3^3 - G_3^3$, the treatment will be similar to that of the $S_1^3 - D_1^3$ case, though here, we will take $R_{ij} = R_{ij,0} + R_{ij,1} + \mathcal{O}(p^2)$ and $g_i(s) = \mathcal{O}(p^0)$, since all the channels start to contribute at the same order p^0 and are free of unnatural scattering lengths.

RESULTS AND DISCUSSION

In this section we consider the phenomenological applications of the previous scheme to the S and higher partial waves. Of particular relevance is to study the triplet S-wave channel, S_1^3, and its mixing with the D_1^3, since here the KSW amplitudes do not converge for $p \gtrsim 100$ MeV although pions are explicitly included. We first discuss the S_0^1 channel and then consider the S_1^3 coupled with the D_1^3 partial wave. After that we turn to discuss the P, D, F_2^3 and G_3^3 waves and compare with other approaches.

$\star$ S_0^1 channel.

We follow the notation of ref.[13]. At NLO we have two counterterms, ξ_1 and ξ_2 together with γ which already appears at LO. The first two counterterms are calculated in terms of γ and the scattering length a_s and effective range r_0 by performing the effective range expansion in eq.(4) with $\mathscr{R}^{NLO}$ given in eq.(16). We then have:

$$\xi_1 = \frac{-g_A^2 M^2 (6\gamma^2 - 8\gamma m_\pi + 3m_\pi^2)(-1+a_s v)^2 + 12 f^2 m_\pi^2 (4 - 8a_s v + a_s^2 M v^2 \pi r_0)}{96 f^2 m_\pi^2 (-1+a_s v)^2 \pi^2},$$

$$(20)$$

$$\xi_2 = \frac{M(g_A^2 \gamma M (\gamma - 2m_\pi) v(-1+a_s v) + 8f^2(-v^2 + \gamma^2(-1+a_s v) + \gamma v(-1+a_s v))\pi)}{32 f^2 m_\pi^2 v(-1+a_s v)\pi^2}.$$

At NNLO there are other counterterms, ξ_3, ξ_4 and ξ_5. The KSW power counting implies that ξ_5 must be small. We either fix it to zero or leave it as a free parameter in the fits, turning out to be negligibly small in the latter case. So that in the following we take $\xi_5 = 0$ in this channel. The counterterms ξ_3 and ξ_4 are then fixed in terms of ξ_1, ξ_2, γ, a_s and r_0 by performing the ERE. At NNLO we perform two kind of fits, either ξ_1 and ξ_2 are free parameters or they are fixed once again by the NLO expression, eq.(20), in the fitting procedure. The expressions of ξ_3 and ξ_4 in terms of ξ_1, ξ_2, γ, a_s and r_0 can be found in ref.[16].

The dashed line in fig.1 represents the NLO fit from eq.(4). At NNLO we indicate by the dash-dotted line the fit with ξ_1, ξ_2 calculated as in NLO, so that only v enters as a free parameter in the fit. The solid line represents the NNLO fit leaving ξ_1 and ξ_2 as free parameters. We also show by the dotted line the NNLO KSW phase shifts from ref.[13]. The values for the parameters that we take are $f = 130.673$ MeV, $g_A = 1.267$, $m_\pi = 138$ MeV, $M = 939$ MeV, $a_s = -23.714$ fm and $r_0 = 2.73$ fm. On the other hand, since a_s is so large, γ can be taken directly zero and we do fix it in that way. We have made a least square minimization process since the errors from the Nijmegen partial wave analysis [23] are in general extremely small. Our NLO results are fitted up to $p < 300$ MeV and the NNLO ones are fitted to the whole energy range, $T_{lab} < 360$ MeV. In this channel the KSW scheme works quite well, as shown in the figure by the dotted line. In our scheme the NLO works worse than both NNLO results which fit the data up to rather high energies. We show in table 2 the fitted and calculated values for the parameters involved in this partial wave. The latter are indicated with a star and their values are the fixed ones taken in the fit.

TABLE 2. Counterterms and subtraction constant in the S_0^1 channel at NLO and NNLO. The stars indicate that the corresponding parameters have been calculated from the ERE in terms of the other ones, a_s and r_0 as explained in the text.

S_0^1	NLO	NNLO	NNLO
v	270 MeV	710 MeV	630 MeV
ξ_1	0.23*	0.24*	0.41
ξ_2	0.03*	0.03*	0.16
ξ_3	...	0.21*	0.20*
ξ_4	...	0.23*	0.07*

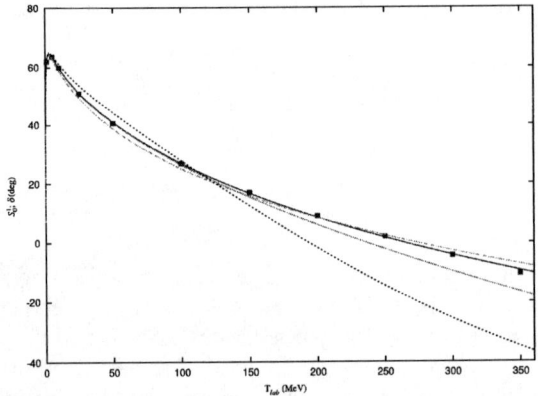

FIGURE 1. Phase shifts for the S_0^1 channel. The dotted line is the NNLO KSW result [13]. The dashed line represents the NLO results from eq.(4). The dash-dotted and solid lines are the NNLO results with ξ_1, ξ_2 fixed as in NLO or taken as free parameters, respectively. The data correspond to the Nijmegen partial wave analysis, ref.[23].

★ $S_1^3 - D_1^3$ coupled channels.

We treat the counterterms in analogous lines as described above for the S_0^1 channel. Nevertheless, we keep now γ as a free parameter for the fit as well as ξ_5 and ξ_6 at NNLO. The ξ_6 counterterm is a new one related to the $S_1^3 - D_1^3$ mixing. One should take into account that although we use the same names for γ and the ξ's counterterms as in the S_0^1 case, they are indeed different [12, 13]. As before ξ_1 and ξ_2 are given at NLO in terms of γ, a_s and r_0 from the ERE. At NNLO we then calculate ξ_3 and ξ_4 from ξ_1, ξ_2, γ, a_s and r_0 with now ξ_1 and ξ_2 either taken as free parameters or fixed as in NLO. The expressions for ξ_1, ξ_2, ξ_3 and ξ_4 determined from the ERE are given in ref.[16].

The results of the fit are presented in figs.2 and 5. In fig.2 we show the phase shifts for the partial wave S_1^3 and the mixing angle ε_1. The latter is defined so that $|S_{11}| = \cos(2\varepsilon_1)$. We take $a_s = 5.425$ fm and $r_0 = 1.749$ fm in the S_1^3 channel. In fig.5 we present the elastic D_1^3 phase shifts together with the phase shifts of the other D-waves. The v_1 and v_2 are

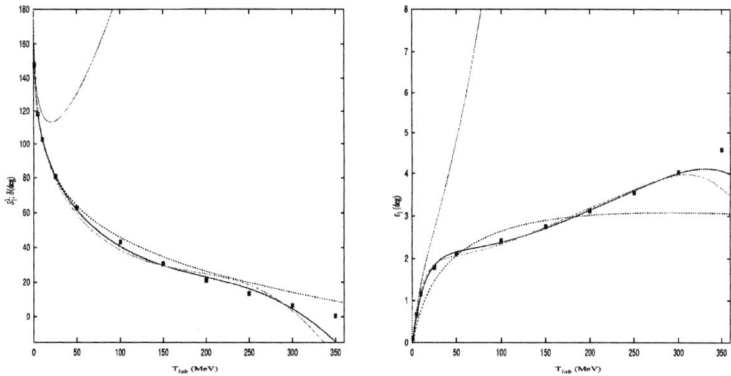

FIGURE 2. Phase shifts for the S_1^3 channel and mixing angle ε_1. The dotted line is the NNLO KSW result [13]. The dashed line represents the NLO results from eq.(4). The dash-dotted and solid lines are the NNLO results with ξ_1, ξ_2 fixed as in NLO or taken as free parameters, respectively. The data correspond to the Nijmegen partial wave analysis, ref.[23].

TABLE 3. Counterterms and subtraction constants in the S_1^3, $S_1^3 - D_1^3$ channels at NLO and NNLO. The stars indicate that the corresponding parameters have been calculated from the ERE in terms of the other ones, a_s and r_0 as explained in the text.

$S_1^3, S_1^3 - D_1^3$	NLO	NNLO	NNLO
v_1	$> \Lambda_{\chi PT}$	170 MeV	190 MeV
v_2	$> \Lambda_{\chi PT}$	590 MeV	560 MeV
γ	0.42 fm^{-1}	0.39 fm^{-1}	0.51 fm^{-1}
ξ_1	0.31*	0.47*	0.28
ξ_2	−0.03*	0.04*	0.06
ξ_3	...	−0.01*	−0.10*
ξ_4	...	−0.08*	−0.06*
ξ_5	...	0.17	0.17
ξ_6	...	0.62	0.47

subject to rather large uncertainties from the fit. At NLO the symbol $> \Lambda_{\chi PT}$ in table 3 indicates that at this order the subtractions constants v_i can take any value arbitrarily large, let us say, above the chiral symmetry breaking scale. At NNLO the value shown for v_1 is indicative, the uncertainty being typically around 100 MeV and one can find other similar fits with v_1 e.g. around 300 MeV. In addition, the minimization program can find other kind of minima with rather different values for the $\xi_1 - \xi_4$ counterterms to those shown in the third column of table 3 and with v_1 as large as M_ρ. Nevertheless, we have chosen the results shown in the third column of table 3 since they come up very similar to those from the more restricted fits of the first and second columns, and furthermore they drive to smaller corrections to R_{11} at NNLO when compared with its value at NLO, see fig.3.

The agreement with data of the new scheme presented here, as shown in figs.2, is remarkably good. At NLO, dashed-lines, the results already follow closely the trend of the data. At NNLO, we distinguish between two types of fits as in the S_0^1 case. They are indicated by the dash-dotted lines, which correspond to the NNLO results when ξ_1 and ξ_2 are given from the ERE at NLO, and by the solid lines when the previous counterterms are taken as free parameters in the fit. Both curves are very similar and agree very well with data. As it is well known, the KSW approach treated in its standard form [13] presents serious problems of convergence, as it is apparent when considering the NNLO KSW results [13] given by the dotted line in fig.2. They only converge for very low energies although the pion fields are explicitly included. Particularly serious is the discrepancy with the mixing angle ε_1. This observable has no counterterms at NLO KSW and only includes one free counterterm, ξ_6, at NNLO. Furthermore, it is free of the large cancellations that affect $\delta(S_1^3)$. It was clarified in ref.[13] that its bad behaviour was due to large corrections from the twice iterated one pion exchange diagrams which are enhanced by large numerical factors. In our power counting the input kernel, $\mathscr{R}$, is infinitely iterated and with it the pion exchange as any other contribution. This ensures that pions are treated non perturbatively which is behind the phenomenological success of figs.2.

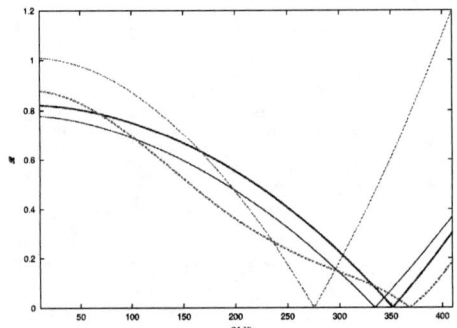

FIGURE 3. $|\mathscr{R}^{NLO}|$ (thin solid line), $|\mathscr{R}^{NNLO}|$ (thick solid line) for the S_0^1 channel and $|R_{11}^{NLO}/4|$ (thin dashed line) and $|R_{11}^{NNLO}/4|$ (thick dashed line) for the S_1^3 channel. In the figure p is the center of mass three-momentum.

We also show in fig.3 the absolute values $|\mathscr{R}^{NLO}|$ and $|\mathscr{R}^{NNLO}|$ for the S_0^1 channel (solid lines) and the same, although divided by four to keep the lines on the same scale, for the matrix element $|R_{11}|$ of the S_1^3 channel. The thick lines refer to the NNLO calculations and the thin ones to NLO. The convergence properties are quite good in a broad range of the center of mass three momentum p, of size about $\Lambda_{NN} \simeq 300$ MeV. This scale is the one expected for the KSW EFT [12], although in the end this EFT does not convergence in the triplet channels for $p \gtrsim 100$ MeV, as already discussed. We see that within our scheme, at the same time that we keep the KSW power counting, we are able to fulfill these expectations. This is the aim of the present investigation.

$\star P, D, F_2^3$ and G_3^3 partial waves.

Up to NNLO in the KSW approach, order p, the physics included in the description of partial waves higher than S waves, except for the mixing between the $S_1^3 - D_3^3$, contains only one pion exchange at order p^0 and the reducible part of the twice iterated one

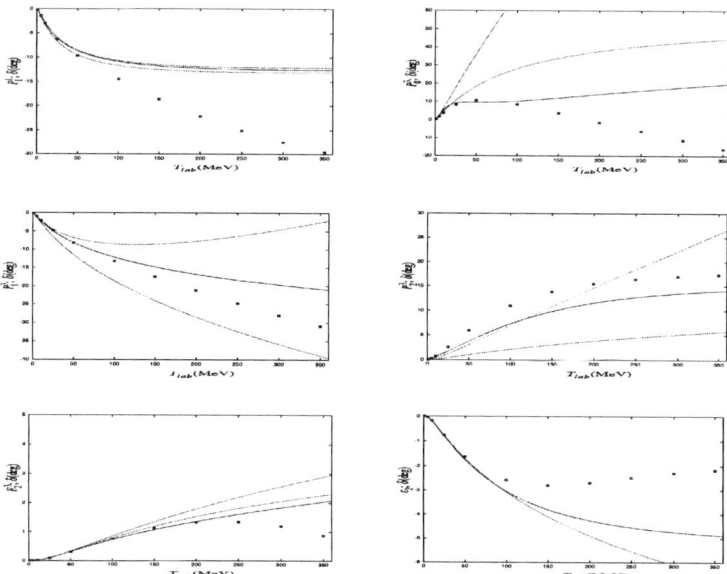

FIGURE 4. Phase shifts for the P_1^1, P_0^3, P_1^3, P_2^3, F_2^3 and ε_2 from left to right and top to bottom, respectively. The dotted line, when present, is the NNLO KSW result [13]. The dashed line represents the NLO results from eq.(4). The solid lines are the NNLO results. The data correspond to the Nijmegen partial wave analysis, ref.[23].

pion exchange, order p. The former contribution constitutes the LO of the Weinberg's scheme [6, 11]. We are then neglecting at this order important contributions, particularly for the P and D waves, from two pion exchange irreducible diagrams, counterterms for the P waves at order p^2, as well as the πN counterterms (which are saturated to a large extend by the Δ isobar), that would appear at order p^3 [11, 24]. Hence a full NNNLO calculation in the KSW scheme should be pursued and then used within our scheme to offer a more complete study of these higher partial waves. For the F and G waves and mixing parameters ε_2 and ε_3, these extra contributions are not so important, see e.g.[24]. This is apparent from our results presented in figs.4 and 5 as well. At the order we are working in these partial waves there are no counterterms and the only free parameters are the subtraction constants v's, one for each partial wave. In most of the channels, they take arbitrarily large or negative values, that is, any value with modulus typically above the expansion scale $\Lambda_{\chi pT} \simeq 0.7 - 1$ GeV gives essentially the same results. Our curves are indeed quite similar to what is obtained in the Weinberg's approach at LO, see [11]. Our results at NNLO, solid lines, improve those of the NLO, dashed lines, except for the D_1^3 where, though the NNLO is better at low T_{lab}, they depart from data more than the NLO ones for higher energies. They also improve the results from the pure KSW amplitudes, dotted lines, although in these cases, as expected, the resummation effects are not so spectacular as in the S-waves.

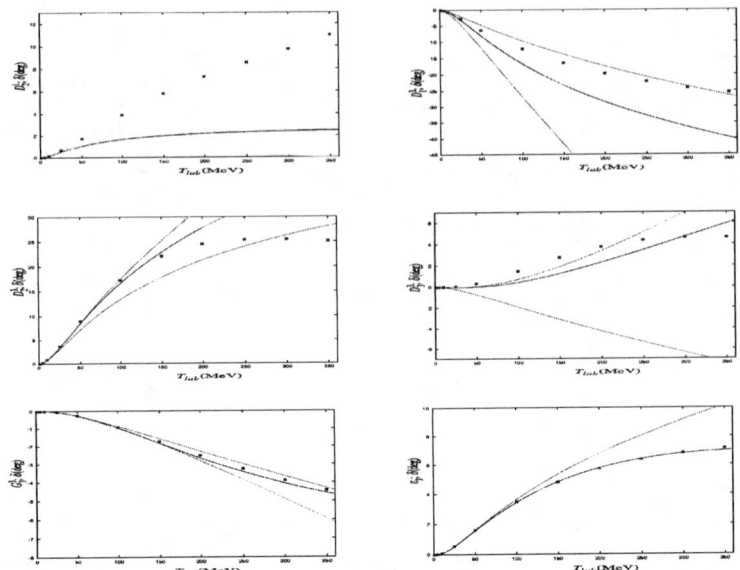

FIGURE 5. Phase shifts for the D_2^1, D_1^3, D_2^3, D_3^3, G_3^3 and ε_3 from left to right and top to bottom, respectively. The dotted line, when present, is the NNLO KSW result [13]. The dashed line represents the NLO results from eq.(4). The solid lines are the NNLO results. The data correspond to the Nijmegen partial wave analysis, ref.[23].

CONCLUSIONS

We have established a new analytic expansion in order to treat systematically nucleon-nucleon interactions. The original KSW amplitudes at NLO [12] and NNLO[13] have been used to fix an interacting on-shell kernel, $\mathscr{R}$, once the unitarity or right hand cut is resumed to all orders. This kernel is fixed by performing a KSW expansion and then matching with the inverses of the pure KSW amplitudes to any arbitrary given order. As a result, a scheme emerges that treats pions non-perturbatively but in harmony with the KSW power counting, and that restores the expected range of convergence of the KSW EFT about Λ_{NN}. It is also worth to remark that for the S-waves a very good agreement with data has been obtained up to the opening of the $NN\pi$ threshold.

Further applications of the present scheme should be pursued and a KSW N^3LO calculation should also be performed and then used in our scheme, particularly for a more complete study of the P and D partial waves.

ACKNOWLEDGMENTS

This work has been partially supported by the spaniard Ministerio de Ciencia y Tecnología under contract FPA2002-03265.

REFERENCES

1. S. Weinberg, Physica **A96**, 327 (1979).
2. J. Gasser and H. Leutwyler, Ann. Phys. (N.Y.) **158**, 142 (1984).
3. U.-G. Meißner, Rep. Prog. Phys. **56**, 903 (1993);
   G. Ecker, Prog. Part. Nucl. Phys. **35**, 1 (1995);
   A. Pich, Rep. Prog. Phys. **58**, 563 (1995);
   H. Leutwyler, *Encyclopedia of Analytic QCD*, edited by M. Shifman, World Scientific; hep-ph/0008124
4. U.-G. Meißner, *Encyclopedia of Analytic QCD*, edited by M. Shifman, World Scientific; hep-ph/0007092.
5. S. Weinberg, Phys. Lett. **B251**, 288 (1990); Nucl. Phys. **B363**, 3 (1991).
6. C. Ordóñez, U. van Kolck, Phys. Lett. **B291**, 459 (1992);
   C. Ordóñez, L. Ray and U. van Kolck, Phys. Rev. Lett. **72**, 1982 (1994); Phys. Rev. **C53**, 2086 (1996);
   U. van Kolck, Phys. Rev. **C49**, 2932 (1994).
7. T.S. Park, D.P. Min and M. Rho, Phys. Rev. Lett. **74**, 4153 (1995); Nucl. Phys. **A596**, 515 (1996).
8. D.B. Kaplan, M.J. Savage and M.B. Wise, Nucl. Phys. **B478**, 629 (1996).
9. D. Eiras and J. Soto, nucl-th/0107009.
10. S.R. Beane, P.F. Bedaque, W.C. Haxton, D.R. Phillips and M.J. Savage, *Encyclopedia of Analytic QCD*, edited by M. Shifman, World Scientific; hep-ph/0008064.
11. E. Epelbaum, W. Glöckle, U.-G. Meißner, Nucl.Phys. **A637**, 107 (1998); Nucl. Phys. **A671**, 295 (2000).
12. D.B. Kaplan, M.J. Savage and M.B. Wise, Phys. Lett. **B424**, 390 (1998); Nucl. Phys. **B534**, 329 (1998).
13. S. Fleming, T. Mehen and I. Stewart, Nucl. Phys. **A677**, 313 (2000); Phys. Rev. **C61**, 044005 (2000).
14. T.D. Cohen and J.M. Hansen, Phys. Rev. **C59**, 13 (1999); Phys. Rev. **C59**, 3047 (1999).
15. M. Lutz, Nucl. Phys. **A677**, 241 (2000).
16. J.A. Oller, nucl-th/0207086.
17. G.F. Chew and S. Mandelstam, Phys. Rev. **119**, 467 (1960).
18. J.A. Oller and E. Oset, Phys. Rev. **D60**, 074023 (1999);
    J.A. Oller, Phys. Lett. **B477**, 187 (2000);
    M. Jamin, J.A. Oller and A. Pich, Nucl. Phys. **B587**, 331 (2000); Nucl. Phys. **B622**, 279 (2002); Eur. Phys. J. **C24**, 237 (2002).
19. J.A. Oller and E. Oset, Nucl. Phys. **A620**, 438 (1997), (E)-ibid **A652**, 407 (1999).
    For a general introduction and discussion see J.A. Oller, Invited talk at YITP Workshop on Possible Existence on the sigma meson and its Implications to Hadron Physics, hep-ph/0007349.
20. U.-G. Meißner and J.A. Oller, Nucl. Phys. **A673**, 311 (2000).
21. J.A. Oller, U.-G. Meißner, Phys. Lett. **B500**, 263 (2001); Phys. Rev. **D64**, 014006 (2001).
22. A. Dobado, M.J. Herrero and T.N. Truong, Phys. Lett. **B235**, 134 (1990).
    A. Dobado and J.R. Peláez, Phys. Rev. **D56**, 3057 (1997).
    T. Hannah, Phys. Rev. **D55**, 5613 (1997).
    J.A. Oller, E. Oset and J.R. Peláez, Phys. Rev. Lett. **80**, 3452 (1998); Phys. Rev. **D59**, 074001 (1999), (E)-ibid **D60**, 099906 (1999).
    F. Guerrero and J.A. Oller, Nucl. Phys. **B537**, 459 (1999).
    A. Gómez Nicola, J. Nieves, J. R. Peláez and E. Ruiz Arriola, Phys. Lett. **B486**, 77 (2000).
    A. Gómez Nicola and J.R. Peláez, Phys. Rev. **D65**, 054009 (2002).
23. V.G.J. Stoks, R.A.M. Klomp, C.P.F. Terheggen and J.J. de Swart, Phys. Rev. **C49**, 2950 (1994).
24. N. Kaiser, R. Brockmann and W. Weise, Nucl. Phys. **A625**, 758 (1997);
    N. Kaiser Phys. Rev. **C65**, 017001 (2002).
25. J.A. Oller, Phys. Rev. **C65**, 025204 (2002).
26. U.-G. Meißner, J.A. Oller and A. Wirzba, Ann. Phys. **297**, 27 (2002).

Criterium for the index theorem on the lattice

Pedro Bicudo

Dep. Física and CFIF, Inst. Sup. Téc., 1049-001 Lisboa, Portugal

Abstract. We study how far the Index Theorem can be extrapolated from the continuum to finite lattices with finite topological charge densities. To examine how the Wilson action approximates the Index theorem, we specialize in the lattice version of the Schwinger model. We propose a new criterion for solutions of the Ginsparg-Wilson Relation constructed with the Wilson action. We conclude that the Neuberger action is the simplest one that maximally complies with the Index Theorem, and that its best parameter in $d = 2$ is $m_0 = 1.1 \pm 0.1$.

1. INTRODUCTION

One of the constraints that lead to the Nielsen-Ninomiya [1] no-go theorem, is the chiral invariance of the Dirac action D for massless fermions on the lattice. Under certain conditions [1], the spectrum of fermions ψ would suffer from doubling, and the axial anomaly would cancel. This has been, from the onset, a recurrent problem of Lattice QCD [2]. In order to recover the axial anomaly, at least at the pertubative level, Ginsparg and Wilson [3] derived a relation, the Ginsparg-Wilson Relation (GWR) which explicitly breaks the standard chiral symmetry,

$$D\gamma_5 + \gamma_5 D = D\gamma_5 R D \, , \tag{1}$$

where R is a matrix proportional to the lattice spacing a (in most of this talk we will consider the case where R is a simple number). Recently Lüsher [4] proved that chiral invariance of an action D which complies with the GWR can be recovered in an extended form,

$$\delta\psi = \gamma_5(I - \frac{1}{2}RD)\psi \, , \quad \delta\bar{\psi} = \bar{\psi}(I - \frac{1}{2}DR)\gamma_5 \, . \tag{2}$$

Thus a conjecture appeared in the literature suggesting that it may be possible to overcome the Nielsen-Ninomiya no-go theorem on the lattice and to fully simulate, without doubling, the chiral symmetry of QCD (see [5] for a recent review). This includes PCAC [6] and the axial anomaly [7]. In particular the Atiyah-Singer [8] Index Theorem has a Hasenfratz-Laliena-Niedermayer [9] version on the lattice,

$$n_- - n_+ = q \, , \tag{3}$$

where $n_{-(+)}$ is the number of zero modes of the Dirac action D with negative (positive) chirality. q is an *integer* topological charge of the gauge field configuration, defined with

CP660, *Hadron Physics: Effective Theories of Low Energy QCD*, edited by A. H. Blin et al.

the charge density, a function of the lattice site $\vec{n}$,

$$\rho(\vec{n}) = \frac{1}{2}tr\Big|_{colour,Dirac} \{\gamma_5 RD(\vec{n},\vec{n})\} \ . \tag{4}$$

Therefore each GWR Dirac action *defines* the density (4) and the index (3). Moreover examples of lattice actions have been found where this index is nontrivial and coincides with a charge independently defined from a topological gauge density. For instance the fixed point topological density [10] coincides, on all gauge configurations, with the density (4) defined with the fixed point Dirac operator. The recent overlap GWR solution of Neuberger [11] also solves the GWR and complies with the Index Theorem. The Neuberger solution was checked in dimension $d = 2$ for smooth gauge configurations [12, 13] , and in dimension $d = 4$ in the continuum limit [14, 15]. Thus the GWR, together with these recent results, essentially solve the problem of the topological Index Theorem on the lattice.

Nevertheless we remark that the definition of the Dirac operator and the definition of the topological charge on the lattice are not unique. Moreover the topological charge on the lattice is not conserved [16]. This suggests that different GWR Dirac operators and topological densities produce different indexes when we depart from the continuum limit, to a finite lattice with finite charge density. In this talk we address the *extremum* problem of finding the lattice Dirac operator which provides the best index (3). Because this is a difficult problem, we specialize in lattice actions which are constructed with the Wilson action [2, 17]. In order to compare the index of different Dirac operators, we first choose a single topological charge density. Although this is an arbitrary definition, let us choose the simplest near neighbor lattice topological charge density which discretizes the continuum density $\varepsilon_{\mu\nu...}F^{\mu\nu}\cdots$ and which sums to an integer total topological charge. Then we search for the GWR Dirac operator which index (3) is closer to this simplest topological charge.

The arbitrariness in the choice of our lattice operators is partly constrained by locality. Locality is a crucial issue to connect the lattice to the Quantum Field Theories in the continuum. Only local actions are universal. Moreover it has been conjectured [5, 18] that local GWR solutions have better topological properties than non-local ones. Only topological configurations which are larger than the range of the couplings in D would produce a number of zero modes n_- different from n_+ in agreement with eq. (3). This conjecture is certainly correct in the case of the Zenkin [19, 20] action. The Zenkin action is non-local [21], and although it complies with eq. (1), it does not comply with eq. (3). Thus we will restrain from using non-local definitions for the lattice action and for the lattice charge density.

In this talk we investigate in detail the eigenvalues of some lattice actions, with the aim to clarify how the Index Theorem is extrapolated from the continuum to finite lattices with finite topological charge densities. We also propose a new criterion for GWR solutions constructed with the Wilson action. In Section 5 we study in detail the effects of topological gauge configurations on the eigenvalues of the Wilson action for the discrete version of the Schwinger model. In Section 2 we review properties of lattice actions. In Section 3 we address the topological Abelian charge in $d = 2$. In Section 4 we review the Wilson action. In Section 6 we present our criterion. Finally in Section 7 we conclude.

2. SOME PROPERTIES OF LATTICE ACTIONS

A plausible property of lattice actions is γ_5 hermiticity. For instance the Wilson action is γ_5 Hermitean. When an action verifies $D^\dagger = \gamma_5 D \gamma_5$, we can show that D and $D^\dagger$ have the same eigenvalues λ,

$$
\begin{aligned}
D v &= \lambda v \\
\Leftrightarrow D^\dagger \gamma_5 v &= \lambda \gamma_5 v,
\end{aligned}
\tag{5}
$$

and that the eigenvalues of D and $D^\dagger$ are mapped by γ_5. When we conjugate eq (5), we find that for every right eigenvalue there is a left eigenvalue λ^* with eigenvector $v^\dagger \gamma_5$, and so the conjugate transformation $\lambda \to \lambda^*$ leaves the spectrum invariant. The spectrum of eigenvalues λ is symmetric with respect to the real axis of the Argand plot.

We proceed to define chirality in the lattice. Eq. (5) can be used to show that the γ_5 is involved in a extended orthogonality condition,

$$
\left(\lambda_2^* - \lambda_1\right) v_2^\dagger \gamma_5 v_1 = 0 .
\tag{6}
$$

A possible definition for the chirality of an eigenvector v is,

$$
\chi = \frac{v^\dagger \gamma_5 v}{v^\dagger v},
\tag{7}
$$

and eq. (6) shows that the complex eigenvalues have a vanishing chirality, $\chi = 0$. Only the real eigenvalues may have a non vanishing chirality.

We now study properties of GWR solutions. The GWR implies that if v is a zero mode of the GWR action $Dv = 0$, then $D\gamma_5 v = 0$. Thus we can use $1 - \gamma_5$ and $1 + \gamma_5$ to decompose the Kernel in a set of left vectors and a set of right vectors. This shows that the zero modes have a chirality $\chi = \pm 1$. Moreover we can verify that if D is a GWR action, then $-D + 2/R$ also complies with the GWR. Thus the eigenvalues $2/R$ of D also have a chirality $\chi = \pm 1$.

It is also convenient to define an intermediate matrix,

$$
V = RD - I \Leftrightarrow D = \frac{1}{R}(I + V)
\tag{8}
$$

because the GWR is equivalent to $V^{-1} = \gamma_5 V \gamma_5$. In the case that V is γ_5 Hermitean, we also obtain that $V^{-1} = V^\dagger$, and this shows that the eigenvectors of V belong to the unitary circle of the complex Argand plane. The eigenvalues of D are in a circle with center $1/R + 0i$ and radius $1/R$. We will be interested in the -1 eigenvalues of V which correspond to zero modes of D.

Moreover the crucial property for the Index Theorem has been show,

$$
\frac{R}{2} tr\{\gamma_5 D\} = n_- - n_+ ,
\tag{9}
$$

both with analytical [4] and algebraic [14] methods.

3. TOPOLOGICAL CONFIGURATIONS IN D=2

The $d = 2$ case is very convenient for the simple study of discrete topology. The Euclidean Dirac matrices $\gamma_{1 \to 5}$ of $d = 4$ are now replaced by the Pauli matrices $\sigma_{1 \to 3}$. The $d = 2$ charge density is similar to the rotational of the vector potential,

$$\rho(x) = F_{12}(x) , \tag{10}$$

it is similar to a magnetic field. On the lattice this magnetic field density is extracted from the plaquette,

$$P(\vec{n}) = U(\vec{n})_1 U(\vec{n} + \hat{1})_2 U(\vec{n} + \hat{2})_1^* U(\vec{n})_2^* \tag{11}$$

and a possible definition of topological charge density is,

$$\rho_1(\vec{n}) = \frac{1}{2\pi} \arg P(\vec{n}) , \tag{12}$$

where $-1/2 < \rho_1(\vec{n}) < 1/2$. This definition of the topological density is not unique, the necessary condition is that it must reproduce the correct continuum limit. The definition (12) of the topological charge is particularly interesting because it is quantized. It is clear that the total topological charge of the lattice, which is defined with,

$$q = \sum_{\vec{n}} \rho(\vec{n}) , \tag{13}$$

is an integer. The charge q_1 is the sum of the decimal part of a set of numbers, and these numbers have a vanishing sum, therefore q_1 is an integer. This is similar to $d = 2$ compact QED, where the magnetic monopoles arise naturally on the lattice. Here the topological charge is equivalent to a magnetic flux through the toroidal $d = 2$ lattice.

We now construct non-trivial gauge configurations. In this work we are particularly interested in extrapolating continuum properties to finite lattices. We specialize to gauge configurations with uniform topological charge density. These configurations are unique, except for gauge transformations. In the $N \times N$ lattice a simple configuration is built with the prescription [12] of Chiu,

$$
\begin{aligned}
U_j(\vec{n}) &= \exp\left[iA_j(\vec{n})\right] , \\
A_1(n_1, n_2) &= -2\pi Q \frac{n_2 - 1}{N^2} , \\
A_2(n_1, N) &= 2\pi Q \frac{n_1 - 1}{N} ,
\end{aligned}
\tag{14}
$$

where the undefined $A_j(n)$ are zero. With this definition (14) and in the particular case of integer Q, the topological density is constant. If $-N^2/2 < Q < N^2/2$ we find that $\rho_1 = Q/N^2$ and that the topological charge is $q_1 = Q$.

The case where Q is an arbitrary continuous parameter can be used when we are interested in an interpolation between the cases of uniform density. Then the density is uniform, except at the single point (N,N). The gauge configuration (14) is periodic in Q, with a period of N^2. In the relevant range $-N^2/2 < Q < N^2/2$ we find that the topological charge q_1 of eq. (12) is a step-like function of Q,

$$q_1 = \text{int}\left(Q\frac{N^2-1}{N^2} + \frac{1}{2}\right).$$ (15)

4. THE WILSON ACTION

The Wilson [2] action is the simplest and most widely used action in lattice field theory. It also provides a good example to observe roots and doubles,

$$
^wD = \frac{1}{a}\sum_j \gamma_j C_j + \frac{2r}{a}\sum_j B_j ,
$$

$$
C_{j\vec{n},\vec{n}'} = \frac{U(\vec{n})_j \delta_{n+\hat{j},n'} - U(\vec{n}')_j^\dagger \delta_{\vec{n},\vec{n}'+\hat{j}}}{2} ,
$$

$$
B_{j\vec{n},\vec{n}'} = \frac{2\delta_{\vec{n},\vec{n}'} - U(\vec{n})_j \delta_{n+\hat{j},\vec{n}'} - U(\vec{n}')_j^\dagger \delta_{\vec{n},\vec{n}'+\hat{j}}}{4} ,
$$ (16)

where r is a parameter of the order of 1. The vector γ_j term of eq. (16) is the naive massless Dirac action on the lattice, where the derivative C_j is computed by means of finite differences. In the limit of free fermions, which corresponds to $U_j = 1$, all these operators commute. In this case the Fourier transform of the above matrices depend on a single momentum,

$$
\begin{aligned}
I(k) &= 1 , \\
C_j(k) &= i\sin k_j , \\
B_j(k) &= \sin^2\frac{k_j}{2} .
\end{aligned}
$$ (17)

In the free case, and in d dimensions, $C_j(k)$ has 2^d roots at all possible combinations of $k_j = 0$ and $k_j = \pi$, see eq. (17). This is the source of the doubler problem because only the root at vanishing k is physical. Wilson [2] included the scalar B_j term in eq. (16) to remove the unwanted doubles.

However this scalar term B_j is not chiral invariant. Chiral invariance can be implemented with the Lüscher transformation. However Horvath showed that the RD operator can at most [22] be exponentially local, therefore the ^wR of the GWR is not a simple number in this case. From the GWR relation, see eq. (1) we get,

$$^wD{}^wR = I + {}^wD\frac{1}{^wD^\dagger},$$ (18)

where the Wilson action $^wD^\dagger$ is in general invertible, see Section 5. We can remark that the operator in eq. (18) is a solution of the GWR, however it is not exponentially local. In this case the nontrivial operator of the Lüscher transformation,

$$\gamma_5 \left(I - \frac{1}{2}\,^wR^wD \right) = \frac{1}{2}\left(\gamma_5 +\,^wD^{-1}\gamma_5^wD \right), \qquad (19)$$

is also non-local. Moreover $^wD^wR$ is not γ_5 Hermitean. This excludes both the Wilson Action and the $^wD^wR$ as the best candidate to study exact chiral symmetry on the lattice.

5. EIGENVALUES OF THE WILSON ACTION

Here we depart from the free case and study the eigenvalues of the Wilson action for topological gauge configurations . This continues reference [23]. The Wilson action ^wD is the simplest and most used lattice action, see Section 4. Our framework is the massless Wilson action with parameters $a = r = 1$, see eqs (16), in a simple 2-dimensional U(1) gauge theory which is a discrete version of the Schwinger model, see Section 3. In the continuum the Wilson action [24] exactly complies with the Index Theorem (3), but for finite lattices it does not. Our aim in this section is to extrapolate the topological charge from the continuum to the lattices of different size $N \times N$, and to study how the Wilson Action deviates from the Index theorem.

We encounter the problem that the topological charge and the topological density on the lattice are not uniquely defined. This problem is not present in the continuum, where the topological density and its derivatives are well defined. This implies that a lattice action close to the continuum limit is only allowed to have a topological charge density both small and nearly constant. However we are interested, both physically and mathematically, in extrapolating the Index Theorem to the case of finite lattices, with finite topological charge. And then we want to investigate how far the topological density can be increased without spoiling the index theorem. To remain as close as possible to the continuum definitions, we momentarily specialize to very smooth gauge configurations. In particular we study the Wilson action in configurations with constant plaquette density, see Section 3, which is equivalent to a constant topological charge density.

First we study some general properties of the Wilson action. Because it is γ_5 Hermitean, the spectrum of its eigenvalues, here denoted λ, is symmetric with respect to the real axis of the Argand plot, see Section 2. Moreover in $d = 2$ it turns out that in the case of even lattices, the spectrum of $^wD - 2I$ in the Argand plot is also symmetric with respect to the imaginary axis. The transformation $(\lambda - 2) \to -(\lambda - 2)^*$ leaves the spectrum invariant (see for instance Figs. 1,2,3). So we will use even $N \times N$ lattices with $N \geq 4$ from now on, for simplicity. We also remark that $0 < \text{real}(\lambda) < 4$ for any non vanishing gauge configuration, thus ^wD is invertible (except in the free quark case). In what concerns $\gamma_5(^wD - 2I)$ we observe that its eigenvalues are exactly symmetric with respect to the origin. This implies for instance that the traces $\text{tr}\left[\gamma_5\,(^wD - 2I)^{-1}\right]$ and $\text{tr}\left[\gamma_5\,^wD\right]$ both vanish for any finite N. This is not the case of the trace $\text{tr}\left[\gamma_5\,^wD^{-1}\right]/\mathscr{Z}$ which is non vanishing and has been proposed as a topological index [25], where $\mathscr{Z}$ is a constant

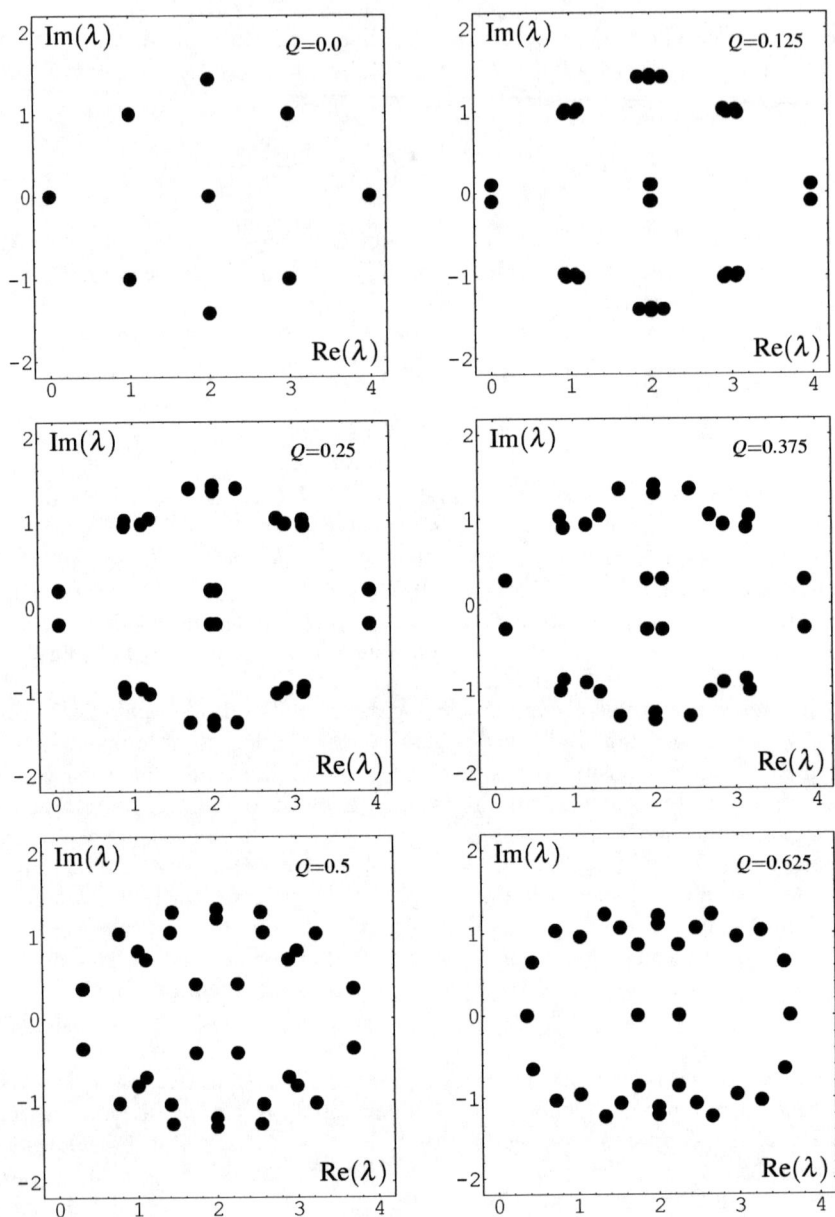

FIGURE 1. Sequence of the eigenvalues of the Wilson action as a function of the topological parameter Q. We show the Argand plot of the eigenvalues λ of the Wilson ^{w}D, on a 4×4 lattice and with parameters $a = r = 1$.

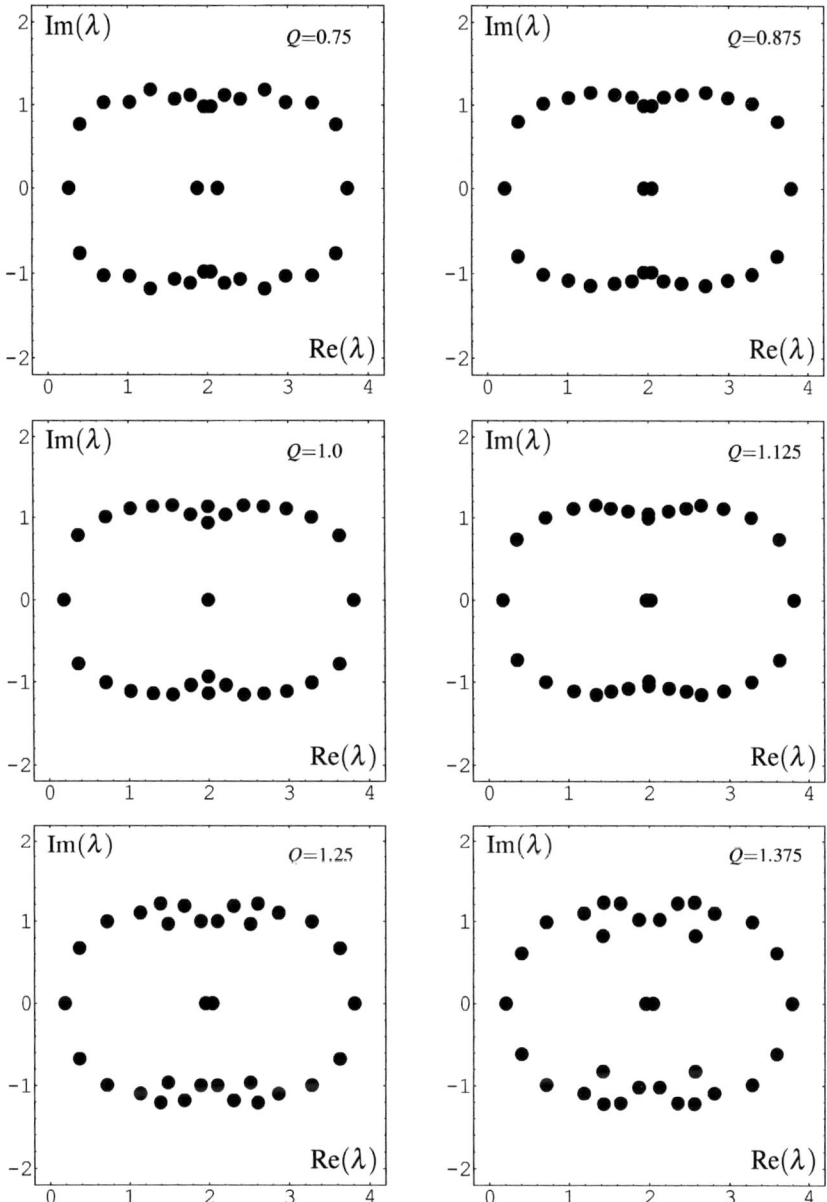

FIGURE 2. Sequence of eigenvalues of the Wilson action, continuing Fig. 2. The Wilson action successfully separates the doubling of the spectrum, not only in the free case of $Q = 0.0$, but also in topologically nontrivial cases.

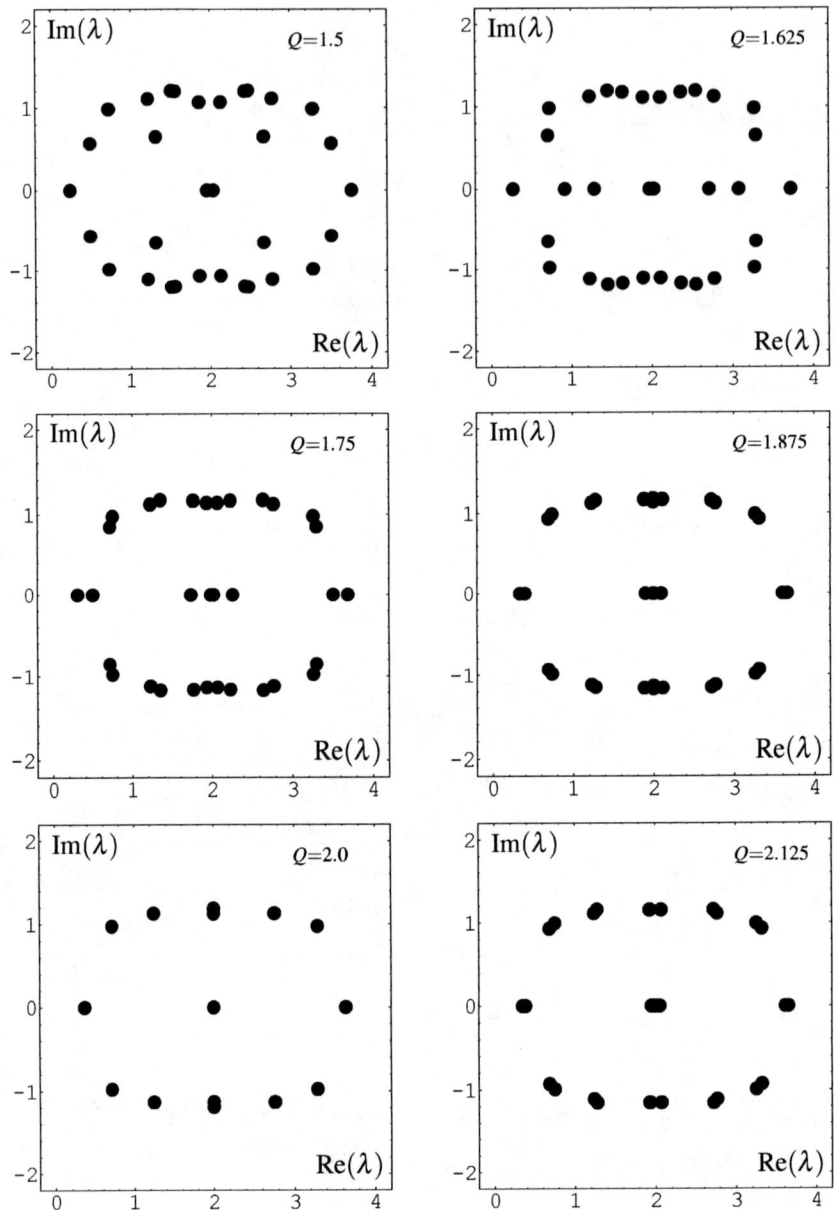

FIGURE 3. Sequence of eigenvalues of the Wilson action, continuing Figs. 1 and 2. For some particular values of Q, two opposite pairs of complex eigenvalues transform into two opposite pairs of real eigenvalues.

normalizing factor. However we will not further study this proposed topological charge because it clearly is neither local nor integer.

We now study quantitatively the real eigenvalues and the topological properties of the Wilson action ^WD. We are interested in eigenvalues $\lambda \simeq 0$ with chirality $\chi \simeq \pm 1$, see Section 2. The non vanishing chirality implies that the relevant eigenvalues are the real eigenvalues. The B_j term was introduced in the free action by Wilson to separate the four roots of the C_j, see Section 4. In the free limit this results in four real eigenvalues: one root at 0, two degenerate eigenvalues equal to 2 and one eigenvalue equal to 4. Here we find that quite a similar separation exists in the quadruplets of real eigenvalues that occur in gauge configurations with a finite topological charge (see Figs. 1,2,3). Moreover the chirality of these real eigenvalues approximately agree with the Index Theorem. In particular we observe that,
- The number of real eigenvalues is always a multiple of four, $4|q_2|$. The integer $q_2 = -\text{sign}(\chi)|q_2|$ is a possible definition of the topological charge of the gauge configuration, where χ is the chirality of the smallest real eigenvalue.
- In the present case of a constant topological density, $q_2 = q_1$ if $|\rho_1|$ is smaller than half of the maximum possible charge density. The integer topological charge q_1 and its topological density ρ_1 are defined in Section 3.
- The real eigenvalues and the corresponding chiralities are close to integers. The difference is proportional to a small number, $\varepsilon = |q_2|/N^2$. The $4|q_2|$ eigenvalues can be divided in the following 3 sets of eigenvectors, with eigenvalues,

$$\lambda = 0 + 3.0\varepsilon + o(\varepsilon^2) \quad , \quad \chi = [-1 + o(\varepsilon^2)]sign(q_2),$$
$$\lambda = 2 + o(\varepsilon^2) \quad , \quad \chi = [1 - 2.0\varepsilon + o(\varepsilon^2)]sign(q_2),$$
$$\lambda = 4 - 3.0\varepsilon + o(\varepsilon^2) \quad , \quad \chi = [-1 + o(\varepsilon^2)]sign(q_2), \tag{20}$$

which contain respectively $|q_2|, 2|q_2|$ and $|q_2|$ eigenvalues.
- When $|q_2|$ approaches the maximum value of the order of $n^2/4$, these 3 sets of eigenvalues spread out to 3 finite intervals, and eventually they overlap. It turns out that right before these intervals overlap, they just leave a gap at $[1.0, 1.2]$.
- In the continuum limit of $N \to \infty$, and for very smooth topological objects, all the different $\varepsilon \to 0$. So we find that in this limit of large N, and from the view point of zero modes, the Index Theorem is verified,

$$n_- - n_+ = q_2 . \tag{21}$$

This result agrees with the proofs of Fujikawa [14] and Adams [15].

If we now leave the case of constant topological density, it occurs that $|\rho_1| < 0.22$ (we mean that the absolute value of the topological density ρ_1 is smaller than 0.22 in any plaquette of the lattice) is a sufficient condition to enforce that $q_2 = q_1$. We verified this crucial result with a large number of randomly generated gauge configurations, for all possible q_1. All the remaining properties, that we just detailed, of the eigenvalues of the Wilson action are also maintained when $|\rho_1| < 0.22$. This suggests that it is possible to extrapolate the continuum limit to a significant and finite subinterval of the range $-1/2 < \rho_1 < 1/2$.

To illustrate how the complex eigenvalues transform in real ones, we arbitrarily interpolate between different constant charge densities with a continuous variation of the topological parameter Q (see Section 3). It turns out that for some particular values of Q, which produce a large $|\rho_1|$ at the point (N,N), two opposite pairs of complex eigenvalues (with 0 chirality) quite suddenly transform into two opposite pairs of real eigenvalues, and then these real eigenvalues continuously increase their chirality $|\chi|$. This transition point is continuous, but at this precise value of Q the velocity of the eigenvalues in the Argand plot is infinite. At this point the topological number q_2 steps up (down). The opposite also happens, at some other particular values of Q, where 4 real eigenvalues suddenly transform into complex eigenvalues. The different topological charges can also be computed (see Section 3). and the observed pattern is quite general. The charge q_2, defined with the real eigenvalues of the Wilson action, and the charge q_1, defined with the plaquette density P in Section 3, are quite close only up to half of the maximum possible topological q_1. This verifies our empirical rule of $|\rho_1| < 0.22$.

In Figs. 1,2,3 we show a sequence of the eigenvalues of the Wilson action for consecutive values of the interpolating topological parameter Q. A simple 4×4 lattice is used and the eigenvalues are displayed in the Argand plan.

6. CRITERION

We aim to understand how the Index Theorem extrapolates from the continuum limit to finite lattices with finite topological charge densities. In d=2, and providing that the topological density $|\rho_1| < 0.22$, we show that the Wilson action ^wD has the correct number of small real eigenvalues to comply with the index theorem. The problem is that these eigenvalues are not exactly vanishing and that their chirality is not exactly ± 1. The deviations to the index theorem are of order $\varepsilon = |q_1|/N^2$, where q_1 is the integer topological charge. We also prove that the GWR Neuberger solutions either project these approximate zero modes of ^wD on the origin, with correct chirality, or send them close to the remote other end $2/R$ of the unitary circle.

This motivates a new criterion for GWR solutions ^{gwr}D with index identical to q_1 (and γ_5 Hermitean, constructed with the Wilson action), which relies in the function $f(\lambda)$. This function $f(\lambda)$ is straightforwardly defined by replacing ^wD and $^wD^\dagger$ by the same real number λ in the expression for ^{gwr}D. We map $^{gwr}D(^wD,^wD^\dagger) \to f(\lambda) = {}^{gwr}D(\lambda,\lambda)$.

The $f(\lambda)$ is a trivial case of a GWR solution. Moreover the γ_5 hermiticity implies that $f(\lambda)$ is real. We thus conclude that $f(\lambda) = 0$ or $f(\lambda) = 2/R$.

In the free case ^wD and $^wD^\dagger$ commute, and have the correct real eigenvalues at 0, while the doubles correspond to the eigenvalues 2 or 4. We remark that the doubles correspond to pathological eigenvectors, with alternating signs in neighboring lattice sites. We conclude that, unless the fermion fields are redefined, a correct GWR solution must produce $f(0) = 0$, $f(2) = 2/R$ and $f(4) = 2/R$.

Studying small topological densities ρ_1 with finite q_1, we can further show that $f([0,3.0\varepsilon]) = 0$, $f(2) = 2/R$, $f([4-3.0\varepsilon,4]) = 2/R$. In the case of small topological densities, ^wD and $^wD^\dagger$ have the same eigenvalues, but they do not exactly have the same eigenvectors. Nevertheless the relevant small real eigenvalues λ of the Wilson action, have a chirality $\chi = [-1 + o(\varepsilon^2)]\mathrm{sign}(q_2)$ and this implies that the difference between the ^wD eigenvector v and the $^wD^\dagger$ eigenvector $\gamma_5 v$ is very small, of order ε^2. This shows that $^{gwr}Dv = f(\lambda)v + o(\varepsilon^2)$, the correct eigenvector v' of ^{gwr}D is quite close to v (and to $\gamma_5 v$). If ^{gwr}D complies with the index theorem, we can assume that this eigenvector v' is a zero mode, because v is already quite close to the correct fermionic zero mode. Therefore $f(\lambda) \simeq 0$. Moreover we can only have $f(\lambda) = 0$ or $2/R$. This implies that $f(\lambda) = 0$. Because $\lambda = 0 + 3.0\varepsilon + o(\varepsilon^2)$ we find that f projects the interval $[0, 3.0\varepsilon]$ on the origin 0. Inversely, if $f(\lambda) = 0$, ^{gwr}D has $|q_1|$ eigenvectors close to $f(\lambda) = 0$. Moreover they all have similar chirality $\chi = [-1 + o(\varepsilon^2)]\mathrm{sign}(q_2)$. Even if these eigenvectors mix, which is natural because they are very close, their chirality will remain close to $\mathrm{sign}(q_2)$, and finite. *Therefore these $|q_1|$ eigenvalues are real, vanishing, with the correct chirality, and in the right number to comply with the index theorem*. Similarly we can show that f projects on $2/R$ both a small neighborhood of the point 2 and the interval $[4 - 3.0\varepsilon, 4]$. If we assume that the complex eigenvalues of the Wilson action are irrelevant to the index because they have vanishing chirality, we conclude that the condition $f([0, 3.0\varepsilon]) = 0$, $f(2) = 2/R$, $f([4 - 3.0\varepsilon, 4]) = 2/R$ is both a necessary and a sufficient condition for any γ_5 Hermitean GWR solution, constructed with the Wilson action, and free from doubling, to comply with the index theorem in the case of small topological density.

In the case of a finite topological charge density the three sectors of real eigenvectors of the Wilson that include 0, 2 and 4 spread out. Analytically it is hard to find how far the exact criterion that we just derived for small ε can be extended to finite ε. Nevertheless our numerical study of the Wilson action shows that up to moderate densities $|\rho_1| < 0.22$, the sectors including 0, 2 and 4 remain separated. For instance the sector of 0, with chiralities $\chi \simeq - \mathrm{sign}(q_1)$, and the sector of 2, with chiralities $\chi \simeq + \mathrm{sign}(q_1)$, are separated by a gap located at $\lambda_0 = 1.1 \pm 0.1$. It is thus quite plausible that a correct GWR solution should project the smaller $|q_2|$ real eigenvalues λ of ^wD on 0, and that the remaining $3|q_2|$ real eigenvalues should be projected onto the other end of the circle, close to $2/R$. We may now determine which is the function $f(\lambda)$ that maximally complies with the index theorem, up to the highest possible topological density 0.22 . This $f(\lambda)$ is a step function that jumps precisely at the gap $\lambda_0 = 1.1 \pm 0.1$.

Finally when the density is large, $|\rho_1| > 0.22$ or more at one or many plaquettes of the lattice, the index theorem is spoiled. This is natural on the lattice, because for large densities the different definitions of topological charge density diverge and the topological charge is not conserved. In what concerns the three sectors of real eigenvectors of the Wilson action, that include respectively 0, 2 and 4, they swell so much that they start to overlap. The real eigenvalues either mix and transform into complex eigenvalues (with vanishing chirality), or they carry their chirality to the wrong sector. We remark that this is not a physical problem, providing the gauge configurations with large plaquette densities are suppressed by the gauge part of the action.

We now present our criterion, which is very simple to use. " **A γ_5 Hermitean GWR solution ^{gwr}D with constant R and constructed with the Wilson ^wD action complies maximally with the index theorem if and only if $^{gwr}D \to \frac{2}{R}\theta(\lambda - \lambda_0)$ when we replace $^wD, ^wD^\dagger \to \lambda$, where λ is a real number, $0 \leq \lambda \leq 4$ and λ_0 belongs to a well determined and narrow subinterval of $]0, 2[$ ".** Our criterion also applies to GWR solutions constructed with any other action ^oD (other than the Wilson action), that

141

approximately complies with the index theorem and that is γ_5 hermitian, ${}^0D^\dagger = \gamma_5 {}^0D\gamma_5$.

7. CONCLUSION

This talk is devoted to clarify how far the Index Theorem can be extrapolated from the continuum to finite lattices with finite topological charge densities.

The $d = 2$ Wilson action ${}^wD$ is examined, and we find that it approximately complies with the Index Theorem for finite topological charge densities lower $|\rho_1| = 0.22$. We also study the GWR solution of Neuberger which exactly complies with the Index Theorem. Finally we produce a criterion for GWR solutions, constructed with the Wilson action, that maximally comply with the Index Theorem.

With our criterion, it is simple to show that locality constitutes a sufficient condition for a Dirac action to comply with the index theorem together with the simplest topological charge q_1. It is clear that any local action is differentiable in the free limit, and therefore the condition $f(0) = 0$, $f(2) = 2/R$, $f(4) = 2/R$ can be extended to a neighborhood of these points. Using our criterion, this proves that any local action (GWR solution and with the correct free limit) complies with the index theorem at least in the case of small topological charge densities. The locality conjecture is correct in the case of actions built from the Wilson action, and it is probably correct in the general case. Moreover the locality condition can be improved. We aim at finding the action with the highest convergence radius in the neighborhood of the points $0, 2, 4$, and in a sense this is comparable with finding the action with the highest locality.

Our criterion complies with the Chiu and Zenkin criterion [26] which states that the correct GWR solutions must have at least one eigenvalue at $2/R$. However we observe that it is possible to find a non-local action with eigenvalues at $2/R$, which does not have zero modes outside the free limit. Therefore, unlike the locality condition, the Chiu-Zenkin criterion is not a sufficient condition to force an action to comply with the Index Theorem, although it certainly constitutes a necessary condition.

Our criterion can be applied to the Neuberger GWR solution defined with $R\,{}^nD = I + ({}^wD - m_0)/\sqrt{({}^wD - m_0)^\dagger\,({}^wD - m_0)}$. With the λ substitution we find that ${}^nD \to \frac{2}{R}\theta(\lambda - m_0)$. This complies with our criterion when $m_0 = \lambda_0$, and it is quite evident that this is the simplest action which complies with it. The criterion shows that the Neuberger overlap action is the simplest GWR solution constructed with the Wilson action that maximally complies with the Index Theorem.

Moreover, in the Schwinger model and using the dimensionless units of $a = r = 1$, we find that $1.0 < \lambda_0 < 1.2$. This determines $m_0 = \lambda_0$ with a value which is not very far from the $m_0 \simeq 0.8$ that provides the highest locality in the free limit. This choice of m_0 increases the precision of the observation of Chandrasekharan [13], stating that the Neuberger action only complies with the Index Theorem if $0 < m_0 < 2$, $(0 < m_0 < 2$ can also be derived from the root and pole structure of the free Neuberger action). The choice of m_0 completely fixes the GWR parameter R because the free continuum limit of the lattice implies that $m_0 = a/R$. So the best choice seems to be $R \simeq 0.9a$.

Finally we outline possible continuations of this study. We are researching the analytical extension of the proof of our criterion to finite topological densities $0 << |\rho_1| < 0.22$.

The repetition of the present study in four dimensions would also be physically relevant. I acknowledge Misha Polikarpov for explaining computing techniques and topology on the lattice. I am also grateful to Herbert Neuberger for explaining the locality conjecture and for pointing to a numerical error. I thank Dimitri Diakonov, Isabel Salavessa and Emilio Ribeiro for discussions on the topological index.

REFERENCES

1. Nielsen, H., and Ninomiya, M., *Nuc. Phys.* **B185**, 20 (1981);
   Nuc. Phys. **B193**, 173 (1981).
2. Wilson, K. G., *Phys. Rev.* **D10** 2445 (1974);
   "Quarks and Strings on a Lattice", in *New Phenomena in Subnuclear Physics, Erice lectures-1975*, edited by Zichichi, A., plenum, New York, 1977.
3. Ginsparg, P.H., and Wilson,K.G. *Phys. Rev.* **D25**, 2649 (1982).
4. Lüscher, M., *Nucl. Phys.* **B538**, 515 (1999) [arXiv:hep-lat/9808021];
   Phys. Lett. **B428**, 342 (1998) [arXiv:hep-lat/9802011].
5. Niedermayer, F., *Nucl. Phys. Proc. Suppl.* **73**, 105 (1999) [arXiv:hep-lat/9810026].
6. Chandrasekharan, S., *Phys. Rev.* **D60**, 074503 (1999) [arXiv:hep-lat/9805015].
7. Zinn-Justin, J. "The Regularization Problem and Anomalies in Quantum Field Theory" , in *Topology of Strongly Correlated Systems, XVIII Lisbon Autumn School- 2000*, edited by Bicudo, P., Ribeiro, J., Sacramento, P., Seixas, J., Vieira, V., World Scientific, Singapore, 2001 ;
   H. Neuberger, " Regulated Chiral Gauge Theory " , ibid.
8. Atiyah, M. F., Singer, I. M., *Annals Math.* **87**, 485 (1968);
   Annals Math. **87**, 546 (1968);
   Atiyah, M.F., and Segal, G. B., *Annals Math.* **87**, 531 (1968);
   Nakahara, M., *Geometry, Topology and Physics*, Graduate Student Series in Physics, Institute of Physics Publishing, Bristol and Philadelphia, 1990.
9. Hasenfratz, P., Laliena, V., and Niedermayer, F., *Phys. Lett.* **B427**, 125 (1998) [arXiv:hep-lat/9801021].
10. Hasenfratz, P., and Niedermayer, F., *Nucl. Phys.* **B414**, 785 (1994) [arXiv:hep-lat/9308004];
    Bietenholz, W., and Wiese, U. J., *Nucl. Phys.* **B464**, 319 (1996) [arXiv:hep-lat/9510026].
11. Neuberger, H., *Phys. Lett.* **B417**, 141 (1998), [arXiv:hep-lat/9707022];
    Phys. Lett. **B427**, 353 (1998), [arXiv:hep-lat/9801031].
12. Chiu, T., *Phys. Rev.* **D58**, 074511 (1998) [arXiv:hep-lat/9804016].
13. Chandrasekharan, S., *Phys. Rev.* **D59**, 094502 (1999) [arXiv:hep-lat/9810007].
14. Fujikawa, K., *Phys. Rev.* **D60**, 074505 (1999) [arXiv:hep-lat/9904007];
    Nucl. Phys. **B546**, 480 (1999) hep-th/9811235.
15. Adams, D. H., [arXiv:hep-lat/9812003].
16. t Hooft, G., *Phys. Lett.* **B349**, 491 (1995) [arXiv:hep-th/9411228].
17. For a recent study of other classes of GWR solutions see,
    Gattringer, G., and Hip, I., *Phys. Lett.* **B480**, 112 (2000) [arXiv:hep-lat/0002002].
18. Hernandez, P., Jansen, K., and Luscher, M., *Nucl. Phys.* **B552**, 363 (1999) [arXiv:hep-lat/9808010].
19. S.V. Zenkin, Bull. Lebedev Phys. Inst. (1988) NO.910;
    Zenkin, S. V., *Mod. Phys. Lett.* **A6**, 151 (1991);
20. Chiu, T., Wang, C., and Zenkin, S. V., *Phys. Lett.* **B438**, 321 (1998) [arXiv:hep-lat/9806031];
21. Bicudo, P., *Phys. Lett.* **B478**, 379 (2000) [arXiv:hep-lat/9909157].
22. Horvath, I., *Phys. Rev.* **D60**, 034510 (1999) [arXiv:hep-lat/9901014].
23. Farchioni, F., Hip, I., and Lang, C. B., *Phys. Lett.* **B443**, 214 (1998) [arXiv:hep-lat/9809016].
24. Hernández, P., *Nucl. Phys.* **B536**, 345 (1998) [arXiv:hep-lat/9801035].
25. Smit, J. Vink, J., *Nucl. Phys.* **B286**, 485 (1987);
    Allés, B. et al., *Phys. Rev.* **D58**, 071503 (1998).
26. Chiu, T., and Zenkin, S.V., Phys. Rev. **D59**, 074501 (1999) [arXiv:hep-lat/9806019].

HOT AND DENSE MATTER

Off-shell effects in nuclear matter from an EFT point of view

B. V. Krippa[*†], M. C. Birse[*], J. A. McGovern[*] and N. R. Walet[†]

[*]Theoretical Physics Group, Department of Physics and Astronomy, University of Manchester,
Manchester M13 9PL, UK
[†]Department of Physics, UMIST, P.O. Box 88, Manchester M60 1QD, UK

Abstract. Effective field theory requires all observables to be independent of the representation used for the quantum field operators. Off-shell parts of the in-medium vertex functions depend on the partucular representation so that off-shell properties of the interactions should not lead to any observable effects. We analyse this issue in the context of many-body approaches to nuclear matter, where it should be possible to shift into three-body force the contributions from purely off-shell two-body interactions. We show that none of the commonly used truncations of the two-body scattering amplitude such as the ladder, Brueckner-Hartree-Fock or parquet approximations respect this requirement.

There has been much activity in recent years applying the methods of Effective Field Theory (EFT) to nuclear and hadron systems. The EFT approach is based on the separation of low and high energy scales and makes use of the fact that the dynamics at low energy is only weakly dependent on the high-energy degrees of freedom.

Hence a detailed knowledge of the interaction at short distances is not required. The low-energy physics can then be described using a local effective Lagrangian [1]. The physical amplitudes can be obtained from this in the form of expansions in powers of the low-energy scales involved, thus providing the possibility of estimating the accuracy of the calculation at every step. An EFT should be formulated in a way respecting the symmetries of the underlying theory, in our case QCD, so that the hadron EFT can be viewed as the low energy representation of QCD. The EFT program has been successfully implemented for meson-meson and meson-baryon scattering (see Ref. [2] and references therein). The natural extension of the EFT approach would be to apply the method to nuclear forces which cannot yet be derived from QCD.

Most of the calculations done so far were based on phenomenological nucleon-nucleon potentials [3]. There is a variety of these potentials, all of which give similarly good descriptions of the available low-energy two-body scattering data but which have different off-shell behaviours. These different off-shell behaviours lead to different predictions when applied to calculate nuclear observables. It led to a lot of speculation in the past on what is the "correct" off-shell behaviour of the nucleon-nucleon potential.

As we discuss below, from the EFT point of view even the issue of "correct" off-shell behaviour of the nucleon-nucleon forces is ill-defined one since off-shell effects are unobservable. In recent years it has become clear that a consistent description of nuclear systems requires also three-body forces both for systems consisting of a small number of nucleons [4], as well as for nuclear matter [5]. Part of the role of the three-body force

CP660, *Hadron Physics: Effective Theories of Low Energy QCD*, edited by A. H. Blin et al.
© 2003 American Institute of Physics 0-7354-0120-9/03/$20.00

is to compensate for the different off-shell behaviours of the two-body forces. This is as expected because, as we argued above, the physics should not depend on the off-shell behaviour of the interactions. Exact calculations with phase-equivalent two-body forces and their corresponding three- (and higher-) body forces should therefore give identical results. However this leaves open the question of whether the approximate many-body techniques used in actual calculations respect this property.

Application of the EFT approach to the nucleon-nucleon (NN) system turns out to be rather complicated due to the large s-wave scattering length, which gives rise to an additional small scale in the problem. If the scattering length were similar to the range of the NN interaction (usually referred to as "natural") then it would be possible to use a perturbative treatment. Instead, the unnaturally large scattering length means that the leading term in the expansion of the short range NN potential should be iterated to all orders, as shown by Weinberg [6] and van Kolck [7]. The remaining terms in the potential can be treated as perturbations, organised according to the power counting elucidated by Kaplan, Savage and Wise (KSW) [8]. This can also be thought of as an expansion around a renormalization-group fixed point which corresponds to a bound state at threshold [9].

Since the off-shell properties depend on the representation chosen for the field operator in EFT's, independence of observables on representation therefore means that off-shell effects should cancel in the final answer. This cancellation is a consequence of a property known as "reparametrization invariance". In formal field theory the corresponding result is known as the equivalence theorem [10] which states that S-matrix elements are not affected by the choice of interpolating fields featuring in a Lagrangian.

This issue has been considered in several recent papers [11, 12, 13, 14, 15]. (Further references can be found in Ref. [14].) In Ref. [11] it was shown that the determination of the off-shell part of the NN scattering amplitude using the bremsstrahlung process is not possible because one can always to redefine the field operators and shift contribution between off-shell terms and contact interactions. In Ref.[12] it was pointed out that any redundant interaction which can be removed by the equation of motion does not affect for the deuteron electromagnetic form-factors but may contribute to the off-shell amplitude. The authors of Ref. [13] studied the violation of the equivalence theorem in the approach based on the Bethe-Salpeter equation with a tree level kernel. Furnstahl et al. [14] have demonstrated that for a model consisting of a dilute Fermi system with a *natural* two-body scattering length three-body counterterms can indeed cancel the effects of off-shell part of the two-body interaction.

Here we explore the extension of these ideas to dense systems with an unnatural scattering length, such as nuclear matter. It is worth noting that, in spite of being greatly reduced in nuclear matter due to the Pauli principle [16], the in-medium analog $a^{(m)}$ of the vacuum NN-scattering length is still large enough to make an expansion in dimensionless quantity $p_F a^{(m)}$ (where p_F is the Fermi-momentum) completely useless.

We find that the effects of off-shell two-body interactions can indeed be cancelled by three-body forces. This is unsurprising given the general nature of the equivalence theorem. Of more importance are the sets of diagrams that need to be included to obtain this cancellation, and the implications for commonly used many-body truncation schemes. In other words, the equivalence theorem can be utilized as a diagnostic tool to

148

determine the minimal set of graphs needed to satisfy the reparametrization invariance requirement. We shall use our analysis to show inconsistency of various many-body methods.

At low enough energies we can work with EFT's for the NN system where the interaction can be treated as purely short-ranged. In principle these should be improved by by including one-pion exchange explicitly, but there is still some debate about the best way to do this. (See, for example, Ref. [17] and references therein.) We consider first an EFT which leads to a purely energy-dependent NN potential. To next-to-leading order (NLO) in the small scale this spin- and isospin-independent potential has the form

$$V_1 = C_0 + C_2' p^2, \tag{1}$$

where p^2/M is the relative kinetic energy of the nucleons. The LO coupling constant C_0 is of order Q^{-1} (where Q is a generic low-energy scale) in KSW counting and so it should be treated nonperturbatively [8]. The NLO coupling C_2' is proportional to the effective range, and is thus of order Q^0 and can be treated perturbatively in this counting scheme. Nonetheless, to show an example of the principles discussed above, we solve the Lippmann-Schwinger (LS) equation with this potential to all orders in C_2', and find the vacuum T-matrix

$$T_1 = \frac{C_0 + p^2 C_2'}{1 + \frac{M}{4\pi}(C_0 + p^2 C_2')(ip + \mu)}. \tag{2}$$

Here we have used a subtractive renormalization procedure [8, 18]. (All coupling constants here should be understood as renormalized ones which depend on μ to ensure that the scattering amplitude is μ-independent.)

More generally, the effective Lagrangian can also include interactions with space derivatives of the nucleon fields and this leads to a potential that depends on momentum as well as energy. The most general NLO potential has the form

$$V_2 = C_0 + C_2' p^2 + \frac{1}{2} C_2 (\mathbf{k}^2 + \mathbf{k}'^2 - 2p^2), \tag{3}$$

where $\mathbf{k}$ and $\mathbf{k}'$ denote the initial and final relative momenta of the nucleons. The coupling C_2 describes a purely off-shell interaction. Solving the LS equation we get

$$T_2 = T_1 \left[1 + \frac{1}{2(C_0 + p^2 C_2')} \left(C_2(\mathbf{k}^2 + \mathbf{k}'^2 - 2p^2) \right. \right.$$
$$\left. \left. - \frac{M}{8\pi} C_2^2 (p^2 - \mathbf{k}^2)(p^2 - \mathbf{k}'^2)(ip + \mu) \right) \right], \tag{4}$$

where T_1 is given by Eq. (2). From this we can see that the two T-matrices coincide on-shell ($\mathbf{k}^2 = \mathbf{k}'^2 = p^2$) and so the scattering observables are indeed independent of the off-shell behaviour of the potential as required by the equivalence theorem.

The situation becomes much less trivial in the presence of the nuclear medium. An in-medium T-matrix [19] can be obtained by solving the Feynman-Galitskii (FG) equation,

$$T^m = V + V G^F T^m, \tag{5}$$

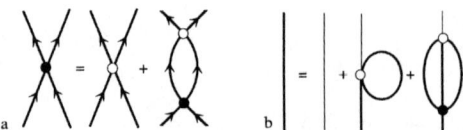

FIGURE 1. Hugenholtz diagrams representing the equation for (a) the T matrix and (b) the dressed propagator. The open circle denotes the LO potential C_0, and the solid circle the T matrix.

where G^F denotes the two-nucleon propagator which contains both particle-particle (pp) and hole-hole (hh) states. This T-matrix can be thought of as an extension of the more familiar G-matrix [20, 21] to include hh as well as pp ladders. It is convenient to represent the FG equation graphically in terms of the Hugenholtz diagrams of Fig. 1(a). These are versions of Feynman diagrams which explicitly incorporate antisymmetry of the interactions. Internal lines represent Feynman propagators which describe both particles and holes. The arrows represent the flow of quantum numbers such as baryon number. Each topologically distinct diagram should be multiplied by a symmetry factor to take account of the number of ways it can be constructed from the antisymmetric vertices. More details of these diagrams and the rules for evaluating them can be found in the textbooks [20, 21].

The solution of the FG equation is rather straightforward in the case of zero total momentum of the nucleons. For the potential V_1 it takes the form

$$T_1^m = \frac{1}{\frac{1}{T_1} + \frac{M}{4\pi^2}[p\log\frac{p+p_F}{p-p_F} - 2p_F]},\tag{6}$$

where p_F is the Fermi momentum. In the same way we can solve the FG equation for the potential V_2. Note that all the renormalisation issues are dealt with on the level of the vacuum T-matrix so that no divergencies appear in the T_1^m by construction. We shall assume that the C_2 term can be treated as a perturbation. For simplicity we omit the energy-dependent term C_2' from now on. Although this term makes a physically important contribution to the energy of the two-particle amplitude, it does not take part in the cancellation of off-shell effects which is of interest here. To first order in C_2 the in-medium T-matrix can be written

$$T_2^m = T_1^m - T_1^m \frac{C_2}{C_0}(2p^2 - \mathbf{k}^2 - \mathbf{k'}^2) - 2(T_1^m)^2 \frac{C_2}{C_0}\frac{M}{6\pi^2}p_F^3.\tag{7}$$

If we now evaluate T_2^m at the on-shell point, we see that it does not agree with T_1^m since the last term does not vanish. This indicates that calculations of nuclear matter based on the in-medium T-matrix (or similarly the G-matrix) do not satisfy the requirement of reparametrization invariance. Alternatively, in more traditional nuclear physics language, results for in-medium observables depend on the off-shell behaviour assumed for the NN potential. Such a dependence is unphysical and should not be present. A clue to how the dependence may be removed comes from the form of the final term in Eq. (7), which is proportional to the density. Its structure is thus similar to that arising from a three-body contact interaction. This suggests that it may be possible

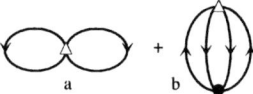

FIGURE 2. Hugenholtz diagrams for the ground state energy at first order order in C_2 (the open triangle).

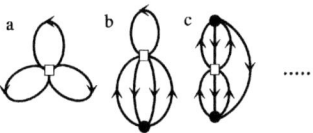

FIGURE 3. Hugenholtz diagrams for the ground state energy at first order in the three-body force D_0 (the open square).

to trade off the off-shell dependence against a three-body force. As shown below, this can be done, provided our approach includes more than just ladder diagrams.

Before exploring what additional physics is needed to remove the off-shell dependence, we should note that there is no clear separation of scales in strongly interacting, dense matter. This is an unsolved problem for the application of EFT's: no power counting has been found which leads to a consistent expansion. One should really solve the many-body theory for C_0 exactly, by constructing the full in-medium NN vertex, Γ. Nonetheless simpler approximations are commonly used in nuclear physics, typically replacing the full NN vertex by a G- or T-matrix. Including ph rings as well as pp and hh ladders leads to the parquet approximation [22, 20]. We examine here the consistency of these approximations with reparametrization invariance.

At LO in C_2 The contributions at LO in C_2 to the ground-state energy of matter are shown in Fig. 2, where the solid dot denotes an in-medium NN vertex. If C_0 were weak enough we could expand these diagrams perturbatively to get a contribution of order MC_0C_2. The resulting diagrams would have an identical structure to those in Fig. 4 below, dut with bare propagators. As shown in Ref. [14], they can be exactly cancelled against the LO contribution of a contact three-body interaction with strength $D_0 = 12MC_0C_2$. This is as required by the equivalence theorem, since the off-shell term and three-body force with this strength are both generated from a Lagrangian which contains neither by the same field redefinition. The details are given in Ref. [14]. For definiteness we repeat the relevant Feynman rules here: the two-body vertices, represented by an open circle and a triangle respectively, are

$$-iC_0S_2 \quad \text{and} \quad iC_2(\Delta_i + \Delta_{i'} + \Delta_j + \Delta_{j'})S_2, \tag{8}$$

where $\Delta_i = Mp_i^0 - (\mathbf{p}_i)^2/2$, p_i being the four-momentum of the ith nucleon, and the spin-isospin structure is given by $S_2 = \delta_{ii'}\delta_{jj'} - \delta_{ij'}\delta_{ji'}$. The three-body vertex (an open square) is

$$-iD_0\left[\delta_{ii'}(\delta_{jj'}\delta_{kk'} - \delta_{jk'}\delta_{kj'}) + \text{cyclic}(i',j',k')\right]. \tag{9}$$

151

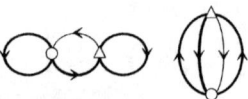

FIGURE 4. The contributions from Fig. 2 proportional to the integral I_0. Diagrams which can be obtained from those shown by simply reversing all the arrows are not shown separately.

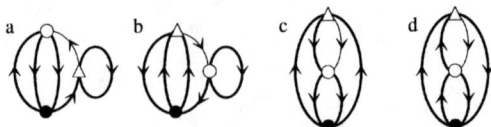

FIGURE 5. (a-c) Contributions proportional to the integral I_1 obtained from Fig. 2 in the BHF approximation. (d) Extra contribution from the parquet approximation. Diagrams which can be obtained from these by simply reversing all the arrows are not shown separately.

When the leading two-body vertex is resummed we get an effective vertex $\Gamma(p_i, p_j; p_{i'}, p_{j'})S_2$ which is denoted by a filled circle [1].

Treating C_0 nonperturbatively, the same three-body force gives rise to the diagrams shown in Fig. 3. Each of these diagrams gives a contribution equal to a distinct integral multiplied by D_0 and a degeneracy factor. The detailed forms of these integrals, which we denote by $I_0, I_1, I_2, \ldots$ are not needed here. However we can use these integrals to classify the structures which arise from the diagrams of Fig. 2. To evaluate them, we note that the off-shell vertex can be written as a sum of four pieces, each of which can cancel a bare propagator on one "leg":

$$G_0(q)(Mq_0 - \mathbf{q}^2/2)C_2 = iMC_2, \tag{10}$$

where $G_0(q)$ is the bare single-particle propagator. The diagrams in Fig. 2 give rise to many different contributions, which can be identified by iterating the equations for the in-medium NN vertex and dressed propagator to pull out a bare propagator ending on a lowest-order vertex C_0 on any of the lines in the original diagrams. When the bare propagator is cancelled against the off-shell vertex, as in Eq. (10), the result is one of the integrals I_n multiplied by MC_0C_2 and a numerical factor. Thus we can examine the cancellation of off-shell dependence for each integral in turn.

We consider first the Brueckner-Hartree-Fock approximation (BHF) [20, 21], in which propagators are dressed and the in-medium NN vertex is obtained by iterating the potential in the pp and hh channels, as shown in Fig. 1. The contributions proportional to I_0 from Figs. 2(a) and (b) are shown in Fig. 4. Except for the dressing of the propagators, these have the same structures as the perturbative diagrams considered in Ref. [14] and they can be shown to cancel with Fig. 3(a) in the same way.

---

[1] Beyond the ladder approximation the NN vertex also has a piece which is symmetric in spin-isospin and antisymmetric in momenta. In the integrals we are considering the rest of the momentum dependence is symmetric, and so the other structure does not contribute.

Fig. 2(a) gives one other contribution, shown in Fig. 5(a), which is proportional to I_1. In the ladder approximation to the NN vertex, Fig. 2(b) also gives contributions proportional to I_1, shown in Fig. 5(b-c). The sum of Figs. 5(a-c) is $-2g(g-1)(2g-3)MC_0C_2I_1$, where g is the spin-isospin degeneracy factor ($g = 4$ for symmetric nuclear matter). In contrast, the three-body force gives $g(g-1)(g-2)D_0I_1/2$ from Fig. 3(b). We see that the degeneracy coefficients do not agree and cancellation does not occur. For example, the off-shell dependence is nonzero for neutron matter ($g = 2$) where the Pauli principle forbids a contact three-body force.

There is one other structure proportional to I_1, Fig. 5(d). However this cannot be generated from the diagrams of Fig. 2 if the potential is iterated in the pp and hh channels only; it requires iteration in the ph channel as well. When this contribution is included, the degeneracy factors agree and the off-shell dependence proportional to I_1 is indeed cancelled by the three-body force with $D_0 = 12MC_0C_2$.

The crucial point to note is that the cancellation requires diagrams which can only be obtained by iterating the two-body potential in the ph channel. These are not contained in the ladder or BHF approximations and so any approach based on a G- or T-matrix cannot satisfy the equivalence theorem. Observables calculated in these approaches will have an unphysical off-shell dependence which cannot be absorbed into a three-body force.

The need for diagrams with iteration in the ph channels suggests that one should try a more complete approach. One such is the parquet approximation [22, 20], which iterates the interaction in all (pp, hh and both ph) two-body channels. In this approximation the goal is to obtain a system of integral equations to generate all reducible diagrams contributing to the many-body observables in terms of the set of 2-body irreducible diagrams.

In its simplest variant the parquet summation is a simultaneous summation of the ph rings and pp ladders with both direct and exchange terms leading to an in-medium two-body vertex. One notes that, due to its complexity parquet theory has never, to our knowledge, been consistently applied to nuclear matter even in its simplest form. However, as follows from our analysis, the inclusion of both ph rings and pp ladders is required to satisfy the reparameterization invariance and so both types of graphs should present in any consistent nuclear EFT. If we now interpret the solid circles in Fig. 2 as parquet NN vertices constructed from C_0, then all of the contributions in Fig. 5 can be generated by iterating the parquet equations. (Note that the parquet self energy can still be expressed in the form of Fig. 1(b) [23], and so the discussion of I_0 above is unchanged.) It is worth mentioning that just like the standard approach based on the G-matrix where the propagator dressing and two-body vertex must be determined in a self-consistent manner the parquet approximation should also be formulated in a way allowing for the self-consistent treatment of the one-particle self-energy which itself is determined by the two-body vertex. It means that the complete parquet theory should be represented by a coupled system of equations consisting of the parquet equations for the NN-vertex and Dyson equation for the one-particle self-energy.

Turning now to terms proportional to I_2, which ought to cancel with the three-body graph Fig. 3(c), we find one self-energy contribution, Fig. 6(a). This would be present even in the BHF approximation for the T-matrix but, not unexpectedly, this does not provide the cancellation. In the parquet approximation, Fig. 2(b) gives additional contri-

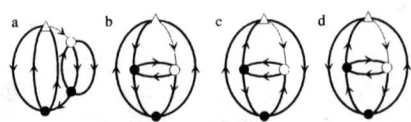

FIGURE 6. (a) Contribution from Fig. 2(b) proportional to the integral I_2 in the BHF approximation. (b-c) Extra contribution in the parquet approximation. (d) Diagram containing a non-parquet contribution. Diagrams which can be obtained from those shown by simply reversing all the arrows are not shown separately.

butions, shown in Fig. 6(b,c), and part of (d). Only when the non-parquet contributions are included (see Table 3 of Ref. [22]), i.e. with a full set of diagrams, do we match the result from Fig. 3(c). Thus we conclude that the parquet approximation also violates reparametrization invariance!

It may well be possible to include the necessary structures by extending the parquet approximation along the lines discussed in Ref. [20], starting from a basic vertex which is a sum of diagrams which are two-particle irreducible in all channels. However if, as suggested there, these structures are simply added in perturbatively, they will not generate the full in-medium vertices needed for the diagrams of Fig. 5. Of course, the comment made above about self-consistent determination of the one-particle self-energy is applicable in this (more complicated!) case too. One cannot also exclude the possibility that the sets of graphs with an larger number of NN-vertices would require diagrams with more complicated topologies to satisfy reparametrization invariance. It is worth emphasizing that the main obstacle plaguing the consistent formulation of a nuclear EFT is the problem of establishing a power counting scheme. If such a counting were found one could then determine the minimal set of graphs required to satisfy reparametrization invariance at any given order up to truncation errors.

In summary, our results demonstrate that the requirement of reparametrization invariance, which would require the effects of off-shell dependence of the two-body interaction to be cancelled by a three-body force, are not satisfied by any of the commonly used truncations of the two-body scattering amplitude such as ladder, BHF or parquet approximations. The violations show up in structures with higher numbers of insertions of the in-medium NN vertices for the more sophisticated truncations, but as the interaction is strong this does not provide a consistent expansion scheme. Finding such a scheme remains essential for practical applications of EFT's to dense, strongly interacting matter.

This work was supported by the EPSRC.

REFERENCES

1. S. Weinberg, Physica A **96**, 32 (1979).
2. D. Phillips, nucl-th **0203040**.
3. R. Machleidt and I. Slaus, J. Phys. G: Nucl. Part. Phys. **27** R69 (2001).
4. S. C. Pieper, V. R. Pandharipande, R. B. Wiringa, and J. Carlson, Phys. Rev. C **64**, 014001 (2001).
5. H.Q. Song, M. Baldo, G. Giansiracusa and U. Lombardo, Phys. Rev. Lett. **81**, 1584 (1998).
6. S. Weinberg, Nucl. Phys. **B363**, 3 (1991).
7. U. van Kolck, Nucl. Phys. **A645**, 273 (1999).

8. D. B. Kaplan, M. J. Savage and M. B. Wise, Nucl. Phys. **B534**, 329 (1998).
9. M. C. Birse, J. A. McGovern and K. G. Richardson, Phys. Lett. **B464**, 169 (1999).
10. R. Haag, Phys. Rev. **B112**, (1958) 669; S. Kamefuchi, L. O'Raifeartaigh and A. Salam, Nucl. Phys. **28**, 529 (1961).
11. H. Fearing, Phys. Rev. Lett. **81**, 4 (1998); H. Fearing and S. Scherer, Phys. Rev. **C62**, 054006 (2000).
12. D. B. Kaplan, M. J. Savage and M. B. Wise, Phys. Rev. C **59**, 617 (1999).
13. S. Kondratyuk, Phys. Lett. **B521**, 204 (2001).
14. R. J. Furnstahl, H.-W. Hammer and N. Tirfessa, Nucl. Phys. **A689**, 846 (2001).
15. R. J. Furnstahl and H.-W. Hammer, Phys. Lett. **B531**, 203 (2002).
16. B. Krippa, nucl-th **9910072**, in Proceedings of International Workshop "Hadron99", 1999, Coimbra, Portugal.
17. S. R. Beane, P. F. Bedaque, M. J. Savage and U. van Kolck, Nucl. Phys. **A700**, 377 (2002).
18. J. Gegelia, Phys. Lett. **B429**, 227 (1998).
19. P. Bozek, Phys. Rev. C **65**, 054306 (2002).
20. J. P. Blaizot and G. Ripka, *Quantum theory of finite systems* (MIT Press, Cambridge, 1985).
21. J. W. Negele and H. Orland, *Quantum many-particle systems* (Perseus, New York, 1988).
22. A. D. Jackson, A. Lande and R. A. Smith, Phys. Rep. **86**, 55 (1982).
23. A. D. Jackson and R. A. Smith, Phys. Rev. **A36**, 2517 (1987).

Thermal Meson properties within Chiral Perturbation Theory

A.Gómez Nicola*, J.R.Peláez†*, A.Dobado**‡ and F.J.Llanes-Estrada**

*Departamento de Física Teórica II, Universidad Complutense, 28040 Madrid, Spain.
†Dip. di Fisica. Universita' degli Studi and INFN, Firenze, Italy.
**Departamento de Física Teórica I, Universidad Complutense, 28040 Madrid, Spain.
‡Theory Group, Lawrence Berkeley National Laboratory, Berkeley, CA 94720, USA.

Abstract. We report on our recent work about the description of a meson gas below the chiral phase transition within the framework of Chiral Perturbation Theory. As an alternative to the standard treatment, we present a calculation of the quark condensate which combines the virial expansion and the meson-meson scattering data. We have also calculated the full one-loop elastic pion scattering amplitude at finite temperature and we have unitarized the amplitude using the Inverse Amplitude Method in order to reproduce the temperature effects on the mass and width of the ρ and σ resonances. Our results show a clear increase of the thermal ρ width, as expected from previous analysis. The results for the σ are consistent with Chiral Symmetry Restoration. We comment on the relevance of our results within the context of Relativistic Heavy Ion Collisions.

INTRODUCTION AND MOTIVATION

The recent development of the Relativistic Heavy Ion Collision (RHIC) program is one of the main motivations to study hadronic matter under extreme conditions of temperature T and density. Here we will consider the meson gas formed when the plasma created after one such RHIC has expanded and hadronized, being in a state where the chiral symmetry is restored. There are strong indications that one could observe the medium effects on such a system. For instance, the dilepton spectrum shows an anomalous behaviour for invariant masses near the ρ mass [1]. The flatness of the spectrum is consistent with a modification of the mass and width of the ρ's which have time to decay inside the plasma, so that their spectral function acquires thermal corrections due to the interaction with the hot and dense hadron gas [2, 3, 4, 5, 6, 7, 8, 9].

In such a system, external momenta and temperature are small compared to the chiral symmetry breaking scale $\Lambda_\chi \simeq 1$ GeV. The relevant degrees of freedom are then the lightest mesons and the interactions among them are best described by a low-energy QCD effective theory based on chiral symmetry. The most general framework comprising the QCD Chiral Symmetry Breaking pattern $SU_L(N_f) \times SU_R(N_f) \rightarrow SU_V(N_f)$ is Chiral Perturbation Theory (ChPT) [10, 11, 12, 13] where any observable can be calculated as an expansion in p/Λ_χ, p denoting a meson mass, momenta or the temperature. Thus, ChPT provides model-independent predictions, just by fixing a few low-energy constants (LEC). This program has included the calculation of thermodynamic observables such as the free energy density and the quark condensate, as we will briefly review below. Throughout this paper, we will neglect finite baryon density effects, as ideally in

CP660, Hadron Physics: Effective Theories of Low Energy QCD, edited by A. H. Blin et al.
© 2003 American Institute of Physics 0-7354-0120-9/03/$20.00

the central rapidity region formed after a RHIC.

Since ChPT is intended to provide a systematic low-energy and low-T perturbative expansion, we do not expect it to reproduce resonances. This is strongly related to the fact that ChPT only satisfies unitarity in a perturbative fashion. Over the last few years, there has been a lot of work devoted to enlarge the ChPT applicability range by using unitarization methods, i.e, imposing exact unitarity requirements, like the Inverse Amplitude Method (IAM) [14] or approaches based on Lippmann-Schwinger or Bethe-Salpeter equations [15]. These methods provide a good agreement with the experimental phase shifts and they are able to generate resonances, like the ρ and the σ for the $SU(2)$ chiral symmetry. In addition, they can be extended to include coupled channels, describing successfully all the meson-meson data and resonances in the $SU(3)$ case, up to 1.2 GeV [16, 17].

What we will show below is that only requiring chiral symmetry and unitarity one can also describe successfully the thermal behaviour of the ρ and σ resonances. For that purpose, one needs first to calculate the thermal pion scattering in ChPT, which has been done in [18]. We shall discuss the main features of such thermal amplitude below. Then, by using the IAM extended to finite T, one can construct a nonperturbative thermal unitarized amplitude which, in particular, describes the behaviour of the thermal ρ and σ [19]. We will present the results for the thermal mass and width of the ρ and σ in that approach, as well as for the T-dependence of the effective $g_{\rho\pi\pi}$ vertex. The main implications of our results in the context of RHIC and chiral symmetry restoration will be also discussed below.

THE MESON GAS AND CHIRAL SYMMETRY AT FINITE T WITHIN CHPT

For the reasons explained above, it is important to provide an accurate description of the low-T meson gas in thermal equilibrium. For instance, the signature of chiral symmetry restoration at $T_c \simeq 150\text{-}200$ MeV should be observed in the thermal evolution of the order parameter, the quark condensate $\langle \bar{q}q \rangle(T)$ from below the transition point. As we have just discussed, ChPT provides a model independent description of the thermodynamic observables, based only on chiral symmetry. The only extra ingredient is the temperature, which is treated as an $\mathcal{O}(p)$ parameter.

The first calculations of the pion gas within ChPT go back to [20], where $\langle \bar{q}q \rangle(T)$ and the thermal dependence of $f_\pi(T)$ were calculated up to $\mathcal{O}(T^2)$ (one loop) in the chiral limit. That result already showed a behaviour compatible with chiral symmetry restoration. In [21] a thorough analysis up to $\mathcal{O}(T^6)$ was performed, including the free energy, the quark condensate and an estimate of the thermal effects of *free* kaons and etas. The $\mathcal{O}(T^4)$ corrections to $f_\pi(T)$ have been analyzed in [22] where it has been shown that beyond $\mathcal{O}(T^2)$ one has to consider separately the space and time components of the axial current, so that there are two independent $f_\pi^{s,t}$, which, in addition, can be complex. In fact, their imaginary part is proportional to the in-medium pion damping rate while their real parts are related to the deviations of the pion velocity from the speed of light. Other analysis of the thermal pion dispersion law can be found in [23, 24].

The analysis of typical nonequilibrium effects such as explosive pion production after a RHIC can be also studied within the ChPT context [25]. It should be pointed out that many of these properties have also been investigated using specific models for low-energy QCD. The most popular is the $O(4)$ model, which reproduces a critical behaviour already in the mean field approach. Apart from introducing the σ explicitly, conventional perturbation theory in the $O(4)$ model breaks down, which has been dealt with at finite T using different nonperturbative approaches [26, 27].

When calculating thermodynamic quantities such as the pressure or the quark condensate from ChPT, the usual approach is to use the Feynman rules of Thermal Field Theory [28] to the order considered. This is particularly cumbersome in the three flavor case if one wishes to include the full dependence on temperature, quark masses and $SU(3)$ interactions. An alternative [21, 29, 30] is to perform a virial expansion of the pressure as [31]

$$\beta P = \sum_i B_i(T)\xi_i + \sum_i \left(B_{ii}(T)\xi_i^2 + \frac{1}{2}\sum_{j\neq i} B_{ij}(T)\xi_i\xi_j \right)..., \tag{1}$$

where $i = \pi, K, \eta$. This is a dilute gas expansion in the fugacities $\xi_i = \exp(-m_i/T)$. The binary interactions between the different species show up in [21, 29, 30]:

$$B_{ij}^{(int)} = \frac{\xi_i^{-1}\xi_j^{-1}}{2\pi^3} \int_{m_i+m_j}^{\infty} dE\, E^2 K_1(E/T) \sum_{I,J,S} (2I+1)(2J+1)\delta_{I,J,S}^{ij}(E), \tag{2}$$

where K_1 is the first modified Bessel function and $\delta_{I,J,S}^{ij}$ are the elastic scattering $ij \to ij$ phase shifts *at T=0* (chosen so that $\delta = 0$ at threshold) of a state ij with isospin I, angular momentum J and strangeness S. What makes the virial expansion useful is that the T dependence on thermodynamical observables up to $T \simeq 200 - 250$ MeV can be obtained just from the $T = 0$ phase shifts, which have been calculated for all possible meson-meson interactions in $SU(3)$ ChPT to one loop. They can be found for instance in [17]. Note that for the pressure one could even use the experimental phase shifts directly. However, the quark condensate is given by the derivative of the pressure with respect to the quark mass and therefore the analytic dependence of the δ^{ij} with the different meson masses is needed.

For temperatures much below 150 MeV, massive states like kaons and eta are thermally suppressed, typically by the Boltzmann factors $\exp(-M_K/T)$ and, in principle, the suppression is even stronger for the interactions among them and with pions, as (1) shows. For low T it is then reasonable to treat those states as free particles, as it was done in [21]. Moreover, at low T the integrals in (2) are dominated by the phase shifts at threshold. However, for higher temperatures, the effect of massive states becomes increasingly important and, furthermore, the dependence of the interactions with the pion mass can be large so that their contribution to the quark condensate becomes sizable [30]. When the effect of the strange states is taken into account, there are two main questions that can be analyzed: on the one hand, the effect on the non-strange condensate $\langle \bar{u}u \rangle = \langle \bar{d}d \rangle = \langle \bar{q}q \rangle/2$ (in the isospin limit) of adding another flavor. Since the number of degrees of freedom increases, so it does the entropy and one expects that the

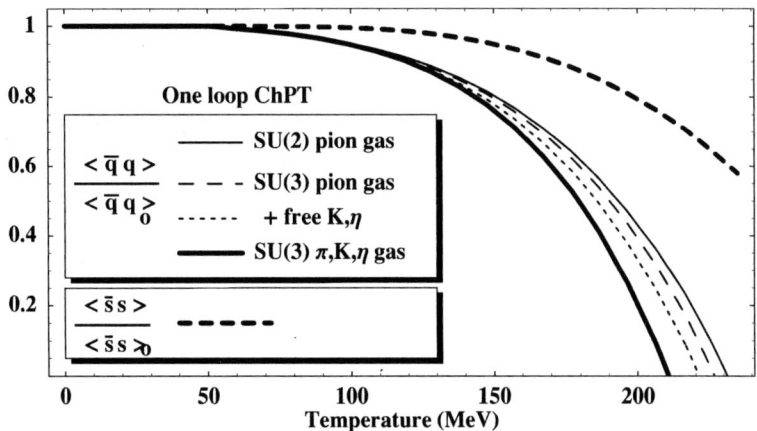

FIGURE 1. The thermal evolution of the $\langle \bar{q}q \rangle$ and $\langle \bar{s}s \rangle$ condensates in ChPT

collective state is closer to "disorder". This would imply a decrease of the critical temperature, as it is indeed observed in lattice calculations [32], although with gluons and massless quarks. On the other hand, one can study the thermal evolution of the strange condensate $\langle \bar{s}s \rangle$. Since $m_s \gg m_{u,d}$, one expects that the thermal evolution for $\langle \bar{s}s \rangle(T)$ is slower than for the non-strange condensate. Note that the quark masses play a similar role as external magnetic fields in ferromagnets, so that increasing the external field means that it takes more thermal energy to restore the symmetry. Let us remark that, in contrast to the lattice, the physical masses are easily incorporated in our approach.

We report here on the virial expansion calculation in $SU(3)$ ChPT that has been done in [30]. The main results for the quark condensates are summarized in Figure 1. First, it should be pointed out that the curves are plotted at most up to the point where the condensates vanish. The ChPT condensates do not vanish above those "critical" values, since they have been obtained from a perturbative series in the temperature. In fact, the results should be trusted only in the region showed in the graphs, as explained above. The thermal evolution of $\langle \bar{q}q \rangle(T)$ is shown in different approximations, namely, considering only pions in $SU(2)$ or $SU(3)$ (their tiny difference comes from the $\mathcal{O}(p^4)$ phase shifts), adding free kaons and etas [21] and, finally, considering the full $SU(3)$ interactions [30]. Note also that these interactions (basically only πK and $\pi \eta$ are important at these temperatures) yield a larger effect than naively expected. The reason is that they depend strongly on m_π, which is more sensitive to $m_{u,d}$ than m_K or m_η. Taking into account the numerical values of the LEC's with their errors, one gets a reduction of the melting temperature of $T_m^{\langle \bar{q}q \rangle SU(2)} - T_m^{\langle \bar{q}q \rangle SU(3)} = 21^{+14}_{-7}$ MeV, in remarkable agreement with the chiral limit lattice calculations [32], taking into account that we have used the actual meson masses. In addition, we estimate $T_m^{\langle \bar{q}q \rangle SU(3)} = 211^{+19}_{-7}$ MeV [30]. The second effect, is also clearly seen in Figure 1: The thermal evolution of the strange condensate from the broken phase is much slower than the non-strange one. From Figure 1, we see that there is still about 80 % left for $\langle \bar{s}s \rangle(T)/\langle \bar{s}s \rangle(0)$ when $\langle \bar{q}q \rangle(T)/\langle \bar{q}q \rangle(0)$ has already melted. Finally, one may wonder about the effect of calculating the integrals in (2) with

ChPT, whose applicability does not extend to infinity. Indeed, when (2) is evaluated [30] with the phase shifts unitarized with the coupled channel IAM [17] , which have a much larger applicability range, the numerical results only change very slightly. Thus, the main conclusions remain the same, since, as we have already commented, the main contribution to the integrals comes from the low-energy region and the IAM agrees with ChPT at low energy, improving only the high energy behavior.

PION SCATTERING AT $T \neq 0$ IN ONE LOOP CHPT

If the pion gas is dilute enough, i.e., at very low temperatures, it is reasonable to assume that the only dependence of pion scattering with T can be accounted for through the Bose-Einstein distribution functions and one can ignore the T-dependence of the scattering amplitudes. However, that dependence could be important in several contexts. For instance, it has been suggested that an enhancement of pion scattering in the scalar channel near threshold could be an indication of chiral symmetry restoration [27]. In addition, a previous calculation [9, 33] of thermal $\pi\pi$ scattering in the Nambu-Jona-Lasinio model, shows a singular behaviour of the scattering lengths at some critical temperature, which may be related to a Mott transition. Furthermore, if one wishes to extend to finite T the fruitful unitarization program in ChPT to describe resonances, the full calculation of the $\pi\pi$ scattering amplitude to one loop and the extension of perturbative unitarity are essential ingredients.

For the above reasons, it is important to provide a model-independent description of pion scattering for temperatures T well below the chiral scale Λ_χ. This can be achieved in ChPT. The calculation of the thermal scattering lengths in ChPT to one loop has been done in [34] whereas the calculation of the full thermal amplitude in one loop ChPT has been recently carried out [18]. We will report here the main results and features of that work.

There are two formal aspects regarding the calculation of the scattering amplitude at finite T that are worth pointing out. The first one is that we are considering the thermal amplitude defined by taking $T = 0$ asymptotic pion sates, the T dependence coming from the four-point function, calculated using the Feynman rules of Thermal Field Theory in the imaginary-time formalism [28]. Then, one can perform an analytic continuation from discrete frequencies $\omega_n = 2\pi nT$ to real energies E as $i\omega_n \to E + i\varepsilon$. Such analytic continuation corresponds to the retarded four-point function in the so called real-time formalism [35]. The retarded functions have a causal and analytic structure [36] suitable to extend perturbative unitarity at $T \neq 0$. The same definition of thermal amplitude has been used in [9, 34].

The second point is that the loss of Lorentz covariance inherent to the thermal formalism (due to the choice of the thermal bath rest frame) means that any two-body scattering amplitude with four-momenta $k_1 k_2 \to k_3 k_4$ will depend separately on the variables $\mathbf{S_0}$, $|\vec{\mathbf{S}}|$, $\mathbf{T_0}$, $|\vec{\mathbf{T}}|$, $\mathbf{U_0}$ and $|\vec{\mathbf{U}}|$, where $\mathbf{S} = k_1 + k_2$, $\mathbf{T} = k_1 - k_3$ and $\mathbf{U} = k_1 - k_4$. At $T = 0$, the amplitude depends only on the Mandelstam variables $s = \mathbf{S}^2, t = \mathbf{T}^2, u = \mathbf{U}^2$ and any $\pi\pi$ scattering amplitude can be related to that of $\pi^+\pi^- \to \pi^0\pi^0$, called $A(s,t,u)$, by isospin and crossing transformations. At $T \neq 0$ and since the temperature does not

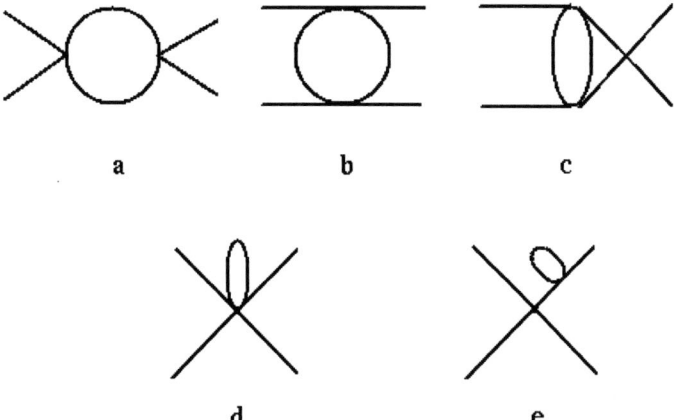

FIGURE 2. One loop diagrams contributing to $\pi\pi$ scattering

modify the interaction vertices, the crossing symmetry still holds, but now in terms of **S**, **T** and **U**. Therefore, any $\pi\pi$ thermal amplitude can be written in terms of the thermal $A(\mathbf{S},\mathbf{T},\mathbf{U};\beta)$. The loop diagrams are the same as for $T=0$ and are given in Figure 2.

The final result for the thermal amplitude can be written as:

$$A(\mathbf{S},\mathbf{T},\mathbf{U};\beta) = A_2(\mathbf{S},\mathbf{T},\mathbf{U}) + A_4^{pol}(\mathbf{S},\mathbf{T},\mathbf{U}) + A_4^{tad}(\mathbf{S},\mathbf{T},\mathbf{U};\beta) + A_4^{uni}(\mathbf{S},\mathbf{T},\mathbf{U};\beta), \quad (3)$$

where A_2 is tree level contribution coming from the $\mathcal{O}(p^2)$ lagrangian (the nonlinear sigma model) while A_4^{pol} contains both the tree level $\mathcal{O}(p^4)$ contributions plus polynomials coming from the renormalization of the loop integrals. Both A_2 and A_4^{pol} are temperature independent. The A_4^{uni} term represents those contributions from diagrams a,b,c in Figure 2. They yield the correct analytic structure and will ensure perturbative unitarity in all channels. Finally, the contribution A_4^{tad} accounts for tadpoles like those in diagrams d,e in Figure 2 plus terms coming from diagrams a,b,c proportional to the tadpole integral.

The detailed results for the different contributions above can be found in [18] and we do not give them here for brevity. As a first check of consistency, we recover the $T \to 0^+$ limit of [11]. Furthermore, when the thermal amplitude is projected into partial waves a_{IJ} of definite isospin I and angular momentum J (defined in the center of mass frame where the pions are at rest with the thermal bath) we also agree with the results given in [34] for the scattering lengths.

Another important check of consistency, which will be crucial for our analysis in the next section, concerns the imaginary part of the partial waves and perturbative unitarity. At $T=0$, unitarity constraints the partial waves, for $s > 4m_\pi^2$ and below other inelastic thresholds, to satisfy

$$\text{Im}\,a(s) = \sigma(s)|a(s)|^2, \quad (4)$$

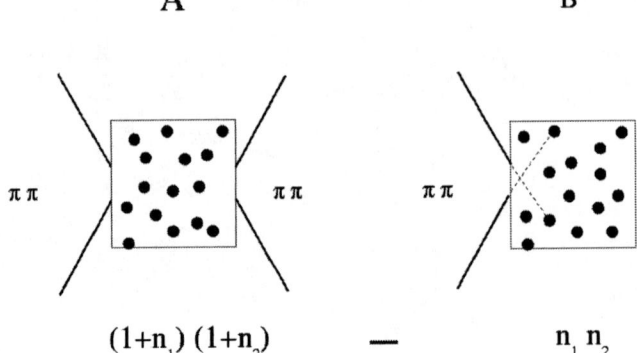

FIGURE 3. Interpretation of thermal bath contributions to the pion scattering phase space. In the process A, the medium enhances the two-pion production, while process B represents the absorption of the initial pions by the thermal bath.

where $\sigma(s) = \sqrt{1 - 4m_\pi^2/s}$ is the two-pion phase space, whereas the ChPT series only satisfies unitarity perturbatively, i.e.,

$$\operatorname{Im} a_2(s) = 0, \quad \operatorname{Im} a_4(s) = \sigma(s) \left| a_2(s) \right|^2, \quad .... \tag{5}$$

It is possible to generalize the relation (5) to the case of any one-loop elastic scattering amplitude at finite T. The derivation is given in [18] and follows closely the analysis in [37] of the discontinuity in the self-energy of a particle decaying in the thermal bath. The result for the thermal perturbative unitarity relation is:

$$\operatorname{Im} a_2(s) = 0, \quad \operatorname{Im} a_4(s; \beta) = \sigma_T(S_0) \left| a_2(s) \right|^2, \quad S_0 > 2m_\pi, \tag{6}$$

where

$$\sigma_T(E) = \sigma(E^2) \left[1 + \frac{2}{\exp(\beta |E|/2) - 1} \right] \tag{7}$$

and it has been assumed that only $\pi\pi$ states are available in the thermal bath. Thus, σ_T is nothing but the thermal phase space, whose origin can be physically interpreted by writing $1 + 2n_B(E/2) = [1 + n_B(E/2)]^2 - n_B^2(E/2)$ where $n_B(x) = (\exp(x/T) - 1))^{-1}$ is the Bose-Einstein distribution function. Written in this way, the first term represents the enhancement of phase space due to the presence of pions in the medium, while the second term accounts for the absorption of the two initial pions by pions in the bath. This is schematically depicted in Figure 3. We have checked explicitly that the relation (6) holds for our thermal $\pi\pi$ scattering amplitude. Moreover, the partial waves at $T \neq 0$

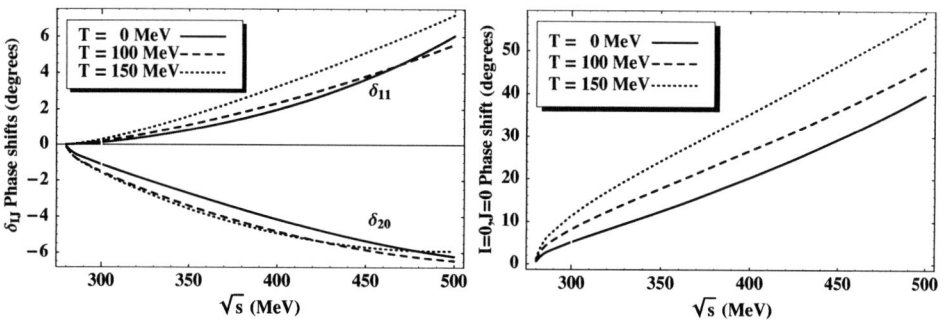

FIGURE 4. Temperature evolution of the phase shifts δ_{IJ} for $IJ = 11, 00, 20$.

can be analytically continued to the s complex plane [19] and they display the same analytic structure as the $T = 0$ one, i.e, they have a right unitarity cut in the real axis above $s > 4m_\pi^2$ (the discontinuity across the cut is given by (6)) and a left one for $s < 0$. We remark that our results remain valid when the density of states with more than two pions is small. This is equivalent to neglecting higher powers of n_B and is therefore similar to the dilute gas approach. For small energies and temperatures, that is, strictly within ChPT, the dilute gas approach is consistent, as we have also seen in the previous section. When the range of energies is extended, as we will do in the next section, one should bear in mind that the range of T where our approach is valid is such that the Bose-Einstein factors still remain small.

Finally, we have plotted the thermal phase shifts in Figure 4. These results deserve some comments. First, we observe that the absolute value of the phase shifts in all channels increases with T, while their sign (i.e, the attractive or repulsive nature of each channel) is preserved. Recall that the phase shifts are related to the real part of the amplitude as $\delta_T \simeq \sigma_T(\sqrt{s})\left[a_2(s) + \mathrm{Re}\, a_4(s, \beta)\right]$. The dominant contribution to the phase shifts thermal enhancement is given by the thermal phase space factor σ_T, while the T-dependence of the real part of the amplitude is rather weak at low T. In particular, we do not observe any significant thermal enhancement for the real part of a_{00}, which would be interpreted as chiral symmetry restoration [27]. However, it must be stressed that we are just considering the *perturbative* amplitude, valid only for low T. In the next section, we will consider a nonperturbative extension of the amplitude, which does show a behaviour compatible with chiral symmetry restoration in the 00 channel.

Another important comment regarding our results shown in Figure 4 concerns the 11 channel, i.e, the ρ channel. Recall that, following the hypothesis of resonance saturation, increasing δ_{11} is equivalent to increasing the ratio $\Gamma_\rho f_\pi^4 / M_\rho^5$ [11, 38]. Therefore, our results are consistent with a thermal increase of the rho width coming mostly from thermal phase space and an almost constant rho mass, at very low T [3, 4, 8]. This behaviour will be confirmed by our analysis in the next section, where we will find also important corrections at higher temperatures.

THE THERMAL ρ AND σ

The purpose of this section is to show that only from chiral symmetry and unitarity one finds a thermal evolution of the masses and widths of the ρ and σ at rest consistent with previous analysis and with the dilepton spectrum observed in RHIC. Unlike the models where the resonances are included as explicit degrees of freedom, we will start from the model-independent pion scattering amplitude in one loop ChPT calculated in the previous section and, imposing exact unitarity, we will construct a nonperturbative amplitude whose poles in the second Riemann sheet in the 00 and 11 channels correspond to the σ and ρ respectively.

The exact unitarity relation (4) implies that any partial wave satisfies $a = 1/(\operatorname{Re} a^{-1} - i\sigma)$ on the real axis below inelastic thresholds. A unitarization method is just one way of approximating $\operatorname{Re} a^{-1}$. The IAM uses the one-loop ChPT and thus ensures that exact unitarity is exactly satisfied and, at the same time, the low-energy predictions of ChPT are preserved. The IAM unitarized amplitude for one channel reads $a^{IAM} = a_2^2/(a_2 - a_4)$ [14] and coincides formally with the [1,1] Padé approximant in the f_π^{-2} expansion.

We have already sketched in the introduction the virtues of the IAM formula at $T = 0$ for the description of resonances and the data for higher energies. At $T \neq 0$, we have seen in the previous section that, perturbatively in ChPT and for a dilute gas, the partial waves satisfy thermal unitarity (6)-(7). Therefore, following the same steps as for the $T = 0$ case and motivated by the success of the IAM approach, we will consider the unitarized IAM thermal amplitude:

$$a^{IAM}(s;T) = \frac{a_2^2(s)}{a_2(s) - a_4(s;T)},$$ (8)

which satisfies the exact elastic unitarity condition

$$\operatorname{Im} a^{IAM}(s;T) = \sigma_T(s) \left| a^{IAM}(s;T) \right|^2.$$ (9)

The first hint that (8) provides a proper description of thermal resonances comes from the following simple argument. The behaviour of the 11 partial wave in the real axis near $s = M_T^2$, the ρ mass squared, can be described by a Breit-Wigner parametrization (valid for narrow resonances $\Gamma_T \ll M_T$):

$$a^{BW}(s;T) = \frac{R_T(s)}{s - M_T^2 + i\Gamma_T M_T}$$ (10)

where $R_T(s)$ can be related to the $\rho\pi\pi$ effective vertex (see below). Now, compare (8) with (10) near $s = M_T^2$. First, one gets $\operatorname{Re} a_4(M_T^2) = a_2(M_T^2)$, which defines the resonance mass and, second, from the unitarity relation (9), we have $\Gamma_T M_T = -R_T(M_T^2)\sigma_T(M_T^2)$. Therefore, if the thermal corrections to R_T and to M_T were much smaller than those to Γ_T ($R_T \simeq R_0$ and $M_T \simeq M_0$) we would simply get

$$\Gamma_T \simeq \Gamma_0 \left[1 + 2n_B \left(M_0/2 \right) \right],$$ (11)

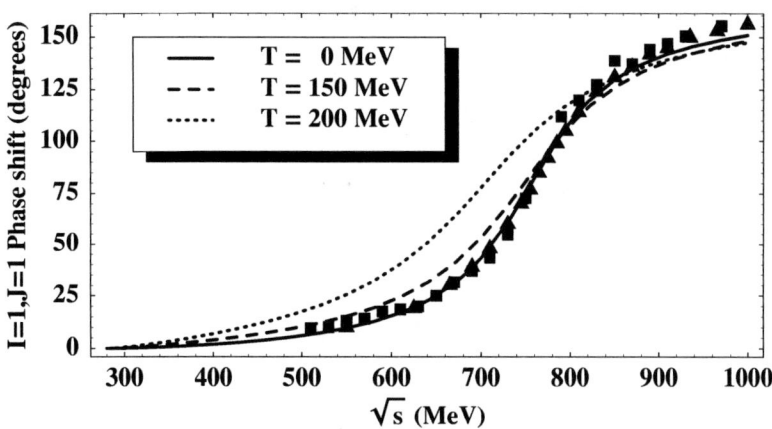

FIGURE 5. δ_{11} for different temperatures. For the data see [14].

which is the behaviour expected at very low T for a ρ at rest [4, 3, 8] that we had already anticipated in the previous section. As discussed in [37, 4] and also in the previous section, this result accounts for the difference between the direct decay $\rho \rightarrow \pi\pi$ and the inverse one $\pi\pi \rightarrow \rho$ which is allowed in the thermal bath and is responsible for the dilepton production. Note that in order to derive (11) we have neglected the T-dependence in $\mathrm{Re}\,a_4(S;T)$. Therefore, by using the complete result for the thermal one-loop amplitude discussed in the previous section, we will be able to find, for higher temperatures, the T dependence on M_T, R_T, possible deviations from the low-T behaviour (11), and even more importantly, the thermal evolution of both the ρ and σ poles in the complex plane, which is the consistent way to generate resonances within ChPT. Recall that a Breit-Wigner description is not valid for the σ. According to our previous discussion, the upper limit in T to which our approach is valid will be dictated by the condition $n_B(M/2,T) < 1$ where M is the mass of the resonance described (ρ or σ). This gives roughly $T < 300$ MeV for the ρ and $T < 180$ MeV for the σ.

In the above discussion, it is crucial to obtain the analytic continuation of the thermal amplitude to the complex s-plane. The details can be found in [19]. The result shows the same right and left cuts structure as the $T = 0$ amplitude, the discontinuities across the cuts being T-dependent. Once such analytic continuation has been obtained, it can be continued to the second Riemann sheet, using (9), as $a^{II}(s;T) = a^{IAM}(s;T)/[1 - 2i\sigma_T(s)a^{IAM}(s;T)]$.

Let us first show the results for δ_{11} for different temperatures, depicted in Figure 5. The $SU(2)$ LEC (see [11] for their definition) we have used are $l_1 = -0.3$, $l_2 = 5.6$, $l_3 = 3.4$ and $l_4 = 4.3$ and are obtained by fitting the $T = 0$ scattering data, which yields $M_0 = 770$ MeV and $\Gamma_0 = 159$ MeV. We see clearly the broadening of the ρ as T increases. This is confirmed by the evolution of the thermal poles, which is shown in Figure 6. The ρ width increases with T while the ρ mass decreases slightly. The σ pole deserves some comments. We see that for low T the σ width increases and its mass decreases slightly, following a similar behaviour as the ρ, i.e, mostly dictated by the

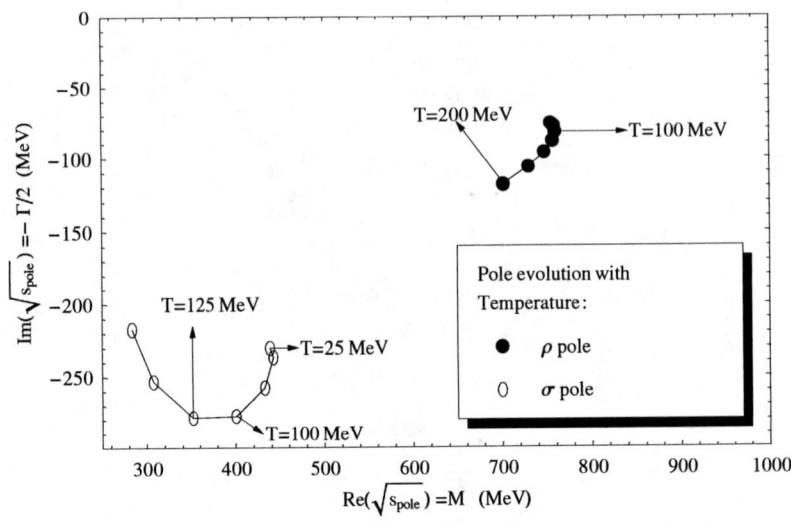

FIGURE 6. Evolution of the σ and ρ poles with the temperature.

thermal space factor σ_T. However, for $T > 125$ MeV, M_σ decreases more rapidly and Γ_σ starts to decrease. This behaviour has an interesting explanation in terms of chiral symmetry restoration. On the one hand, since the T-dependence of $m_\pi(T)$ is rather weak up to $T \simeq 200$ MeV [24, 27], the decrease of $M_\sigma(T)$ points towards $M_\sigma \to m_\pi$ and that decreasing is stronger as T increases, unlike the ρ mass. On the other hand, as T increases and M_σ approaches $2m_\pi$ from above, both the direct $\sigma \to 2\pi$ and inverse $2\pi \to \sigma$ decays become kinematically disallowed, so that Γ_σ is reduced. At low T, the decrease of the mass is much weaker and the phase space contribution dominates, making Γ_σ grow. Similar results and discussion for the σ can be found in [27, 39].

Finally, in Figure 7 we have collected our results for the ρ. First, we clearly observe a significant deviation (for $T > 100$ MeV) between the full result (obtained either from the pole position in Figure 6 or from the phase shift in Figure 5) and the naive thermal phase space prediction (11), stressing the importance of having a full T-dependent ChPT description of $\pi\pi$ scattering. Moreover, although M_ρ changes little, consistently with Vector Meson Dominance [2, 4], our results show a sizable slight decrease of the ρ mass for $T > 150$ MeV, which seems to be favored by phenomenological analysis of RHIC dilepton data [6, 8]. In Figure 7 we have also plotted the effective $\rho\pi\pi$ vertex, defined from R_T in (10) as $R_T = g_T^2 \left(4m_\pi^2 - M_T^2\right)/48\pi$, from the VMD $\rho\pi\pi$ coupling [4, 7] with a thermal ρ ($g_0 \simeq 6.2$). At low temperatures $g_T \lesssim g_0$ ($g_{50}/g_0 \simeq 0.9991$) in agreement with the chiral analysis in [7]. For higher temperatures g_T grows, although the thermal corrections are much smaller than those at finite baryon density [40].

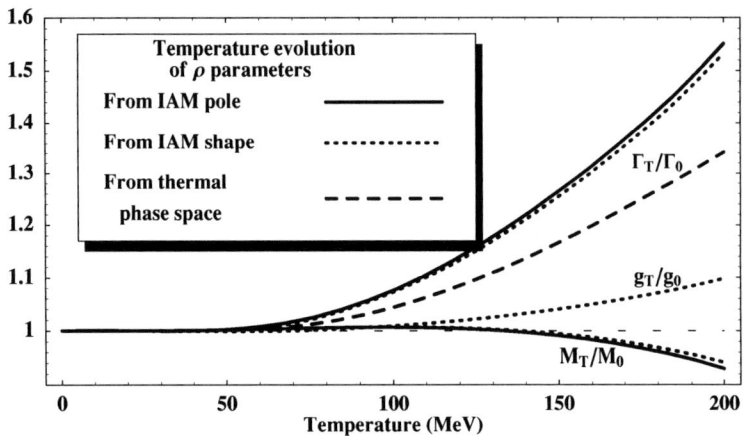

FIGURE 7. Thermal evolution of the ρ mass, width and $\rho\pi\pi$ effective vertex

CONCLUSIONS

We have reviewed some basic aspects about the ChPT description of the meson gas formed after a Relativistic Heavy ion Collision. In particular, we have shown that one can combine the virial expansion with ChPT in order to obtain the thermal evolution of the quark condensate below the chiral phase transition in the dilute gas regime. That approach allows to use directly all the available information on meson-meson scattering for three light flavors in ChPT. Contrary to lattice studies, in this approach physical masses are easily accounted for. The main effects of considering the strange sector in the condensate are, first, that the strange condensate melts much more slowly than the non-strange one and, second, that the melting temperature for the non-strange condensate is sizably smaller with three flavors than with two, and that, surprisingly, the πK and $\pi\eta$ interactions provide a largest contribution to this effect than free kaons or etas.

We have also shown the results of a recent calculation of the $\pi\pi$ scattering amplitude at finite T in one loop ChPT. The partial waves satisfy an extended version of perturbative unitarity, where the only change with respect to the $T = 0$ case is that the phase space is thermally enhanced by the presence of pions in the thermal bath. Thus, the imaginary part of the thermal amplitude has a neat physical interpretation in terms of absorption and induced emission of pion pairs in the thermal bath. The thermal phase shifts are also enhanced with T, keeping their attractive or repulsive nature, mainly dominated by the phase space factor.

The thermal pion scattering amplitude can be unitarized using the IAM, following similar steps as for $T = 0$. The unitarized partial waves can then be analytically continued to the complex energy plane and their poles in the $I = J = 1$ and $I = J = 0$ channels correspond to the ρ and the σ. This approach provides a description of the thermal effects for a ρ and σ at rest, only from chiral symmetry and unitarity. We have found that Γ_ρ increases significantly with T. For low T, M_ρ and the effective vertex $g_{\rho\pi\pi}$ remain almost constant, which is consistent with neglecting the T-dependence in the real part of the amplitudes and considering only the contribution from the thermal phase space. However, for higher T, the full T-dependence in the amplitude has to be taken into account. Thus, for $T > 100$ MeV, Γ_ρ acquires significant corrections from the pure thermal space approximation and $g_{\rho\pi\pi}$ increases, while for $T > 150$ MeV, M_ρ shows a sizable decreasing behaviour. As for the σ pole, although its interpretation as a resonance is much less clear than the ρ, the thermal behaviour of both M_σ and Γ_σ, as obtained from its associated pole, can be understood from chiral symmetry restoration.

Our results agree with several theoretical analysis and are consistent with phenomenological studies of RHIC dilepton data. We have only used chiral symmetry and unitarity as guiding principles, without including the resonances as explicit degrees of freedom. It would be interesting to apply our chiral unitary approach to the production of dileptons from thermal $\pi\pi$ annihilation near the ρ energy, in order to be able to provide more accurate predictions about the observed dilepton spectrum in RHIC. This is just but one of the possible directions that we will pursue in the near future.

ACKNOWLEDGMENTS

A.G.N and J.R.P wish to thank the organizers of the "2nd International Workshop on Hadron Physics" for their kind invitation. We acknowledge financial support from the Spanish CICYT projects, FPA2000-0956, PB98-0782 and BFM2000-1326. This work was supported in part by the Director, Office of Science, Office of High Energy and Nuclear Physics of the U.S. Department of Energy under Contract DE-AC03-76SF00098. J.R.P. acknowledges support from the CICYT-INFN collaboration grant 003P 640.15. A.D. acknowledges support from the Universidad Complutense del Amo Program. Eqns.(4)-(7) and Fig.4 are reprinted from "Finite-temperature pion scattering to one loop in chiral perturbation theory" [18] with permission from Elsevier Science.

REFERENCES

1. HELIOS-3 collaboration, *Nucl. Phys.* **590**, 127c (1995); CERES (NA45) Collaboration, *Nucl. Phys.* **A661**, 23c (1999).
2. Dey, M., Eletsky, V.L. and Ioffe, B.L. *Phys. Lett.* **B252**, 620 (1990); Gale, C. and Kapusta, J.I. *Nucl. Phys.* **B357**, 65 (1991).
3. Dominguez, C.A., Loewe, M. and Rojas, J.C. *Z. Phys.* **C 59**, 63 (1993).
4. Pisarski, R.D. *Phys. Rev.* **D 52**, 3773 (1995); *Nucl. Phys.* **A 590**, 553c (1995).
5. Haglin, K. *Nucl. Phys.* **A584**, 719 (1995).
6. Li, G.Q., Ko, C.M. and Brown, G.E., *Phys. Rev. Lett.* **75**, 4007 (1995); Ko, C.M., Li, G.Q, Brown, G.E. and Sorge, H., *Nucl. Phys.* **A610**, 342c (1996).
7. Song, C. and Koch, V. *Phys. Rev.* **C54** 3218 (1996).
8. Eletsky, V.L., Belkacem, M., Ellis, P.J. and Kapusta, J.I. *Phys. Rev.* **C64**, 035202 (2001).
9. He, Y.B., Hüfner, J., Klevansky, S.P. and Rehberg, P. *Nucl. Phys.* **A630**, 719 (1998).
10. Weinberg, S. *Physica* **A96**, 327 (1979).
11. Gasser, J. and Leutwyler, H. *Ann.Phys.(N.Y.)* **158**, 142 (1984).
12. Gasser, J. and Leutwyler, H., *Nucl. Phys.* **B250**, 465 (1985).
13. Meissner, U.G., *Rep.Prog.Phys.* **56**, 903 (1993); Pich, A., *Rep.Prog.Phys.* **58**, 563 (1995); Dobado, A., Gómez Nicola, A., Maroto, A.L. and Peláez, J.R., *Effective Lagrangians for the Standard Model*, Springer-Verlag, Berlin, 1997;

14. Truong, T.N. *Phys. Rev. Lett.* **61**, 2526 (1988); *Phys. Rev. Lett.* **67**, 2260 (1991); Dobado, A., Herrero, M.J. and Truong, T.N., *Phys. Lett.* **B235**, 134 (1990); Dobado, A. and Peláez, J.R. *Phys. Rev.* **D47**, 4883 (1993); *Phys. Rev.* **D56**, 3057 (1997).
15. Oller, J.A. and Oset, E., *Nucl. Phys.* **A620**, 438 (1997); Nieves, J. and Ruiz Arriola, E., *Phys. Lett.* **B455**, 30 (1999); *Nucl. Phys.* **A679**, 57 (2000); *Phys. Rev.* **D63**, 076001 (2001).
16. Oller, J.A., Oset, E. and Peláez, J.R., *Phys. Rev. Lett.* **80**, 3452 (1998); *Phys. Rev.* **D59**, 074001 (1999); **60**, 099906(E) (1999). Guerrero, F. and Oller, J.A., *Nucl. Phys.* **B537**, 459 (1999); *Nucl. Phys.* **B602** 641 (E).
17. Gómez Nicola, A. and Peláez, J.R., *Phys. Rev.* **D65**, 054009 (2002).
18. Gómez Nicola, A., LLanes-Estrada, F.J. and Peláez, J.R., *Phys. Lett.* **B550**, 55-64 (2002).
19. Dobado, A., Gómez Nicola, A., LLanes-Estrada, F.J. and Peláez, J.R. *Phys. Rev.* **C66**, 055201 (2002).
20. Gasser, J. and Leutwyler, H., *Phys. Lett.* **B184**, 83 (1987).
21. Gerber, P. and Leutwyler, H., *Nucl. Phys.* **B321**, 387 (1989).
22. Pisarski, R.D. and Tytgat, M. *Phys. Rev.* **D54**, 2989 (1996); Toublan, D. *Phys. Rev.* **D56**, 5629 (1997); Martinez Resco, J.M. and Valle Basagoiti, M.A., *Phys. Rev.* **D58**, 097901 (1998)
23. Goity, J.L. and Leutwyler, H., *Phys. Lett.* **B228**, 517 (1989).
24. Schenk, A., *Phys. Rev.* **D47**, 5138 (1993).
25. Gómez Nicola, A., *Phys. Rev.* **D 64**, 016011 (2001); Gómez Nicola, A. and Galan-González, V., *Phys. Lett.* **B449**, 288 (1999).
26. Pisarski, R.D. and Wilczek, F., *Phys. Rev.* **D29**, 338 (1984); Rajagopal, K. and Wilczek, F., *Nucl. Phys.* **B399**, 395 (1993); Bochkarev, A. and Kapusta, J.I, *Phys. Rev.* **D54**, 4066 (1996); Amelino-Camelia, G., *Phys. Lett.* **B407**, 268 (1997); Petropoulos, N., *J.Phys.***G25**, 2225 (1999); Nemoto, Y., Naito, K. and Oka,M., *Eur.Phys.J.***A9**, 245 (2000); Ayala, A. and Sahu, S. *Phys. Rev.* **D 62**, 056007 (2000).
27. Chiku, S. and Hatsuda, T., *Phys. Rev.* **D57**, R6 (1998), *Phys. Rev.* **D58**, 076001 (1998).
28. Le Bellac, M., *Thermal Field Theory*, Cambridge University Press, London, 1996.
29. Dobado, A. and Peláez, J.R., *Phys. Rev.* **D59**, 034004 (1999).
30. Peláez, J.R., *Phys. Rev.* **D66**, 096007 (2002).
31. Dashen, R., Ma, S., Bernstein, H.J., *Phys. Rev.* **187**, 345 (1969).
32. Karsch, F., Peikert, A. and Laermann, E., *Nucl. Phys.* **B605**, 579 (2001).
33. Quack, E., Zhuang, P., Kalinovsky,Y., Klevansky, S.P. and Hufner,J., *Phys. Lett.* **B348**, 1 (1995).
34. Kaiser, N., *Phys. Rev.* **C59**, 2945 (1999).
35. Baier, R. and Niégawa, A. *Phys. Rev.* **D49**, 4107 (1994).
36. Kobes,R. *Phys. Rev.* **D42**, 562 (1990); **D43**, 1269 (1991).
37. Weldon, H.A. *Phys. Rev.* **D28**, 2007 (1983); *Ann.Phys. (N.Y.)* **214**, 152 (1992).
38. Donoghue, J.F., Ramirez, C. and Valencia, G., *Phys. Rev.* **D39**, 1947 (1989); Ecker, G., Gasser, J., Pich. A. and de Rafael, E., *Nucl. Phys.* **B321**, 311 (1989).
39. Yokokawa, K., Hatsuda, T., Hayashigaki, A. and Kunihiro,T., *Phys. Rev.* **C66**, 022201 (2002).
40. Broniowski, W., Florkowski, W. and Hiller, B., *Nucl. Phys.* **A696** 870 (2001).

Theta Term in QCD sum rules at Finite Temperature and the Neutron Electric Dipole Moment

Mohamed Chabab

LPHEA, Physics Department, Faculty of Science-Semlalia, P.O. Box 2390, Marrakesh, Morocco

Abstract. By using thermal QCD sum rules to investigate the $\bar{\theta}$ induced neutron electric dipole moment d_n, we have examined the behaviour of broken CP symmetry at finite temperature. We find that, below the critical temperature, the ratio $\mid \frac{d_n}{\bar{\theta}} \mid$ slightly decreases but survives at temperature effects, implying T nonrestoration of CP-invariance [1].

INTRODUCTION

To establish connection between particle physics and cosmology, it is essential to study the behaviour of symmetry breaking in the early universe, i.e. at high temperature. The CP symmetry is, without doubt, one of the most fundamental symmetries in nature. It is intimately related to theories of interactions between elementary particles and represents a cornerstone in constructing grand unified and supersymmetric models. However, its breaking still carries a cloud of mystery in particle physics and cosmology since it is necessary to explain baryogenesis and is required by theories with domain walls.

According to the CPT theorem, CP violation implies T violation. The latter is tested through the measurement of the neutron electric dipole moment (NEDM) d_n. The upper experimental limit gives confidence that the NEDM can be another manifestation of CP breaking. In the Standard Model, CP violation is originated from two sources: the first source, which appears in the electroweak sector, is parametrized by a single phase in the Cabbibo-Kobayashi-Maskawa (CKM) quark mixing matrix [2]. The other source is due to the the so called θ-term of QCD. In fact, the the QCD effective lagrangian contains an additional CP-odd four dimensional operator embedded in the following topological term:

$$L_\theta = \theta \frac{\alpha_s}{8\pi} G_{\mu\nu} \tilde{G}^{\mu\nu}, \tag{1}$$

where $G_{\mu\nu}$ is the gluonic field strength, $\tilde{G}^{\mu\nu}$ is its dual and α_s is the strong coupling constant. The $G_{\mu\nu}\tilde{G}^{\mu\nu}$ quantity is a total derivative, consequently it contribute to the physical observables only through non perturbative effects. The NEDM is related to the $\bar{\theta}$-angle by the following relation:

$$d_n \sim \frac{e}{M_n} \left(\frac{m_q}{M_n}\right) \bar{\theta} \sim \left\{ \begin{array}{ll} 2.7 \times 10^{-16}\bar{\theta} & [3] \\ 5.2 \times 10^{-16}\bar{\theta} & [4] \end{array} \right. \tag{2}$$

CP660, *Hadron Physics: Effective Theories of Low Energy QCD*, edited by A. H. Blin et al.
© 2003 American Institute of Physics 0-7354-0120-9/03/$20.00

and consequently, according to the experimental measurements $d_n < 1.1 \times 10^{-25} ecm$ [5], the $\bar{\theta}$ parameter must be less than 2×10^{-10} [6]. The well known strong CP problem consists in explaining the smallness of $\bar{\theta}$. In this regard, several scenarios were suggested. The most popular one is the Peccei and Quinn [7], in which $\bar{\theta}]$ is identified to a very light pseudo goldstone boson called the axion. This particle arises from the spontaneous breakdown of a global $U_A(1)$ symmetry and may well be important to explain the puzzle of dark matter providing a peace of information on the missing mass of the universe [8].

Our aim in this work is to study the thermal behaviour of the CP symmetry breaking and the temperature effects on its restoration. This is motivated by the possibility to restore some broken symmetries by increasing the temperature.

This paper is organized as follows: Section 2 is devoted to the calculations of the NEDM induced by the $\bar{\theta}$ using QCD sum rules. In section 3, we show how one introduces temperature in QCD sum rules calculations. The last section is devoted to a discussion and qualitative analysis of the thermal effects on the CP symmetry.

NEDM FROM QCD SUM RULES

Since the NEDM has essaentially a non perturbative nature, one should any how take into account some effects which escape the perturbative treatement. One such approach, dealing with the strong strong coupling regime and based on first principales of the theory is QCD sum rules. Such approach has been applied successfully to the investigation of hadronic properties at low energies, particularly to certain baryonic magnetic form factor. In order to derive the NEDM through through this approach [9, 10], we consider a lagrangian containing the following P and CP violating operators:

$$L_{P,CP} = -\theta_q m_* \sum_f \bar{q}_f i \gamma_5 q_f + \theta \frac{\alpha_s}{8\pi} G_{\mu\nu} \tilde{G}^{\mu\nu}. \tag{3}$$

θ_q and θ are respectively two angles coming from the chiral and the topological terms and m_* is the quark reduced mass given by $m_* = \frac{m_u m_d}{m_u + m_d}$. Then, as usual, we start from the two points correlation function in QCD background with a nonvanishing θ and in the presence of a constant external electomagnetic field $F^{\mu\nu}$ [10]:

$$\Pi(q^2) = i \int d^4x e^{iqx} < 0|T\{\eta(x)\bar{\eta}(0)\}|0 >_{\theta,F}. \tag{4}$$

where $\eta(x)$ is the neutron interpolating current [12]:

$$\eta = 2\varepsilon_{abc}\{(d_a^T C \gamma_5 u_b)d_c + \beta(d_a^T C u_b)\gamma_5 d_c\}, \tag{5}$$

and β is a mixing parameter. To select the appropriate Lorentz stucture, $\Pi(q^2)$ is expanded in terms of the electromagnetic charge as:

$$\Pi(q^2) = \Pi^{(0)}(q^2) + e\Pi^{(1)}(q^2, F^{\mu\nu}) + O(e^2). \tag{6}$$

The first term $\Pi^{(0)}(q^2)$ is the nucleon propagator which includes only the CP-even parameters, while the second term $\Pi^{(1)}(q^2, F^{\mu\nu})$ is the polarization tensor which may be expanded through Wilson OPE as: $\sum C_n < 0|\bar{q}\Gamma q|0 >_{\theta,F}$, where Γ is an arbitrary Lorentz structure and C_n are the Wilson coefficient functions calculable in perturbation theory [14, 13]. From this expansion, we keep only the CP-odd contribution part. The electromagnetic dependence of these matrix elements is determined in terms of the magnetic susceptibilities κ, χ and ξ, defined as [13]:

$$< 0|\bar{q}\sigma^{\mu\nu}q|0 >_F = \chi e_q F^{\mu\nu} < 0|\bar{q}q|0 > \tag{7}$$

$$g < 0|\bar{q}G^{\mu\nu}q|0 >_F = \kappa e_q F^{\mu\nu} < 0|\bar{q}q|0 > \tag{8}$$

$$2g < 0|\bar{q}\tilde{G}^{\mu\nu}q|0 >_F = \xi e_q F^{\mu\nu} < 0|\bar{q}q|0 > \tag{9}$$

Besides, by considering the anomalous axial current, one obtains the following θ dependence of $< 0|\bar{q}\Gamma q|0 >_\theta$ matrix elements [9]:

$$m_q < 0|\bar{q}\Gamma q|0 >_\theta = im_* \theta < 0|\bar{q}\Gamma q|0 > +O(m_q^2) \tag{10}$$

where m_q and m_* are respectively the quark and reduced masses. $O(m_q^2)$ connection is negligible since $m_\eta >> m_\pi$.

Putting altogether the above ingredients and after a straightforward calculation [10], the following expression of $\Pi^{(1)}(q^2, F^{\mu\nu})$ for the neutron is derived:

$$\Pi(-q^2) = -\frac{\bar{\theta}m_*}{64\pi^2} < 0|\bar{q}q|0 > \{\tilde{F}\sigma, \hat{q}\}[\chi(\beta+1)^2(4e_d - e_u)\ln(\frac{\Lambda^2}{-q^2})$$

$$-4(\beta-1)^2 e_d(1 + \frac{1}{4}(2\kappa+\xi))(\ln(\frac{-q^2}{\mu_{IR}^2}) - 1)\frac{1}{-q^2}$$

$$-\frac{\xi}{2}((4\beta^2 - 4\beta + 2)e_d + (3\beta^2 + 2\beta + 1)e_u)\frac{1}{-q^2}...], \tag{11}$$

where $\bar{\theta} = \theta + \theta_q$ is the physical phase and $\hat{q} = q_\mu \gamma^\mu$.

The QCD expression (2.7) is confronted to the phenomenological parametrisation $\Pi^{Phen}(-q^2)$ written in terms of the Neutron hadronic properties. The latter is given by:

$$\Pi^{Phen}(-q^2) = \{\tilde{F}\sigma, \hat{q}\}(\frac{\lambda^2 d_n m_n}{(q^2 - m_n^2)^2} + \frac{A}{(q^2 - m_n^2)} + ...), \tag{12}$$

where m_n is the neutron mass, e_q is the quark charge. A and λ^2, which originate from the phenomenological side of the sum rule, represent respectively a constant of dimension 2 and the neutron coupling constant to the interpolating current $\eta(x)$. This coupling is defined via a spinor v as $< 0|\eta(x)|n >= \lambda v e^{\alpha\gamma_5}$.

THERMAL NEDM SUM RULES

The introduction of finite temperature effects may provide more precision to the phenomenological values of hadronic observables. Within the framework of QCD sum rules, the Temperature evolution of the correlation functions manifests itself in the thermal average of the Wilson operator expansion[15, 18]. Hence, at relatively low temperature, the system can be regarded as a non interacting gas of bosons. In this approximatyion, the Thermal dependance of the vacuum condensates can be written as :

$$< O^i >_T = < O^i > + \int \frac{d^3 p}{2\varepsilon (2\pi)^3} < \pi(p)|O^i|\pi(p) > n_B(\frac{\varepsilon}{T}) \qquad (13)$$

where $\varepsilon = \sqrt{p^2 + m_\pi^2}$, $n_B = \frac{1}{e^x - 1}$ is the Bose-Einstein distribution and $< O^i >$ is the standard vacuum condensate (i.e. at T=0). In this approximation, we only kept the pion contributions, since in the low temperature region, the effects of heavier resonances ($\Gamma = K, \eta, ..etc$) are dumped by their distibution functions $\sim e^{\frac{-m_\Gamma}{T}}$[17]. To compute the pion matrix elements, we apply the soft pion theorem given by:

$$< \pi(p)|O^i|\pi(p) > = -\frac{1}{f_\pi^2} < 0|[Q_5^a, [Q_5^a, O^i]]|0 > + O(\frac{m_\pi^2}{\Lambda^2}), \qquad (14)$$

where Λ is a hadron scale and Q_5^a is the isovector axial charge defined by:

$$Q_5^a = \int d^3 x \bar{q}(x) \gamma_0 \gamma_5 \frac{\tau^a}{2} q(x). \qquad (15)$$

Direct application of the above formula to the quark and gluon condensates shows the following features[16, 17]:
(i) Only $< \bar{q}q >$ is sensitive to temperature. Its behaviour at finite T is given by:

$$< \bar{q}q >_T \simeq (1 - \frac{\varphi(T)}{8}) < \bar{q}q >, \qquad (16)$$

where $\varphi(T) = \frac{T^2}{f_\pi^2} B(\frac{m_\pi}{T})$ with $B(z) = \frac{6}{\pi^2} \int_z^\infty dy \frac{\sqrt{y^2 - z^2}}{e^y - 1}$ and f_π is the pion decay constant ($f_\pi \simeq 93 MeV$). The variation with temperature of the quark condensate $< \bar{q}q >_T$ results in two different asymptotic behaviours, namely:

$$< \bar{q}q >_T \simeq (1 - \frac{T^2}{8 f_\pi^2}) < \bar{q}q > \quad for \quad \frac{m_\pi}{T} \ll 1 \qquad (17)$$

$$< \bar{q}q >_T \simeq (1 - \sqrt{\frac{\pi m_\pi}{2T}} \frac{T^2}{8 f_\pi^2} e^{\frac{-m_\pi}{T}}) < \bar{q}q > \quad for \quad \frac{m_\pi}{T} \gg 1 \qquad (18)$$

(ii) The gluon condensate is nearly constant at low temperature and a T dependence occurs only at order T^8.

The determination of the ratio $\frac{d_n}{\theta}$ sum rules at non zero temperature is now easily performed by applying Borel operator to both parametrisation of the Neutron correlation function shown in Eqs. (2.7) and (2.8). Then finite temperature effects are introduced via the procedure discussed above. Finally, by invoking the quark-hadron duality, we deduce the final sum rules of the $\bar{\theta}$ induced NEDM at finite temperature:

$$\frac{d_n}{\bar{\theta}}(T) = -\frac{M^2 m_*}{16\pi^2} \frac{1}{\lambda_n^2(T)M_n(T)} \left(1 - \frac{\varphi(T)}{8}\right) < \bar{q}q > \left[4\chi(4e_u - e_d) - \frac{\xi}{2M^2}(4e_u + 8e_d)\right] e^{\frac{M_{\bar{n}}^2}{M^2}},$$

(19)

where M represents the Borel parameter. The single pole contribution entering the sum rules via the constant A has been neglected, as suggested in [9].

The value of β has been set to 1 in (3.5). This choice is more appropriate for us since it suppresses the infrared divergences. The T- evolution of the coupling constant and the mass of the neutron were determined from the thermal nucleon sum rules [17].

Within the dilute pion gas approximation, Eletsky has shown that the contribution induced the pion-nucleon scattering has to be considered [19]. It enters the nucleon sum rules through the coupling constant $g_{\pi NN}$, whose values lie within the range 13.5-14.3 [20].

Numerical analysis is performed with the following input parameters: the Borel mass has been chosen within the values $M^2 = 0.55 - 0.7 GeV^2$ which correspond to the optimal range (Borel window) in the $\frac{d_n}{\bar{\theta}}$ sum rule at $T = 0$ [10]. For the χ and ξ susceptibilities we take $\chi = -5.7 \pm 0.6 GeV^{-2}$ [21] and $\xi = -0.74 \pm 0.2$ [22]. As to the vacuum condensates appearing in (3.5), we use their standard values [11].

ANALYSIS AND CONCLUSION

We have established the relation between the NEDM and $\bar{\theta}$ angle at non zero temperature from QCD sum rules. We find that the behaviour of the ratio $\frac{d_n}{\bar{\theta}}$ is connected to the thermal evolution of the pion parameters f_π, m_π and of $g_{\pi NN}$.

By analysing the ratio as a function of T in the region of validity of thermal sum-rules $[0, T_c]$, we learn that $|\frac{d_n}{\bar{\theta}}|$ decreases smoothly with T (about 16% variation for temperature values up to 200 MeV) but survives at finite temperature. This means that either the NEDM value decreases or $\bar{\theta}$ increases. Consequently, for a fixed value of $\bar{\theta}$ the NEDM decreases but it does not exhibit any critical behaviour. Furthermore, if we start from a non vanishing $\bar{\theta}$ value at $T = 0$, it is not possible to remove it at finite temperature. We also note that $|\frac{d_n}{\bar{\theta}}|$ grows as M^2 or χ susceptibility increases. It also grows with quark condensate rising. However this ratio is insensitive to both the ξ susceptibility and the coupling constant $g_{\pi NN}$. We notice that for higher temperatures, the analysis of $|\frac{d_n}{\bar{\theta}}| = f(\frac{T}{T_c})$ exhibits a brutal increase justified by the fact that for temperatures beyond the critical value T_c, at which the chiral symmetry is restored, the constants f_π and $g_{\pi NN}$ become zero and consequently from Eq(3.5) the ratio $\frac{d_n}{\bar{\theta}}$ behaves as a non vanishing constant. The large difference between the values of the ratio for $T < T_c$ and $T > T_c$ may be a consequence of the fact that other contributions to the the spectral function have ben neglected, like the scattering process $N + \pi \to \Delta$. These contributions which are of the order T^4, are negligible in the low temperature region but become substantial for $T \geq T_c$. Moreover, this difference may also originate from the use of soft pion approximation which is valid essentially for low T ($T < T_c$). Therefore it is clear from this qualitative analysis, which is based on the soft pion approximation, that temperature

174

does not play a fundamental role in the suppression of the undesired θ-term and hence the broken CP symmetry is not restored [24]. Indeed, some exact symmetries can be broken by increasing temperature [23, 24]. The symmetry non restoration phenomenon, which means that a broken symmetry at T=0 remains broken even at high temperature, is essential for discrete symmetries, CP symmetry in particular. Indeed, the symmetry non restoration allows us to avoid wall domains inherited after the phase transition [25] and to explain the baryogenesis phenomenon in cosmology [26]. Furthermore, it can be very useful for solving the monopole problem in grand unified theories [27].

AKNOWLEDGMENTS
The author is deeply grateful to Prof. Brigitte Hiller and to the organization Committe for the invitation to the Second International Conference on Hadron Physics. He also wishes to thank Prof. Joao Providencia for his hospitality during the visit to CPT at Coimbra.
This work is supported by the convention de cooperation between CNRST/ICCTI 681.02/CNR.

REFERENCES

1. Chabab, M., El Biaze, N., and Markazi, R., J. Phys. **G27**, 2275 (2001).
2. Cabibbo, N., Phys. Rev. Lett. **10**, 531 (1963);
   Kobayashi, M and Maskawa, T., Prog. Theor. Phys. **49**, 652 (1973).
3. Baluni, V, Phys. Rev **D19**, 2227 (1979).
4. Crewther, R., Di Vecchia, P., Veneziano, G. and Witten, E., Phys. Letters **B88**, 123 (1979).
5. Barnett, R. M., and al, Phys. Rev. **D54**, 1 (1996).
6. Peccei, R. D., hep-ph/9807516.
7. Peccei, R. D., and Quinn, H. R., Phys. Rev. **D16**, 1791 (1977).
8. Lazarides, G., and Shafi, Q., "Monopoles, Axions and Intermediate Mass Dark Matter", hep-ph/0006202.
9. Pospelov, M. and Ritz, A., Nucl. Phys. **B558**, 243 (1999).
10. Pospelov, M., and Ritz, A., Phys. Rev. Letters **83**, 2526 (1999).
11. Shifman, M. A., Vainshtein, A. I., and Zakharov, V. I., Nucl. Phys. **B147**, 385 (1979).
12. Ioffe, B. L., Nucl. Phys. **B188**, 317 (1981);
    Chung, Y. and al, Phys. Lett.**B102**, 175 (1981); Nucl. Phys. **B197**, 55 (1982)
13. Ioffe, B. L., and Smilga, A. V., Nucl. Phys. **B232**, 109 (1984).
14. Shifman, M.A., Vainshtein, A. I., and Zakharov, V. I., Nucl. Phys. **B166**, 493 (1980).
15. Bochkarev, A. I., and Shaposhnikov, M. E., Nucl. Phys. **B268**, 220 (1986).
16. Gasser, J., and Leutwyler, H., Phys. Letters **B184**, 83 (1987);
    Leutwyler, H., in *QCD 20 years later*, edited by . P.M. Zerwas and H.A. Kastrup, World scientific Proceedings, Singapore, 1992. 1993).
17. Adami, C., and Zahed, I., Phys.Rev **D45**, 4312 (1992);
    Hatsuda, T., Koike, Y., and Lee, S. H., Nucl. Phys. **B394**, 221 (1993).
18. Mallik, S., and Mukherjee, K., Phys. Rev. **D58**, 096011 (1998).
19. Eletsky, V. L., Phys. Lett. **B245**, 229 (1990); Phys. Letters **B352**, 440 (1995) .
20. Blomgten, J., in *A critical issue in the determination of the pion nucleon decay constant*, Phys. Scripta **T87**, 53 (2000).
21. Belyaev, V. M., and Kogan, Y. I., Sov. J. Nucl. Phys. **40**, 659 (1984).
22. Kogan, I. I., and Wyler, D., Phys. Letters **B274**, 100 (1992).
23. Weinberg, S., Phys. Rev **D9**, 3357 (1974).

24. Mohapatra, R. N., and Senjanovic, G., Phys. Rev. **D20**, 3390 (1979) ;
    Dvali, G., Melfo, A., and Senjanovic, G., Phys.Rev. **D54**, 7857 (1996).
25. Zeldovich, Y. B., Kobzarev, I. Y., and Okun, L., JETP. **40**, 1 (1974);
    Kibble, T. W., J. Phys. **A9**, 1987 (1976); Phys. Rep. **67**, 183 (1980).
26. Sakharov, A., JETP Letters **5**, 24 (1967).
27. Dvali, G., Melfo, A., and Senjanovic, G., Phys. Rev. Letters **75**, 4559 (1995).

Thermal model for RHIC, part I: particle ratios and spectra

Wojciech Florkowski and Wojciech Broniowski

The H. Niewodniczański Institute of Nuclear Physics
ul. Radzikowskiego 152, 31-342 Kraków, Poland

Abstract. A simple thermal model with single freeze-out and flow is used to analyze the ratios of hadron yields and the hadron transverse-mass spectra measured in $\sqrt{s_{NN}}$ = 130 GeV Au+Au collisions at RHIC. An overall very good agreement between the model predictions and the data is achieved for all measured hadron species including hyperons.

INTRODUCTION

The main features of the soft hadron production at RHIC, such as the ratios of hadron abundances, the transverse-mass spectra, the elliptic flow, or the HBT radii may be efficiently understood in the framework of a statistical model which combines the standard thermal analysis of the hadron ratios with a suitably parameterized expansion of matter at freeze-out [1, 2, 3]. In the present paper we outline the main assumptions of the model and concentrate on the discussion of the particle ratios [4, 5] and the hadron transverse-mass spectra [6, 7, 8, 9]. The subsequent paper in these Proceedings [10] contains the analysis of the elliptic flow and the pion HBT correlation radii.

The main ingredients of our approach are as follows: i) the chemical freeze-out and the thermal (kinetic) freeze-out occur simultaneously, ii) all hadronic resonances are included in the calculation of both the hadron yields and the spectra, and iii) the freeze-out hypersurface and the flow at freeze-out are defined by the simple expressions inspired by the Bjorken model [11]. Below we discuss in more detail these three points.

Freeze-out

Our approach includes a complete set of hadronic resonances in both the calculation of the hadron abundances and the calculation of the hadron spectra. Then, it turns out that the distinction between the two freeze-outs [12] is not necessary. Our analysis showed [4] that the decays of the resonances, which are initially present in a heat bath at the temperature of 165 MeV, effectively lower the inverse slope parameters of the spectra by about 30-40 MeV. This is just the desired effect which explains the typical difference between T_{chem} and T_{kin} (i.e., between the temperature at the chemical freeze-out and the thermal freeze-out, respectively). As a consequence, we have found, at least for the RHIC data, that no extra elastic rescattering is required in order to describe

CP660, *Hadron Physics: Effective Theories of Low Energy QCD*, edited by A. H. Blin et al.

simultaneously both the ratios and the spectra. In other words, we assume one universal freeze-out taking place at [1]

$$T_{\text{chem}} = T_{\text{kin}} \equiv T. \tag{1}$$

Recently, experimental hints have been found in favor of our assumption (1). A successful reconstruction of the $K^*(892)^0$ states by the STAR experiment [13], together with the very good agreement between the measured yield of $K^*(892)^0$ and the prediction of the thermal model [3, 14] suggests a picture with the short expansion time between the two freeze-outs. Such a picture is natural if the production of particles (hadronization) occurs in such conditions that neither elastic or inelastic processes are effective. An example here is provided by the sudden hadronization model of Ref. [15].

Resonances

For thermal systems the contribution from high-lying (heavy) states is damped by the exponential factor, $\exp(-m_\perp/T)$. This fact, at the first sight, suggests that most of the resonances present in a hadron gas at a moderate temperature may be neglected. However, although the high-lying states are suppressed, their number increases according to the Hagedorn hypothesis [16, 17, 18, 19], such that their net effect turns out to be important. Indeed, hadronic resonances have been included in numerous applications of the statistical models used in the studies of the ratios of hadron multiplicities [14, 20, 21, 22, 23] and the effects connected with their decays are essential for the successful description of the data. We note that only a quarter of the observed pions at RHIC comes from the "primordial" pions present at freeze-out, and the remaining three quarters are produced from the decays of resonances.

In our approach, the same number of the resonances is included in the calculation of the hadron ratios and in the calculation of the hadron spectra. In this way, our theoretical spectra have always the correct relative normalization. Moreover, we have worked out semi-analytic formulas for the treatment of the resonance decays [4]. This allows us to sum up exactly many small contributions, especially, those appearing in the sequential decays. All two- and three-body decays are taken into account with the branching ratios taken from the tables. In the case of the three-body decays, the matrix elements are approximated by a constant, hence only the phase-space effect is included.

Expansion

In order to calculate the spectra we need to specify the expansion of the matter at freeze-out; clearly, the Doppler effect due to flow modifies the spectra and must be properly included. Our choice of the freeze-out hypersurface and the four-velocity at freeze-out has been made in the spirit of Refs. [11, 24, 25, 26, 27, 28, 29, 30] and is defined by the two conditions:

$$\tau = \sqrt{t^2 - r_x^2 - r_y^2 - r_z^2} = \text{const.}, \tag{2}$$

and

$$u^\mu = \frac{x^\mu}{\tau} = \frac{t}{\tau}\left(1, \frac{r_x}{t}, \frac{r_y}{t}, \frac{r_z}{t}\right). \tag{3}$$

The constant in Eq. (2) will be later denoted simply by τ. In order to make the transverse size,

$$\rho = \sqrt{r_x^2 + r_y^2}, \tag{4}$$

finite, we impose the condition

$$\rho < \rho_{max}. \tag{5}$$

We note that the four-velocity (3) defining the hydrodynamic flow at freeze-out is proportional to the coordinate (Hubble-like expansion). This form of the flow and the fact the the coordinates t and r_z are not limited and appear in the boost-invariant combination in (2) mean that our model is boost-invariant. In practical calculations it is convenient to introduce the following parameterization [28]:

$$\begin{aligned} t &= \tau \cosh\alpha_\| \cosh\alpha_\perp, \quad r_z = \tau \sinh\alpha_\| \cosh\alpha_\perp, \\ r_x &= \tau \sinh\alpha_\perp \cos\phi, \quad r_y = \tau \sinh\alpha_\perp \sin\phi, \end{aligned} \tag{6}$$

where $\alpha_\|$ is the rapidity of the fluid element, $v_z = r_z/t = \tanh\alpha_\|$, and $\alpha_\perp$ describes the transverse size, $\rho = \tau \sinh\alpha_\perp$. The transverse velocity is $v_\rho = \tanh\alpha_\perp/\cosh\alpha_\|$. The element of the hypersurface is defined as

$$d\Sigma_\mu = \varepsilon_{\mu\alpha\beta\gamma}\frac{\partial x^\alpha}{\partial\alpha_\|}\frac{\partial x^\beta}{\partial\alpha_\perp}\frac{\partial x^\gamma}{\partial\phi}d\alpha_\|d\alpha_\perp d\phi, \tag{7}$$

where $x^0 = t$, $x^1 = r_x$, $x^2 = r_y$, $x^3 = r_z$, and $\varepsilon_{\mu\alpha\beta\gamma}$ is the Levi-Civita tensor. A straightforward calculation yields

$$d\Sigma^\mu(x) = u^\mu(x)\,\tau^3\sinh(\alpha_\perp)\cosh(\alpha_\perp)\,d\alpha_\perp d\alpha_\|d\phi. \tag{8}$$

Equation (8) shows that the four-vectors $d\Sigma^\mu$ and u^μ are parallel. In this case the spectra may be obtained from the expression analogous to the Cooper-Frye [31, 32] formula

$$\frac{dN}{d^2p_\perp dy} = \int p^\mu d\Sigma_\mu\, f(p\cdot u), \tag{9}$$

but with the distribution f which has collected the products of resonance decays (for details see [3]). With parameterization (6) we can rewrite Eq. (9) in the form

$$\frac{dN}{2\pi m_\perp dm_\perp dy} = \tau^3\int_{-\infty}^{+\infty}d\alpha_\|\int_0^{\rho_{max}/\tau}\sinh\alpha_\perp d\left(\sinh\alpha_\perp\right)\int_0^{2\pi}d\xi\, p\cdot u f(p\cdot u), \tag{10}$$

where

$$p\cdot u = m_\perp\cosh\alpha_\|\cosh\alpha_\perp - p_\perp\cos\xi\sinh\alpha_\perp. \tag{11}$$

One can notice that the spectrum (10) is, as expected from the assumed boost invariance, independent of the rapidity y.

RATIOS OF HADRON ABUNDANCES

In the case of the boost-invariant systems, the ratios of hadron multiplicities at midrapidity, $dN/dy|_{y=0}$, are simply related to the ratios of the local densities, n_i, since

$$\left.\frac{dN_i/dy}{dN_j/dy}\right|_{y=0} = \frac{N_i}{N_j} = \frac{n_i}{n_j}. \tag{12}$$

The first part of this equality follows from the boost invariance, whereas the second part is a consequence of the factorization of the volume of the system (this point is discussed in detail in Ref. [10]). Since the firecylinder formed at RHIC is approximately boost-invariant (at least within one unit of rapidity at $y = 0$, which is sufficient for our considerations which concentrate only on the central region), the ratios at zero rapidity measured for various particles may be used to fit the thermal parameters of the model. This is an important observation indicating that the ratios are not sensitive to the particular form of expansion. Consequently, the parameters of our model can be fixed in two steps: with help of the ratios we first fix the thermodynamic parameters, and later with help of the spectra we fix the two expansion parameters, τ and $\rho_{\max}$.

The density of the ith hadron species is calculated from the ideal-gas expression

$$n_i = g_i \int d^3p \, f^{(i)}(p),$$

$$f^{(i)}(p) = \frac{1}{(2\pi)^3} \left(\exp\left[(E_i(p) - \mu_B B_i - \mu_S S_i - \mu_I I_i)/T\right] \pm 1\right)^{-1}, \tag{13}$$

where g_i is the spin degeneracy, B_i, S_i, and I_i denote the baryon number, strangeness, and the third component of isospin, and $E_i(p) = \sqrt{p^2 + m_i^2}$. The quantities μ_B, μ_S and μ_I are the chemical potentials which enforce the corresponding conservation laws. We recall that Eq. (13) is used to calculate the "primordial" densities of stable hadrons as well as of all resonances at freeze-out, which decay later on.

Initially, the temperature, T, and the baryon chemical potential, μ_B, were fitted with the χ^2 method to the 9 preliminary experimental ratios of the hadron yields (for the list of the ratios used in this fit and for more details of our analysis see Refs. [4, 5]). The μ_S and μ_I were determined with the conditions that the initial strangeness of the system is zero, and the ratio of the baryon number to the electric charge is the same as in the colliding nuclei. This procedure gave us $T = 165 \pm 7$ MeV and $\mu_B = 41 \pm 5$ MeV. In Table I we present the result of the fit with a wider set of the up-to-date available hadronic ratios. It yields $T = 168 \pm 5$ MeV and $\mu_B = 41 \pm 4$ MeV. It is interesting to observe that the newly released data are fully consistent with the thermal picture and only a small change of the thermodynamic parameters follows when the new set of the ratios is used.

A characteristic feature of our fit is that the optimal temperature is consistent with the value for the deconfinement phase transition obtained from the QCD lattice simulations: $T_c = 154 \pm 8$ MeV for three massless flavors and $T_c = 173 \pm 8$ MeV for two massless flavors [33]). This type of the behavior has been also found in other statistical calculations

TABLE 1. Optimal thermal parameters and the ratios of hadron multiplicities at zero rapidity.

	Model	Experiment
Fitted thermal parameters		
T [MeV]	168 ± 5	
μ_B [MeV]	41 ± 4	
μ_S [MeV]	10	
μ_I [MeV]	-1	
χ^2/n	0.6	
Theoretical ratios and the data		
π^-/π^+	1.02	1.00 ± 0.02 [35], $\quad 0.99 \pm 0.02$ [36]
$\bar{p}/\pi^-$	0.09	0.08 ± 0.01 [37]
K^-/K^+	0.92	0.92 ± 0.03 [38], $\quad 0.93 \pm 0.07$ [39] 0.91 ± 0.09 [35], $\quad 0.92 \pm 0.06$ [36]
K^-/π^-	0.16	0.15 ± 0.02 [40]
K_0^*/h^-	0.047	0.042 ± 0.011 [13]
$\overline{K_0^*}/h^-$	0.042	0.039 ± 0.011 [13]
$\bar{p}/p$	0.65	0.61 ± 0.07 [37], $\quad 0.64 \pm 0.08$ [39] 0.60 ± 0.07 [35], $\quad 0.61 \pm 0.06$ [36]
$\bar{\Lambda}/\Lambda$	0.69	0.71 ± 0.04 [38]
$\bar{\Xi}/\Xi$	0.77	0.83 ± 0.06 [38]
ϕ/h^-	0.020	0.021 ± 0.001 [41]
ϕ/K^-	0.15	0.13 ± 0.03 [41]
Λ/p	0.47	0.49 ± 0.03 [42, 43]
Ω^-/h^-	0.0011	0.0012 ± 0.005 [44]
Ξ^-/π^-	0.0072	0.0088 ± 0.0020 [45]
Ω^+/Ω^-	0.86	0.95 ± 0.16 [38]

[14, 34] and is interpreted as the argument for a scenario in which the hadronic ratios are fixed just in the hadronization process.

TRANSVERSE-MASS SPECTRA

Having fixed the two independent thermodynamic parameters of the model, we can use Eq. (10) to fit the transverse-mass spectra and to fix the two remaining geometric parameters of the model. This method has been initially applied in the study of the spectra of pions, kaons and protons, which gave us, for the most central events, the following values of the size parameters:

$$\tau = 7.66 \text{ fm}, \qquad \rho_{\max} = 6.69 \text{ fm}. \qquad (14)$$

In Fig. 1 we show our results for all up-to-now available spectra at $\sqrt{s_{NN}} = 130$ GeV for the most-central collisions. In the upper part of Fig. 1 we show the spectra of pions, kaons, and antiprotons (used earlier to determine the geometric parameters) and

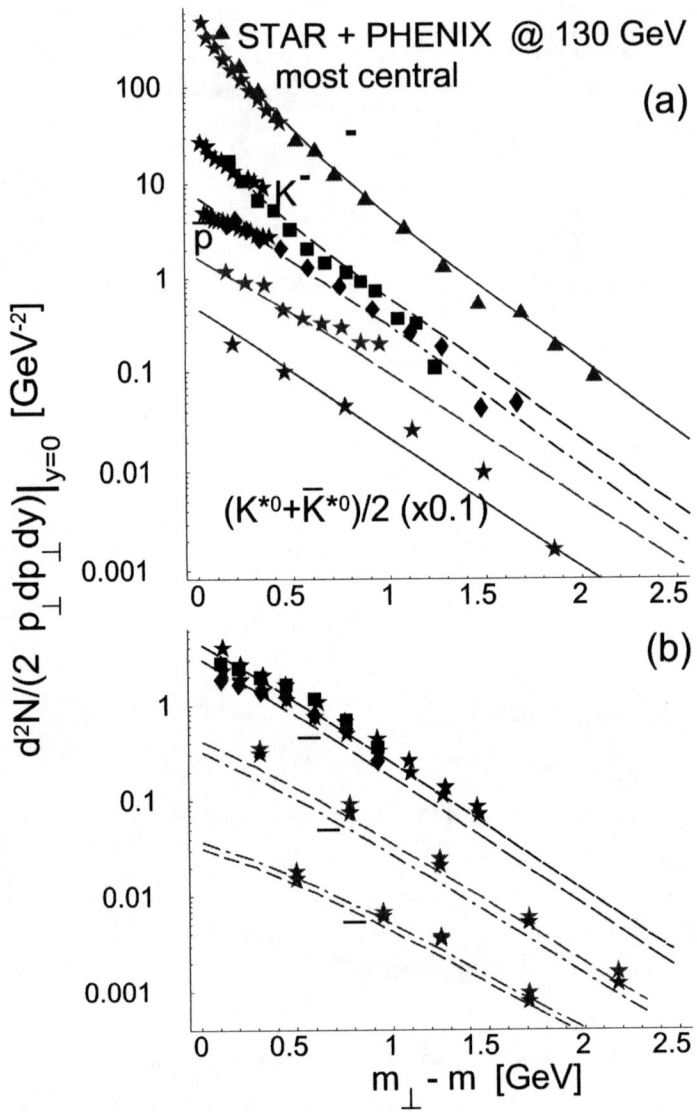

FIGURE 1. The transverse-mass spectra at midrapidity. The data from STAR are denoted by asterisks, other symbols are used to denote the PHENIX data. All spectra are for most central collisions

the predicted spectra of the ϕ and $K^*(892)^0$ mesons. The predicted spectrum of the ϕ mesons agrees very well with the measurement [41], with model curve crossing five out of the nine data points. The ϕ meson production is of a particular interest in relativistic heavy-ion collisions, since its spectrum reflects the initial temperature of the hadronic system. This is because its interaction with the hadronic environment is negligible. Moreover, it does not receive any contribution from resonance decays. Thus, the agreement of the model and the data supports the idea of one universal freeze-out. The upper part of Fig. 1 also shows the averaged spectrum of K^* resonances, with the data from Ref. [13]. Once again we observe a very good agreement between the model curve and the experimental points. As already mentioned above, the successful description of both the yield and the spectrum of $K^*(892)^0$ mesons supports the concept of the thermal description of hadron production at RHIC, and brings evidence for a very small interval between chemical and thermal freeze-outs.

In the lower part of Fig. 1 we show the predictions of the model for the spectra of hyperons. Again, in view of the fact that no extra parameters are introduced here and no refitting has been performed, the agreement is impressive. We note that the preliminary [46] data for the Ξ's used in the figure were subsequently updated, which resulted in the reduction of the data by about a factor of 2. This correction makes our agreement with the data even better. The data accumulated at lower energies at SPS showed that the slope of the Ω hyperon was much steeper than for other particles [47]. On the contrary, in the case of RHIC the model predictions for the Ω are as good as for the other hadrons. Since the Ω contains three strange quarks, it is most sensitive for modifications of the simple thermal model used here, *e.g.* the use of canonical instead of the grand-canonical ensemble. The agreement of Fig. 1 does not support the need for inclusion of these effects.

SUMMARY

The success of our model in reproducing the hadron ratios and the transverse-mass spectra indicates that the particles are indeed produced thermally. The model has only four parameters and describes the data with surprising accuracy. Since our approach uses hadronic degrees of freedom and starts at freeze-out, important theoretical questions concerning the earlier stages of the evolution cannot be addressed in this framework. We think, however, that our results constrain any more microscopic descriptions of the evolution of the matter formed in ultra-relativistic heavy-ion collisions. Further applications of the model aiming at the description of the elliptic flow and the HBT pion radii will be presented in the next contribution to these Proceedings [10].

ACKNOWLEDGMENTS

Supported in part by the Polish State Committee for Scientific Research, grant 2 P03B 09419. WB acknowledges the support of PRAXIS XXI/BCC/429/94.

REFERENCES

1. Broniowski, W., and Florkowski, W., *Phys. Rev. Lett.*, **87**, 272302 (2001).
2. Broniowski, W., and Florkowski, W., *Phys. Rev. C*, **65**, 064905 (2002).

3. Broniowski, W., Baran, A., and Florkowski, W., *Acta Phys. Pol. B*, **33**, 4235 (2002).
4. Florkowski, W., Broniowski, W., and Michalec, M., *Acta Phys. Pol. B*, **33**, 761 (2002).
5. Michalec, M., *Thermal description of particle production in ultra-relativistic heavy-ion collisions*, Ph.D. thesis, Institute of Nuclear Physics, ul. Radzikowskiego 152, 31-342 Kraków, Poland (2001), available as nucl-th/0112044.
6. Broniowski, W., and Florkowski, W., *Acta Phys. Pol. B*, **33**, 1935 (2002).
7. Broniowski, W., and Florkowski, W., "Thermal model at RHIC: particle ratios and $p_\perp$ spectra," in *Ultrarelativistic Heavy-Ion Collisions*, edited by M. Buballa, W. Nörenberg, B.-J. Schaefer, and J. Wambach (GSI, Darmstadt) Hirschegg, Austria, 2002, p. 146, hep-ph/0202059.
8. Florkowski, W., and Broniowski, W., *Acta Phys. Pol. B*, **33**, 1629 (2002).
9. Florkowski, W., and Broniowski, W., "Thermal description of transverse-momentum spectra at RHIC," in *Proceedings of Quark Matter 2002 Conference*, Nucl. Phys. A in print, nucl-th/0208061.
10. Broniowski, W., Baran, A., and Florkowski, W., "Thermal model at RHIC, part II: elliptic flow and HBT radii," following paper.
11. Bjorken, J. D., *Phys. Rev. D*, **27**, 140 (1983).
12. Heinz, U., *Nucl. Phys. A*, **661**, 140c (1999).
13. Xu, Z., STAR Collaboration, *Nucl. Phys. A*, **698**, 607c (2002).
14. Braun-Munzinger, P., Magestro, D., Redlich, K., and Stachel, J., *Phys. Lett. B*, **518**, 41 (2001).
15. Rafelski, J., and Letessier, J., *Phys. Rev. Lett.*, **85**, 4695 (2000).
16. Hagedorn, R., *Suppl. Nuovo Cim.*, **3**, 147 (1965).
17. Broniowski, W., and Florkowski, W., *Phys. Lett. B*, **490**, 223 (2000).
18. Broniowski, W., "Distinct Hagedorn temperatures from the particle spectra: a higher one for mesons, a lower one for baryons," in *Few-Quark Problems*, edited by B. Golli, M. Rosina, and S. Širca, Bled, Slovenia, 2002, p. 14, hep-ph/0008112.
19. Tounsi, A., Letessier, J., and Rafelski, J., "Hadronic matter equation of state and the hadron mass spectrum," in *Hot Hadronic Matter*, Divonne-les-Bains, France, 1994, p. 105.
20. Braun-Munzinger, P., Heppe, I., and Stachel, J., *Phys. Lett. B*, **465**, 15 (1999).
21. Gaździcki, M., and Gorenstein, M. I., *Acta Phys. Pol. B*, **30**, 2705 (1999).
22. Yen, G. D., and Gorenstein, M. I., *Phys. Rev. C*, **59**, 2788 (1999).
23. Becattini, F., Cleymans, J., Keranen, A., Suhonen, E., and Redlich, K., *Phys. Rev. C*, **64**, 024901 (2001).
24. Baym, G., Friman, B., Blaizot, J. P., Soyeur, M., and Czyż, W., *Nucl. Phys. A*, **407**, 541 (1983).
25. Milyutin, P., and Nikolaev, N. N., *Heavy Ion Phys.*, **8**, 333 (1998).
26. Siemens, P. J., and Rasmussen, J., *Phys. Rev. Lett.*, **42**, 880 (1979).
27. Schnedermann, E., Sollfrank, J., and Heinz, U., *Phys. Rev. C*, **48**, 2462 (1993).
28. Csörgő, T., and Lörstad, B., *Phys. Rev. C*, **54**, 1390 (1996).
29. Rischke, D. H., and Gyulassy, M., *Nucl. Phys. A*, **597**, 701 (1996).
30. Scheibl, R., and Heinz, U., *Phys. Rev. C*, **59**, 1585 (1999).
31. Cooper, F., and Frye, G., *Phys. Rev. D*, **10**, 186 (1974).
32. Cooper, F., Frye, G., and Schonberg, E., *Phys. Rev. D*, **11**, 192 (1975).
33. Karsch, F., *Nucl. Phys. A*, **698**, 199c (2002).
34. Becattini, F., *J. Phys. G*, **28**, 1553 (2002).
35. Back, B. B., et al., PHOBOS Collaboration, *Phys. Rev. Lett.*, **87**, 102301 (2001).
36. Bearden, I. G., BRAHMS Collaboration, *Nucl. Phys. A*, **698**, 667c (2002).
37. Harris, J., STAR Collaboration, *Nucl. Phys. A*, **698**, 64c (2002).
38. Adams, J., et al., STAR Collaboration, nucl-ex/0211024.
39. Ohnishi, H., PHENIX Collaboration, *Nucl. Phys. A*, **698**, 659c (2002).
40. Caines, H., STAR Collaboration, *Nucl. Phys. A*, **698**, 112c (2002).
41. Adler, C., et al., STAR Collaboration, *Phys. Rev. C*, **65**, 041901 (2002).
42. Adler, C., et al., STAR Collaboration, *Phys. Rev. Lett.*, **89**, 092301 (2002).
43. Adler, C., et al., STAR Collaboration, *Phys. Rev. Lett.*, **87**, 262302 (2001).
44. Suire, C., STAR Collaboration, nucl-ex/0211017.
45. Castillo, J., STAR Collaboration, nucl-ex/0210032.
46. Castillo, J., STAR Collaboration, *J. Phys. G*, **28**, 1987 (2002).
47. Antinori, C., et al., WA97 Collaboration, *J. Phys. G*, **27**, 375 (2001).

Thermal model for RHIC, part II: elliptic flow and HBT radii

Wojciech Broniowski, Anna Baran, and Wojciech Florkowski

The H. Niewodniczański Institute of Nuclear Physics, PL-31342 Cracow, Poland

Abstract. We continue the analysis of the preceding talk with a discussion of the elliptic flow and the Hanbury-Brown–Twiss pion correlation radii. It is shown that the thermal model can be extended to describe these phenomena. The description of the elliptic flow involves an appropriate deformation of the freeze-out hyper-surface and flow velocity. The obtained results agree reasonably with the data for soft (< 2 GeV) transverse momenta. For the pionic HBT correlation radii the experimental feature that $R_{out}/R_{side} \simeq 1$ is naturally obtained. The reproduction of individual R_{side} and R_{out} can be achieved with the inclusion of the excluded volume corrections, which effectively increase the radii by $\sim 30\%$.

INTRODUCTION

In the preceding talk [1], from now on referred to as (I), it has been shown that the thermal approach is successful in the description of the particle ratios and $p_{\perp}$-spectra at RHIC. Here we continue our investigation, studying azimuthal asymmetry of the spectra and the pionic Hanbury-Brown–Twiss correlation radii.

ELLIPTIC FLOW

When the nuclei collide at non-zero impact parameter, $b \neq 0$, the momentum distribution of the produced particles carries azimuthal asymmetry. In general, at mid-rapidity ($y = 0$) we may write the following Fourier decomposition in the azimuthal angle ϕ, measured from the reaction plane:

$$\frac{dN}{d^2 p_{\perp} dy}\bigg|_{y=0} = \frac{dN}{2\pi p_{\perp} dp_{\perp} dy}\bigg|_{y=0} \left(1 + 2v_2 \cos 2\phi + 2v_4 \cos 4\phi + \dots\right). \qquad (1)$$

The sines are absent due to the symmetry condition $\phi \to -\phi$, which is simply the reflexion with respect to the reaction plane, while the coefficients of cosines with odd multiples of ϕ vanish for the case of symmetric nuclei and at $y = 0$, when the symmetry $\phi \to \pi - \phi$ holds. The elliptic-flow coefficient, v_2, can therefore be computed as

$$v_2 = \frac{\int_0^{2\pi} \frac{dN}{d^2 p_{\perp} dy}\big|_{y=0} \cos 2\phi \, d\phi}{\int_0^{2\pi} \frac{dN}{d^2 p_{\perp} dy}\big|_{y=0} d\phi}. \qquad (2)$$

CP660, *Hadron Physics: Effective Theories of Low Energy QCD*, edited by A. H. Blin et al.

The value of v_2 is an important signature of the physics occurring in heavy-ion collisions. Most importantly, its non-vanishing value indicates that the production mechanism is not a simple composition of nucleon-nucleon collisions, since in that case the asymmetry of production in each such collision would average out practically to zero. Thus, interactions with other particles (rescattering, asymmetric collective flow, ...) are necessary to generate non-vanishing v_2. In hydrodynamical approaches the elliptic flow has been analyzed in many papers, see *e.g.* [2, 3, 4, 5, 6]). The coefficient v_2 depends on the impact parameter, b, on the transverse momentum $p_\perp$, as well as, obviously, on the type of the considered particle. All these dependences are measured at RHIC. The impact parameter, b, is traded for the experimentally more useful centrality parameter, c, which to a very good accuracy is given by [7]

$$c \simeq \frac{b^2}{(2R)^2}.$$ (3)

There are two basic empirical facts from RHIC which we will use in our approach. Firstly, v_2 *is positive* [8, 9, 10, 11, 12, 13, 14, 15, 16, 17], which means that the collective flow is faster in the reaction plane than out-of-plane. Secondly, the measurement [18] of the azimuthal dependence of the R_{side} pion HBT radius shows, that the shape of the system at freeze-out *is elongated out of the reaction plane*. We will now use these two facts in our choice of the parameterization of the hypersurface at freeze-out. A natural extension of Eq. (I.6) is to introduce the azimuthal shape asymmetry,

$$\begin{aligned} r_x &= \rho_{\max}\sqrt{1-\varepsilon}\cos\phi, \\ r_y &= \rho_{\max}\sqrt{1+\varepsilon}\sin\phi, \end{aligned}$$ (4)

with r_z and t kept as in the symmetric case of Eq. (I.6) of paper (I). Our convention is that r_x lies in the reaction plane, and r_y is perpendicular to the reaction plane. For positive ε this produces elongation out of the reaction plane, as seen in the experiment. The four-velocity of Eq. (I.3) is modified as follows:

$$\begin{aligned} u_x &= \frac{1}{N}r_x\sqrt{1+\delta}\cos\phi, \\ u_y &= \frac{1}{N}r_y\sqrt{1-\delta}\sin\phi, \\ u_z &= \frac{1}{N}r_z, \\ u_t &= \frac{1}{N}t. \end{aligned}$$ (5)

The normalization N is such that $u^\mu u_\mu = 1$. Positive δ means faster flow in the reaction plane, which corresponds to positive v_2. Certainly, the choice (4,5) is by no means unique, but it grasps the essential empirical features.

The modified expansion model has four geometric parameters: τ, $\rho_{\max}$, ε and δ. These parameters depend on the centrality parameter, c. Fortunately, the effect of ε and δ on the ϕ-averaged spectra, $dN/(2\pi p_\perp dp_\perp dy)|_{y=0}$, is negligible and enters at the level of

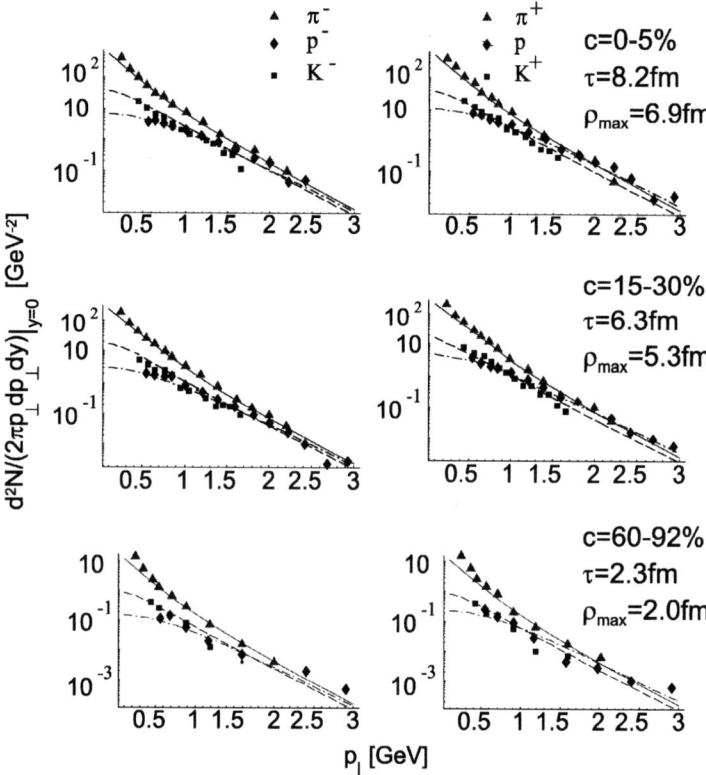

FIGURE 1. The model fit of the pion, kaon, and proton spectra to the PHENIX data for $\sqrt{s_{NN}} =$ 130 GeV [15] at three centrality bins, arranged top to bottom. Negative (positive) hadron are shown in the left (right) side. The optimum values of the size parameters τ and ρ_{max} are given for each centrality bin.

a few percent. Thus we may first fit τ and ρ_{max} to the ϕ-averaged $p_\perp$-spectra at various centrality parameters, assuming for the moment vanishing ε and δ. The result is shown in Fig. 1. We note that the fit works as good as for the most-central case presented in (I). The optimum values of parameters are collected in Table 1, where they are also compared to the minimum-bias fit and the joint fit to the most central PHENIX [15] and STAR [19] data. In fact, the qualitative dependence of τ and ρ_{max} on c is as expected: the larger c, *i.e.* the more peripheral collision, the smaller values of the size parameters. Figure 2 visualizes this dependence. In Table 1 we also list the ratio of ρ_{max}/τ, and the maximum and average values of the flow parameter, β. Interestingly, these quantities depend very weakly on c. They are defined as follows:

$$\beta_\perp^{max} = \frac{\rho_{max}}{\sqrt{\tau^2 + \rho_{max}^2}},$$

$$\langle \beta_\perp \rangle = \int_0^{\rho_{max}} r\,dr \frac{r}{\sqrt{\tau^2 + r^2}} \Big/ \int_0^{\rho_{max}} r\,dr. \tag{6}$$

187

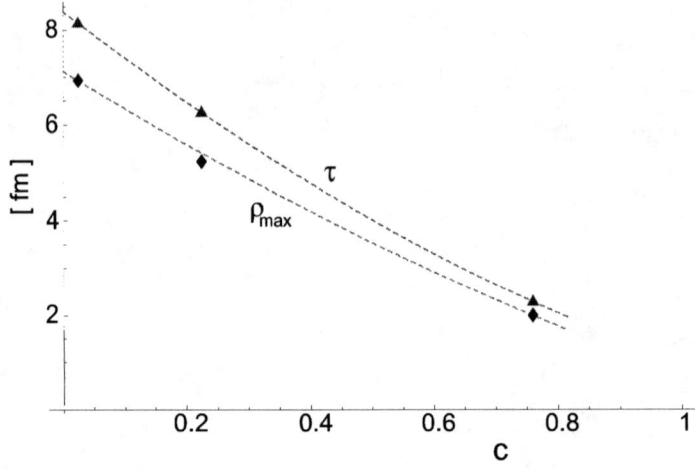

FIGURE 2. The dependence of the size parameters τ and ρ_{max} on the centrality parameter c, as fitted to the PHENIX data on ϕ-integrated $p_\perp$-spectra at $\sqrt{s_{NN}} = 130$ GeV [15].

Ideally, the dependence of the ε parameter on c should come from the measurement of R_{side} at various centralities. Then the model evaluation of this quantity would allow to fit independently $\varepsilon(c)$ to the data. We hope to be able to proceed in such a manner in the future. Unfortunately, no necessary experimental results are available at the moment. In this circumstance we take a reasonable theoretical estimate for $\varepsilon(c)$ based on Ref. [4], which leads to $\varepsilon = 0.1$, 0.21, and 0.35 in the centrality bins $0 - 15\%$, $15 - 30\%$, and $30 - 60\%$, respectively. Finally, the parameter δ is obtained by fitting the model predictions to the v_2 measurements. Our approach includes, as described in detail in (I), the decays of resonances. The calculation is straightforward and very similar to the one discussed in (I), although takes a much longer computer time due to a lower degree of symmetry. The technicalities will be presented elsewhere. The simplified results for

TABLE 1. Comparison of the size parameters obtained by fitting particle spectra at various values of th centrality parameter c. Their ratio, as well as the maximum and average value of the flow parameter, $\beta_\perp$, are also given.

	PHENIX @130GeV				PHENIX+STAR @130GeV
c [%]	min. bias	0-5	15-30	60-92	0-5/0-6
τ [fm]	5.6	8.2	6.3	2.3	7.7
ρ_{max} [fm]	4.5	6.9	5.3	2.0	6.7
ρ_{max}/τ	0.81	0.84	0.84	0.87	0.87
$\beta_\perp^{max}$	0.62	0.64	0.64	0.66	0.66
$\langle \beta_\perp \rangle$	0.46	0.47	0.47	0.48	0.48

v_2 presented here include all resonances up to $m_\Delta = 1.232$ GeV, and do not take into account three-body decays.

Figure 3 shows the result of our calculation for three different centrality bins. The elliptic flow coefficient grows with the momentum. In continues to grow for large momenta, where saturation is seen in the experiment, however the thermal model cannot be trusted at momenta larger than about, say, 2 GeV, where hard dynamics is important. We observe that the effects of resonance decays, large in both the numerator and denominator of Eq. (2), cancel to a large degree in the ratio, and the net effect in v_2 is small. The dependence of v_2 on centrality for various transverse-momentum bins is show in

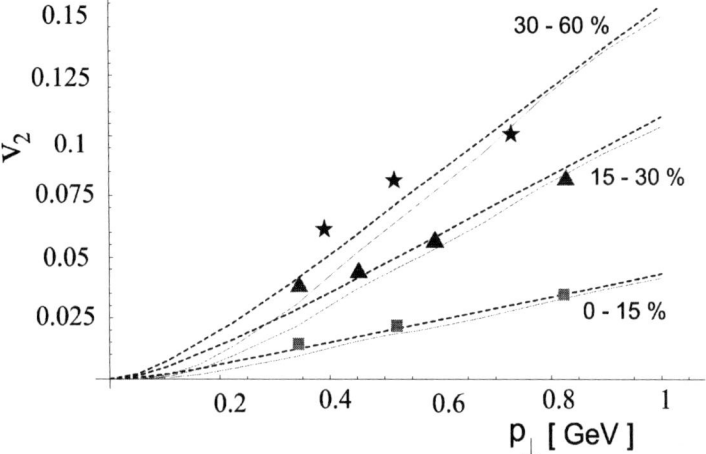

FIGURE 3. Dependence of v_2 (accumulated from all particle species) on the perpendicular momentum, $p_\perp$, for three centrality bins: $0-15\%$ (bottom), $15-30\%$ (middle), and $30-60\%$ (top). Thick (thin) lines correspond to the calculation with (without) resonance decays. Experimental points come from the PHENIX collaboration at $\sqrt{s_{NN}} = 130$ GeV [17]. The taken values for the shape-asymmetry parameter ε are, from the lowest to highest centrality bin, 0.1, 0.21, and 0.35, while the fitted values of the flow-asymmetry parameter δ are 0.145, 0.34, and 0.35, respectively.

Fig. 4. We note that our calculation works well except large c and $p_\perp$, *i.e.* except for high-momentum particles from most peripheral collisions.

In Fig. 5 we compare the model predictions for v_2 integrated over c for identified particles: pions, kaons, and protons. The general characteristics, with lighter particles having larger v_2, are reproduced. The agreement is not satisfactory only for the case of protons at lowest centrality, where the data is compatible with zero.

To summarize this part we note that

1. Elliptic flow can be introduced in the thermal approach by suitably modifying the freeze-out hypersurface.
2. The shape-deformation parameter, $\varepsilon(c)$, should and hopefully will be taken independently from future data on the azimuthal asymmetry of the R_{side} HBT correlation radius at various centralities.
3. The velocity-asymmetry parameter, $\delta(c)$, can be fitted to reproduce v_2. Predictions for identified-particle v_2 follow and are reasonable.

189

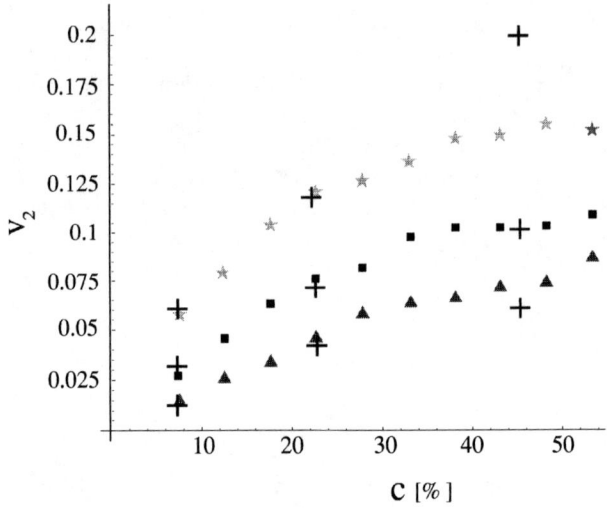

FIGURE 4. Dependence of v_2 (accumulated from all particle species) on the centrality parameter for various transverse-momentum bins: $0.4-0.6\,GeV$ (bottom), $0.6-1\,GeV$ (middle), and $1-2.5\,GeV$ (top). Crosses show the model calculation, while other symbols are the experimental points from the PHENIX collaboration at $\sqrt{s_{NN}} = 130\,GeV$ [17]. The ε and δ parameters are the same as in Fig. 3.

4. The $p_\perp$-dependence shows a monotonic growth, which agrees with the data at lower values of $p_\perp$, but certainly fails to produce saturation at large momenta, where the thermal approach is not applicable.

5. Resonance decays do not have a very large effect on v_2.

HBT RADII

Now we pass to the description of the pionic Hanbury-Brown–Twiss correlation radii (for a review of the problem see, *e.g.*, [20]). The studied object is the two-particle correlation function for identical particles, in the present case $\pi^+\pi^+$ or $\pi^-\pi^-$. It is given by

$$C(\vec{q},\vec{P}) = \frac{\{n_{\vec{p}_1} n_{\vec{p}_2}\}}{\{n_{\vec{p}_1}\}\{n_{\vec{p}_2}\}}, \tag{7}$$

where $\{.\}$ denotes averaging over events, p_1 and p_2 are the momenta of the pions, $\vec{q} = \vec{p}_2 - \vec{p}_1$, and $\vec{P} = \vec{p}_1 + \vec{p}_2$. We use the Bertch-Pratt parameterization [21, 22, 23],

$$C(\vec{q},\vec{P}) = 1 + \lambda e^{-\left(q_{out}^2 R_{out}^2 + q_{side}^2 R_{side}^2 + q_{long}^2 R_{long}^2 + 2q_{out}q_{long}R_{ol}^2\right)}. \tag{8}$$

First, let us briefly recall the experimental highlights. Two facts came as a great surprise with the RHIC data. First, the R_{side} and R_{out} radii practically do not depend

190

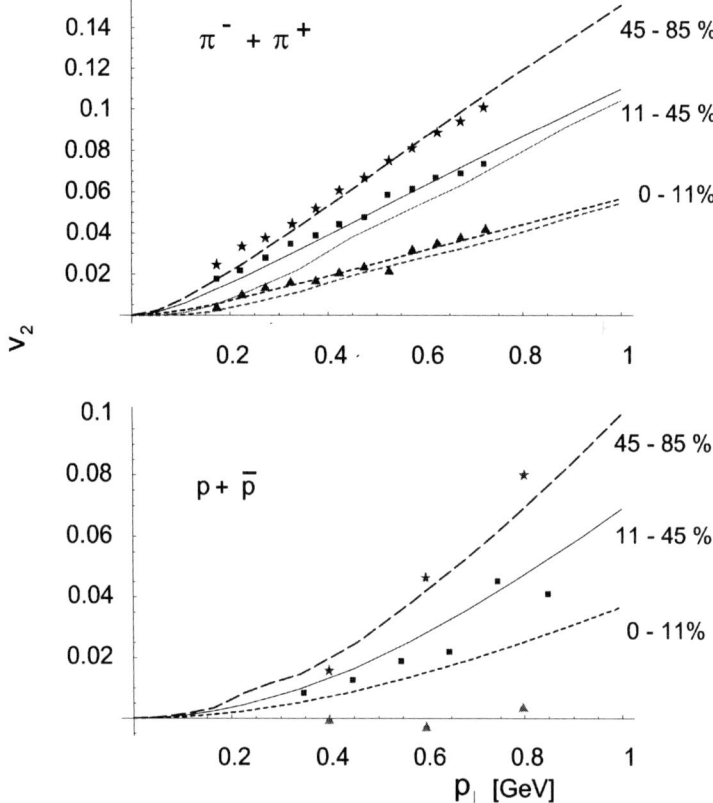

FIGURE 5. Dependence of v_2 integrated over c for identified particles on the transverse momentum, $p_\perp$. Thick (thin) lines correspond to the calculation with (without) resonance decays. Experimental points come from the STAR collaboration at $\sqrt{s_{NN}} = 130$ GeV [12]. The taken values for the shape-asymmetry parameter ε are, from the lowest to highest centrality bin, 0.08, 0.25, and 0.52, while the fitted values of the flow-asymmetry parameter δ are 0.15, 0.35, and 0.52, respectively.

on the collision energy [8, 16], and acquire similar values from AGS to RHIC, despite the increase of the energy by almost two orders of magnitude. Secondly, the ratio of R_{out} to R_{side}, which can be interpreted as a measure of the duration time of the freeze-out, is close or even less than one. This is in contradiction to anticipations from numerous hydrodynamic simulations, which had predicted $R_{\text{out}}/R_{\text{side}}$ significantly larger than one.

Our model evaluation of the correlation function is performed according to the formalism of Ref. [24], with a technical approximation of neglecting the finite life-time of resonances. In that case

$$C(\vec{q},\vec{P}) = 1 + \frac{\left| \int d\Sigma(x) \cdot u(x) e^{iq \cdot x} S(P \cdot u(x)) \right|^2}{\int d\Sigma \cdot u S((P + \frac{q}{2}) \cdot u(x)) \int d\Sigma \cdot u S((P - \frac{q}{2}) \cdot u(x))}, \tag{9}$$

where the source function is

$$S(p \cdot u) = \frac{1}{(2\pi)^3} e^{-(p \cdot u - \mu)/T} + \text{contribution from resonances.} \qquad (10)$$

As discussed in Ref. [25], our model values for the geometric parameters τ and ρ_{max} are low, of the order of the size of the colliding nuclei. As a result, the values of the R_{side} and R_{out} HBT radii obtained with the procedure described above are about 30% too low compared to the experiment. The problem can be alleviated with the inclusion of the excluded-volume (Van der Waals) corrections. Such effects have been realized to be important since the early thermal studies of the particle production in relativistic heavy-ion collisions [26, 27, 28, 29], where they led to a significant dilution of system. In the case of the Boltzmann statistics, which is a very good approximation in our case [30], the excluded volume corrections bring in a factor [28]

$$\frac{e^{-Pv_i/T}}{1 + \sum_j v_j e^{-Pv_j/T} n_j}, \qquad (11)$$

into the phase-space integrals, where P is the pressure, $v_i = 4\frac{4}{3}\pi r_i^3$ is the excluded volume for the particle of species i, and n_i is the density of particles of species i. The pressure can be calculated self-consistently from the equation

$$P = \sum_i P_i^0(T, \mu_i - Pv_i/T) = \sum_i P_i^0(T, \mu_i) e^{-Pv_i/T}, \qquad (12)$$

where P_i^0 denotes the partial pressure of the ideal gas of hadrons of species i. With the simplest assumption that the excluded volumes for all particles are equal, $r_i = r$, the excluded-volume correction manifests itself as a common scale factor, which we may denote by S^{-3}. The Frye-Cooper formula can then be written in the form [31, 32]

$$\frac{dN_i}{d^2p_\perp dy} = \tau^3 \int_{-\infty}^{+\infty} d\alpha_\| \int_0^{\rho_{\text{max}}/\tau} \sinh\alpha_\perp d\left(\sinh\alpha_\perp\right)$$
$$\times \int_0^{2\pi} d\xi \, p \cdot u S^{-3} f_i(p \cdot u), \qquad (13)$$

where $p \cdot u = m_\perp \cosh\alpha_\| \cosh\alpha_\perp - p_\perp \cos\xi \sinh\alpha_\perp$. As can be immediately seen from this expression, the presence of the factor S^{-3} in Eq. (13) may be compensated by rescaling ρ and τ by the factor S. That way the system becomes more dilute and larger in such a way, that the *particle multiplicities and the spectra are left intact*.

With our values of the thermodynamic parameters $\sum_i P_i^0(T, \mu_i) = 80$ MeV/fm^3, and we find $S = 1.3$ with $r = 0.6$ fm. Such a value of the excluded volume is compatible with values typically obtained in other calculations. The increase of the size parameters by 30% is what we need to bring the values of R_{side} and R_{out} up to the experimental ball park.

Our results are shown in Fig. 6. We note that the agreement with data is very reasonable. In particular, the ratio of $R_{\text{out}}/R_{\text{side}}$ is close to one, and drops below one at

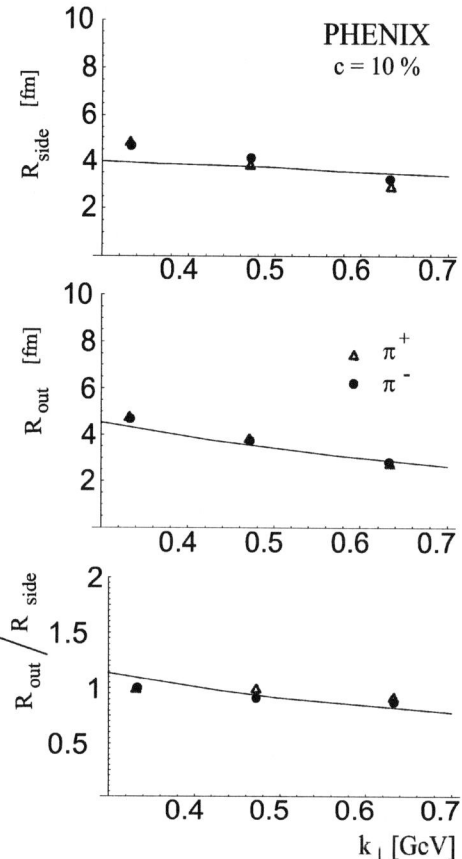

FIGURE 6. Model predictions for the pionic R_{side} and R_{out} HBT correlation radii (top two panels), and their ratio (bottom panel), confronted with the PHENIX data $Au + Au$ data at $\sqrt{s_{NN}} = 130$ GeV and average centrality 10%. The quantity $k_\perp$ is the total momentum of the pion pair.

larger values of the pair momentum $k_\perp$. The plots of R_{side} and R_{out} include the excluded-volume correction factor $S = 1.3$. We observe that these radii decrease with the pair momentum $k_\perp$, although somewhat slower than indicated by the data. We note that the radius R_{long} cannot be reliably evaluated in our model. This is due to the assumption of the boost invariance, *cf.* Eq. (I.2), which leads to too large values of R_{long}.

CONCLUSION

To conclude, we list the main results of our approach:

1. The thermal model works for the particle ratios, see (I).

193

2. Supplied with expansion, it works for the $p_\perp$-spectra, see (I). The complete treatment of resonances is essential, and the assumption of single freeze-out leads to good predictions. Moreover, the strange particles, including Ω, are described properly with no need for extra parameters.

3. Supplied with azimuthal asymmetry, the model can be used to describe the elliptic flow at moderate transverse momenta, up to $p_\perp \sim 2$ GeV. The fitted values of the velocity asymmetry parameter, δ, are reasonable.

4. Supplied with the excluded-volume corrections, the model works also for the HBT radii R_{side} and R_{out}. In particular, the ratio of $R_{\text{out}}/R_{\text{side}}$ is close to 1.

5. The description is efficient, involving two thermal parameters T and μ_B, two size parameters τ and ρ_{max}, and in the case of azimuthal asymmetry, two deformation parameters, ε and δ.

6. Finally, we note that the model also works for the case of RHIC at $\sqrt{s_{NN}} = 200$ GeV A as well as for SPS at $\sqrt{s_{NN}} = 17$ GeV [33, 34].

ACKNOWLEDGMENTS

This research has been supported by PRAXIS XXI/BCC/429/94, and by the Polish State Committee for Scientific Research (KBN), grants 2 P03B 11623 and 2 P03B 09419.

REFERENCES

1. Florkowski, W., and Broniowski, W. (2002), preceding talk.
2. Kolb, P. F., Sollfrank, J., Ruuskanen, P. V., and Heinz, U. W., *Nucl. Phys.*, **A661**, 349–352 (1999).
3. Voloshin, S. A., and Poskanzer, A. M., *Phys. Lett.*, **B474**, 27–32 (2000).
4. Kolb, P. F., Huovinen, P., Heinz, U. W., and Heiselberg, H., *Phys. Lett.*, **B500**, 232–240 (2001).
5. Hirano, T., *Phys. Rev. Lett.*, **86**, 2754–2757 (2001).
6. Magas, V. K., Csernai, L. P., and Strottman, D. D. (2000), nucl-th/0009049.
7. Broniowski, W., and Florkowski, W., *Phys. Rev.*, **C65**, 024905 (2002).
8. Adler, C., et al., *Phys. Rev. Lett.*, **87**, 082301 (2001).
9. Ackermann, K. H., et al., *Phys. Rev. Lett.*, **86**, 402–407 (2001).
10. Snellings, R. J. M., *Nucl. Phys.*, **A698**, 193–198 (2002).
11. Lacey, R. A., *Nucl. Phys.*, **A698**, 559–563 (2002).
12. Adler, C., et al., *Phys. Rev. Lett.*, **87**, 182301 (2001).
13. Adler, C., et al., *Phys. Rev.*, **C66**, 034904 (2002).
14. Adcox, K., et al., *Phys. Rev. Lett.*, **86**, 3500–3505 (2001).
15. Adcox, K., et al., *Phys. Rev. Lett.*, **88**, 242301 (2002).
16. Adcox, K., et al., *Phys. Rev. Lett.*, **88**, 192302 (2002).
17. Adcox, K., *Phys. Rev. Lett.*, **89**, 212301 (2002).
18. Snellings, R. (2001), hep-ph/0111437.
19. Adler, C., et al., *Nucl. Phys.*, **A698**, 64–77 (2002).
20. Baym, G., *Acta Phys. Polon.*, **B29**, 1839–1884 (1998).
21. Pratt, S., *Phys. Rev. Lett.*, **53**, 1219–1221 (1984).
22. Pratt, S., *Phys. Rev.*, **D33**, 72–79 (1986).
23. Bertsch, G., Gong, M., and Tohyama, M., *Phys. Rev.*, **C37**, 1896–1900 (1988).
24. Bolz, J., Ornik, U., Plumer, M., Schlei, B. R., and Weiner, R. M., *Phys. Rev.*, **D47**, 3860–3870 (1993).
25. Broniowski, W., and Florkowski, W., *Phys. Rev. C*, **65**, 064905 (2002).
26. Braun-Munzinger, P., Stachel, J., Wessels, J. P., and Xu, N., *Phys. Lett.*, **B344**, 43–48 (1995).

27. Braun-Munzinger, P., Stachel, J., Wessels, J. P., and Xu, N., *Phys. Lett.*, **B365**, 1–6 (1996).
28. Yen, G. D., Gorenstein, M. I., Greiner, W., and Yang, S.-N., *Phys. Rev.*, **C56**, 2210–2218 (1997).
29. Yen, G. D., and Gorenstein, M. I., *Phys. Rev.*, **C59**, 2788–2791 (1999).
30. Michalec, M., *Thermal description of particle production in ultra-relativistic heavy-ion collisions*, Ph.D. thesis, Institute of Nuclear Physics, Kraków, Poland (2001), nucl-th/0112044.
31. Broniowski, W., and Florkowski, W., *Phys. Rev. Lett.*, **87**, 272302 (2001).
32. Broniowski, W., Baran, A., and Florkowski, W., *Acta Phys. Pol. B*, **33**, 4235 (2002).
33. Broniowski, W., and Florkowski, W., "Thermal model at RHIC: particle ratios and $p_\perp$ spectra," in *Ultrarelativistic Heavy-Ion Collisions*, edited by M. Buballa, W. Nörenberg, B.-J. Schaefer, and J. Wambach, (GSI, Darmstadt), Hirschegg, Austria, 2002, p. 146.
34. Broniowski, W., and Florkowski, W., *Acta Phys. Pol. B*, **33**, 1935 (2002).

195

Diquark Condensation in Electrically and Color Neutral Quark Matter

Michael Buballa[*†], Frederik Neumann[*] and Micaela Oertel[**]

[*]Institut für Kernphysik, TU Darmstadt, Schlossgartenstr. 9, 64289 Darmstadt, Germany
[†]Gesellschaft für Schwerionenforschung (GSI), Planckstr.1, 64291 Darmstadt, Germany
[**]IPN-Lyon, 43 Bd du 11 Novembre 1918, 69622 Villeurbanne Cédex, France

Abstract.
We discuss the possible phase structure of strongly interacting matter in the regime of high densities and low temperatures where the quarks are expected to be in a color superconducting state. We focus on the additional constraints imposed by the requirement of electric and color neutrality which is essential for the description of neutron star interiors. To that end we employ a 3-flavor NJL-type quark model, which treats the diquark condensates and the quark-antiquark condensates on an equal footing. The resulting phase diagram at zero temperature and the various independent chemical potentials turns out to be very rich, containing at least five different color superconducting phases. We search for regions of electrically and color neutral matter and also discuss the possibility of mixed phases with zero net charge. Neglecting surface and Coulomb effects we find nine different mixed phases with up to four components. Preliminary estimates indicate, however, that the mixed phases become unstable if surface and Coulomb effects are included.

INTRODUCTION

The structure of the QCD phase diagram is one of the most exciting topics in the field of strong interactions (For reviews see, e.g. [1, 2, 3, 4, 5]). For a long time the discussion was restricted to two phases: the hadronic phase and the quark-gluon plasma (QGP). The former contains "our" world, where quarks and gluons are confined to color-neutral hadrons and chiral symmetry is spontaneously broken due to the presence of a non-vanishing quark condensate $\phi = \langle \bar{\psi}\psi \rangle$. In the QGP quarks and gluons are deconfined and chiral symmetry is (almost) restored, $\phi \simeq 0$.

Although color-superconducting phases were discussed already in the '70s [6, 7, 8] and '80s [9], until quite recently not much attention was payed to this possibility. This changed dramatically after it was discovered that due to non-perturbative effects, the gaps which are related to these phases could be of the order of $\Delta \sim 100$ MeV [10, 11], much larger than expected from the early perturbative estimates. Since in standard weak-coupling BCS theory the critical temperature is given by $T_c \simeq 0.57\Delta(T = 0)$, this also implies a sizable extension of the color-superconducting phases into the temperature direction [12]. It was concluded that color-superconducting phases could be relevant for neutron stars [13, 14] and – in very optimistic cases – even for heavy-ion collisions [15].

Rather soon after the beginning of this new era, it was noticed that there is probably more than one color superconducting phase in the QCD phase diagram. At large chemical potential, where up, down, and strange quarks can condense, matter is expected to

CP660, *Hadron Physics: Effective Theories of Low Energy QCD*, edited by A. H. Blin et al.
© 2003 American Institute of Physics 0-7354-0120-9/03/$20.00

be in the so-called color-flavor locked phase [16], whereas at intermediate densities, just above the deconfinement phase transition, we might have a two-flavor color superconductor (2SC). Other phases have been suggested more recently, like crystalline phases [17, 18, 19] or a CFL phase accompanied by a kaon condensate (CFL + K) [20, 21]. In addition, those color or flavor degrees of freedom which do not participate in the "standard" condensates could pair in different, usually more fragile, channels, thus forming additional phases [22, 23, 24, 25].

Most of the calculations so far have been performed using a common chemical potential for all quarks (see, e.g., [3, 4]). As we will see, in that case a large region of the phase diagram is occupied by a 2SC phase [26, 27]. For the interior of neutron stars, it is however important, to impose β-equilibrium as well as electric and color neutrality[1]. Consider a system of quarks, electrons and muons with the number densities $n_{f,c}, n_e$, and n_μ, respectively. Here $f = u, d, s$ ("up", "down", "strange") refers to the flavor and $c = r, g, b$ ("red", "green", "blue") refers to the color. The total flavor and color densities are then given by $n_f = \sum_c n_{f,c}$ and $n_c = \sum_f n_{f,c}$, respectively. In compact stars older than a few minutes neutrinos can freely leave the system. In this case lepton number is not conserved and we have four independent conserved charges, namely the total electric charge

$$n_Q = \frac{2}{3} n_u - \frac{1}{3} n_d - \frac{1}{3} n_s - n_e - n_\mu \qquad (1)$$

and the three color charges n_r, n_g, and n_b or, equivalently, their linear combinations

$$n = n_r + n_g + n_b, \qquad n_3 = n_r - n_g, \qquad n_8 = \frac{1}{\sqrt{3}} (n_r + n_g - 2n_b). \qquad (2)$$

Here n corresponds to the total quark number density, while n_3 and n_8 describe color asymmetries. Note that $n/3$ also describes the conserved baryon number. The four conserved charges are related to four independent chemical potentials, μ, μ_3, μ_8, and μ_Q. The individual chemical potentials "felt" by each particle can then be expressed in terms of these four chemical potentials according to their electric and color charges. In this way one finds

$$\mu_{d,c} = \mu_{s,c} = \mu_{u,c} + \mu_e \quad \text{for all } c, \qquad (3)$$

which is usually referred to as β-equilibrium. On the other hand electric and color neutrality means

$$n_Q = n_3 = n_8 = 0. \qquad (4)$$

It was argued by Alford and Rajagopal that these constraints strongly disfavor or even rule out the 2SC phase in compact stars [29]. Recently this question was analyzed in great detail by Steiner, Reddy, and Prakash [30], and by us [31] within an NJL-type model. In this model the various diquark condensates can be treated on the same footing as the quark-antiquark condensates. This is rather important because the latter are not

---

[1] Strictly speaking, color neutrality is not sufficient, but color singletness has to be imposed. This does, however, not induce a large cost in energy [28], such that we can consider matter which is only color neutral.

only responsible for spontaneous chiral symmetry breaking in vacuum, but also lead to density and phase dependent constituent quark masses. In this article we summarize the main result of these investigations.

DIQUARK CONDENSATES IN 2- AND 3-FLAVOR SYSTEMS

According to Cooper's theorem any arbitrarily weak interaction leads to an instability at the Fermi surface which is cured by the formation of Cooper pairs. At very large densities, where asymptotic freedom allows to perform the analysis in terms of a single gluon exchange it can easily be shown that there are indeed attractive channels and hence QCD matter must be a color superconductor at these densities. However, because of the large number of possible channels related to the quantum numbers of spin, flavor and color, we can almost be sure that also in the non-perturbative regime just above the deconfinement phase transition some of them will be attractive.

In general, a diquark condensate may be written as

$$\langle \psi^T \mathcal{O} \psi \rangle , \tag{5}$$

where ψ is a quark field and $\mathcal{O}$ is an operator, acting in color, flavor and Dirac space,

$$\mathcal{O} = \mathcal{O}_{Dirac} \otimes \mathcal{O}_{flavor} \otimes \mathcal{O}_{color} . \tag{6}$$

It can also contain derivatives, but we will not consider this possibility here. A priori, the only restriction to $\mathcal{O}$ is provided by the Pauli principle, which requires that $\mathcal{O}$ must be totally antisymmetric. This still leaves many possibilities, and thus the interaction must decide about the actual condensation pattern.

As already mentioned, at very large chemical potentials, $\mu \gg \Lambda_{QCD}$, $\alpha_s(\mu)$ is small and the problem can be (and has been) attacked from first principles [12, 15, 32, 33, 34]. To estimate the range of validity of these calculations we assume (quite optimistically) that the perturbative regime begins at $\mu \approx 1.5$ GeV. For two massless flavors this corresponds to a baryon density $\rho_B = 2/(3\pi^2)\mu^3 \approx 30$ fm^{-3}, which is about 175 times nuclear saturation density. It turns out that the situation is even worse: In a numerical study Rajagopal and Shuster [35] found that (gauge dependent) higher-order terms can only be neglected if $\mu \gg 10^5$ GeV!

Hence, asymptotic studies, although interesting by themselves, cannot be trusted down to densities which are present, e.g., in the interiors of neutron stars. In this regime one has to rely on effective interactions, like instanton interactions [36, 37], or (local or nonlocal) 4-point interactions ("NJL-type models") with quantum numbers also abstracted from the instanton vertex [10, 11] or derived in a more phenomenological way [27, 30, 31, 38]. This is quite analogous to the Landau-Migdal interaction used to describe nuclear matter. However, we should be aware of the fact that there are presently no data to constrain the parameters in the deconfined phase itself. They are therefore usually fixed in vacuum, which is clearly a source of big uncertainties.

Nevertheless, there are good reasons to believe, that a dominant role is played by the Lorentz-invariant scalar ($J = 0^+$) condensate,

$$s_{AA'} = \langle \psi^T C \gamma_5 \tau_A \lambda_{A'} \psi \rangle , \tag{7}$$

which corresponds to the most attractive channel, both for interactions with the quantum numbers of a single gluon exchange as well as for instanton induced interactions. Here C is the matrix of charge conjugation, and τ_A and $\lambda_{A'}$ are the antisymmetric generators of flavor-$SU(N_f)$ and color-$SU(N_c)$, respectively. Throughout this article, we will restrict ourselves to the physical number of colors, $N_c = 3$. Then the $\lambda_{A'}$ denote the three antisymmetric Gell-Mann matrices, λ_2, λ_5 and λ_7, i.e., $s_{AA'}$ is a color anti-triplet.

Concerning the number of flavors we begin with two idealized cases: If the strange quark is very heavy it decouples and we are left with two light flavors, u and d. In this case, there is only one antisymmetric $SU(N_f)$ generator, namely τ_2, i.e., the flavor index in Eq. (7) is restricted to $A = 2$. In the limit of massless up and down quarks $s_{2A'}$ is invariant under chiral $SU(2)_L \times SU(2)_R$ transformations. The three condensates s_{22}, s_{25}, and s_{27}, form a vector in color space, which can always be rotated into the $A' = 2$-direction. Hence the two-flavor superconducting state (2SC) state can be characterized by

$$s_{22} \neq 0 \quad \text{and} \quad s_{AA'} = 0 \quad \text{if} \quad (A, A') \neq (2, 2). \tag{8}$$

Since only the first two colors ("red" and "green") participate in the s_{22}, while the third one ("blue") does not, color $SU(3)$ is spontaneously broken down to $SU(2)$. As a result five of the eight gluons acquire a mass [39].

For $N_f = 3$ there are three antisymmetric $SU(N_f)$-generators, τ_2, τ_5, and τ_7, describing ud-, us-, and ds-pairing, respectively. The two-flavor condensation pattern, Eq. (8) is still possible, but now there are several other combinations which cannot be transformed into s_{22} via color or flavor rotations.

In the case of three degenerate light flavors, the most favored pattern at high density is the so-called color-flavor locked (CFL) state [16], characterized by the situation

$$s_{22} = s_{55} = s_{77} \neq 0 \quad \text{and} \quad s_{AA'} = 0 \quad \text{if} \quad A \neq A'. \tag{9}$$

In this state color $SU(3)$ as well as the chiral $SU(3)_L \times SU(3)_R$ and the $U(1)$-symmetry related to baryon-number conservation are broken down to a common $SU(3)_{color+V}$ subgroup where color and flavor rotations are locked. As a consequence all gluons receive a mass and there is a gap in the dispersion laws of all nine (3 flavors, 3 colors) quark quasiparticle states.

The situations discussed above are idealizations of the real world, where the strange quark mass M_s is neither infinite, such that strange quarks can be completely neglected nor degenerate with the masses of the up- and down quarks. For sufficiently large quark chemical potentials $\mu \gg M_s$, the s quark mass can be neglected against μ, and matter is expected to be in the CFL phase. It is not clear, however, whether this CFL phase is directly connected to the hadronic phase [40] at low densities, or whether an intermediate 2SC phase exists, where only up and down quarks are paired. It is obvious that the answer to this question depends on the strange quark mass. This has first been analyzed by Alford, Berges and Rajagopal [41] who have studied the color-flavor unlocking phase transition in a model calculation with different values of M_s. Assuming that the region below $\mu \simeq 400$ MeV belongs to the hadronic phase, these authors came to the conclusion that a 2SC-phase exists if $M_s \gtrsim 250$ MeV. Here M_s is an effective "constituent"mass of the strange quark, which could be considerably larger than the current quark mass

$m_s \sim 100$ to 150 MeV in the Lagrangian. It is in general T- and μ-dependent and can depend on the presence of quark-antiquark and diquark condensates. In particular, it can be discontinuous along a first-order phase transition line. This means, not only the phase structure depends on the effective quark mass, but also the quark mass depends on the phase. In Refs. [26, 27] we have studied these interdependencies within an NJL-type quark model. Below we will discuss the main results of these investigations.

QUARK MODEL

We consider the Lagrangian

$$\mathscr{L}_{eff} = \bar{\psi}(i\partial\!\!\!/ - \hat{m})\psi + \mathscr{L}_{q\bar{q}} + \mathscr{L}_{qq}, \tag{10}$$

where ψ denotes a quark field with three flavors and three colors. The mass matrix $\hat{m}$ has the form $\hat{m} = diag(m_u, m_d, m_s)$ in flavor space. Throughout this paper we will assume $m_u = m_d$, whereas the strange quark mass m_s will be different. To study the interplay between the color-superconducting diquark condensates and the quark-antiquark condensates we consider an NJL-type interaction with a quark-quark part

$$\mathscr{L}_{qq} = H \sum_{A=2,5,7} \sum_{A'=2,5,7} (\bar{\psi} i\gamma_5 \tau_A \lambda_{A'} C \bar{\psi}^T)(\psi^T C i\gamma_5 \tau_A \lambda_{A'} \psi). \tag{11}$$

and a quark-antiquark part

$$\mathscr{L}_{q\bar{q}} = G \sum_{a=0}^{8} \left[(\bar{\psi}\tau_a\psi)^2 + (\bar{\psi} i\gamma_5 \tau_a\psi)^2 \right] - K \left[\det_f \left(\bar{\psi}(1+\gamma_5)\psi \right) + \det_f \left(\bar{\psi}(1-\gamma_5)\psi \right) \right]. \tag{12}$$

Here $\tau_0 = \sqrt{\frac{2}{3}} \, \mathbb{1}_f$ is proportional to the unit matrix in flavor space. Eq. (12) corresponds to a typical 3-flavor NJL-model Lagrangian which has often been used to study meson spectra [42, 43] or properties of quark matter at finite densities or temperatures [44, 45, 46]. It consists of a $U(3)_L \times U(3)_R$-symmetric 4-point interaction and a 't Hooft-type 6-point interaction which breaks the $U_A(1)$ symmetry. The above interaction terms might arise from some underlying more microscopic theory and are understood to be used at mean-field level in Hartree approximation. In particular we do not consider any contribution from the 6-point interaction to the diquark condensate.

In order to calculate the mean-field thermodynamic potential Ω_q at temperature T and chemical potentials μ_{fc} (corresponding to quarks of flavor f and color c) we linearize $\mathscr{L}_{eff}$ in the presence of the three quark-antiquark condensates $\phi_f = \langle \bar{f}f \rangle$, $f = u, d, s$, and the three diquark condensates s_{AA}, $A = 2, 5, 7$. Introducing the constituent quark masses

$$M_i = m_i - 4G\phi_i + 2K\phi_j\phi_k, \qquad (i,j,k) = \text{any permutation of } (u,d,s), \tag{13}$$

and the diquark gaps

$$\Delta_{ud} = -2H s_{22}, \Delta_{us} = -2H s_{55}, \Delta_{ds} = -2H s_{77}, \tag{14}$$

and employing Nambu-Gorkov-formalism the result can be written in the following way:

$$\Omega_q(T, \{\mu_i\}; \chi) = -T \sum_n \int \frac{d^3p}{(2\pi)^3} \frac{1}{2} \mathrm{Tr} \ln \left(\frac{1}{T} S^{-1}(i\omega_n, \vec{p})\right)$$
$$+ 2G(\phi_u^2 + \phi_d^2 + \phi_s^2) - 4K\phi_u\phi_d\phi_s + H\left(|s_{22}|^2 + |s_{55}|^2 + |s_{77}|^2\right).$$
$$(15)$$

Here $\chi = \{\phi_u, \phi_d, \phi_s, s_{22}, s_{55}, s_{77}\}$ denotes the set of condensates and $\omega_n = (2n-1)\pi T$ are fermionic Matsubara frequencies. The inverse fermion propagator reads

$$S^{-1}(p) = \begin{pmatrix} \not{p} - \hat{M} + \hat{\mu}\gamma^0 & \gamma_5(\Delta_{ud}\tau_2\lambda_2 + \Delta_{us}\tau_5\lambda_5 + \Delta_{ds}\tau_7\lambda_7) \\ -\gamma_5(\Delta_{ud}^*\tau_2\lambda_2 + \Delta_{us}^*\tau_5\lambda_5 + \Delta_{ds}^*\tau_7\lambda_7) & \not{p} - \hat{M} - \hat{\mu}\gamma^0 \end{pmatrix},$$
$$(16)$$

where $\hat{M}$ and $\hat{\mu}$ are diagonal matrices containing the different constituent masses M_f and chemical potentials μ_{fc}. Note that including color, flavor, and Dirac degrees of freedom in addition to the Nambu-Gorkov components shown explicitly, S^{-1} is a 72×72 matrix. The evaluation of the corresponding trace in Eq. (15) is therefore tedious but straight forward work and will not be presented here.

Later we will also consider a system of quarks and charged leptons. The total thermodynamic potential is then simply taken to be the sum of Ω_q and a lepton contribution Ω_l describing non-interacting massless electrons and non-interacting massive muons.

Our model is incomplete without specifying the parameters to be used in the numerical calculations. To this end we adopt the parameters of Ref. [45], which have been determined by fitting the pseudoscalar meson masses and decay constants in vacuum: a sharp 3-momentum cutoff $\Lambda = 602.3$ MeV to regularize the integrals, the bare quark masses $m_u = m_d = 5.5$ MeV and $m_s = 140.7$ MeV, and the coupling constants $G = 1.835\Lambda^{-2}$ and $K = 12.36\Lambda^{-5}$. Similarly, one could fix the quark-quark coupling constant H by fitting baryon masses within a Fadeev approach [47, 48]. For simplicity, however, we take $H = G$. The main justification of this somewhat arbitrary choice should be seen in the fact, that at $\mu \sim 400$ MeV we find $\Delta_{ud} \sim 100$ MeV, i.e., of the same order which is usually quoted in the literature. (For further arguments see Ref. [31]). It follows that none of our quantitative results should be taken literally. On the other hand we do not expect big qualitative differences if H is varied within a certain range.

EQUAL CHEMICAL POTENTIALS

Let us first assume that all quark chemical potentials are equal, i.e., $\mu_Q = \mu_3 = \mu_8 = 0$. Then, because of isospin symmetry, we have $M_u = M_d$ and $\Delta_{us} = \Delta_{ds}$. The values of these quantities as functions of μ for $T = 0$ are displayed in Fig. 1. In the left panel the diquark gaps Δ_{ij} are shown. One can clearly distinguish three regimes, separated by first-order phase transitions: At low chemical potentials we find a normal phase with vanishing diquark condensates, followed by the 2SC phase with $\Delta_{ud} \neq 0$ and finally the CFL phase where also Δ_{us} and Δ_{ds} are nonzero. At the phase boundaries we also observe strong discontinuities in the constituent quark masses (right panel of Fig. 1).

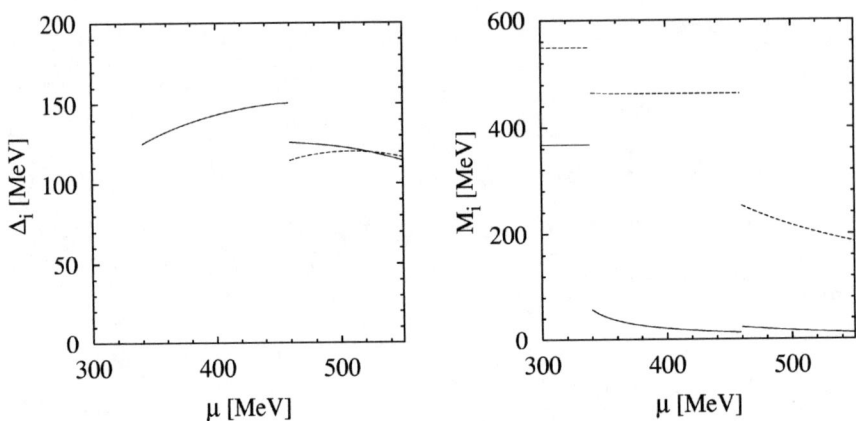

FIGURE 1. Various quantities at zero temperature as functions of a common quark chemical potential μ. Left: diquark gaps Δ_{ud} (solid) and $\Delta_{us} = \Delta_{ds}$ (dashed). Right: constituent masses $M_u = M_d$ (solid) and M_s (dashed).

For instance at the 2SC-CFL transition point at $\mu = 460$ MeV the strange quark mass drops from 463 MeV to 252 MeV, i.e., from a value larger than μ to a value considerably smaller than μ. Hence any simple criterion which relates the position of the color-flavor (un-)locking transition to the strange quark mass can at best give a very rough estimate. This demonstrates the importance of the simultaneous treatment of diquark and quark-antiquark condensates [26, 27].

In Fig. 2 we show the corresponding densities. The total quark number density n is displayed in the left panel, the ratios n_Q/n and n_8/n in the right panel. Note that all densities are zero in the "normal phase", i.e., here this phase corresponds to the vacuum. Therefore the ratios n_i/n are not well defined in this regime[2]. As one can see in the figure, the two color superconducting phases carry both, electric and color charges. The electric charge of the 2SC phase is easily understood. Since $\mu_Q = 0$, there are no leptons and the densities of up and down quarks are equal. Moreover, the strange quarks are too heavy to be populated in this regime. Hence the total electric charge density is given by $n_Q = n/6$. The nonvanishing color density n_8 reflects the fact that for equal chemical potentials the densities of the paired (red and green) quarks are larger than the density of the unpaired (blue) quarks [49].

At the transition to the CFL phase n_8/n does not change very much, whereas the electric charge density drops significantly due to a strong increase of the density of strange quarks. To a large extent, this is caused by the sudden drop of the strange quark mass, but also by the CFL pairing pattern which relates color and flavor densities in a

---

[2] One could argue, however, that in the normal phase with one common chemical potential μ one should always have $n_u = n_d$ and $n_r = n_g = n_b$, and therefore the correct limits should be $n_Q/n = 1/6$ and $n_8/n = 0$.

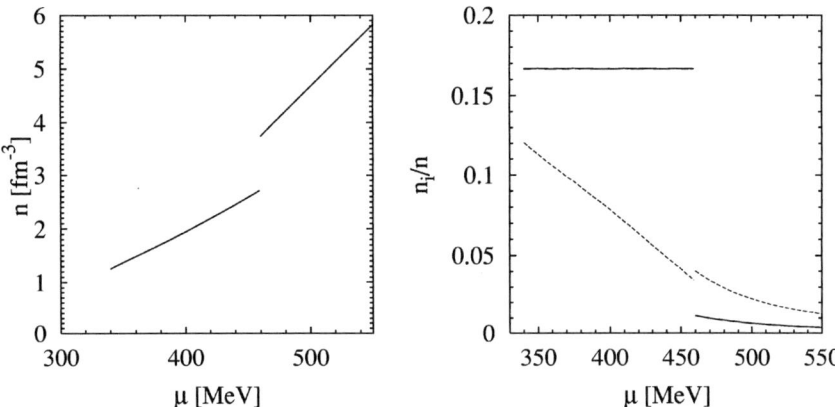

FIGURE 2. Densities at $T = 0$ as functions of a common quark chemical potential μ. Left: total quark number density n. Right: ratios n_Q/n (solid) and n_8/n (dashed).

peculiar way [30]:

$$n_u = n_r , \qquad n_d = n_g , \qquad n_s = n_b \qquad \text{(CFL)} . \qquad (17)$$

For the present case that means $n_Q = 1/(2\sqrt{3}) \, n_8$, in agreement with our numerical results.

NON-EQUAL CHEMICAL POTENTIALS

The above results show that in order obtain electrically and color neutral quark matter solutions we have to choose non-equal quark chemical potentials. In the CFL phase the neutralization is relatively easy. As one can see from Eq. (17), color neutral CFL matter is automatically electrically neutral as long as there are no leptons. This means we should keep μ_Q equal to zero and only vary μ_8 until color (and hence electric) neutrality is reached. The corresponding solutions have been constructed in Refs. [30, 31].

The situation is much more difficult for the 2SC phase. Here we have no relation which links electric neutrality to color neutrality and we have to choose nonvanishing values of both, μ_8 and μ_Q. Whereas color neutrality is achieved quite easily [49] the main problem is to obtain *electrically* neutral 2SC matter. Therefore let us first study the influence of a nonvanishing chemical potential μ_Q.

In Fig. 3 we show the phase diagram in the $\mu - \mu_Q$-plane for $\mu_8 = \mu_3 = 0$. Since we are interested in neutralizing the electrically positive 2SC phase, we choose μ_Q to be negative. As long as this is not too large, we find again the normal phase at lower values of μ, the 2SC phase in the intermediate region and the CFL phase for large μ. This changes dramatically around $\mu_Q \simeq 180$ MeV where both, the 2SC phase and the CFL phase disappear and a new phase emerges. This phase is analogous to the 2SC phase but

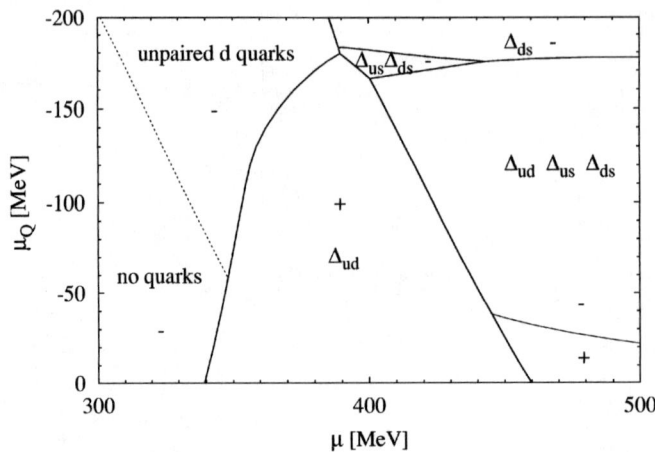

FIGURE 3. Phase diagram in the $\mu - \mu_Q$ plane for $T = \mu_3 = \mu_8 = 0$ (adapted from Ref. [31]). The various phases separated by the solid lines are characterized by different non-vanishing diquark gaps Δ_{ij} as indicated in the figure. In the non-superconducting phase quarks are present only above the dashed line. The "+" and "-" signs indicate the sign of the electric charge density in the corresponding region. The dotted line corresponds to electrically (but not color) neutral matter in the CFL phase.

with ds pairing, instead of ud pairing ("2SC$_{ds}$"). In a small intermediate regime there is even another phase which contains us and ds but no ud pairs ("SC$_{us+ds}$").

Qualitatively, the existence of these phases is quite plausible: At low values of $|\mu_Q|$ the Fermi momenta of the up and down quarks are relatively similar to each other, whereas the strange quarks are suppressed because of their larger mass. With increasing negative μ_Q, however, the up quarks become more and more disfavored and eventually the Fermi momenta are ordered as $p_F^u < p_F^s < p_F^d$. It is then easy to imagine that only ds pairing or – in some intermediate regime – us and ds pairing is possible.

In the phase diagram, Fig. 3, we also indicated the sign of the electric charge density for the various regions, and the line of electrically neutral matter in the CFL phase (dotted line). Note that there is no other electrically neutral regime in this diagram (apart from the vacuum at small μ and $\mu_Q = 0$). In the normal phase, there are no quarks below the dashed line, corresponding to the line $\mu - 1/3\mu_Q = M_d$. This region is nevertheless negatively charged due to the leptons which are present for any $\mu_Q < 0$. Above the dashed line there are also d quarks rendering the matter even more negative. (In the right corner of this phase there is also a very small fraction of u quarks.) The "new" phases, 2SC$_{ds}$ and SC$_{us+ds}$, are also negatively charged. On the contrary, the entire 2SC phase is positively charged, even at the largest values of $|\mu_Q|$.

The difficulty to obtain electrically neutral 2SC matter can be traced back to the fact that the pairing of two quark species forces their densities to be equal [50]. For the 2SC pairing pattern, Eq. (8), this means that the sum of red and green u quarks is equal to the sum of red and green d quarks. As long as no strange quarks are present, the related positive net charge can only be compensated by the blue quarks and the leptons, which would require a very large negative μ_Q. However, before this point is reached a phase transition takes place and the 2SC phase is no longer stable. Again the selfconsistent treatment is very important in this context: Similar to what we have seen in Fig. 1 the phase transition is triggered by a sudden drop of the strange quark mass. This leads to a strong increase of the strange Fermi momentum such that the strange quarks can also participate in a condensate. At the same time the up quark mass is increased which in some regions causes a break-up of the *ud* condensates.

MIXED PHASES

So far we have restricted ourselves to study homogeneous phases. Now suppose we have a mixed phase consisting of two components, 1 and 2, with charge densities $n_i^{(1)}$ and $n_i^{(2)}$, respectively. As pointed out by Glendenning, in a compact star it is sufficient to demand that the *average* charge density vanishes [51]. If the two components occupy the volume fractions x_1 and $x_2 = 1 - x_1$, respectively, the average densities are given by

$$n_i = x_1 n_i^{(1)} + (1 - x_1) n_i^{(2)} . \tag{18}$$

This is zero for

$$x_1 = \frac{n_i^{(2)}}{n_i^{(2)} - n_i^{(1)}} . \tag{19}$$

To be meaningful the solution must be in the interval $0 < x_1 < 1$. This is fulfilled when the charge densities $n_i^{(1)}$ and $n_i^{(2)}$ have opposite signs, which is an obvious prerequisite for a charge neutral mixture.

Glendenning has discussed this problem in the context of the quark-hadron phase transition. Since he did not consider color-superconducting phases, he only had to care about electric neutrality. We now have to generalize this procedure to construct electrically and color neutral mixed phases. This is more difficult because we have to require that the result of Eq. (19) is the same for $i = Q$, 3, and 8 at the same time. This is the case when

$$n_Q^{(1)} : n_Q^{(2)} = n_3^{(1)} : n_3^{(2)} = n_8^{(1)} : n_8^{(2)} . \tag{20}$$

In the four-dimensional space spanned by the four independent chemical potentials μ, μ_Q, μ_3 and μ_8 the phase boundaries are three-dimensional hypersurfaces. Since Eq. (20) imposes two additional constraints, electrically and color neutral mixed phases can be constructed along a one-dimensional line. As discussed in Ref. [31] there are also electrically and color neutral mixed phases consisting of three or even four components. All these solutions exist along one-dimensional lines in the space of chemical potentials.

In Ref. [31] we have constructed the neutral mixed phase solutions for the model discussed above. We have found nine different mixed phases consisting of two, three or four components. The volume fractions of the various components as functions of the quark number chemical potential are shown in the left panel of Fig. 4, the corresponding average densities are given in the right panel. We see that the "normal", 2SC and CFL components are dominant at lower, intermediate and higher chemical potentials, respectively, more or less in those regimes where we have found these phases

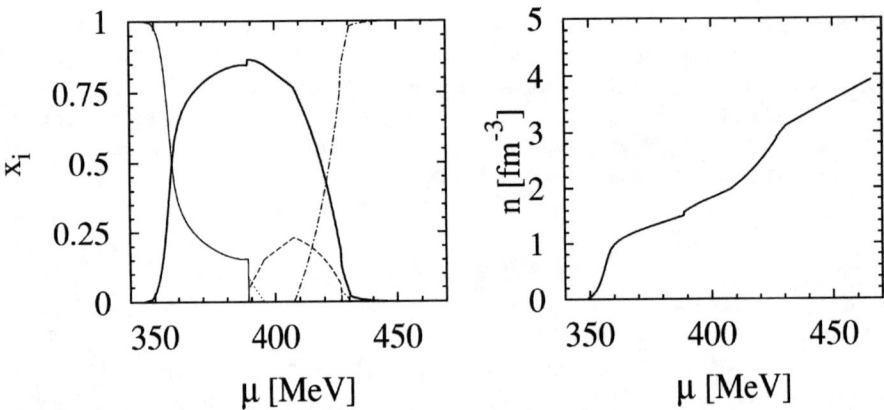

FIGURE 4. Left: Volume fractions of the various components in the mixed phase region as functions of μ (adapted from Ref [31]): normal (thin solid), 2SC (thick solid), CFL (dashed-dotted), SC_{us+ds} (dashed), $2SC_{us}$ (dotted). Right: Corresponding quark number density averaged over the components.

being stable if no neutrality constraints are imposed (cf. Fig. 1). We also have "exotic" components, like an SC_{us+ds} component which fills about 25% of the volume at $\mu \simeq$ 410 MeV.

We should remark that in our analysis we have completely neglected the surface energy and the energies of the electric and color-electric fields, which arise in mixed phases consisting of non-neutral components. Of course the mixed phases are only stable if the gain in bulk free energy is larger than these so-far neglected contributions. Unfortunately, in order to calculate these contributions we need to know the surface tensions between the various phases which are so-far unknown quantities. However, preliminary estimates along lines of the analysis performed in Ref. [52] seem to indicate that for "reasonable" surface tensions the mixed phases are likely to be unstable [31]. In this case it is possible that neutral 2SC matter that we found to be unstable in the previous section, is rendered stable and might exist in a compact star. In fact, the authors of Ref. [30] found that there is a certain interval of the chemical potential μ, where homogeneous neutral 2SC matter has a lower free energy than homogeneous neutral quark matter in the CFL phase or in the normal phase. This interval corresponds roughly to the regime where we found a dominant 2SC component.

SUMMARY

We have studied the phase structure of strongly interacting matter in the regime of high densities and low temperatures where the quarks are in a color superconducting state. Our calculations have been performed within an NJL-type model, which allows to treat diquark condensates on an equal footing with quark-antiquark condensates, leading to density (and phase-) dependent effective quark masses. This was shown to be very important for the description of the phase transition from the two-flavor color superconductor (2SC) where only up and down quarks are paired to the so-called color-flavor locked (CFL) phase, which also incorporates strange quarks [26].

In this article we have focussed on the conditions which are relevant for the interior of neutron stars older than a few minutes, i.e., dense, electrically and color neutral matter in β-equilibrium. To that end we had to introduce four independent chemical potentials, related to the total quark number, electric charge and two color charges. The resulting phase diagram turned out to be very rich. In addition to the familiar 2SC and CFL phases it contains at least three further color-superconducting phases, like the SC_{us+ds}-phase, where u and d quarks are paired with s quarks, but not with each other. However, apart from the CFL phase (and of course the vacuum) we did not succeed to find homogeneous neutral matter in the stable regimes of this phase diagram. In particular the 2SC-phase remained always positive. On the other hand we were able to construct mixed phases with zero net charge. Neglecting surface and Coulomb effects we found nine different mixed phases with up to four components. Preliminary estimates indicate, however, that these phases might become unstable if surface and Coulomb effects are included.

A lot of work remains to be done. First, the influence of surface of Coulomb effects on the stability of the mixed phases has to be investigated more carefully. We should also take into account the effect of Goldstone bosons which could condense in the CFL phase [53, 54]. Moreover, we neglected the possibility of the formation of crystalline (LOFF) phases [18, 19]. These might become favored over phases with BCS pairing if the Fermi surfaces are pulled apart. As we are mostly dealing with pairing of quarks with unequal Fermi momenta, we suppose that especially in the vicinity of phase boundaries, the possible formation of a crystalline phase should be taken into account. Finally, our model is certainly unrealistic at low densities, where confinement and hadronic degrees of freedom have to be taken into account.

ACKNOWLEDGMENTS

M.B. thanks the organizers of the workshop, in particular Maria Ruivo, for their warm hospitality in Coimbra and financial support. M.O. acknowledges financial support from the Alexander von Humboldt-foundation as a Feodor-Lynen fellow.

REFERENCES

1. Halasz, M.A., Jackson, A.D., Shrock, R.E., Stephanov, M.A., and Verbaarschot, J.J.M., *Phys. Rev.* **D 58**, 096007 (1998).
2. Rajagopal, K., *Nucl. Phys.* **A 661**, 150–161 (1999).
3. Rajagopal K. and Wilczek F., "The Condensed Matter Physics of QCD", in B.L. Ioffe Festschrift *At the Frontier of Particle Physics / Handbook of QCD*,**vol. 3**, edited by M. Shifman, World Scientific, Singapore, 2001, pp. 2061–2151.
4. Alford, M. *Ann. Rev. Nucl. Part. Sci.* **51**, 131–160 (2001).
5. Sannino, F., *arXiv.org e-Print archive*, **hep-ph/0205007** (2002).
6. Collins, J.C. and Perry, M.J., *Phys. Rev. Lett.* **34**, 1353–1356 (1975).
7. Barrois, B., *Nucl. Phys.* **B 129**, 390–402 (1977).

8. Frautschi, S.C., "Asymptotic freedom and color superconductivity in dense quark matter", in *Proc. of the Workshop on Hadronic Matter at Extreme Energy Density*, edited by N. Cabibbo, Erice (1978).
9. Bailin, D. and Love, A., *Phys. Rep.* **107**, 325–385 (1984).
10. Alford, M., Rajagopal, K., and Wilczek, F., *Phys. Lett.* **B 422**, 247–256 (1998).
11. Rapp, R., Schäfer, T., Shuryak, E.V., and Velkovsky, M., *Phys. Rev. Lett.* **81**, 53–56 (1998).
12. Pisarski, R.D. and Rischke, D.H., *Phys. Rev.* **D 60**, 094013 (1999).
13. Weber, F., *Acta Phys. Polon.* **B 30**, 3149–3169 (1999).
14. Blaschke, D., Glendenning, N.K., and Sedrakian A. (eds.), *Physics of Neutron Star Interiors*, Lecture Notes in Physics, **vol. 578**, Springer, Berlin, Heidelberg (2001).
15. Pisarski, R.D. and Rischke, D.H., *Phys. Rev.* **D 61**, 051501; ibd., 074017 (2000).
16. Alford, M., Rajagopal, K., and Wilczek, F., *Nucl. Phys.* **B 537**, 443–458 (1999).
17. Rapp, R., Shuryak, E.V., and Zahed, I., *Phys. Rev.* **D 63**, 034008 (2001).
18. Alford, M., Bowers, J., and Rajagopal, K., *Phys. Rev.* **D 63**, 074016 (2001).
19. Bowers, J. and Rajagopal, K., *Phys. Rev.* **D 66**, 065002 (2002).
20. Schäfer, T., *Phys. Rev. Lett.* **85**, 5531–5534 (2000).
21. Bedaque, P.F. and Schäfer, T. *Nucl. Phys.* **A 697**, 802-822 (2002).
22. Schäfer, T., *Phys. Rev.* **D 62**, 094007 (2000).
23. Buballa, M., Hošek, J., and Oertel, M., *arXiv.org e-Print archive* **hep-ph/0204275** (2002).
24. Sannino, F. and Schäfer, W. *Phys. Lett.* **B 527**, 142–148 (2002).
25. Alford, M.G., Bowers, J.A., Cheyne, J.M., and Cowan, G.A., *arXiv.org e-Print archive* **hep-ph/0210106** (2002).
26. Buballa, M. and Oertel, M., *Nucl. Phys.* **A 703**, 770–784 (2002).
27. Oertel, M. and Buballa, M., "Color-Flavor (Un-)locking", in *Ultrarelativistic heavy ion collisions*, edited by M. Buballa, W. Nörenberg, B.-J. Schaefer, and J. Wambach, Proceedings of the International workshop XXX on gross properties of nuclei and nuclear excitations, Hirschegg/Austria, GSI, Darmstadt 2002, pp. 110–115.
28. Amore, P., Birse, M.C., McGovern, J.A., and Walet, N.R., *Phys. Rev.* **D 65**, 074005 (2002).
29. Alford, M. and Rajagopal, K., *JHEP* **0206**, 031 (2002).
30. Steiner, A., Reddy, S., and Prakash, M., *arXiv.org e-Print archive* **hep-ph/0205201** (2002).
31. Neumann, F., Buballa, M., and Oertel, M., *Nucl. Phys.*, A, in press [*arXiv.org e-Print archive*, **hep-ph/0210078** (2002)].
32. Son, D.T., *Phys. Rev.* **D 59**, 094019 (1999).
33. Schäfer, T. and Wilczek, F., *Phys. Rev.* **D 60**, 114033 (1999).
34. Hong, D.K., Miransky, V.A., Shovkovy, I.A., and Wijewardhana, L.C.R., *Phys. Rev.* **D 61**, 056001 (2000), err. ibid. **D 62** 059903 (2000).
35. Rajagopal, K., Shuster, E., *Phys. Rev.* **D 62**, 085007 (2000).
36. Rapp, R., Schäfer, T., Shuryak, E.V., and Velkovsky, M., *Annals Phys.* **280**, 35–99 (2000).
37. Carter, G.W. and Diakonov, D., *Phys. Rev.* **D60**, 016004 (1999).
38. Schwarz, T.M., Klevansky, S.P., and Papp, G., *Phys. Rev.* **C60**, 055205 (1999).
39. Rischke, D.H., *Phys. Rev.* **D 62**, 034007 (2000).
40. Wilczek, F. and Schäfer, T., *Phys. Rev. Lett.* **82**, 3956–3959 (1999).
41. Alford, M., Berges, J., and Rajagopal, K., *Nucl. Phys.* **B 558**, 219–242 (1999).
42. Takizawa, M., Tsushima, K., Kohyama, Y., and Kubodera, K., *Nucl. Phys.* **A 507**, 611–648 (1990).
43. Klimt, S., Lutz, M., Vogl, U., and Weise, W., *Nucl. Phys.* **A 516**, 429–468 (1990); Vogl, U., Lutz, M., Klimt, S., and W. Weise, *ibid.*, 469–495 (1990).
44. Lutz, M., Klimt, S., and Weise, W., *Nucl. Phys.* **A 542**, 521–558 (1992).
45. Rehberg, P., Klevansky, S.P., and Hüfner, J., *Phys. Rev.* **C 53**, 410–429 (1996).
46. Buballa, M. and Oertel, M., *Phys. Lett.* **B 457**, 261–267 (1999).
47. Ishii, N., Bentz, W., and Yazaki, K., *Phys. Lett.* **B 318**, 26–31 (1993).
48. Hanhart, C. and Krewald, S., *Phys. Lett.* **B 344** 55–60 (1995).
49. Buballa, M., Hošek, J., and Oertel, M., *Phys. Rev.* **D 65**, 014018 (2002).
50. Rajagopal, K. and Wilczek, F., *Phys. Rev. Lett.* **86**, 3492–3495 (2001).
51. Glendenning, N.K., *Phys. Rev.* **D 46**, 1274–1287 (1992).
52. Alford, M., Rajagopal, K., Reddy, S., and Wilczek, F., *Phys. Rev.* **D 64**, 074017 (2001).
53. Bedaque, P.F. and Schäfer, T., *Nucl. Phys.* **A 697**, 802–822 (2002).
54. Kaplan, D.B. and Reddy, S., *Phys. Rev.* **D 65 054042 (2002)**.

Effects of quark matter and color superconductivity in compact stars

D. Blaschke*[†], H. Grigorian*[**], D.N. Aguilera*[‡], S. Yasui*[§] and H. Toki[§]

*Fachbereich Physik, Universität Rostock, D-18051 Rostock, Germany
[†]Bogoliubov Laboratory of Theoretical Physics,
Joint Institute for Nuclear Research, 141980, Dubna, Russia
**Department of Physics, Yerevan State University, 375025 Yerevan, Armenia
[‡]Instituto de Física Rosario, Bv. 27 de febrero 210 bis, 2000 Rosario, Argentina
[§]Research Center for Nuclear Physics, Osaka University, Ibaraki 567 - 0047, Japan

Abstract. The equation of state for quark matter is derived for a nonlocal, chiral quark model within the mean field approximation. We investigate the effects of a variation of the form factors of the interaction on the phase diagram of quark matter under the condition of β- equilibrium and charge neutrality. Special emphasis is on the occurrence of a diquark condensate which signals a phase transition to color superconductivity and its effects on the equation of state. We calculate the quark star configurations by solving the Tolman- Oppenheimer- Volkoff equations and obtain for the transition from a hot, normal quark matter core of a protoneutron star to a cool diquark condensed one a release of binding energy of the order of $\Delta Mc^2 \sim 10^{53}$ erg. We study the consequences of antineutrino trapping in hot quark matter for quark star configurations with possible diquark condensation and discuss the claim that this energy could serve as an engine for explosive phenomena. A "phase diagram" for rotating compact stars (angular velocity-baryon mass plane) is suggested as a heuristic tool for obtaining constraints on the equation of state of QCD at high densities. It has a critical line dividing hadronic from quark core stars which is correlated with a local maximum of the moment of inertia and can thus be subject to experimental verification by observation of the rotational behavior of accreting compact stars.

INTRODUCTION

Color superconductivity in quark matter [1] is one interesting aspect of recent discussions devoted to the physics of compact star interiors [2]. Since calculations of the energy gap of quark pairing predict a value $\Delta \sim 100$ MeV and corresponding critical temperatures for the phase transition to the superconducting state are expected to follow the BCS relation $T_c = 0.57\ \Delta$, the question arises whether diquark condensation can lead to remarkable effects on the structure and evolution of compact objects. If positively answered, color superconductivity of quark matter could provide signatures for the detection of a deconfined phase in the interior of compact objects (pulsars, Low-mass X-ray binaries) via observations. Hong, Hsu and Sannino have conjectured [3] that the release of binding energy due to Cooper pairing of quarks in the course of protoneutron star evolution could provide an explanation for the unknown source of energy in supernovae, hypernovae or gamma-ray bursts, see also [4]. Their estimate of energy release did not take into account the change in the gravitational binding energy due to the change in the structure of the stars quark core. We will reinvestigate the question of a

CP660, *Hadron Physics: Effective Theories of Low Energy QCD*, edited by A. H. Blin et al.

possible binding energy release due to a color supercondcutivity transition by taking into account changes in the equation of state (EoS) and the configuration of the quark star selfconsistently and by including the effects of antineutrino trapping [5].

As a first step in this direction we will discuss here the two flavor color supercon-ducting (2SC) quark matter phase which occurs at lower baryon densities than the color-flavor-locking (CFL) one, see [6, 7]. We will investigate the influence of the formfactor of the interaction on the phase diagram and the EoS of dense quark matter under the conditions of charge neutrality and isospin asymmetry due to β-equilibrium relevant for compact stars.

Finally we consider the question whether the effect of diquark condensation which occurs in the earlier stages of the compact star evolution ($t \simeq 100$ s) [8, 9] at temperatures $T \sim T_c \sim 20 - 50$ MeV can be considered as an engine for exposive astrophysical phenomena like supernova explosions due to the release of a binding energy of about $10^{52} \div 10^{53}$ erg, as has been suggested before [3, 4].

THERMODYNAMICS OF A NONLOCAL CHIRAL QUARK MODEL

The phase structure of electrically and color neutral quark matter in β-equilibrium has been studied in [7] and it has been shown that at densities relevant for compact star interiors the 2SC phase should be dominant over the CFL phase. The latter one could only be stable in the very inner core and thus does not occupy a large enough volume in order to cause observable effects. Therefore, we consider in the present work two flavor quark matter in with 2SC superconductivity only. The more general case which includes the CFL phase does not invalidate the scenario developed in the following and will be studied in a subsequent work. We consider the grand canonical thermodynamical potential for 2SC quark matter within a nonlocal chiral quark model [10] where in the mean field approximation the mass gap ϕ and the diquark gap Δ appear as order parameters which can be expressed as in [11] by

$$
\begin{aligned}
\Omega_q(\phi, \Delta; \mu_q, \mu_I, T) = {} & \frac{\phi^2}{4G_1} + \frac{\Delta^2}{4G_2} - \frac{1}{\pi^2} \int_0^\infty dq q^2 \{ \omega \left[\varepsilon_r(-\mu_q - \mu_I), T \right] \\
& + \omega \left[\varepsilon_r(\mu_q - \mu_I), T \right] + \omega \left[\varepsilon_r(-\mu_q + \mu_I), T \right] + \\
& + \omega \left[\varepsilon_r(\mu_q + \mu_I), T \right] \} - \frac{2}{\pi^2} \int_0^\infty dq q^2 \{ \omega \left[\varepsilon_b(E(q) - \mu_q) - \mu_I, T \right] \\
& + \omega \left[\varepsilon_b(E(q) + \mu_q) - \mu_I, T \right] + \omega \left[\varepsilon_b(E(q) - \mu_q) + \mu_I, T \right] \\
& + \omega \left[\varepsilon_b(E(q) + \mu_q) + \mu_I, T \right] \} + \Omega_{vac} ,
\end{aligned}
\tag{1}
$$

where we have introduced the quark chemical potential $\mu_q = (\mu_u + \mu_d)/2$ and the chemical potential of the isospin asymmetry $\mu_I = (\mu_u - \mu_d)/2$ instead of the chemical potentials of up and down quark flavors. We neglect here a possible difference between the chemical potentials of paired and unpaired colors for the same quark flavor. For a more general approach see [7, 12]. The factor 2 in the last integral comes from the degeneracy of the blue and green colors ($\varepsilon_b = \varepsilon_g$). We have introduced the notation

$$\omega[\varepsilon_c, T] = T \ln\left[1 + \exp\left(-\frac{\varepsilon_c}{T}\right)\right] + \frac{\varepsilon_c}{2}, \tag{2}$$

where the first argument is given by

$$\varepsilon_c(x) = x\sqrt{1 + \Delta_c^2/x^2}, \tag{3}$$

and we assume that the green and blue colors are paired and the red one remains unpaired, so that we have

$$\Delta_c = g(q)\Delta(\delta_{c,b} + \delta_{c,g}). \tag{4}$$

The dispersion relation for unpaired quarks with dynamical mass function $m(q) = m + g(q)\phi$ is given by

$$E_f(q) = \sqrt{q^2 + m^2(q)}, \tag{5}$$

where $g(q)$ denotes the formfactor of the quark interaction, for which we employ the following models

$$g_L(q) = [1 + (q/\Lambda_L)^{2\alpha}]^{-1}, \quad \alpha > 1, \tag{6}$$
$$g_G(q) = \exp(-q^2/\Lambda_G^2), \tag{7}$$
$$g_{NJL}(q) = \theta(1 - q/\Lambda_{NJL}). \tag{8}$$

The Lorentzian (L) momentum distribution can interpolate between a soft (Gaussian-type, $\alpha \sim 2$) and a hard cutoff (NJL, $\alpha > 30$) depending on the parameter α. The parametrization of the model can be found in Refs. [13, 10]. The contribution from the leptons should be added to the quark thermodynamical potential Ω_q in order to obtain the total one

$$\Omega(\phi, \Delta; \mu_q, \mu_I, \mu_e, \mu_{\bar{\nu}_e}, T) = \Omega_q(\phi, \Delta; \mu_q, \mu_I, T) + \sum_{l \in \{e, \bar{\nu}_e\}} \Omega^{id}(\mu_l, T), \tag{9}$$

where

$$\Omega^{id}(\mu, T) = -\frac{1}{12\pi^2}\mu^4 - \frac{1}{6}\mu^2 T^2 - \frac{7}{180}\pi^2 T^4 \tag{10}$$

is the thermodynamical potential for an ideal gas of massless fermions. The stellar matter in the quark core of compact stars consists of u and d quarks, electrons e and antineutrinos $\bar{\nu}_e$ under the conditions of β-equilibrium: $d \longleftrightarrow u + e^- + \bar{\nu}_e$, which in terms of chemical potentials reads $\mu_e + \mu_{\bar{\nu}_e} = -2\mu_I$, and charge neutrality: $\frac{2}{3}n_u - \frac{1}{3}n_d - n_e = 0$, which also could be written as $n_q + 3n_I - 6n_e = 0$. The number densities n_j, $j \in \{u, d, e, \bar{\nu}_e, q, I\}$ are defined as

$$n_j = -\frac{\partial\Omega}{\partial\mu_j}\bigg|_{\phi_0, \Delta_0; T}. \tag{11}$$

The conditions for the local extrema of Ω_q, correspond to coupled gap equations for the two order parameters ϕ and Δ

$$\left.\frac{\partial \Omega}{\partial \phi}\right|_{\phi=\phi_0,\Delta=\Delta_0} = \left.\frac{\partial \Omega}{\partial \Delta}\right|_{\phi=\phi_0,\Delta=\Delta_0} = 0 . \tag{12}$$

The global minumum of Ω_q represents the state of thermodynamical equilibrium from which all equations of state can be obtained by derivation. In the following subsections we want to comment on aspects of this stellar matter model which turn out to be essential for the discussion of quark matter in compact stars.

Effect of formfactors

The nonvanishing of the order parameters ϕ or Δ signals the presence of a phase with broken chiral symmetry or color superconductivity, respectively. In Fig. 1 we show the resulting phase diagram of quark star matter under the above constraints and neglecting the CFL phase which should appear only at such high densities that it will at best occupy a negligible volume in the very inner core of a compact star configuration. From Fig. 1

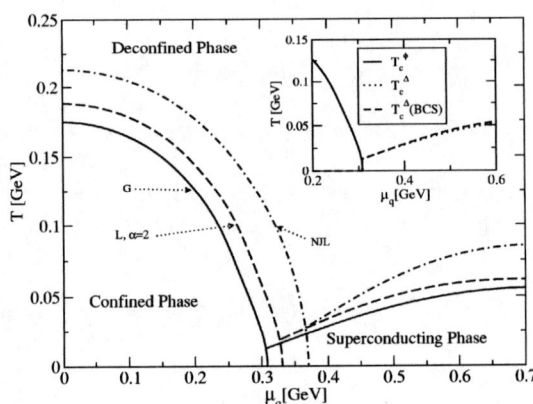

FIGURE 1. Phase diagrams for different form factors: Gaussian (solid lines), Lorentzian $\alpha = 2$ (dashed lines) and NJL (dash-dotted). In the upper panel the comparison with the BCS formula for $T_c = 0.57\,\Delta(T = 0,\mu_q)\,g(\mu_q)$ is shown for the Gaussian model.

we see that the softer the formfactor $g(q)$ the lower the critical temperatures and chemical potentials for the phase transition to quark matter with vanishing order parameters or to color superconducting quark matter at low temperatures. It is remarkable that a modified BCS relation for the critical 2SC temperature holds [10].

Antineutrino trapping

The results for the solution of the gap equations (12) are shown in Fig. 2 (left panel) for different values of the antineutrino chemical potential which is a measure for the density of trapped antineutrinos in a hot, young protoneutron star. For values $\mu_{\bar{\nu}_e} > 72$ MeV the flavor asymmetry becomes large enough to prevent diquark pairing and therefore color superconductivity at low densities. Simultaneously, the onset density for quark matter

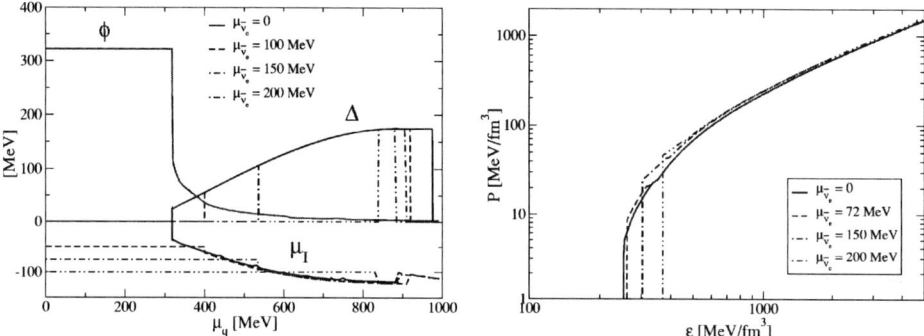

FIGURE 2. Left panel: Mass gap ϕ, diquark gap Δ and isospin chemical potential μ_I as a function of the quark chemical potential μ_q for different values of the antineutrino chemical potential $\mu_{\bar{\nu}_e}$. Solutions obey β-equilibrium and charge neutrality conditions. Right panel: Pressure vs. energy density for different values of the antineutrino chemical potential $\mu_{\bar{\nu}_e}$. The onset of the appeareance of quark matter is shifted to higher energy densities due to antineutrino trapping.

occurence is shifted to higher densities, see Fig. 2 (right panel). These solutions will be used for the discussion of a scenario of hot quark star evolution in the next Section. Before that we have to consider the question whether the quark-hadron phase transition would occur at too high energy densities so that no quark core could exist in a hybrid compact star.

Quark hadron phase transition

We construct a quark hadron phase transition at zero temperature using a linear and a nonlinear Walecka model for the hadronic phase and perform a Maxwell construction for the phase transition. Although it has been claimed that this procedure might be in contradiction with the Gibbs conditions for phase equilibrium when more than one conserved charge exists in the system [14], recent reinvestigation including charge screening has revealed the opposite [15]. The resulting EoS with a deconfinement transition confirms that the lowest critical phase transition (energy) density is obtained for the softest (Gaussian) formfactor model, see Fig. 3.

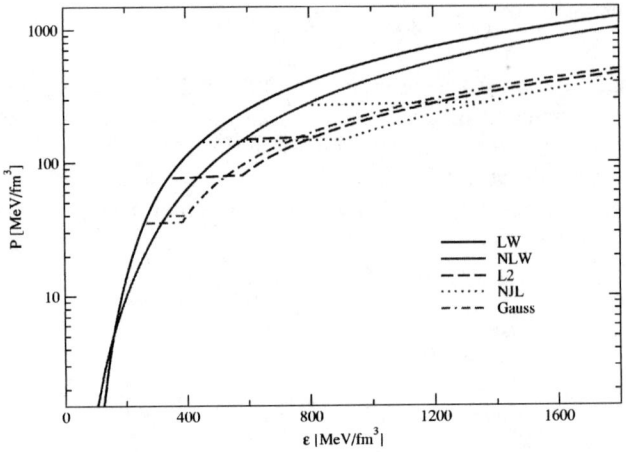

FIGURE 3. EoS of compact star matter with quark hadron phase transition. Hadronic EoS: linear (LW) and nonlinear (NLW) Walecka model, quark matter: nonlocal separable model with 2SC quark matter and different form factors: Gaussian (Gauss), Lorentzian $\alpha = 2$ (L2) and cutoff (NJL).

COMPACT STAR CONFIGURATIONS

What compact star configurations will result from the use of the above EoS? Will a stable compact star with quark matter core be obtained? We will answer these questions in the present Section by solving the Tolman-Oppenheimer-Volkoff equations for the case without rotation and by applying the perturbative method of solution of Einstein equations for stationary rigid rotation. First we consider quark stars without hadronic shell which can be thought of as the simplified models for a compact star interior and after that we discuss hybrid stars.

Quark stars - engine for explosive phenomena?

The engine which drives supernova explosions and gamma ray bursts being among the most energetic phenomena in the universe remains still puzzling. The phase transiton to a quark matter phase may be a mechanism that could release such an amount of energy [16, 17]. It has been proposed that due to the Cooper instability in dense Fermi gas cold dense quark matter shall be in the color superconducting state with a nonvanishing diquark condensate [18, 2]. The consequences of diquark condensation for the cooling of compact stars due to changes in the transport properties and neutrino emissivities have been investigated much in detail, see [19, 20, 8, 10], and may even contribute to the explanation of the relative low temperature of the pulsar in the supernova remmant 3C58

[21]. Unlike the case of normal (electronic) superconductors, the pairing energy gap in quark matter is of the order of the Fermi energy so that diquark condensation gives considerable contributions to the equation of state (EoS) of the order of $(\Delta/\mu)^2$. Therefore, it has been suggested that there might be scenarios which identify the unknown source of the energy of 10^{53} erg with a release of binding energy due to Cooper pairing of quarks in the core of a cooling protoneutron star [3]. In that work the total diquark condensation energy released in a bounce of the core is estimated as $(\Delta/\mu)^2 M_{core}$ corresponding to a few percent of a solar mass, that is 10^{52} erg. It has been shown in [10] by solving the selfconsistent problem of the star configurations, however, that these effects due to the softening of the EoS in the diquark condensation transition lead to an increase in the gravitational mass of the star contrary to the naive estimates and that no explosion occurs.

Therefore, we have suggested a new mechanism of energy release [5] which involves a first order phase transition induced by antineutrino trapping. This phenomenon occurs in hot compact star configurations at temperatures $T \geq 1$ MeV where the mean free path of (anti-)neutrinos becomes smaller than the typical size of a star [22]. During the

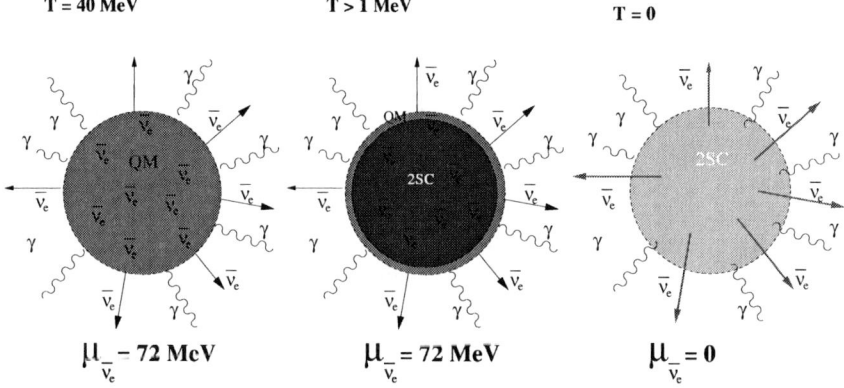

FIGURE 4. Left two graphs: Quark star cooling by neutrino and photon emission from the surface in the case of antineutrino trapping when $T > 1$ MeV. Right graph: antineutrino untrapping and burst-type release.

collapse in the hot era of protoneutron star evolution, antineutrinos are produced due to the β-processes. Since they have a small mean free path, they cannot escape and the asymmetry in the system is increased. This entails that the formation of the diquark condensate is shifted to higher densities or even inhibited depending on the fraction of trapped antineutrinos. As the quark star cools, a two-phase structure will occur. Despite of the asymmetry, the interior of the quark star (because of its large density) could consist of color superconducting quark matter, whereas in the more dilute outer shell, diquark condensation cannot occur and quark matter is in the normal state, opaque to antineutrinos for $T \geq 1$ MeV. When in the continued cooling process the antineutrino mean free path increases above the size of this normal matter shell, an outburst of neutrinos occurs and gives arise to a released energy of the order of $10^{53} - 10^{54}$ erg. This untrapping transition is of first order and could lead to an explosive phenomenon. Three stages of this scenario of hot quark star evolution are illustrated in Fig. 4. In Fig. 5 these

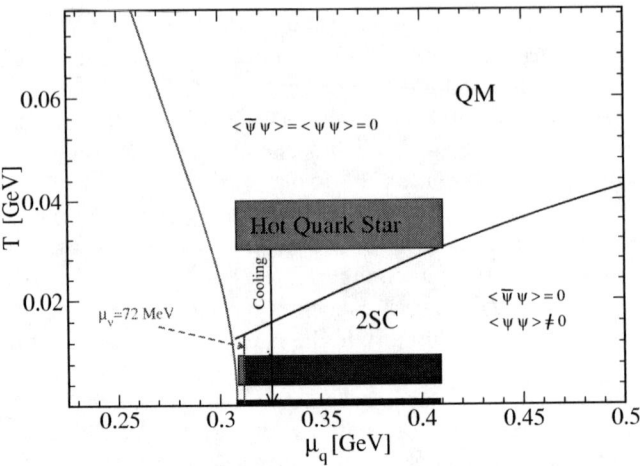

FIGURE 5. Three stages of quark star cooling in the phase diagram corresponding to Fig. 4.

stages are shown schematically in the quark matter phase diagram. From Fig. 2 (right panel) we can see that the EoS without antineutrinos is softer than with antineutrinos (it has a lower pressure at a given energy density) and therefore allows more compact configurations (Fig. 6, left part of left panel). The presence of antineutrinos tends to increase the mass of the star for a given central density (Fig. 6, right part of left panel).

To estimate the effect of antineutrinos on star configurations we choose a reference configuration without antineutrinos with the mass of a typical neutron star $M_f = 1.4\,M_\odot$ (see Fig. 6). The corresponding radius is $R_f = 10.29$ km and the central density $n_q = 10.5\,n_0$, where $n_0 = 0.16$ fm^{-3} is the saturation density. The configurations with trapped antineutrinos and nonvanishing $\mu_{\bar{\nu}_e}$ to compare with we choose to have the same total baryon number as the reference star: $N_B = 1.48\,N_\odot$, where $N_\odot$ is the total baryon number of the sun. For $\mu_{\bar{\nu}_e} = 72$ MeV we obtain $M_A = 1.47\,M_\odot$ and for $\mu_{\bar{\nu}_e} = 150$ MeV, $M_B = 1.72\,M_\odot$. The differences in the radii are $R_A - R_f = 0.1$ km and $R_B - R_f = 0.3$ km and in the central densities $n_q^A - n_q^f = 0.6\,n_0$ and $n_q^B - n_q^f = 2.1\,n_0$, respectively. This is a consequence of the hardening of the EoS due to the presence of antineutrinos. The mass defect $\Delta M_{if} = M_i - M_f$ can be interpreted as an energy release if there is a process which relates the configurations with M_i and M_f being the initial and final states, respectively. In the right panel of Fig. 6 we show limits for the binding energy release as a function of the final state mass when conservative values for the antineutrino chemical potential during the trapping are chosen. The antineutrino untrapping transition results in a first order phase transition which gives rise to an explosive release of the binding energy, as required for a scenario which should explain the engine of supernovae or gamma-ray bursts [5].

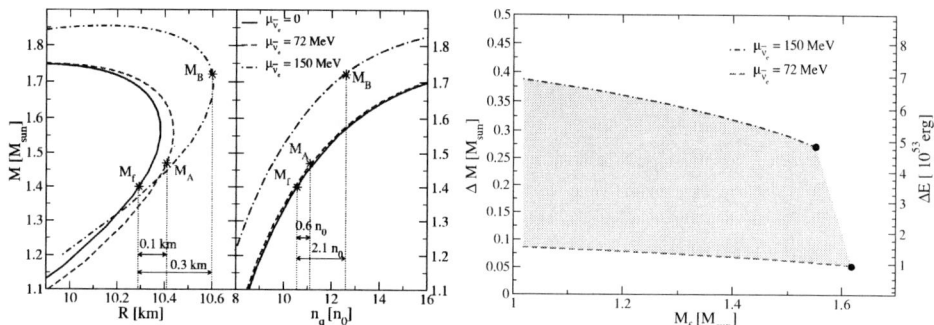

FIGURE 6. Left panel: Quark star configurations for different antineutrino chemical potentials $\mu_{\bar{\nu}_e} = 0$, 72, 150 MeV. The total mass M in solar masses M_{sun} is shown as a function of the radius R (left panel) and as a function of the central number density n_q in units of the nuclear number density n_0 (right panel). Asterisks denote configurations with the same total baryon number. Right panel: Mass defect ΔM and corresponding energy release ΔE due to antineutrino untrapping as a function of the mass of the final state M_f. The shaded region is defined by the estimates for the upper and lower limits of the antineutrino chemical potential in the initial state $\mu_{\bar{\nu}_e} = 150$ MeV (dashed-dotted line) and $\mu_{\bar{\nu}_e} = 72$ MeV (dashed line), respectively.

Hybrid stars

Once the EoS with a quark-hadron phase transition is defined and the constraints of beta equilibrium and charge neutrality are obeyed (see previous Section), the corresponding hybrid star configurations are obtained from the solution of Einsteins equations [23].

Is a quark core possible?

The answer to this question depends cruically on the employed EoS. Within the setting of the present model, stable quark matter cores are possible for a Gaussian formfactor model but not for a cut-off one (NJL). This confirms on the one hand conclusions previously obtained within other approaches using an NJL model [24, 25] for quark matter but on the other hand presents an alternative quark matter model for which quark cores are possible, see Fig. 7.

Cooling curves and SNR 3C58

In order to obtain the cooling behavior of the presented hybrid star model, we employ the program code developed recently [19, 8] so as to describe 2SC quark matter and investigate the dependence on the compact star mass [21]. The result shown in Fig. 8

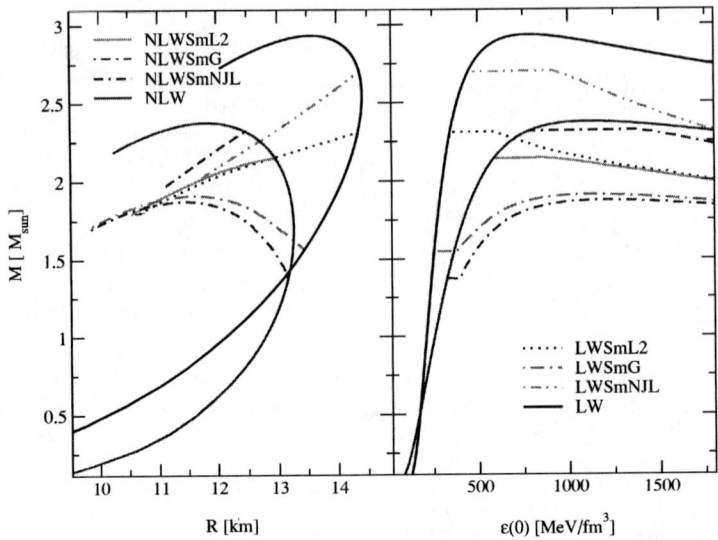

FIGURE 7. Mass - radius and mass - central energy density curves for hybrid stars. Stable quark matter cores are possible for a Gaussian formfactor model but not for a cut-off one (NJL).

demonstrates that a massive star may have a large enough quark core to entail an enhaced cooling behavior in accordance with the recent data point reported by CHANDRA measurements of the pulsar in the supernova remnant 3C58. This is, however, not a unique feature of a quark matter interior and could also be explained by other exotic phases of dense matter [26].

Population gap for accreting LMXBs

Our focus is on the elucidation of qualitative features of signals from the high density phase transition in the pulsar timing, therefore we use a generic form of an equation of state (EoS) with such a transition. We use the polytropic type equation of state for different values of the incompressibility [23] $K_{L,H}(n) = 9 \, dP/dn$ at the saturation density, see Ref. [27]. The phase transition between the lower and higher density phases is made by the Maxwell construction and compared to a relativistic mean field model consisting of a linear Walecka plus dynamical quark model EoS with a Gibbs construction, Fig. 9.

We introduce a classification of rotating compact stars in the plane of their angular frequency Ω and mass (baryon number N) which we will call *phase diagram*. In this diagram, configurations with high density matter cores are separated from conventional ones by a critical phase transition line. The position and the form of these lines are

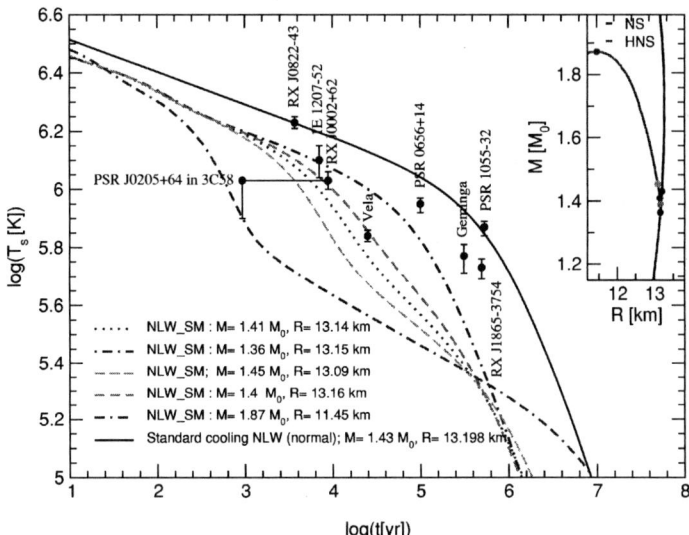

FIGURE 8. Cooling evolution of the hybrid stars with different masses. The plot in right corner on the mass - radius plain with dots fingers that configurations for which the cooling evolution is shown. The oservational data are taken from [26].

sensitive to changes in the equation of state of stellar matter [2].

In Fig. 10 we display the phase diagrams for the rotating star configurations, which correspond to the three model EoS of Fig. 9. These phase diagrams have four regions: (i) the region above the maximum frequency $\Omega > \Omega_K(N)$ where no stationary rotating configurations are found, (ii) the region of black holes $N > N_{max}(\Omega)$, and the region of stable compact stars which is subdivided by the critical line $N_{crit}(\Omega)$ into (iii) the region of hybrid stars for $N > N_{crit}(\Omega)$ where configurations contain a core with a second, high density phase and (iv) the region of mono-phase stars without such a core.

From the comparison of the regional structure of these three different phase diagrams in Fig. 10 with the corresponding EoS in Fig. 9 we arrive at the main result of this paper that there are the following correlations between the topology of the lines $N_{max}(\Omega)$ and $N_{crit}(\Omega)$ and the properties of two-phase EoS:

- The hardness of the high density EoS determines the maximum mass of the star, which is given by the line $N_{max}(\Omega)$. Therefore $N_{max}(0)$ is proportional to the parameter $K_H(n_H)$, where n_H is the density of the transition to high density phase.
- The onset of the phase transition line $N_{crit}(0)$ depends on the density n_H and $K_L(n_L)$ where n_L is the density of the transition to the low density phase.

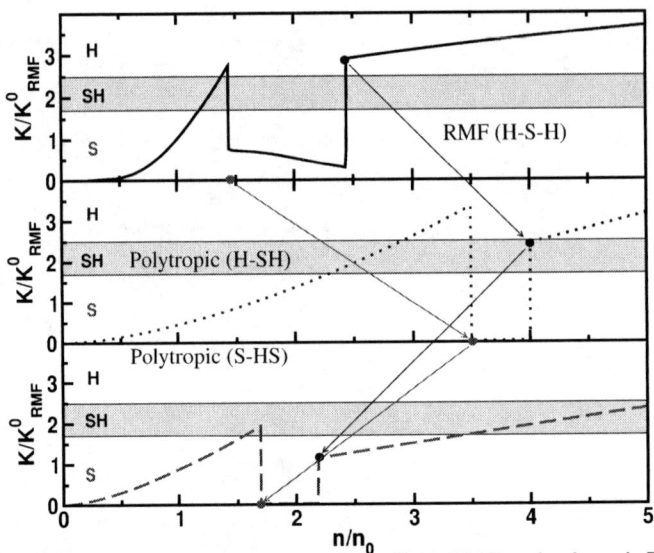

FIGURE 9. Incompressibilities for Relativistic Mean Field (RMF) and polytropic EoS models with a phase transition, see [27].

- The curvature of the lines $N_{max}(\Omega)$ and $N_{crit}(\Omega)$ is proportional to the compressibility of the high and low density phases, respectively.

Therefore, a verification of the existence of the critical lines $N_{crit}(\Omega)$ and $N_{max}(\Omega)$ by observation of the rotational behavior of compact objects would constrain the parameters of the EoS for neutron star matter. We have investigated different trajectories of rotating compact star evolution in the phase diagram in order to identify scenarios, which result in signatures of the deconfinement phase transition [27]. A key evolutionary track is accretion with strong magnetic fields [28]. For this case the $\dot{\Omega}$ first decreases as long as the moment of inertia monotonously increases with N. When passing the critical line $N_{crit}(\Omega)$ for the phase transition, the moment of inertia starts decreasing and the internal torque term K_{int} changes sign. This leads to a narrow dip for $\dot{\Omega}(N)$ in the vicinity of this line. As a result, the phase diagram gets overpopulated for $N \overset{<}{\sim} N_{crit}(\Omega)$ and depopulated for $N \overset{>}{\sim} N_{crit}(\Omega)$ up to the second maximum of $I(N,\Omega)$ close to the black-hole line $N_{max}(\Omega)$. A *population gap* in the phase diagram of compact stars appears as a detectable indicator for hybrid star configurations.

CONCLUSIONS

We have investigated the influence of the diquark condensation on the thermodynamics of the quark matter under the conditions of β-equilibrium and charge neutrality relevant for the discussion of compact stars. The EoS has been derived for a nonlocal chiral

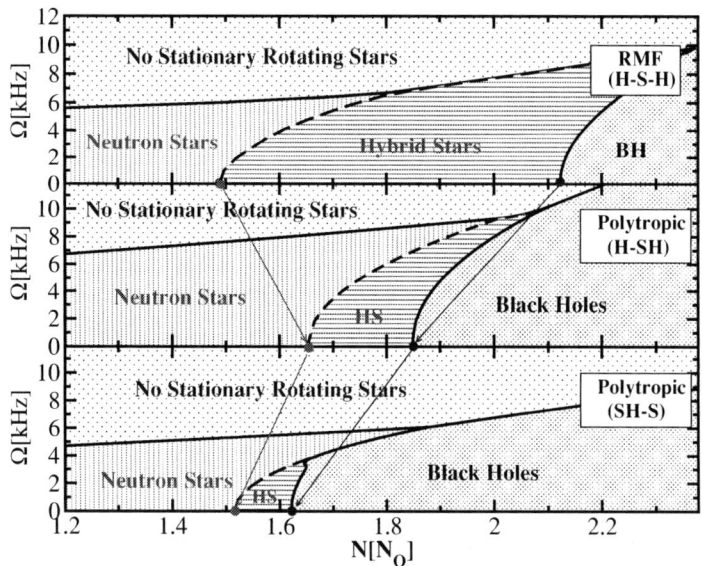

FIGURE 10. Phase diagrams for rotating star configurations corresponding to three model EoS discussed in the text.

quark model in the mean field approximation, and the influence of different formfactors (Gaussian, Lorentzian, NJL) has been studied. We have shown that the smoothness of the interaction changes the critical temperatures and chemical potentials for the onset of the phase transition to lower values.

The phase transition to color superconducting quark matter from the lower density regions at small temperatures ($T < 5 \div 10$ MeV) is of first order, while the melting of the diquark condensate and the corresponding transition to normal quark matter at high temperatures is of second order. The presence of flavor asymmetry due to β-equilibrium in quark matter does not destroy the diquark condensate since the electron fraction $n_e/n_{total} < 0.01$ is too small. The masses of the quark core configurations could be up to 1.7 M_{sun} and the radii could be up to 11 km.

We have investigated the effects of trapped antineutrinos on the asymmetry and diquark condensates in a quark star configurations. By comparing configurations with fixed baryon number the release of energy in an antineutrino untrapping transition is estimated to be of the order of 10^{53} erg. Such a transition is of first order so that antineutrinos can be released in a sudden process (burst). This scenario could play an important role to solve the problem of the engine of supernova explosions and gamma ray bursts. A detailed neutrino transport and cooling calculation should be taken into account in a future work. A second antineutrino pulse is suggested as an observable characteristics of the present scenario.

As there are still many unknowns in the picture we have drawn in this contribution for the possible effects of quark matter and color superconductivity on compact stars we would like to point out that ther will not be one "smoking gun" type signal of quark matter but rather the different facets of the picture which emerges for characteristic features of compact stars containing quark matter should all match in the big puzzle. To bring the pieces together is a task very similar to that of quark gluon plasma search in heavy-ion collision experiments.

ACKNOWLEDGMENTS

The research of D.N. Aguilera has been supported by DFG Graduiertenkolleg 567 "Stark korrelierte Vielteilchensysteme", by the CONICET (Argentina) and by DAAD grant No. A/01/17862. H.G. acknowledges support by DFG under grant No. 436 ARM 17/5/01. S.Y. was supported by DAAD HOST program during his stay at Rostock University. Part of the results reported in these Proceedings have been obtained with our collaborators S. Fredriksson, A.M. Öztas, G. Poghosyan and D.N. Voskresensky.

REFERENCES

1. Rajagopal, K., and Wilczek, F., *At the Frontier of Particle Physics / Handbook of QCD*, World Scientific, Singapore, 2000, vol. 3, chap. The Condensed Matter Physics of QCD, p. 2061.
2. Blaschke, D., Glendenning, N. K., and Sedrakian, A., editors, *Physics of neutron star interiors*, Springer, Heidelberg, 2001.
3. Hong, D. K., Hsu, S. D. H., and Sannino, F., *Phys. Lett.*, **B516**, 362 (2001).
4. Ouyed, R., *eConf*, **C010815**, 209 (2002).
5. Aguilera, D. N., Blaschke, D., and Grigorian, H., *arXiv:astro-ph/0212237* (2002).
6. Steiner, A. W., Reddy, S., and Prakash, M., *arXiv:hep-ph/0205201* (2002).
7. Neumann, F., Buballa, M., and Oertel, M., *arXiv:hep-ph/0210078* (2002).
8. Blaschke, D., Grigorian, H., and Voskresensky, D. N., *Astron. Astrophys.*, **368**, 561 (2001).
9. Carter, G. W., and Reddy, S., *Phys. Rev.*, **D62**, 103002 (2000).
10. Blaschke, D., Fredriksson, S., Grigorian, H., and Oztas, A. M., *arXiv:nucl-th/0301002* (2003).
11. Kiriyama, O., Yasui, S., and Toki, H., *Int. J. Mod. Phys.*, **E10**, 501 (2001).
12. Huang, M., Zhuang, P.-f., and Chao, W.-q., *arXiv:hep-ph/0207008* (2002).
13. Schmidt, S. M., Blaschke, D., and Kalinovsky, Y. L., *Phys. Rev.*, **C50**, 435 (1994).
14. Glendenning, N. K., *Phys. Rev.*, **D46**, 1274 (1992).
15. Voskresensky, D. N., Yasuhira, M., and Tatsumi, T., *Phys. Lett.*, **B541**, 93 (2002).
16. Drago, A., and Tambini, U., *J. Phys.*, **G25**, 971 (1999).
17. Berezhiani, Z., Bombaci, I., Drago, A., Frontera, F., and Lavagno, A., *arXiv:astro-ph/0209257* (2002).
18. Alford, M. G., Bowers, J. A., and Rajagopal, K., *J. Phys.*, **G27**, 541 (2001).
19. Blaschke, D., Klahn, T., and Voskresensky, D. N., *Astrophys. J.*, **533**, 406 (2000).
20. Page, D., Prakash, M., Lattimer, J. M., and Steiner, A., *Phys. Rev. Lett.*, **85**, 2048 (2000).
21. Grigorian, H., Blaschke, D., and Voskresensky, D. N., *MPG-VT-UR 231/02* (2002).
22. Prakash, M., Lattimer, J. M., Sawyer, R. F., and Volkas, R. R., *Ann. Rev. Nucl. Part. Sci.*, **51**, 295 (2001).
23. Glendenning, N. K., *Compact stars: Nuclear physics, particle physics, and general relativity*, Springer, New York, 1997.
24. Schertler, K., Leupold, S., and Schaffner-Bielich, J., *Phys. Rev.*, **C60**, 025801 (1999).
25. Baldo, M., et al., *arXiv:nucl-th/0212096* (2002).
26. Yakovlev, D. G., Kaminker, A. D., Haensel, P., and Gnedin, O. Y., *Astron. Astrophys.*, **389**, L24 (2002).
27. Blaschke, D., Grigorian, H., and Poghosyan, G., *arXiv:astro-ph/0208332* (2002).
28. Poghosyan, G. S., Grigorian, H., and Blaschke, D., *Astrophys. J.*, **551**, L73 (2001).

Effects of the $\rho - \omega$ mixing interaction in relativistic models

D.P.Menezes[*] and C. Providência[†]

[*]Dep. de Física - CFM - Universidade Federal de Santa Catarina - Florianópolis - SC - CP. 476 - CEP 88.040 - 900 - Brazil
[†]Centro de Física Teórica - Dep. de Física - Universidade de Coimbra - P-3004 - 516 Coimbra - Portugal

Abstract. The effects of the $\rho - \omega$ mixing term in infinite nuclear matter and in finite nuclei are investigated with the non-linear Walecka model in a Thomas-Fermi approximation. For infinite nuclear matter the influence of the mixing term in the binding energy calculated with the NL3 and TM1 parametrizations can be neglected. Its influence on the symmetry energy is only felt for the TM1 with a unrealistically large value for the mixing term strength. For finite nuclei the contribution of the isospin mixing term is very large as compared with the expected value to solve the Nolen-Schiffer anomaly.

INTRODUCTION

The determination of the properties of nuclear matter as functions of density, temperature and the neutron-proton composition is an important problem in contemporary nuclear and astrophysics. One can study problems related with the liquid-gas phase transition and possible droplet formation in a vapor system of asymmetric nuclear matter at $T = 0$ and also at finite temperature within the framework of relativistic models [1]. The properties of the arising droplets as neutron skin thickness, surface energy, central density, etc, can be evaluated. In the present work we show the importance of the $\rho - \omega$ mixing in asymmetric nuclear matter and also in finite nuclei in the context of the non-linear Walecka model within the Thomas-Fermi approximation for the TM1 and NL3 parameter sets.

On the other hand, the Nolen-Schiffer anomaly (NSA) [2] is known as the failure of nuclear theory in explaining the experimental mass differences between mirror nuclei. Some years ago, a Hartree-Fock approach was employed to obtain the binding energy and the self-energies for proton and neutron in the Walecka model [3]. The influence of the $\rho - \omega$ mixing term was investigated and shown to be of crucial importance in explaining the NSA in a qualitative way. In what follows we also check whether our formalism can explain the discrepancy between mirror nuclei masses.

The non-linear Walecka model lagrangian density reads:

$$\mathscr{L} = \bar{\psi} \left[\gamma_\mu \left(i\partial^\mu - g_v V^\mu - \frac{g_\rho}{2}\vec{\tau}\cdot\vec{b}^\mu - eA^\mu \frac{(1+\tau_3)}{2} \right) - (M - g_s\phi) \right] \psi$$

CP660, Hadron Physics: Effective Theories of Low Energy QCD, edited by A. H. Blin et al.

$$+\frac{1}{2}(\partial_\mu\phi\partial^\mu\phi - m_s^2\phi^2) - \frac{1}{3!}\kappa\phi^3 - \frac{1}{4!}\lambda\phi^4$$

$$-\frac{1}{4}\Omega_{\mu\nu}\Omega^{\mu\nu} + \frac{1}{2}m_v^2 V_\mu V^\mu + \frac{1}{4!}\xi g_v^4(V_\mu V^\mu)^2$$

$$-\frac{1}{4}\vec{B}_{\mu\nu}\cdot\vec{B}^{\mu\nu} + \frac{1}{2}m_\rho^2\vec{b}_\mu\cdot\vec{b}^\mu - \frac{1}{4}F_{\mu\nu}F^{\mu\nu}$$

$$+\Lambda \mathbf{b}_\mu V^\mu.$$

where $\Omega_{\mu\nu} = \partial_\mu V_\nu - \partial_\nu V_\mu$, $\vec{B}_{\mu\nu} = \partial_\mu\vec{b}_\nu - \partial_\nu\vec{b}_\mu - g_\rho(\vec{b}_\mu\times\vec{b}_\nu)$ and $F_{\mu\nu} = \partial_\mu A_\nu - \partial_\nu A_\mu$. The equation of state (EOS) of this model depends on the correct parametrization of three coupling constants g_s, g_v and g_ρ of the mesons to the nucleons, the nucleon mass M, the masses of the mesons m_s, m_v, m_ρ, the electromagnetic coupling constant $e = \sqrt{\frac{4\pi}{137}}$, the self-interacting coupling constants κ, λ and ξ and $\Lambda = -2m_\rho < V_0|H|b_0 >= -4500 \text{ MeV}^2$ [4], where $b_3^\mu = (b_0, \mathbf{b})$. In what follows, we use two parameter sets, TM1 [5] and NL3 [7].

Looking for a static solution requires the minimization of the thermodynamic potential:

$$\Omega = E - \sum_{i=p,n} \mu_i B_i,$$

with appropriate solutions for the fields coming from

$$\nabla^2\phi = m_s^2\phi + \frac{1}{2}\kappa\phi^2 + \frac{1}{3!}\lambda\phi^3 - g_s\rho_s, \tag{1}$$

$$\nabla^2 b_0 = m_\rho^2 b_0 - \frac{g_\rho}{2}\rho_3 + \Lambda V_0, \tag{2}$$

$$\nabla^2 V_0 = m_v^2 V_0 + \frac{\xi g_v^4}{6}V_0^3 - g_v\rho_B + \Lambda b_0 \tag{3}$$

$$\nabla^2 A_0 = -e\rho_p, \tag{4}$$

where

$$\rho_s = 2\sum_{i=p,n}\int_0^{k_{Fi}}\frac{d^3p}{(2\pi)^3}\frac{M^*}{\varepsilon},$$

$$\varepsilon = \sqrt{p^2 + M*^2}, \qquad M* = M - g_s\phi,$$

$$\mu_p = \sqrt{k_{Fp}^2 + M*^2} + g_v V_0 + \frac{g_\rho}{2}b_0 + eA_0 \qquad \mu_n = \sqrt{k_{Fn}^2 + M*^2} + g_v V_0 - \frac{g_\rho}{2}b_0,$$

where μ_p (μ_n) is the chemical potential for protons (neutrons) and is related with the Fermi momentum and

$$\rho_i = \frac{k_{Fi}^3}{3\pi^2}, \quad i = p, n,$$

with

$$\rho_B = \rho_p + \rho_n \qquad \rho_3 = \rho_p - \rho_n.$$

Infinite matter

We perform a mean field approximation, where the fields are taken as their expectation values, in order study the influence of the $\rho - \omega$ mixing in nuclear matter properties. The equations of motion become

$$\phi_0 = <\phi> \rightarrow \phi_0 = -\frac{\kappa}{2m_s^2}\phi_0^2 - \frac{\lambda}{6m_s^2}\phi_0^3 + \frac{g_s}{m_s^2}\rho_s,$$

$$V_0 = <V_0> \rightarrow V_0 = -\frac{\xi g_v^4}{6m_v^2}V_0^3 + \frac{g_v}{m_v^2}\rho_B - \Lambda b_0,$$

$$b_0 = <b_0> \rightarrow b_0 = \frac{g_\rho}{2m_\rho^2}\rho_3 - \Lambda V_0.$$

and the energy density can be written as:

$$\mathscr{E} = \frac{1}{\pi^2}\sum_{i=p,n}\int_0^{k_{Fi}}p^2 dp\sqrt{p^2+M^{*2}}$$

$$+\frac{m_v^2}{2}V_0^2 + \frac{\xi g_v^4}{8}V_0^4 + \frac{m_\rho^2}{2}b_0^2 + \frac{m_s^2}{2}\phi_0^2 + \frac{\kappa}{6}\phi_0^3 + \frac{\lambda}{24}\phi_0^4 - \Lambda V_0 b_0.$$

Plotting the binding energy for various values of Λ, we have observed that the mixing term can really be neglected for both parameter sets [6] in infinite nuclear matter and for any proton fraction. The nuclear bulk symmetry energy is defined [8] as

$$\mathscr{E}_{sym} = \frac{1}{2}\frac{\partial^2\mathscr{E}}{\partial\delta^2}\bigg|_{\delta=0}, \tag{5}$$

with $\delta = \frac{\rho_3}{\rho_B}$ and which can be analitically rewritten as

$$\mathscr{E}_{sym} = \frac{k_F^2}{6\varepsilon}\rho_B + \mathscr{E}_{sym1}$$

where

$$k_{Fp} = k_F(1+\delta)^{1/3}, \qquad k_{Fn} = k_F(1-\delta)^{1/3},$$

with $k_F = (1.5\pi^2\rho_B)^{1/3}$. For both parametrizations NL3 and TM1 the first order correction coming from the $\rho - \omega$ mixing term is quadratic in Λ^2. For the NL3 parametrization and to second order in Λ we get a simple expression linear in the barionic density

$$\mathscr{E}_{sym1} = \frac{g_\rho^2}{8m_\rho^2}\rho_B\left(1+\Lambda'^2\right) + \mathscr{O}(\Lambda'^4),$$

where $\Lambda' = \frac{\Lambda}{m_v m_\rho}$. For this force, the influence of the $\rho - \omega$ term on the symmetry energy is negligible, even if we consider a value of Λ ten times larger than the presently accepted

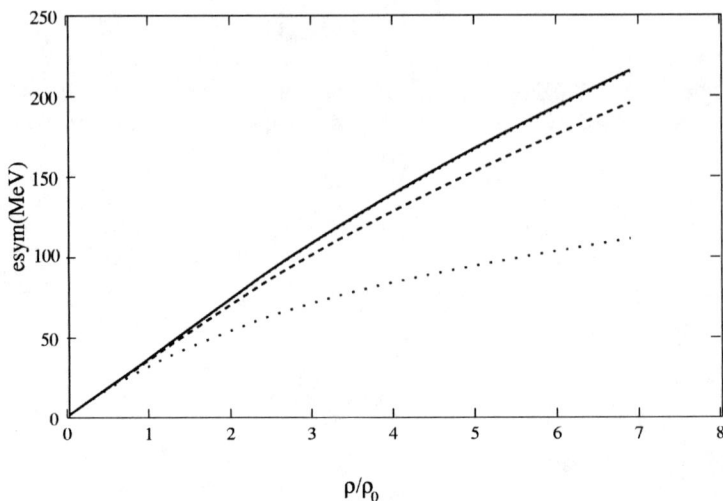

FIGURE 1. Symmetry energy with $\Lambda = 0$ (solid line), $\Lambda = -4500$ MeV^2 (long dashed line) $\Lambda = -20000$ MeV^2 (dashed line) and $\Lambda = -45000$ MeV^2 (dotted line)

experimental value. In fact for $\Lambda = -4500$ MeV^2 [4], $\Lambda' = -0.0075$. Something different happens with the TM1 parametrization since the lowest order correction in Λ has a non linear dependence in ρ_B. It can be seen that the effect of the non-linear terms is already non-negligible for the saturation density.

In fig. 1 the bulk symmetry energy is plotted as a function of the density for different values of Λ, i.e., 0, -4500, -20000 and -45000 MeV^2 . For $\Lambda = -4500$ MeV^2 the effect of the $\rho - \omega$ mixing term is indeed very small. However for a 4 or 10 times greater value the effect would be important at high densities, but these large strenght for the mixing term are not supported by the experiments [4].

In order to obtain the boundary conditions for the coupled differential equations we study the liquid-gas phase coexistence in infinite matter, where mechanical stability

$$\left(\frac{\partial P}{\partial \rho_B} \right)_{Y_p} \geq 0,$$

where P is the pressure, $Y_p = \rho_p/\rho_B$ is the proton fraction and diffusive stability

$$\left(\frac{\partial \mu_p}{\partial Y_p} \right)_{P,T} \geq 0 \qquad \left(\frac{\partial \mu_n}{\partial Y_p} \right)_{P,T} \leq 0$$

are required. The geometrical construction is performed and the binodal section is then built [1], from where points with the same pressure and different proton fractions are obtained and used as boundary conditions for the second order differential equations.

Finite nuclei

As a second step, the coupled differential equations (1)-(4) are solved numerically and all relevant quantities which depend on the fields are then calculated.

The binding energy per nucleon reads:

$$\frac{B}{A} = \varepsilon - M , \quad A = Z+N , \quad \varepsilon = \int \frac{\mathscr{E}(r)}{A} d^3r$$

where

$$\mathscr{E}(r) = 2\sum_i \int_0^{k_{Fi}} \frac{d^3p}{(2\pi)^3} \left(\sqrt{p^2+M^{*2}} + \mathcal{V}_{i0} \right) + \mathscr{E}_{fields},$$

$$\mathscr{E}_{fields} = \frac{1}{2} \left[(\nabla \phi)^2 - (\nabla V_0)^2 - (\nabla b_0)^2 - (\nabla A_0)^2 \right] +$$

$$+ \frac{1}{2} \left[m_s^2 \phi^2 + \frac{2}{3!} \kappa \phi^3 + \frac{2}{4!} \lambda \phi^4 - m_v^2 V_0^2 - \frac{2}{4!} \xi g_v^4 V_0^4 - m_\rho^2 b_0^2 - 2\Lambda V_0 b_0 \right].$$

In analogy with the above definition, we have defined the proton binding energy per proton and the neutron binding energy per neutron as

$$\frac{B_p}{Z} = (\varepsilon_p + \varepsilon_{coul.}) - M , \quad \frac{B_n}{N} = \varepsilon_n - M$$

$$\varepsilon_p = \int \frac{\mathscr{E}_p(r)}{Z} d^3r , \quad \varepsilon_n = \int \frac{\mathscr{E}_n(r)}{N} d^3r$$

where

$$\mathscr{E}_i(r) = 2 \int_0^{k_{Fi}} \frac{d^3p}{(2\pi)^3} \left(\sqrt{p^2+M^{*2}} + \mathcal{V}_{i0} \right) + \frac{\rho_i}{\rho_B} \mathscr{E}_{fields}$$

and

$$\mathscr{E}_{coul.} = 2 \int \frac{d^3p}{(2\pi)^3} \left(eA_0 - \frac{1}{2}(\nabla A_0)^2 \right).$$

RESULTS AND CONCLUSIONS

The binding energy in terms of the baryon density for the TM1 and NL3 parametrizations for $Y_p = 0$, (the equation of state which is most affected by the mixing term) shows a very small influence of the mixing term in pure neutron matter. No influence at all is detected in symmetric nuclear matter ($b_0 = 0$).

TABLE 1: ^{41}Ca and its mirror nucleus - $A_0 = 0$

Force	TM1				NL3			
nucleus	^{41}Ca		^{41}Sc		^{41}Ca		^{41}Sc	
Λ (MeV^2)	0	-4500	0	-4500	0	-4500	0	-4500
R_p (fm)	3.935	3.942	3.967	3.975	3.938	3.946	3.972	3.981
Θ (fm)	0.032	0.018	-0.032	-0.046	0.034	0.018	-0.034	-0.049
B/A	-11.24	-11.30	-11.24	-11.18	-11.02	-11.08	-11.02	-10.96
B_p/Z	-12.35	-11.85	-10.19	-9.59	-12.07	-11.57	-10.02	-9.43
B_n/N	-10.19	-10.78	-12.35	-12.84	-10.02	-10.62	-12.07	-12.57
$B_p/Z - B_n/N$	-2.16	-1.07	2.16	3.25	-2.05	-0.95	2.05	3.14

TABLE 2: ^{41}Ca and its mirror nucleus - $A_0 \neq 0$

Force	TM1			
nucleus	^{41}Ca		^{41}Sc	
Λ (MeV^2)	0	-4500	0	-4500
R_p (fm)	4.018	4.025	4.054	4.061
Θ (fm)	-0.002	-0.003	-0.009	-0.102
B/A	-9.07	-9.00	-8.97	-8.90
B_p/Z	-8.21	-7.52	-6.52	-5.92
B_n/N	-9.88	-10.41	-11.54	-12.04
$B_p/Z - B_n/N$	1.67	2.89	5.02	6.12

TABLE 3: ^{81}Zr and its mirror nucleus - $A_0 = 0$

Force	TM1				NL3			
nucleus	^{81}Zr		^{81}Nb		^{81}Zr		^{81}Nb	
Λ (MeV^2)	0	-4500	0	-4500	0	-4500	0	-4500
R_p (fm)	4.957	4.965	4.976	4.983	4.936	4.945	4.955	4.964
Θ (fm)	0.018	0.003	-0.018	-0.033	0.019	0.018	-0.019	-0.036
B/A	-12.53	-12.57	-12.53	-12.44	-12.28	-12.33	-12.28	-12.21
B_p/Z	-13.16	-12.57	-11.93	-11.27	-12.84	-12.31	-11.73	-10.08
B_n/N	-11.93	-12.57	-13.16	-13.65	-11.73	-12.36	-12.84	-12.36
$B_p/Z - B_n/N$	-1.23	0	1.23	2.38	-1.11	0.05	1.11	2.28

228

TABLE 4: ^{81}Zr and its mirror nucleus - $A_0 \neq 0$

Force	TM1			
nucleus	^{81}Zr		^{81}Nb	
Λ (MeV^2)	0	-4500	0	-4500
R_p (fm)	5.127	5.135	5.150	5.156
Θ (fm)	-0.086	-0.101	-0.124	-0.138
B/A	-9.04	-8.96	-8.92	-8.87
B_p/Z	-6.37	-5.74	-5.75	-5.15
B_n/N	-11.65	-12.11	-12.17	-12.69
$B_p/Z - B_n/N$	5.28	6.37	6.42	7.54

From tables 1 to 4, one can see that the influence of the mixing term on the binding energy is not negligible. For both parametrizations, the $\rho - \omega$ mixing term gives more binding to the neutrons and less to the protons, reinforcing the effect of the Coulomb field. The difference $B_p/Z - B_n/N$ calculated with the inclusion of the mixing term, for both parameter sets, always increases in mirror nuclei whether the Coulomb contribution is included or not. In all tables the binding energy per nucleon is given in units of MeV/A.

The effect of the mixing term on properties such as the surface energy, charge radius or neutron skin is either zero or very small. The charge radius increases slightly and the neutron radius decreases by the same amount. As a result the neutron skin always decreases.

In all nuclei studied, the proton fraction is very close to 0.5. We conclude that the $\rho - \omega$ mixing is much more important in finite nuclei than in infinite nuclear matter where the influence of the mixing term for $Y_p \sim 0.5$ is negligible. It is, therefore, important to stress that finite nuclei are very sensitive to the contribution from the $\rho - \omega$ mixing term despite the fact that it is only a small isospin symmetry breaking term.

TABLE 5: Mirror nuclei energy difference

nucleus	$^{41}Ca/^{41}Sc$		$^{67}As/^{67}Se$		$^{79}Y/^{79}Zr$		$^{81}Zr/^{81}Nb$	
Force	TM1	NL3	TM1	NL3	TM1	NL3	TM1	NL3
Δ_{4500}/A	0.13	0.13	0.13	0.13	0.13	0.13	0.13	0.13
$\Delta_{4500}/(|A\delta|)$	5.13	5.14	8.62	8.63	10.27	10.25	10.20	10.41
Δ_∞	0.037	0.040	0.023	0.025	0.019	0.020	0.019	0.020
$\Delta_\infty/|\delta|$	1.53	1.65	1.53	1.65	1.53	1.65	1.53	1.65
Δ_{GSI}/A	0.177		0.163		0.149		0.150	

TABLE 6: Ficticious large mirror nuclei energy difference

nucleus	$^{499}X/^{499}X'$		$^{1999}X/^{1999}X'$	
Force	TM1	NL3	TM1	NL3
Δ_{4500}/A	0.03	N.E.	0.01	N.E.
$\Delta_{4500}/(|A\delta|)$	13.05	N.E.	12.26	N.E.
Δ_{∞}	0.003	0.003	0.0008	0.0008

(N.E. = not estimated)

Observing the effect of the mixing term in ficticious large mirror nuclei displayed in table 6 with respectively 499 and 1999 nucleons we confirm that the energy difference per particle due to the mixing term is decreasing with the increase of the mass number but it remains larger than the values we obtain for infinite matter. We then conclude that the boundary conditions necessary to solve the field equations for finite systems introduce finite size effects which do not disappear for very large systems.

When the Coulomb interaction is included, the effect of the mixing term becomes much smaller due to the fact that the mixing term is bringing more repulsion to the nucleus with $N > Z$ than to the other. Thus, within our formalism, the values coming from the $\rho - \omega$ interaction are of the correct order in explaining the NSA only in infinite nuclear matter. The overall effect of the mixing term in the nucleus is different according to whether the Coulomb interaction is present or not.

We believe that the $\rho - \omega$ term, although representing a small isospin symmetry-breaking term, should be kept in the parametrization of the nuclear force within the relativistic mean-field framework. This term has the same magnitude as the non-linear ω term, which is used in the TM1 force.

ACKNOWLEDGMENTS

This work was partially supported by CNPq - Brazil and CFT - Portugal under the contract POCTI/35308/FIS/2000. D.P.M. ackowledges the warm hospitality received at the Depto de Física of the University of Coimbra during the Workshop where this work was presented.

REFERENCES

1. C. Providência, D.P. Menezes and L. Brito, Nucl. Phys. **A 703** (2002) 188; D.P. Menezes and C. Providência, Phys. Rev. **C 64** (2001) 044306; D.P. Menezes and C. Providência, Nucl. Phys. **A 650** (1999) 283, D.P. Menezes and C. Providência, Phys. Rev. **C 60** (1999) 024313.
2. J. A. Nolen and J. P: Shiffer, Ann. Rev. Nucl. Sci. **19** (1969) 471.
3. G. Krein, D.P. Menezes and M. Nielsen, Phys. Lett. **B 294** (1992) 7.
4. G.A. Miller, B.M.K. Nefkens and I. Slaus, Phys. Rep. 194 (1990) 1; G.A. Miller, Nucl. Phys. **A 518** (1990) 345.
5. K. Sumiyoshi, H. Kuwabara, H. Toki, Nucl. Phys. **A 581** (1995) 725.
6. D.P. Menezes and C. Providência, Phys. Rev. **C 66** (2002) 015206.
7. G. A. Lalazissis, J. König and P. Ring, Phys. Rev. **C 55** (1997) 540.
8. B.-A. Li, C. M. Ko and W. Bauer, Int. J. Mod. Phys. E **7** (1997) 147.
9. Jonghwa Chang, http://sutekh.nd.rl.ac.uk/CoN/

NUCLEAR MATTER MEAN FIELD WITH EXTENDED NJL MODEL

C. Providência*, J. M. Moreira*, J. da Providência* and S. A. Moszkowski†

*Centro de Física Teórica, Universidade de Coimbra, 3004-516 Coimbra, Portugal
† UCLA, Los Angeles, CA 90095, USA

Abstract. An extended version of the Nambu-Jona-Lasinio (NJL) model is applied to describe both nuclear matter and surface properties of finite nuclei. Several parameter sets are discussed and a comparison of the saturation properties and equation of state (EOS) with the NL3 parametrization of the non-linear Walecka model is made. The confinement-deconfinement phase transition and the properties of asymmetric matter are discussed. A full bosonization of the model is performed.

INTRODUCTION

Strong interaction dynamics of mesons and baryons is believed to be described by QCD, which exhibits a non perturbative behaviour at low energies. This circumstance renders the analytic study of the theory rather difficult. The NJL model is a popular substitute which has in common with QCD important symmetries of the quark-flavour dynamics. This model has been very successful in the description of the meson sector. The question arises: does the model allow also for soliton solutions (non-homogeneous solutions) which are of interest for the description of the baryon sector and of hadronic matter?

The NJL model[1] was originally developed for the purpose of understanding hadron physics. In this model, hadron masses are generated by spontaneous symmetry breaking of the vacuum. In the modern form of the NJL model, we start with essentially massless quarks interacting via zero range interactions, but with a cutoff in momentum space [2, 3, 4, 5]. NJL is thus only an effective theory, in which form factors and other finite range effects have been ignored, and it does not take into account quark confinement. However, it may work quite well in the region of interest in nuclear physics, i.e. for excitations less than the scalar meson mass.

The paper has been organized as follows: It is shown that with the generalization of the NJL model, referred to as the Extended NJL (ENJL) model, it is possible to get a reasonable nuclear equation of state and behaviour of the effective mass. Then we list some numerical results, concerning both nuclear matter and quark matter, and the hadron to quark phase transition at zero temperature. The surface properties of finite nuclei are calculated. The full bosonization of the model is presented, as well as some remarks concerning the connection of the ENJL model to other relativistic chiral models [6, 7]. Finally, some properties of asymmetric nuclear matter are discussed and some brief conclusions are drawn.

CP660, *Hadron Physics: Effective Theories of Low Energy QCD*, edited by A. H. Blin et al.
© 2003 American Institute of Physics 0-7354-0120-9/03/$20.00

EXTENDED NJL MODEL

The NJL model [3] is defined by the Lagrangian density

$$\mathcal{L} = \bar{\psi}(i\gamma^{\mu}\partial_{\mu})\psi + G[(\bar{\psi}\psi)^2 + (\bar{\psi}i\gamma_5\vec{\tau}\psi)^2]. \tag{1}$$

A regularizing momentum cut-off Λ is part of the model. The Lagrangian is equivalent to the Hamiltonian

$$\mathcal{H}_{NJL} = \sum_{k=1}^{N} \vec{p}_k \cdot \vec{\alpha}_k + G \sum_{k,l=1}^{N} \delta(\vec{r}_k - \vec{r}_l)\beta_k\beta_l(1 - \gamma_k^5\gamma_l^5\vec{\tau}_k \cdot \vec{\tau}_l). \tag{2}$$

The vacuum is described by a Slater Determinant $|\Phi_0\rangle$ constructed from plane waves which are negative energy eigenfunctions of the single particle Hamiltonian $h = \vec{p} \cdot \vec{\alpha} + \beta m$. The "constituent mass" m is a variational parameter.

If moreover positive energy eigenfunctions with momentum $\vec{p}$ satisfying $|\vec{p}| < p_F$ are occupied, so that p_F is the Fermi momentum, the energy expectation value $E = \langle\Phi_0|\mathcal{H}_{NJL}|\Phi_0\rangle$ is given by

$$E = -v \sum_{p=p_F}^{\Lambda} \frac{p^2}{\sqrt{m^2 + p^2}} - \frac{m^2 G}{4} v^2 \left[\sum_{p=p_F}^{\Lambda} \frac{1}{\sqrt{m^2 + p^2}}\right]^2. \tag{3}$$

For quark matter, the degeneracy is $v = 2N_cN_f$ and Λ is such that $m = 313$ MeV is the constituent quark mass in the vacuum. The coupling constant G is determined by the value of $F_\pi = 93$ MeV.

The NJL model can be extended [8] to yield more reasonable saturation properties of nuclear matter. An effective density dependent coupling constant is obtained if the following extended NJL Lagrangian density, which actually pushes chiral symmetry restoration to higher densities, is considered,

$$\begin{aligned}
\mathcal{L} &= \bar{\psi}(i\gamma^{\mu}\partial_{\mu})\psi + G_s[(\bar{\psi}\psi)^2 + (\bar{\psi}i\gamma_5\vec{\tau}\psi)^2] - G_v(\bar{\psi}\gamma^{\mu}\psi)^2 \\
&\quad - G_{sv}[(\bar{\psi}\psi)^2 + (\bar{\psi}i\gamma_5\vec{\tau}\psi)^2](\bar{\psi}\gamma^{\mu}\psi)^2.
\end{aligned} \tag{4}$$

For nuclear matter the degeneracy is $v = 2N_f$ and Λ is such that $m = 939$ MeV is the nucleon mass in the vacuum.

We assume that G_s for nuclear matter is either determined by the saturation properties of nuclear matter or is nine times bigger than the quark matter value, G. The philosophy behind the last assumption is that the NN interaction is a manifestation of the instanton interaction between quarks predicted by QCD in the weak coupling regime. Following Guichon [9], we assume that the scalar field emanating from the quark scalar density in moving nucleons, due to the instanton force, acts on the quarks of another nucleon, and produces in this way the nucleon-nucleon force.

The terms in G_v, G_{sv} are supposed to simulate a chiral invariant short range repulsion between nucleons which, due to the overlap of the bags, is increasingly felt as the density increases. Thus, for quark matter we set $G_v = G_{sv} = 0$.

232

TABLE 1. The coupling constants and properties of different equations of state

	nuclear-matter		
	EOS-I	EOS-II	[8]
$G_s(\text{fm}^2)$	3.880	1.746	1.16
$G_v(\text{fm}^2)$	3.952	3.387	3.40
$G_{sv}(\text{fm}^8)$	-4.901	-1.839	-0.92
$G_\rho(\text{fm}^2)$	2.794	-	-
m_0 (MeV)	939	939	939
Λ (MeV)	418.9	553	639.9
ρ/ρ_0 ($m^* = 0$)	3.0	8.0	13.4
$E_{Bq} = E_{BN}$ (MeV)	111	80	120
ρ/ρ_0 ($E_{Bq} = E_{BN}$)	.4.3	3.5	4.5

TABLE 2. Nuclear matter saturation properties and surface properties

	EOS-I	EOS-II	[8]	NL3
$E/A - m_0$(MeV)	16.12	-15.93	-16.50	-16.3
ρ (fm^{-3})	0.148	0.148	0.171	0.148
m^*/m_0	0.75	0.89	0.92	0.60
K (MeV)	295	339	349	272
W_{surf} (MeV)	19.43	12.92	11.07	19.38
R fm	5.33	5.35	5.11	5.328
t (MeV)	2.64	1.32	1.12	2.65

The thermodynamical potential per volume corresponding to (4) is

$$\omega(p_F, \mu) = \langle \bar{\psi}(\vec{\gamma} \cdot \vec{p}) \psi \rangle - G_s \langle \bar{\psi}\psi \rangle^2 + G_v \langle \psi^\dagger \psi \rangle^2 + G_{sv} \langle \bar{\psi}\psi \rangle^2 \langle \psi^\dagger \psi \rangle^2 - \mu \langle \psi^\dagger \psi \rangle \quad (5)$$

where exchange terms have been neglected. By $\langle \bar{\psi}\Gamma\psi \rangle$ we denote the following expectation value per volume $\langle \bar{\psi}\Gamma\psi \rangle = \frac{1}{V}\langle \Phi_0 | \Sigma_k \beta_k \Gamma_k | \Phi_0 \rangle$. The condition $\partial\omega/\partial m = 0$ implies

$$m = -2G_s \langle \bar{\psi}\psi \rangle + 2G_{sv} \langle \bar{\psi}\psi \rangle \langle \psi^\dagger \psi \rangle^2.$$

The condition $\partial\omega/\partial p_F = 0$ implies

$$E_{p_F} = \mu - 2G_v \langle \psi^\dagger \psi \rangle - 2G_{sv} \langle \psi^\dagger \psi \rangle \langle \bar{\psi}\psi \rangle^2,$$

with $E_{p_F} = \sqrt{m^2 + p_F^2}$. These conditions fix the values of p_F, m for given μ.

The properties of the extended NJL model are now easily computed.

PROPERTIES OF NUCLEAR MATTER

In Table 1, numerical results pertaining to several different realizations of the model are presented. As inputs, we use the saturation properties of nuclear matter fitted empirically

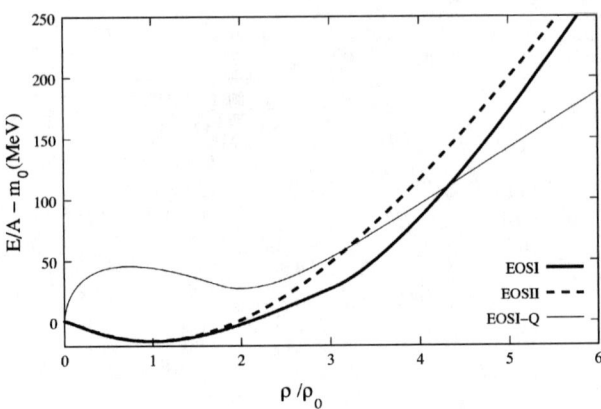

FIGURE 1. Energy vs density in ENJL Model for symmetric nuclear matter (EOSI and II) and for quark matter (EOSI-Q)

in Ref.[10]. In model I, the three coupling constants G_s, G_v and G_{sv} are adjusted to fit the effective mass $m^* = 0.75$ and the saturation energy and density. In model II the inputs are F_π and the saturation energy and density. We have assumed in model II that G_s for nuclear matter is nine times bigger than the quark matter value determined by the value of F_π, G. We have taken $E/A = -15.8$ MeV and $\rho_0 = 0.148$ fm^{-3} according to [10, 11].

Model I is the most realistic of our models, since it fits well both saturation properties and the nuclear surface properties. We also compare our results with those of [8], which give results similar to those of our model II, and [11] in which a relativistic mean field theory (RMF) is used, and which describes well both stable and unstable nuclei. The RMF parametrization introduced in [11] is known as NL3 . Model I gives results quite close to NL3 especially for the surface properties (see Table 2)..

The crucial role played by clustering must be stressed. It should be pointed out that without clustering there is no real binding and that, moreover, the incompressibility becomes unacceptable. This is well illustrated by the curves in Fig 1. At some density, the nucleon matter curve intercepts the quark matter curve, so it is clear that, at high density, quark matter prevails and may be found, for instance, in the core of neutron stars. Fig. 2 shows that clustering pushes chiral symmetry restoration to higher densities.

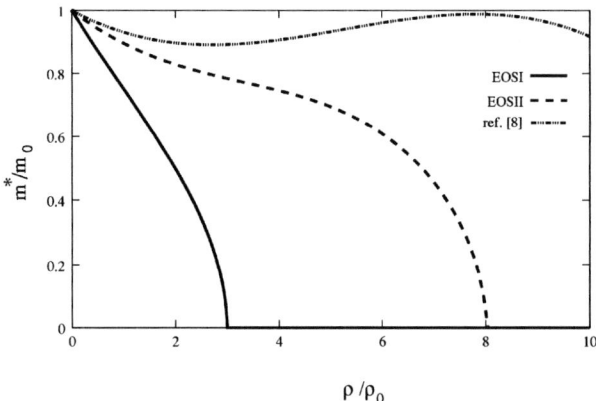

FIGURE 2. Effective Mass vs density in ENJL Model for models I and II and ref.[8]

SURFACE PROPERTIES

Next we will calculate some nuclear surface properties within a Thomas-Fermi approximation. The surface properties are mainly defined by the scalar meson, the meson with the smallest mass, and, therefore, we will only bosonize the corresponding terms of the Lagrangian.

The bosonized Lagrangian density may be formally written as

$$
\begin{aligned}
\mathscr{L} &= \bar{\psi}(i\gamma^{\mu}\partial_{\mu})\psi - G_{v}(\bar{\psi}\gamma^{\mu}\psi)(\bar{\psi}\gamma_{\mu}\psi) \\
&- g(\sigma(\bar{\psi}\psi) + \vec{\pi}\cdot(\bar{\psi}i\gamma_{5}\vec{\tau}\psi))\left(1 - \frac{G_{sv}}{G_{s}}(\bar{\psi}\gamma_{\mu}\psi)(\bar{\psi}\gamma^{\mu}\psi)\right)^{1/2} \\
&+ \frac{1}{2}(\partial_{\mu}\sigma\partial^{\mu}\sigma + \partial_{\mu}\vec{\pi}\cdot\partial^{\mu}\vec{\pi}) - \frac{1}{2}m_{\sigma}^{2}(\sigma^{2} + \vec{\pi}\cdot\vec{\pi}),
\end{aligned}
\tag{6}
$$

where the parameters g, m_{σ} are related to the coupling contanst G_{s}. The bosonized thermodynamical potential per volume is

$$
\begin{aligned}
\omega(p_{F},\mu) &= \langle\bar{\psi}(\vec{\gamma}\cdot\vec{p})\psi\rangle + G_{v}\langle\psi^{\dagger}\psi\rangle^{2} - \mu\langle\psi^{\dagger}\psi\rangle \\
&+ g\sigma\langle\bar{\psi}\psi\rangle\left(1 - \frac{G_{sv}}{G_{s}}\langle\psi^{\dagger}\psi\rangle^{2}\right)^{1/2} + \frac{1}{2}(\partial_{i}\sigma\partial_{i}\sigma) + \frac{1}{2}m_{\sigma}^{2}\sigma^{2}.
\end{aligned}
\tag{7}
$$

Minimization with respect to σ yields eq. (5) for infinite matter, provided we choose $G_{s} = g^{2}/(2m_{\sigma}^{2})$. Variation with respect to m, p_{F} and σ gives:

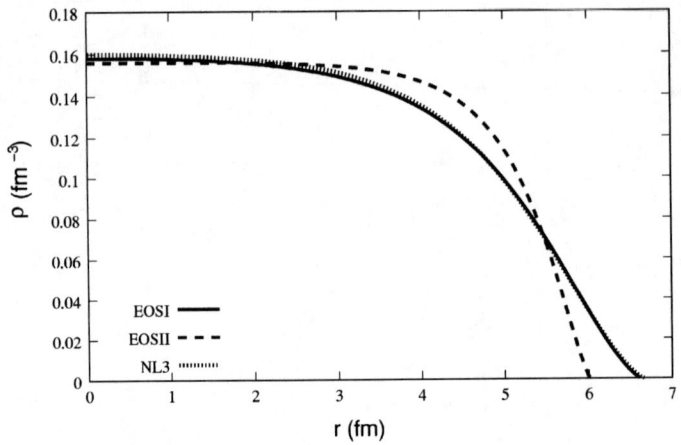

FIGURE 3. Density profile for nuclei with A = 100

$$m = g\sigma \left(1 - \frac{G_{sv}}{G_s}\langle\psi^\dagger\psi\rangle^2\right)^{1/2},$$

$$E_{P_F} = \mu - 2G_v\langle\psi^\dagger\psi\rangle + g\sigma\frac{G_{sv}}{G_s}\langle\bar\psi\psi\rangle\langle\psi^\dagger\psi\rangle\left(1 - \frac{G_{sv}}{G_s}\langle\psi^\dagger\psi\rangle^2\right)^{-1/2},$$

$$\nabla^2\sigma = m_\sigma^2\sigma + g\langle\bar\psi\psi\rangle\left(1 - \frac{G_{sv}}{G_s}\langle\psi^\dagger\psi\rangle^2\right)^{1/2}.$$

Next, we look for droplet solutions within models I and II and compare their surface properties with the results obtained with the parametrization proposed in [8] and the NL3 parametrization [11]. In the small surface thickness approximation ($\nabla^2\sigma \sim \frac{d^2\sigma}{dr^2}$) the free energy F can be rewritten as [13]:

$$F = \int 4\pi r^2 dr\left(\left(\frac{d\sigma}{dr}\right)^2 - C\right) + \mu A, \tag{8}$$

where C is a constant and A is the number of particles. For droplets with radius R and volume V,

$$F(R) = W_{surf}A^{2/3} - CV + \mu A. \tag{9}$$

236

The surface energy per unit area of these droplets in the small thickness approximation is

$$W_{surf} = \frac{4\pi R^2}{A^{2/3}} \int_0^\infty dr \left(\frac{d\sigma}{dr}\right)^2. \tag{10}$$

The surface thickness t is defined as the width of the region where the density drops from $0.9\rho_{B0}$ to $0.1\rho_{B0}$, where ρ_{B0} is the baryonic density at $r = 0$. For a nucleus with $A = 100$ we obtain for model I and II, and from ref. [8] and NL3 parametrization the values of W_{surf}, R, t displayed in Table 2. In the present approach the mass of the σ meson appears as an extra parameter. We choose : $m_\sigma = 2m_q = 626$ MeV as in the NJL model. Taking $m_\sigma = 400$ MeV the surface properties of model II get closer to the NL3 values, namely $W_{surf} = 19.5$ MeV and $t = 2.27$ fm. In Fig. 3 we plot the density profile of a nucleus with 100 particles.

CONFINEMENT-DECONFINEMENT PHASE TRANSITION

As discussed in the last section, at some density, depending on the parameterisation used, the nucleon matter curve intercepts the quark matter curve. At high density the quark matter prevails. In order to study the confinement-deconfinement phase transition we make a Maxwell construction: at the phase transition the temperature, pressure and chemical potential are equal in both phases:

$$\mu_H = \mu_Q, \qquad T_H = T_Q, \qquad p_H = p_Q.$$

In the present simplified approach we only require barionic charge conservation. For a single conserved charge the transition occurs at constant pressure [12]. In fig. 4 we plot the pressure versus the baryonic density. The letters H and Q denote the end of the confined (hadronic) and beginning of the deconfined (quark) phases. We conclude that the density at which the confinement-deconfinement phase transition occurs is quite sensitive to the parameterisations used. We will not discuss the properties of the family of neutron stars predicted by the models discussed above. In order to do this study, it is important to implement both barionic and charge conservation in order not to get a region of instability between the minimum and maximum mass configurations. The kink in EOSI corresponds to the chiral symmetry restoration which occurs at a lower density than the confinement-deconfinement phase transition.

BOSONIZATION OF EXTENDED NJL MODEL

The bosonized ENJL Lagrangian density may be formally written as

$$\begin{aligned}
\mathscr{L} &= \bar{\psi}(i\gamma^\mu \partial_\mu)\psi + g_s\left(\sigma\,\bar{\psi}\psi + \bar{\psi}i\gamma_5\vec{\pi}\cdot\vec{\tau}\,\psi\right)\left[1 + a_1(\bar{\psi}\gamma_\mu\psi)(\bar{\psi}\gamma^\mu\psi)\right]^{1/2} \\
&\quad - g_v\,\bar{\psi}V^\mu\gamma_\mu\psi\left\{1 + a_2\left[(\bar{\psi}\psi)^2 + (\bar{\psi}i\gamma_5\vec{\tau}\psi)^2\right]\right\}^{1/2} \\
&\quad - \frac{1}{2}m_s^2(\sigma^2 + \vec{\pi}\cdot\vec{\pi}) + \frac{1}{2}m_v^2 V^\mu V_\mu
\end{aligned} \tag{11}$$

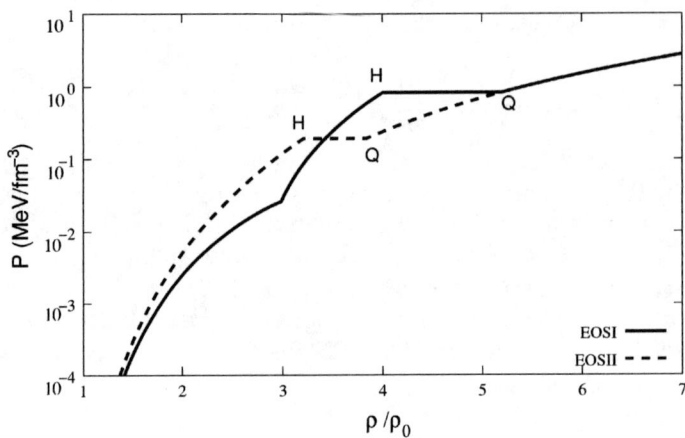

FIGURE 4. Hadron to Quark Matter Phase Transition

from where we get the following equations for the meson fields

$$\sigma = \frac{g_s}{m_s^2} \bar{\psi}\psi \left[1 + a_1(\bar{\psi}\gamma^\mu\psi)^2\right]^{1/2} \tag{12}$$

$$\vec{\pi} = \frac{g_s}{m_s^2} \bar{\psi} i\gamma_5\vec{\tau}\psi \left[1 + a_1(\bar{\psi}\gamma^\mu\psi)^2\right]^{1/2} \tag{13}$$

$$V^\mu = \frac{g_v}{m_v^2} \bar{\psi}\gamma^\mu\psi \left\{1 + a_2\left[(\bar{\psi}\psi)^2 + (\bar{\psi} i\gamma_5\vec{\tau}\psi)^2\right]\right\}.^{1/2} \tag{14}$$

Replacing (12)-(14) in the Lagrangian density we recover the original ENJL model if the following relations are satisfied:

$$\frac{g_s^2}{2m_s^2} = G_s, \qquad \frac{g_v^2}{2m_v^2} = G_v, \qquad \frac{a_1 g_s^2}{2m_s^2} - \frac{a_2 g_v^2}{2m_v^2} = -G_{sv}. \tag{15}$$

There is a certain arbitrariness in the choice of the constants a_1 and a_2. In the bosonized ENJL the meson fields are introduced as auxiliary fields. In order to transform them in new dynamical degrees of freedom we must add the corresponding kinetic terms:

$$\frac{1}{2}(\partial_\mu\sigma\partial^\mu\sigma + \partial_\mu\vec{\pi}\cdot\partial^\mu\vec{\pi}) - \frac{1}{4}\left(\partial_\mu V_\nu - \partial_\nu V_\mu\right)^2.$$

In the present form the Lagrangian density (16) contains terms coupling the meson fields to the fermion fields in a non-linear way. Using eqs. (12)-(14) and keeping only quartic

238

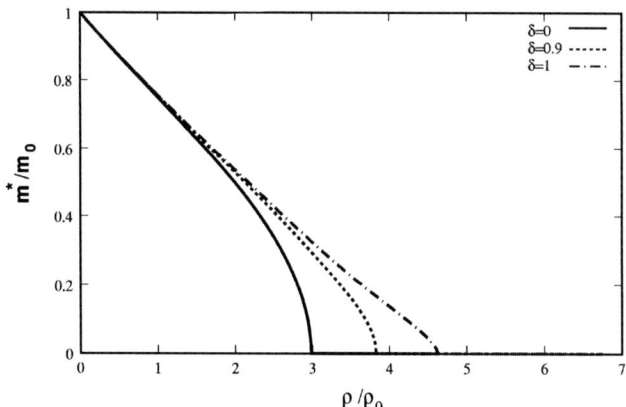

FIGURE 5. Effective mass vs density for EOSI and different values of δ

terms in the meson fields we get, including the kinetic terms,

$$
\begin{aligned}
\mathcal{L} &= \bar{\psi}(i\gamma^{\mu}\partial_{\mu})\psi + g_{s}\left(\sigma\,\bar{\psi}\psi + \bar{\psi}i\gamma_{5}\vec{\pi}\cdot\vec{\tau}\,\psi\right) - g_{v}\,\bar{\psi}V^{\mu}\gamma_{\mu}\psi \\
&+ \frac{1}{2}(\partial_{\mu}\sigma\partial^{\mu}\sigma + \partial_{\mu}\vec{\pi}\cdot\partial^{\mu}\vec{\pi}) - \frac{1}{2}m_{s}^{2}(\sigma^{2}+\vec{\pi}\cdot\vec{\pi}) - \frac{1}{4}\left(\partial_{\mu}V_{v}-\partial_{v}V_{\mu}\right)^{2} + \frac{1}{2}m_{v}^{2}V^{\mu}V_{\mu} \\
&+ \frac{1}{2}\left(\frac{g_{s}m_{v}^{4}}{g_{v}^{2}}a_{1} - \frac{g_{v}m_{s}^{4}}{g_{s}^{2}}a_{2}\right)(\sigma^{2}+\vec{\pi}\cdot\vec{\pi})V^{\mu}V_{\mu}. \tag{16}
\end{aligned}
$$

This Lagrangian density is similar to the one introduced by Boguta [6] which, by including a scalar-vector coupling was able to reproduce nuclear matter saturation properties. However, in [6] an effective mesonic self-interaction corresponding to the "mexican hat" is present. It was shown in [14] that the Dirac-sea in the linear sigma model with valence and Dirac-sea quarks but no "mexican hat" provides the above effective mesonic self-interaction.

In [7] a generalization of the model proposed in [6], including a "bare" vector mass, a quartic term in vector field and allowing for different vector-scalar and vector-nucleon couplings was studied. It was argued that a linear realization of chiral symmetry was too restricitve and although nuclear matter properties may be reproduced there are problems with finite nuclei properties. With our choice of parameters we have shown that not only nuclear matter saturation properties are reasonable but also surface properties.

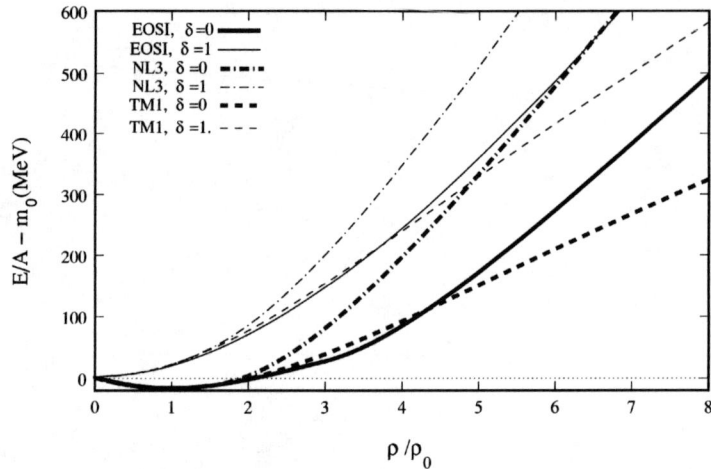

FIGURE 6. Binding energy versus density for EOSI, NL3 and TM1 for symmetric nuclear matter (thick lines) and neutron matter (thin lines)

ISOSPIN ASYMMETRIC NUCLEAR MATTER

The recent advance in unstable nuclear-beam experiments has been providing us with new knowledge on unstable nuclei away from the stability line. We hope to get information on the properties of dense matter under extreme conditions as the neutron-rich enviroment in neutron stars.

In order to describe isospin asymmetric nuclear matter within the present formalism it is important to include in the lagrangian density the isovector-vector term

$$\mathscr{L}_\rho = -G_\rho \left[(\bar{\psi}\gamma^\mu \vec{\tau}\psi)^2 + (\bar{\psi}\gamma_5\gamma^\mu \vec{\tau}\psi)^2 \right]$$

G_ρ is chosen in such a way that the experimental value of the symmetry energy $a_{sym} = 35$

MeV is reproduced. In terms of the asymmetry parameter $\delta = (\rho_p - \rho_n)/(\rho_p + \rho_n)$, where ρ_i, $i = p, n$ is the density of particles of type i, we conclude that the chiral symmetry restoration occurs at larger densities when δ increases, Fig. 5.

In Fig. 6 we compare EOSI with two different parametrizations of the non-linear Walecka model, NL3 [11] and TM1 [15], for symmetric ($\delta = 1$, thick lines) and neutron matter ($\delta = 0$, thin lines). The RMF parametrizations where fitted to both stable and unstable nuclei. However TM1 contains a self-coupling term in the vector meson which weakens the contribution of the vector meson at high densities. We conclude that EOSI is softer than NL3 but stiffer than TM1.

CONCLUSIONS

We have studied an extension of the NJL model which yields reasonable saturation of nuclear matter. An effective density dependent coupling constant is obtained which pushes chiral symmetry restoration to higher densities.

Two variants of the model defined by appropriate sets of the parameters, G_s, G_v, G_{sv}, and Λ, have been studied numerically. The values of the defining parameters, the vacuum properties, the properties at saturation are given in Tables 1 and 2. Model I fits not only the properties at saturation but also the empirical nuclear surface energy and thickness. The term in G_{sv}, responsible for the density dependence of the effective coupling constant, plays an important role in pushing to higher energies the restoration of chiral symmetry and in lowering the incompressibility.

We have discussed the confinement-deconfinement phase transition and have shown that it could occur at larger or smaller densities than the density at which the restoration of chiral symmetry occurs, depending on the choice of the parameters.

A full bosonization of the model was performed and it was shown the relation of the present model and the relativistic chiral model proposed by Boguta [6].

Finally we have discussed asymmetric matter within the present model and have shown that it gives rise to a equation of state softer than the NL3 parametrization of the non-linear Walecka model and stiffer than the TM1 parametrization.

In order to further test the model it is important to calculate other finite nuclei properties. Application of the model to neutron-star matter requires the inclusion of strangeness and beta-equilibrium.

ACKNOWLEDGMENTS

This work was partially supported by FCT and FEDER under the projects POCTI/FIS/451/94 and POCTI/35308/FIS/2000.

REFERENCES

1. Y. Nambu and G. Jona-Lasinio, Phys. Rev. **122**, 345 (1961);**124**, 246 (1961).
2. U. Vogl and W. Weise, Prog. Part. and Nucl. Phys. **27**, 95 (1991).
3. S. P. Klevansky, Revs. of Mod. Physics **64**, 649 (1992).
4. T. Hatsuda and T. Kunihiro, Prog. Theo. Phys. **74**, 765 (1985).
5. T. Hatsuda and T. Kunihiro, Phys. Rep. **247**, 221 (1994).
6. J. Boguta, *Phys. Lett.* **B 120**, 34 (1983).
7. R. J. Furnstahl and B. D. Serot, *Phys. Rev.* **C47**, 2338 (1993).
8. V. Koch, T.S. Biro, J. Kunz, and U. Mosel, Phys. Lett. **B185**, 1 (1987).
9. P.A.M. Guichon, K. Saito, E. Rodionov, and A.W. Thomas, Nucl. Phys. A **601**, 349, 1996).
10. R. J. Furnstahl, Brian D. Serot, Hua-Bin Tang, Nucl. Phys. **A615**, 441 (1997).
11. G.A. Lalazissis, J. König, P. Ring, Phys. Rev. **C55**, 540 (1997).
12. N. K. Glendenning, Compact Stars, Springer-Verlag, New-York, 1997.
13. M. Nielsen and J. da Providencia, J. Phys. G **16**, 649 (1990).
14. M. Fiolhais, J. da Providencia, M. Rosina, and C. A. de Sousa, Phys. Rev.**C56**, 3311 (1997).
15. K. Sumiyoshi, H. Kuwabara, H. Toki, Nucl. Phys. A581 (1995) 725.

$\eta - \eta'$ in a hot and dense medium

P. Costa*, M. C. Ruivo* and Yu. L. Kalinovsky†

*Centro de Física Teórica, Departamento de Física, Universidade, P3004-516 Coimbra, Portugal
†Laboratory of Information Technologies, Joint Institute for Nuclear Research, Dubna, Russia

Abstract. The behavior of η and η' in hot strange quark matter in weak equilibrium with temperature, is investigated within the SU(3) Nambu-Jona-Lasinio [NJL] model. Possible manifestations of restoration of symmetries, by temperature or density, in the behavior of η and η' are discussed. The role played by the combined effect of temperature and density in the nature of the phase transition and meson behavior is also analyzed.

INTRODUCTION

Understanding the QCD phase diagram is one of the great challenges in the physics of strong interactions. It is expected that under extreme conditions (high density and or temperature) phase transitions occur leading to a plasma of deconfined quarks and gluons [QGP] [1, 2, 3]. While the phase transition at zero chemical potential and finite temperature is accepted to be second order or crossover, there are indications that the phase transition with finite chemical potential and zero temperature is first order. Experimental and theoretical efforts have been done in order to explore the $\mu - T$ phase boundary. Recent Lattice results indicate a critical "endpoint", connecting the first order phase transition with the crossover region at $T_E = 160 \pm 35 \mathrm{MeV}, \mu_E = 725 \pm 35 \mathrm{MeV}$ [4]. Besides restoration of chiral symmetry and deconfinement, the rich content of the QCD phase diagram has been recently explored in other directions. For instance, in high density matter, for which neutron stars provide an excellent laboratory, a color locked phase with kaon condensates is expected to occur.

Much attention has also been paid to the question of which symmetries are restored. In the limit of vanishing quark masses the QCD Lagrangian has 8 Goldstone bosons, associated with the dynamical breaking of chiral symmetry. In order to give a finite mass to the mesons chiral symmetry is broken *ab initio* by giving current masses to the quarks, therefore in the high density/temperature regime an indication of the restoration of chiral symmetry would be that the mesons became again almost Goldstone bosons. However, the mass of η' has a different origin than the masses of the other pseudoscalar mesons and it cannot be regarded as the remanent of a Goldstone boson. The problem of the non-existence of the ninth Goldstone boson, predicted by the quark model, was solved by assuming that the QCD Lagragian has a $U_A(1)$ anomaly. The mass of the η' is obtained by explicitly breaking the $U_A(1)$ symmetry, for instance by instantons.

The study of the behavior of mesons in hot and dense matter is an important issue since they might provide a signature of the phase transition and give indications about which symmetries are restored. There is a long standing question [5] about whether only

CP660, *Hadron Physics: Effective Theories of Low Energy QCD*, edited by A. H. Blin et al.

SU(3) chiral symmetry is restored or both SU(3) and $U_A(1)$ symmetries are restored. The behavior of η' in medium or of related observables like the topological susceptibility might help to decide between these scenarios. A decrease of the η' mass in medium could manifest in the increase of the η' production cross section, as compared to that for pp collisions [6].

This paper is devoted to investigate the in medium behavior of the η and η' mesons in hot and dense matter in the framework of the SU(3) Nambu-Jona-Lasinio [NJL] [7] model including the 't Hooft interaction which breaks the $U_A(1)$ symmetry. At variance with the other pseudoscalar SU(3) mesons there is not much information about the behavior of these mesons in medium. Concerning the η meson, the recent discovery of mesic atoms might provide useful information in this concern [8].

The NJL model is an effective quark model where the gluonic degrees of freedom are supposed to be frozen and, besides its simplicity, has the advantage of incorporating important symmetries of QCD, namely chiral symmetry. Since the model has no confining mechanism, several drawbacks are well known. In what concerns our present problem, it should be noticed that the η' mass lies above the $\bar{q}q$ threshold and the model describes this meson as $\bar{q}q$ resonance, which would have the unphysical decay in $\bar{q}q$ pairs. For this reason, it has been proposed that in order to investigate the possible restoration of $U_A(1)$ symmetry, instead of $m_{\eta'}$, the topological susceptibility χ, a more reliable quantity in this model, should be used [9].

The behavior of SU(3) pseudoscalar mesons, in NJL models, with temperature has been studied in [10, 11]. Different studies have been devoted to the behavior of pions and kaons at finite density in flavor symmetric [12] or asymmetric matter [13, 14, 15] . The combined effect of temperature and density on kaons in quark matter simulating symmetric nuclear matter has been investigated in [16].

In this work we investigate the phase transition and the behavior of η, η' in strange quark matter in weak equilibrium at zero and finite temperature. We present the model and formalism in the vacuum (Section 2) and at finite density and temperature (Section 3). In Section 4 we discuss present our results for the phase transition in the $T - \rho$ plane and in Section 5 we present our results and conclusions for the η, η' behavior.

MODEL AND FORMALISM

We will consider in this paper the $U(1)_A$ anomaly in the three flavor Nambu - Jona - Lasinio model. The Lagrangian we employ here can be used as an effective theory of QCD and has the following form

$$
\begin{aligned}
L = {} & \bar{q}(i\partial \cdot \gamma - \hat{m})q + \frac{g_S}{2}\sum_{a=0}^{8}\left[(\bar{q}\lambda^a q)^2 + (\bar{q}(i\gamma_5)\lambda^a q)^2\right] \\
& - \frac{g_V}{2}\sum_{a=0}^{8}\left[(\bar{q}\gamma_\mu\lambda^a q)^2 + (\bar{q}(\gamma_\mu\gamma_5)\lambda^a q)^2\right] \\
& + g_D\left[\det\left[\bar{q}(1-\gamma_5)q\right] + \det\left[\bar{q}(1-\gamma_5)q\right]\right].
\end{aligned}
\tag{1}
$$

Here $q = (u, d, s)$ is the quark field with three flavors, $N_f = 3$, and three colors, $N_c = 3$. λ^a are the Gell - Mann matrices, a=0, 1, ..., 8, $\lambda^0 = \sqrt{\frac{2}{3}}\mathbf{I}$. The explicit symmetry breaking part (1) contains the current quark masses $\hat{m} = \text{diag}(m_u, m_d, m_s)$. Note, that $m_{u,d} \sim 10$ MeV, $m_s \sim 100$ MeV at the scale of QCD (~ 1 GeV). The last term in (1) is the lowest dimensional operator and that has the $SU_L(3) \otimes SU_R(3)$ invariance but breaks the $U_A(1)$ symmetry. This term is a reflection of the axial anomaly in QCD. The model is fixed by the coupling constants g_S, g_V, g_D, the cutoff parameter Λ, which regularizes momentum space integrals and current quark masses. In order to fix parameters we will use the values of pseudoscalar meson masses, decay constants and the quark condensates at zero temperature and baryon number density.

Using the definition of the determinant we can rewrite the Lagrangian (1) as

$$L = \bar{q}(i\partial \cdot \gamma - \hat{m})q + \frac{g_S}{2}\sum_{a=0}^{8}\left[(\bar{q}\lambda^a q)^2 + (\bar{q}(i\gamma_5)\lambda^a q)^2\right]$$

$$- \frac{g_V}{2}\sum_{a=0}^{8}\left[(\bar{q}\gamma_\mu\lambda^a q)^2 + (\bar{q}(\gamma_\mu\gamma_5)\lambda^a q)^2\right]$$

$$+ \frac{g_D}{2}\left[\frac{1}{3}D_{abc}(\bar{q}\lambda^a q)(\bar{q}\lambda^b q)(\bar{q}\lambda^c q) - D_{abc}(\bar{q}\lambda^a q)(\bar{q}(i\gamma_5)\lambda^b q)(\bar{q}(i\gamma_5)\lambda^c q)\right] \quad (2)$$

with summation over $a, b, c = (0, 1, 2, ..., 8)$. The structure constants D_{abc} are coincident with SU(3) structure constants d_{abc} and $D_{0ab} = -\frac{1}{\sqrt{6}}\delta_{ab}$ and $D_{000} = \sqrt{\frac{2}{3}}$. Making the shift

$$(\bar{q}\lambda^a q) \longrightarrow (\bar{q}\lambda^a q) + <\bar{q}\lambda^a q>, \quad (3)$$

where $<\bar{q}\lambda^a q>$ is the vacuum expectation value, and keeping second order terms of $(\bar{q}\lambda^a q)$, we get for pseudoscalar and scalar interactions (for simplicity, we omit vector and axial - vector parts)

$$L = \bar{q}(i\partial \cdot \gamma - \hat{m})q + \frac{1}{2}\left\{(\bar{q}\lambda^a q)S_{ab}(\bar{q}\lambda^b q) + (\bar{q}(i\gamma_5)\lambda^a q)P_{ab}(\bar{q}(i\gamma_5)\lambda^b q)\right\}, \quad (4)$$

in terms of the projectors

$$S_{ab} = g_S\delta_{ab} + g_D D_{abc}<\bar{q}\lambda^c q>, \quad (5)$$
$$P_{ab} = g_S\delta_{ab} - g_D D_{abc}<\bar{q}\lambda^c q> . \quad (6)$$

The hadronization procedure is made via the functional integral

$$Z \sim \int dq d\bar{q} \exp\left\{iW_{eff}[q, \bar{q}]\right\} \quad (7)$$

with the quark effective action

$$W_{eff}[q, \bar{q}] = \int dx L(x) = \int dx\left[\bar{q}(i\partial \cdot \gamma - \hat{m})q\right.$$

$$\left. + \frac{1}{2}(\bar{q}\lambda^a q)S_{ab}(\bar{q}\lambda^b q) + \frac{1}{2}(\bar{q}(i\gamma_5)\lambda^a q)P_{ab}(\bar{q}(i\gamma_5)\lambda^b q)\right]. \quad (8)$$

After integration over quark fields in (7), we obtain the effective meson action

$$W_{eff}[\varphi,\sigma] = -\frac{1}{2}\left(\sigma^a S_{ab}^{-1}\sigma^b\right) - \frac{1}{2}\left(\varphi^a P_{ab}^{-1}\varphi^b\right)$$
$$-i\mathrm{Tr}\ln\left[i(\gamma_\mu\partial_\mu) - \hat{m} + \sigma_a\lambda^a + (i\gamma_5)(\varphi_a\lambda^a)\right]. \tag{9}$$

Here the symbol Tr means the summation over discrete indices and integration over continuous variables. The fields σ^a and φ^a are scalar and pseudoscalar meson nonets, respectively.

The first variation of the action (9) leads to the Dyson - Schwinger equation,

$$M_i = m_i - 2g_S < \bar{q}_i q_i > -2g_D < \bar{q}_j q_j > < \bar{q}_k q_k >, \tag{10}$$

with $i, j, k = u, d, s$ cyclic. M_i are constituent quark masses. The equation (10) is a system of three coupled equations and the flavor mixing occurs through the coupling constant g_D. It is an effect of anomaly contribution. The quark condensates in (10) are determined by

$$< \bar{q}_i q_i >= -i\mathrm{Tr}\frac{1}{\hat{p} - M_i} = -i\mathrm{Tr}\left[S_i(p)\right]. \tag{11}$$

Here $S_i(p)$ is the quark Green function.

To consider the meson mass spectrum, we expand the effective action (9) over meson fields. Keeping the pseudoscalar mesons only, we write the second order effective action as

$$W_{eff}^{(2)}[\varphi] = -\frac{1}{2}\varphi^a\left[P_{ab}^{-1} - \Pi_{ab}(P)\right]\varphi^b \tag{12}$$

with the polarization operator $\Pi_{ab}(P)$, which in the momentum space has the form

$$\Pi_{ab}(P) = iN_c \int \frac{d^4p}{(2\pi)^4}\mathrm{tr}_D\left[S_i(p+\frac{P}{2})(\lambda^a)_{ij}(i\gamma_5)S_j(p-\frac{P}{2})(\lambda^b)_{ji}(i\gamma_5)\right], \tag{13}$$

where now tr_D is the trace over Dirac matrices. In the result, we have obtained an unnormalized inverse meson propagator $D_{ab}^{-1} = P_{ab}^{-1} - \Pi_{ab}(P)$ or

$$D_{ab} = \frac{1}{P_{ab}^{-1} - \Pi_{ab}(P)} \tag{14}$$

which is the 3×3 matrix in the flavor space. We will neglect $\pi^0 - \eta$ and $\pi^0 - \eta'$ mixing terms because they are proportional to $< \bar{q}_u q_u > - < \bar{q}_d q_d >$. We will do the approximation of neglecting also this contribution in the medium. In the basis of $\eta - \eta'$ mesons we may find P_{ab} and $\Pi_{ab}(P)$

$$P_{ab} = \begin{pmatrix} P_{00} & P_{08} \\ P_{08} & P_{88} \end{pmatrix} \longrightarrow P_{ab}^{-1} = \frac{1}{\Delta}\begin{pmatrix} P_{88} & -P_{08} \\ -P_{08} & P_{00} \end{pmatrix} \tag{15}$$

with the determinant $\Delta = P_{00}P_{88} - P_{08}^2$. Then the inverse meson propagator (14) can be presented in the explicit form

$$
\begin{aligned}
D_{ab}^{-1} &= \frac{1}{\Delta} \begin{pmatrix} P_{88} - \Delta\Pi_{00} & -P_{08} - \Delta\Pi_{08} \\ -P_{08} - \Delta\Pi_{08} & P_{00} - \Delta\Pi_{88} \end{pmatrix} \\
&= \frac{1}{\Delta} \begin{pmatrix} A & B \\ B & C \end{pmatrix} \equiv \frac{1}{\Delta} O^{-1} \begin{pmatrix} \underline{A} & 0 \\ 0 & \underline{C} \end{pmatrix} O
\end{aligned}
\tag{16}
$$

with the orthogonal transformation matrix O. Let us suppose that the matrix O has the form

$$
O = \begin{pmatrix} \cos\theta & \sin\theta \\ -\sin\theta & \cos\theta \end{pmatrix}.
\tag{17}
$$

The value of the angle θ can be fixed from the condition

$$
\tan 2\theta = \frac{2B}{C - A}
\tag{18}
$$

which guarantees us a diagonal form of D_{ab}^{-1}. From (17) and (18) we may find $\underline{A}$ and $\underline{C}$ and write D_{ab} as

$$
D_{ab}^{-1} = \frac{1}{\Delta} O^{-1} \begin{pmatrix} \underline{A} & 0 \\ 0 & \underline{C} \end{pmatrix} O = \frac{1}{2\Delta} O^{-1} \begin{pmatrix} D_\eta^{-1} & 0 \\ 0 & D_{\eta'}^{-1} \end{pmatrix} O
\tag{19}
$$

where

$$
D_\eta^{-1} = (A + C) - \sqrt{(C - A)^2 + 4B^2}
\tag{20}
$$

and

$$
D_{\eta'}^{-1} = (A + C) + \sqrt{(C - A)^2 + 4B^2}.
\tag{21}
$$

The masses of the η and η' meson can now be determined by the conditions

$$
D_\eta^{-1}(M_\eta, \mathbf{0}) = 0,
\tag{22}
$$

$$
D_{\eta'}^{-1}(M_{\eta'}, \mathbf{0}) = 0.
\tag{23}
$$

To calculate the coupling constants, we will express D_{ab} directly in the following form

$$
D_{ab} = \Delta \begin{pmatrix} A & B \\ B & C \end{pmatrix}^{-1} = \frac{\Delta}{D} \begin{pmatrix} C & -B \\ -B & A \end{pmatrix}.
\tag{24}
$$

Here

$$
\begin{aligned}
\Delta &= P_{00}P_{88} - P_{08}^2, \\
D &= AC - B^2 \\
&= \Delta(1 - \mathrm{tr}(P\Pi) + \det P \det \Pi)
\end{aligned}
$$

246

and the effective action (12) now is

$$W_{eff}^{(2)}[\varphi] = -\frac{1}{2}\varphi^a D_{ab}\varphi^b \qquad (25)$$

with

$$D_{ab} = \frac{1}{D}\begin{pmatrix} C & -B \\ -B & A \end{pmatrix}. \qquad (26)$$

It has singularities when D equals zero. Making the pole approximation, we expand D over P^2:

$$D(P^2) = D(P^2 = M^2) + \frac{\partial D}{\partial P^2}\Big|_{P^2 = M^2}(P^2 - M^2). \qquad (27)$$

The first term equals zero when $P^2 = M_\eta^2$ (or $P^2 = M_{\eta'}^2$). The second term can be rewritten in the rest frame as

$$D(P^2) = \frac{1}{2M}\frac{\partial D}{\partial P_0}\Big|_{P_0 = M}(P^2 - M^2). \qquad (28)$$

In the case $P^2 = M_\eta^2$ we have the following expression for the meson propagator

$$D_{ab} = \frac{2M_\eta}{\left(\frac{\partial D}{\partial P_0}\right)_{P_0 = M_\eta}}\frac{1}{P^2 - M_\eta^2}\begin{pmatrix} C & -B \\ -B & A \end{pmatrix}. \qquad (29)$$

The matrix

$$M_{ab} = -g_{a\eta}\frac{1}{P^2 - M_\eta^2}g_{b\eta} \qquad (30)$$

allows to parameterize the meson propagator in terms of coupling constants $g_{i\eta}$ with $i = 0, 8$. Using (30), we obtain

$$g_{0\eta}^2 = -\frac{2M_\eta}{\left(\frac{\partial D}{\partial P_0}\right)_{P_0 = M_\eta}}C, \qquad (31)$$

$$g_{8\eta}^2 = -\frac{2M_\eta}{\left(\frac{\partial D}{\partial P_0}\right)_{P_0 = M_\eta}}A, \qquad (32)$$

$$g_{0\eta}g_{8\eta} = \frac{2M_\eta}{\left(\frac{\partial D}{\partial P_0}\right)_{P_0 = M_\eta}}B. \qquad (33)$$

From these coupling constants we may calculate

$$g_{\eta q \bar{q}} = \sqrt{\frac{2}{3}} g_{0\eta} + \frac{1}{\sqrt{3}} g_{8\eta}, \tag{34}$$

$$g_{\eta s \bar{s}} = \sqrt{\frac{2}{3}} g_{0\eta} - \frac{2}{\sqrt{3}} g_{8\eta}. \tag{35}$$

The coupling constants $g_{0\eta'}, g_{8\eta'}, g_{\eta' q \bar{q}}, g_{\eta' s \bar{s}}$ can be evaluated in the same way by the replacement $M_\eta \to M'_\eta$.

Now we consider the properties of η and η' mesons.

First we solve the gap equation (10). Then the matrices A, B, C, D depend on the solutions of M_i and contain the integrals:

$$I_1^i = iN_c \int \frac{d^4 p}{(2\pi)^4} \frac{1}{p^2 - M_i^2} \tag{36}$$

and

$$I_2^{ii}(P) = iN_c \int \frac{d^4 p}{(2\pi)^4} \frac{1}{(p^2 - M_i^2)((p+P)^2 - M_i^2)}, \tag{37}$$

which are divergent. To calculate the physical observables we introduce the cut - off parameter Λ. Putting the integrals to (22) we may calculate M_η. But, since NJL does not confine the quarks and the η' is above the $\bar{q}_u q_u$ and $\bar{q}_d q_d$ continuum, the meson propagator (23) has complex poles. We assume (see details in [11]) that the equation for the poles in the propagator has solutions of the form:

$$P_0 = M_{\eta'} - \frac{1}{2} i\Gamma, \tag{38}$$

where $M_{\eta'}$ is the mass and Γ is the width of the η' - resonance.

In order to avoid the complexity of the analytical continuation, we calculate the integrals (37) in an approximate way, which means that we neglect Γ that occurs in the denominator of $I_2^{ii}(P_0)$ and consider the small value of the width. Since this last integral is complex, a straightforward calculation leads to the replacement in A, B, C, D:

$$P_0^2 I_2^{ii}(P_0) \longrightarrow \left[P_0^2 \mathrm{Re} I_2^{ii}(P_0) + P_0 \Gamma \mathrm{Im} I_2^{ii}(P_0) \right] + i \left[P_0^2 \mathrm{Im} I_2^{ii}(P_0) - P_0 \Gamma \mathrm{Re} I_2^{ii}(P_0) \right] \Big|_{P_0^2 = M_{\eta'}}. \tag{39}$$

Therefore from the zeros of the real and imaginary parts of the propagator we get the mass and the width of η'.

For our numerical calculations we use the parameter set:
$m_u = m_d = 5.5$ MeV, $m_s = 140.7$ MeV, $g_S \Lambda^2 = 3.67$, $g_D \Lambda^5 = -12.36$ and $\Lambda = 602.3$ MeV
that has been determined by fixing the conditions:
$M_\pi = 135.0$ MeV, $M_K = 497.7$ MeV, $f_\pi = 92.4$ MeV and $M_{\eta'} = 960.8$ MeV.

Here f_π is the pion decay constant [10, 11].

Also, we have:
$$M_\eta = 514.8 \text{ MeV}, \ \theta(M_\eta^2) = -11.6°, \ g_{\eta\bar{u}u} = 2.29, \ g_{\eta\bar{s}s} = -3.71$$
and
$$M_{\eta'} = 960.8 \text{ MeV}, \ \theta(M_{\eta'}^2) = -86.8°, \ g_{\eta'\bar{u}u} = 13.4, \ g_{\eta'\bar{s}s} = -6.72.$$

Note that the η' - meson always lies above the quark - antiquark threshold. Since this is an artifact due to the lack of confinement it is a resonant state. Nevertheless, we use $M_{\eta'}$ as an input parameter, because as mentioned in [9], the chiral susceptibility is related with $M_{\eta'}$ by Witten - Veneziano mass formula [17, 18]

$$\frac{2N_f}{f_\pi^2}\chi = M_\eta^2 + M_{\eta'}^2 - 2M_K^2 \tag{40}$$

and takes the value of $M_{\eta'}$ which is closed to the experimental value. In this formula $N_f = 3$ and the chiral susceptibility equals $\chi = (177.2\text{MeV})^4$. The lattice calculations give the number $\chi = (175 \pm 5\text{MeV})^4$.

For the quark condensates we have $< \bar{q}_u q_u > = < \bar{q}_d q_d > = -(241.9\text{MeV})^3$ and $< \bar{q}_s q_s > = -(257.7\text{MeV})^3$.

THE FINITE TEMPERATURE AND DENSITY CASE

The generalization of the NJL model to the finite temperature and density case can be done if we introduce the thermal Green function, which for a system of quarks q_i at given temperature T and chemical potential μ_i is

$$S_i(\vec{x} - \vec{x}', \tau - \tau') = \frac{i}{\beta}\sum_n e^{-i\omega_n(\tau - \tau')} \int \frac{d^3p}{(2\pi)^3} \frac{e^{-i\vec{p}(\vec{x} - \vec{x}')}}{\gamma_0(i\omega_n + \mu_i) - \vec{\gamma}.\vec{p} - M_i}, \tag{41}$$

where $\beta = 1/T$, T is the temperature and the sum is made over the Matsubara frequencies $\omega_n = (2n+1)\pi T$, $n = 0, \pm 1, \pm 2, \ldots$, so that $p_0 \longrightarrow i\omega_n + \mu_i$ with a chemical potential μ_i. Instead of integration over p_0 we have now the sum over Matsubara frequencies. $E_i = \sqrt{\vec{p}^2 + M_i^2}$ is the quark energy.

Because in the gap equation (10) instead of $S_i(p)$ we put the thermal Green functions (41), the current quark masses M_i will now depend on the temperature and chemical potential

$$M_i = m_i - 2g_S << \bar{q}_i q_i >> - 2g_D << \bar{q}_j q_j >> << \bar{q}_k q_k >> \tag{42}$$

where $<< \bar{q}_i q_i >>$ are the quark condensates at finite T and chemical potential μ_i. The condensates are expressed in the terms of the integral $I_1^i(T, \mu_i)$ which is calculated by the substitution of (41) in (36)

$$I_1^i(T, \mu_i) = -\frac{N_c}{4\pi^2} \int \frac{\vec{p}^2 d\vec{p}}{E_i} \left(n_i^+ - n_i^-\right), \tag{43}$$

with the Fermi distribution functions

$$n_i^{\mp} = \left[1 + \exp\left[\beta(E_i \pm \mu_i)\right]\right]^{-1} \tag{44}$$

The integral $I_2^{ii}(P_0, T, \mu_i)$ may be found in the same fashion

$$I_2^{ii}(P_0, T, \mu_i) = -\frac{N_c}{2\pi^2} \mathscr{P} \int \frac{p^2 dp}{E_i} \frac{1}{P_0^2 - 4E_i^2} \left(n_i^+ - n_i^-\right)$$
$$-i\frac{N_c}{4\pi} \sqrt{1 - \frac{4M_i^2}{P_0^2}} \left(n_i^+ \left(\frac{P_0}{2}\right) - n_i^- \left(\frac{P_0}{2}\right)\right). \tag{45}$$

Using these integrals we solve equations (22) and (23) at given temperature and chemical potential taking into account also the relations

$$P_{00}(T, \mu_i) = g_S - \frac{2}{3}g_D \left(<<\bar{q}_u q_u>> + <<\bar{q}_d q_d>> + <<\bar{q}_s q_s>>\right), \tag{46}$$

$$P_{88}(T, \mu_i) = g_S + \frac{1}{3}g_D \left(2 <<\bar{q}_u q_u>> + 2 <<\bar{q}_d q_d>> - <<\bar{q}_s q_s>>\right), \tag{47}$$

$$P_{08}(T, \mu_i) = P_{80} = \frac{1}{3\sqrt{2}}g_D \left(<<\bar{q}_u q_u>> + <<\bar{q}_d q_d>> - 2 <<\bar{q}_s q_s>>\right) \tag{48}$$

and

$$\Pi_{00}(P_0, T, \mu_i) = \frac{2}{3}\left[J_{uu}(P_0, T, \mu_i) + J_{dd}(P_0, T, \mu_i) + J_{ss}(P_0, T, \mu_i)\right], \tag{49}$$

$$\Pi_{88}(P_0, T, \mu_i) = \frac{1}{3}\left[J_{uu}(P_0, T, \mu_i) + J_{dd}(P_0, T, \mu_i) + 4J_{ss}(P_0, T, \mu_i)\right], \tag{50}$$

$$\Pi_{08}(P_0, T, \mu_i) = \frac{\sqrt{2}}{3}\left[J_{uu}(P_0, T, \mu_i) + J_{dd}(P_0, T, \mu_i) - 2J_{ss}(P_0, T, \mu_i)\right], \tag{51}$$

where

$$J_{ii}(P_0, T, \mu_i) = 4(I_1^i + \frac{P_0^2}{2}I_2^{ii}). \tag{52}$$

PHASE TRANSITION

To consider the density effects, in accordance with [13, 14, 16, 19, 20, 21], we concentrate here on the case of asymmetric quark matter with strange quarks in chemical equilibrium maintained by weak interactions and with charge neutrality, by imposing the following constraints on the chemical potentials of quarks and electrons and on its densities:

$$\mu_d = \mu_s = \mu_u + \mu_e \quad \text{and} \quad \frac{2}{3}\rho_u - \frac{1}{3}(\rho_d + \rho_s) - \rho_e = 0, \tag{53}$$

with

$$\rho_i = \frac{1}{\pi^2}(\mu_i^2 - M_i^2)^{3/2}\theta(\mu_i^2 - M_i^2) \quad \text{and} \quad \rho_e = \frac{\mu_e^3}{3\pi^2}. \tag{54}$$

As discussed by several authors, this version of the NJL model exhibits, at $T = 0$ MeV a first order phase transition [15, 16, 22].

For the numerical calculations and discussions of the nature of the phase transition we use the expression for the pressure at given temperature which is connected with the thermodynamical potential as

$$P(\rho, T) = -[\Omega(\rho, T) - \Omega(0, T)], \tag{55}$$

where the thermodynamic potential has the following form

$$\Omega(\rho, T) = \mathscr{E} - TS - \sum_{i=u,d,s} \mu_i N_i \tag{56}$$

and contains the internal energy $\mathscr{E}$, the entropy S and the number of the ith quark N_i.

As a matter of fact, the energy per particle of the quark system has two minimum, corresponding to the zeros of the pressure, the minimum at $\rho_n = 2.25\rho_0$ being an absolute minimum. This implies that the phase transition in this model is a first order one. The results are presented in Figure 1 (left panel) where we have plotted the pressure in terms of ρ_n/ρ_0, where ρ_n is the neutron matter density and $\rho_0 = 0.17$ fm^{-3} is the normal nuclear density.

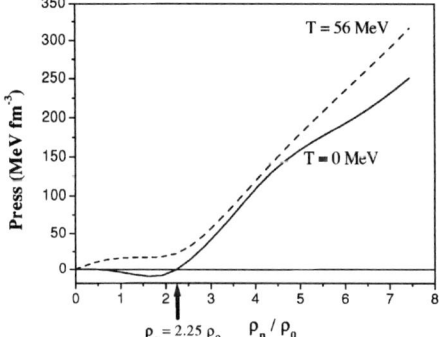 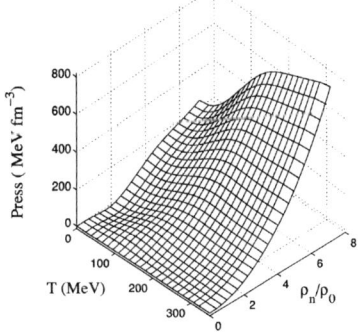

FIGURE 1. Left panel: Pressure $T = 0$ MeV (1st order phase transition) and $T = 56$ MeV (2nd order phase transition); Right panel: Pressure as function of the density and temperature. At $T < 56$ MeV we have a first order phase transition. After that, the phase transition becomes of second order.

At finite temperature we conclude that for $T < 56$ MeV we still have a first order phase transition, but for $T > T_{cl} = 56$ MeV the phase transition is of second order.

Below this temperature there is a range of densities ($\rho_n < 2.25\rho_0$) where the pressure and/or the compressibility are negative meaning that the system is in a mixing phase: a low density (ρ_n^l) phase with massive quarks and a high density (ρ_n^{cr}) phase with light quarks of (partially) restored chiral symmetry (see also [15, 16, 22]).

With this parameterization [22], $\rho_n^l \simeq 0$, and at zero temperature, the model may be interpreted as having a hadronic phase - droplets of light u, d quarks with a density

251

$\rho_n^{cr} = 2.25\rho_0$ surrounded by a non-trivial vacuum - and, above the critical density, a quark phase with partially restored chiral symmetry.

The right panel in Figure 1 shows the combined effect of the temperature and density dependencies of the pressure.

As the temperature increases the pressure becomes positive but the compressibility is still negative. At $T = 56$ MeV and $\rho = 1.53\rho_0$, the compressibility has only on zero and we identify this point as the critical endpoint, the connects the first order and second order phase transition regions.

$\eta - \eta'$ BEHAVIOR IN THE MEDIUM

Figure 2 shows the temperature and density dependencies of constituent quark masses M_i (left panel: $M_u = M_d$, right panel: M_s). As we see at low temperature masses are almost constant and then they decrease with increasing temperature. The same situation can be observed in the density direction.

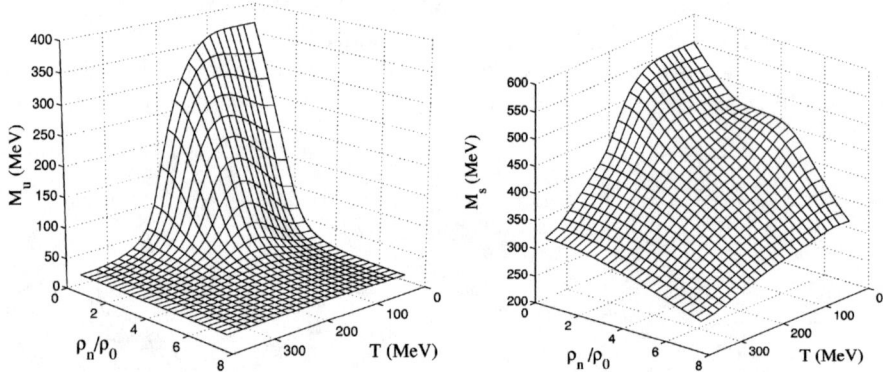

FIGURE 2. Masses of the u and s quarks as function of the density and temperature.

The combined effect of density and temperature shows, as it was expected, that chiral symmetry is partially restored for the light quarks and is badly restored in the strange sector. There is a clear separation between the regions of restored and chiral symmetry in the diagram of the left panel of Figure 2, but the same does not happen in the right panel.

In what concerns the mesonic behavior, before discussing our new findings, we reproduce here the behavior at finite temperature and zero density obtained, for instance in [23]. Figure 3 a) shows the temperature dependence of the M_π, M_η and $M_{\eta'}$ mesons with vanishing chemical potential. For comparison, the curve of $2M_u$ is also indicated. One see that at low temperature the M_π, M_η masses are lower than masses of their constituents. In this case the integrals $I_2^{ii}(P_0,T)$ are real. The crossing of the π and η lines with $2M_u$ line indicates the respective Mott transition temperature T_M for these mesons. For our set of parameters the mesons become unbound at $T_M \approx 200$ MeV. One can also see that $T_{M\eta} < T_{M\pi}$. Above the Mott temperature we have taken into account the imaginary parts of the integrals $I_2^{ii}(P_0, T)$ and used a finite width approximation [11, 23]. As we mentioned before the η' - meson is unbound.

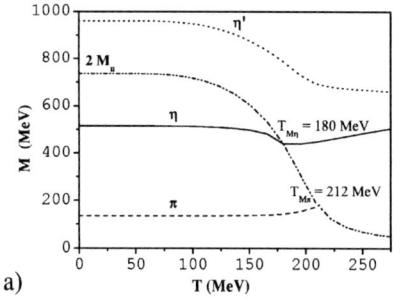 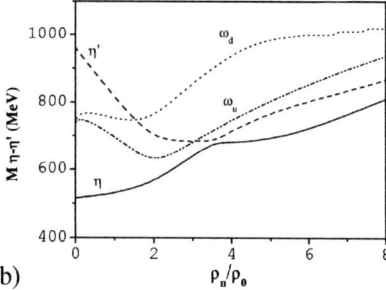

a)   b)

FIGURE 3. a) Temperature dependence of the π (dashed line), η (solid line) and η' (dotted line) masses. The curve $2M_u$ shows the temperature dependence of the quark threshold. Respective Mott temperatures $T_{M\pi}$ and $T_{M\eta}$ are indicated; b) Density dependence of η (solid line) and η' (dotted line) masses.

We start the discussions of the η and η' in matter from the properties of dynamical quark masses and chemical potentials. They are plotted in Figure 4 as functions of ρ_n.

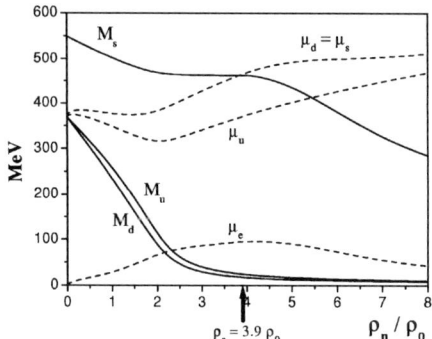

FIGURE 4. Constituent quark masses (solid lines) and chemical potentials, $\mu_s = \mu_u, \mu_e$ (dashed lines) at $T = 0$ MeV.

An important point to be noticed is that, due to the 't Hooft contribution in the gap equations, the mass of the strange quark decreases smoothly and becomes lower than the chemical potential at densities above $3.9\rho_0$, which we will denote from now on as ρ_s'. A more pronounced decrease is then observed, which is due to the presence of s quarks in this regime. As it will be discussed below, this fact is related to a change in the behavior of different observables, as compared to the region of lower densities. At $\rho = \rho_s'$ (Figure 4) the difference between the chemical potentials $\mu_d = \mu_s$ has a maximum and, as the density increase there is a tendency to the restoration of flavor symmetry [21].

Now, we analyze the results for the masses of η and η'.

Having in mind the constraints (53) and (54) we have calculated the density dependence of M_η and $M_{\eta'}$ masses. The results are shown on Figure 3-b.

253

In order to understand these results, it is useful to study the limits of the Dirac sea of this mesons. They can be obtained by looking the limits of the regions of poles in the integrals $I_2^{ii}(P_0)$ (45) and are plotted in Figure 3 (dashed point lines): $\omega_u = 2\mu_u$, $\omega_d = 2\mu_d$ and $\omega_s = 2\mu_s$. (At finite temperature, we will have $\omega_u = 2M_u$, $\omega_d = 2M_d$ and $\omega_s = 2M_s$).

As we can see, the η' - meson lies above the quark - antiquark threshold for $\rho_n < 2.5\rho_0$ and it is a resonant state. After that density, the η' becomes a bound state.

Concerning the masses of η and η' they exhibit a tendency to became degenerate but there are no clear indications of restoration of $U_1(A)$ symmetry. In neutron matter with beta equilibrium the systems shows a tendency to the restoration of flavor symmetry, which is related with the presence of strange quarks in the medium that occurs at about $\rho \sim 3.8\rho_0$.

Concerning the behavior in a hot and dense medium, a feature to be noticed is that the η meson, that has a Goldstone boson like nature, shows more clearly the difference between the chiral symmetric and asymmetric phase, as shown in Figure 5.

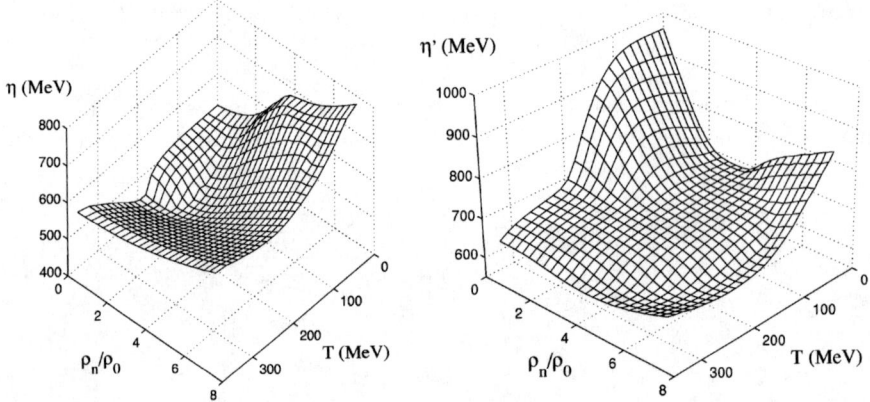

FIGURE 5. Left panel: η mass as function of the density and temperature; Right panel: η' mass as function of the density and temperature.

In conclusion, in the framework of our model, if the behavior of η' at high temperatures and zero density suggests a possible restoration of the $U_A(1)$ symmetry, the same does not happen when we explore the high density region of a dense or hot and dense medium.

ACKNOWLEDGMENTS

Work suported by grant SFRH/BD/3296/2000 (P. Costa), Centro de Física Teórica, FCT and GTAE (Yu. L. Kalinovsky).

REFERENCES

1. F. Karsch and E. Laermann, Phys. Rev. D 50 (1994) 6954;
   K. Kanaya, Prog. Theor. Phys. Sup. 129 (1997) 197.
2. C. Roland, (PHOBOS Collaboration), Nucl. Phys. A 698 (2002) 54.
3. C. Lourenço, Nucl. Phys. A 698 (2002) 13.
4. Z. Fodor and S. D. Katz, Phys. Lett. B 534 (2002) 87, hep-lat/0104001; hep-lat/0204029.
5. E. Shuryak, Comm. Nucl. Part. Phys. 21 (1994) 235.
6. J. Kapusta, D. Kharzeev and L. McLerran, Phys. Rev. D 53 (1996) 5028.
7. Y. Nambu and G. Jona-Lasinio, Phys. Rev. 122 (1961) 345; Phys. Rev. 124 (1961) 246;
   M. K. Volkov, Ann. Phys. 157 (1984) 282.
8. G. A. Sokol, L. N. Pavlyuchenko, nucl-ex/0111020 (2001).
9. K. Ohnishi, K. Fukushima, K. Ohta, Phys. Rev. C 63 (2001) 045203; Phys. Lett. B 514 (2001) 200.
10. U. Vogl and W. Weise, Prog. Part. Nucl. Phys. 27 (1991) 195;
    S. P. Klevansky, Rev. Mod. Phys. 64 (1992) 649;
    T. Hatsuda and T. Kunihiro, Phys. Reports 247 (1994) 221;
    P. Rehberg, S. P. Klevansky, Ann. Phys. 252 (1996) 422.
11. D. Blaschke et al., Nucl. Phys. A 592 (1995) 561.
12. M. C. Ruivo, C. A. de Sousa, B. Hiller and A. H. Blin, Nucl. Phys. A 575 (1994) 460.
13. M. C. Ruivo and C. A. Sousa, Phys. Lett. B 385 (1996) 39.
14. C. A. Sousa and M. C. Ruivo, Nucl. Phys. A 625 (1997) 713. W. Broniowski and B. Hiller, Nucl. Phys. A 643 (1998) 161.
15. M.C. Ruivo, P. Costa and C. A. Sousa, in *Quark Matter in Astro and Particle Physics*, Eds. D. Blaschke, Rostock University, 2001, p. 218, hep-ph/0109234.
16. M. C. Ruivo, C. A. Sousa and C. Providência, Nucl. Phys. A 651 (1999) 59.
17. E. Witten, Nucl. Phys. B 156 (1979) 269.
18. G. Veneziano, Nucl. Phys. B 159 (1979) 213.
19. W. Broniowski and B. Hiller, Nucl. Phys. A 643 (1998) 161.
20. M. C. Ruivo, Hadron Physics - Effective Theories of Low Energy QCD, ed. A. Blin et al, AIP, New York, 2000, p. 237.
21. P. Costa and M. C. Ruivo, Europhysics Letters 60(3) (2002) 356.
22. M. Buballa and M. Oertel, Nucl. Phys. A 642 (1998) 39c; Phys. Lett. B 457 (1999) 261.
23. P. Rehberg, S. P. Klevansky and J. Hüfner, Phys. Rev. C 53 (1996) 410.

J/Ψ PHYSICS

When does anomalous J/ψ suppression happen in nuclear collisions?

Jörg Hüfner* and Pengfei Zhuang†

* Institute for Theoretical Physics, University of Heidelberg, D-69120 Heidelberg, Germany
† Physics Department, Tsinghua University, Beijing 100084, PR China

Abstract. The data for $\langle p_t^2 \rangle(E_t)$ of J/ψ produced in Pb-Pb collisions at the CERN-SPS contain information about when anomalous suppression is active. A general transport equation which describes transverse motion of J/ψ in the absorptive medium is proposed and solved for a QGP and a comover scenario of suppression. While inconsistencies appear in the QGP, the comover approach accounts for the data fairly well. For central collisions the bulk of anomalous suppression happens rather late, 3-4 fm/c after the nuclear overlap.

The discovery in 1996 of anomalous J/ψ suppression in Pb-Pb collisions at the SPS has been one of the highlights of the research with ultrarelativistic heavy ions at CERN [1]. Does it point to the discovery of the predicted quark gluon plasma (QGP)? Six years later the situation is still confused, since several models - with and without the assumption of a QGP - describe the observed suppression, after at least one parameter is adjusted. The data on the mean squared transverse momentum $\langle p_t^2 \rangle(E_t)$ [2] for the ψ (this symbol stands for J/ψ and ψ') in the regime of anomalous suppression and as a function of transverse energy E_t have received less attention - for no good reason. We claim: $\langle p_t^2 \rangle(E_t)$ contains additional information, namely about the time structure of anomalous suppression. This idea has already been considered more than 10 years ago [3] (c.f. also more recent works [4, 5]) and is based on the following phenomenon: Anomalous suppression is not an instantaneous process, but takes a certain time t_A. During this time ψ's produced with high transverse momenta may leak out of the parton/hadron plasma and escape suppression. As a consequence, low p_t ψ's are absorbed preferentially. The $\langle p_t^2 \rangle$ of the surviving (observed) ψ's shows an increase $\delta\langle p_t^2 \rangle$, which grows monotonically with t_A[5]. In this letter we propose a general formalism how to incorporate the effect of leakage into the various models, which have been proposed to describe anomalous suppression. We extract information about the time t_A from a comparison with experiment.

It has become customary to distinguish between normal and anomalous values of suppression $S(E_t) = \sigma^\psi(E_t)/\sigma^{DY}(E_t)$ and values of $\langle p_t^2 \rangle(E_t)$ for ψ's produced in nuclear collisions, c.f. reviews [6, 7]. Here, σ^ψ is the production cross section of a ψ and σ^{DY} is the Drell-Yan cross section in an AB collision. By definition, ψ's produced in pA collisions show normal suppression via inelastic ψN collisions in the final state and normal increase of $\langle p_t^2 \rangle$ (above $\langle p_t^2 \rangle_{NN}$ in NN collisions) via gluon rescattering in the initial state. These normal effects are also present in nucleus-nucleus collisions and happen, while projectile and target nuclei overlap. Anomalous values of S and $\langle p_t^2 \rangle$ are attributed to the action on the ψ by the mostly baryon free phase of partons and/or hadrons (we

CP660, Hadron Physics: Effective Theories of Low Energy QCD, edited by A. H. Blin et al.

call it parton/hadron plasma) which is formed <u>after</u> the nuclear overlap. It may lead to deconfinement of the ψ via colour screening, dissociation via gluon absorption or inelastic collisions by the comovers during the later period of the plasma evolution. In this letter we describe anomalous ψ suppression within a transport theory and apply it to two rather different scenarios: (I) Absorption involving a threshold in the energy density (QGP), (II) continuous absorption via comovers.

We denote by $d\sigma^\psi/d\vec{p}_t(\vec{p}_t, E_t)$ the cross section for the production of a ψ with given p_t and in an event with fixed transverse energy E_t. It can be related to the phase space density f^ψ via

$$\frac{d\sigma^\psi}{d\vec{p}_t}(\vec{p}_t, E_t) = \lim_{t \to \infty} \int d\vec{b} P(E_t; b) \int d\vec{s} f^\psi(\vec{s}, \vec{p}_t, t; \vec{b}). \tag{1}$$

Here, $P(E_t; b)$ describes the distribution of transverse energy E_t in events with a given impact parameter $\vec{b}$ between projectile A and target B. We follow ref. [8] in notation for $P(E_t; \vec{b})$ and the values of the numerical constants. The function $f^\psi(\vec{s}, \vec{p}_t, t; \vec{b})$ is the distribution of ψ's in the transverse phase space $(\vec{s}, \vec{p}_t)$ at time t for given $\vec{b}$.

We define $t = 0$ as the time, when the process of normal suppression and normal generation of $\langle p_t^2 \rangle$ has ceased and denote by $f_N^\psi(\vec{s}, \vec{p}_t; \vec{b})$ the distribution of ψ's at this time. We take f_N^ψ as initial condition for the motion and absorption of the ψ's due to anomalous interactions. The evolution of the ψ is described by a transport equation

$$\frac{\partial}{\partial t} f^\psi + \vec{v}_t \cdot \vec{\nabla}_s f^\psi = -\alpha f^\psi. \tag{2}$$

The time dependence arises from the free streaming of the ψ with transverse velocity $\vec{v}_t = \vec{p}_t/\sqrt{m_\psi^2 + p_t^2}$ (l.h.s.) and an absorptive term on the r.h.s., where the function $\alpha(\vec{s}, \vec{p}_t, t; \vec{b})$ contains all details about the surrounding matter and the absorption process. We have left out effects from a mean field, because the elastic ψN cross section is very small and have also neglected a gain term on the r.h.s., because recombination processes $c\bar{c} + N \to \psi + X$ seem unimportant at SPS energies since at most one $c\bar{c}$ is created per event.

Eq. (2) can be solved analytically with the result

$$f^\psi(\vec{s}, \vec{p}_t, t; \vec{b}) = \exp\left(-\int_0^t dt' \alpha(\vec{s} + \vec{v}_t(t - t'), \vec{p}_t, t'; \vec{b})\right) f_N^\psi(\vec{s} + \vec{v}_t t, \vec{p}_t; \vec{b}), \tag{3}$$

which for $t = 0$ reduces to $f^\psi = f_N^\psi$. If we denote by t_f the time when anomalous suppression has ceased, $\alpha(\vec{s}, \vec{p}_t, t; \vec{b}) = 0$ for $t > t_f$, the limit $t \to \infty$ in eq. (1) can be replaced by setting $t = t_f$, since the distribution in p_t does not change for larger t's.

There is little controversy about ψ production and suppression in the normal phase: The gluons, which fuse to the $c\bar{c}$, collide with nucleons before fusion and gain additional p_t^2. The ψ on its way out is suppressed by inelastic ψN collisions. Neglecting effects of formation time[9], one has

$$f_N^\psi(\vec{s}, \vec{p}_t; \vec{b}) = \sigma_{NN}^\psi \int dz_A dz_B \rho_A(\vec{s}, z_A) \rho_B(\vec{b} - \vec{s}, z_B) \langle p_t^2 \rangle_N^{-1} \exp\left(-p_t^2/\langle p_t^2 \rangle_N\right) \times$$
$$\times \exp\left(-\sigma_{abs}^\psi \left[T_A(\vec{s}, z_A, \infty) + T_B(\vec{b} - \vec{s}, -\infty, z_B)\right]\right), \tag{4}$$

where

$$\langle p_t^2 \rangle_N (\vec{b}, \vec{s}, z_A, z_B) = \langle p_t^2 \rangle_{NN}^{\psi} + a_{gN} \rho_0^{-1} \left[T_A(\vec{s}, -\infty, z_A) + T_B(\vec{b} - \vec{s}, z_B, +\infty) \right] \quad (5)$$

with the thickness function $T(\vec{s}, z_1, z_2) = \int_{z_1}^{z_2} dz \rho(\vec{s}, z)$. All densities ρ_A, ρ_B are normalized to the number of nucleons (ρ_0 is the nuclear matter density). We shortly explain eqs. (4) and (5): For given values $\vec{b}$ and $\vec{s}$ in the transverse plane the ψ is produced at coordinates z_A and z_B in nuclei A and B, respectively. On its way out, the ψ experiences the thicknesses $T_A(\vec{s}, z_A, \infty)$ and $T_B(\vec{b} - \vec{s}, -\infty, z_B)$ in nuclei A and B, respectively and is suppressed with an effective absorption cross section σ_{abs}^{ψ}. The two gluons which fuse carry transverse momentum from two sources. (i) Intrinsic p_t, because they had been confined to a nucleon. The intrinsic part is observable in $NN \to \psi$ collisions and leads to $\langle p_t^2 \rangle_{NN}^{\psi}$ in eq. (5). (ii) In a nuclear collision, the gluons traverse thicknesses $T_A(\vec{s}, -\infty, z_A)$ and $T_B(\vec{b} - \vec{s}, z_B, +\infty)$ of nuclear matter in A and B, respectively, and acquire additional transverse momentum via gN collisions. This is the origin of the second term in eq. (5).

The constants $\sigma_{abs}^{J/\psi}$ and a_{gN} are usually adjusted to the data from pA collisions, before one investigates anomalous suppression. Fig. 1 shows (dashed curves) the results for normal suppression $S^{\psi}(E_t)$ and normal $\langle p_t^2 \rangle^{\psi}(E_t)$ calculated with f_N^{ψ} eq. (4) in eq. (1). While the difference between calculation and data is enormous for the suppression, it is rather small for $\langle p_t^2 \rangle^{\psi}(E_t)$.

Since the physical origin of anomalous suppression is not yet settled, we investigate suppression $S^{\psi}(E_t)$ and $\langle p_t^2 \rangle^{\psi}(E_t)$ for two models, which have rather contradictory assumptions.

   I. QGP scenario: ψ's is totally and rapidly destroyed, when they are in a medium whose energy density above a critical one, and nothing happens elsewhere. As a representative model we use the approach by Blaizot et al. [8].

  II. Comover scenario: The plasma of comovers (partons and/or hadrons) leads to a continuous absorption of long duration due to inelastic collisions with the comoving particles. As a representative model, we use the approaches by Capella et al. [10] and Kharzeev et al. [11].

We begin with model I: In their schematic approach Blaizot et al. [8] include anomalous suppression via

$$f^{\psi}(\vec{s}, \vec{p}_t; \vec{b}) = \theta(n_c - n_p(\vec{b}, \vec{s})) f_N^{\psi}(\vec{s}, \vec{p}_t; \vec{b}). \quad (6)$$

Here n_c is a critical density and $n_p(\vec{b}, \vec{s})$ is the density of participant nucleons

$$n_p(\vec{b}, \vec{s}) = T_A(\vec{s}, -\infty, +\infty) \left[1 - \exp\left(-\sigma_{in}^{NN} T_B(\vec{b} - \vec{s}, -\infty, +\infty) \right) \right] + (A \leftrightarrow B). \quad (7)$$

According to eq. (6) all ψ's are destroyed if the energy density (which is directly proportional to the participant density) at the location $\vec{s}$ of the ψ is larger than a critical density n_c. The other ψ's survive. While the prescription eq.(6) successfully describes the data in the full E_t range of anomalous suppression after the only one free parameter,

n_c, is adjusted, the predictions for $\langle p_t^2 \rangle (E_t)$ are significantly below the data, especially at large E_t (see below).

The expression eq. (6) for the phase space distribution f^ψ including anomalous suppression within the QGP model is recovered within our transport approach eq. (3) by setting

$$\alpha(\vec{s},\vec{p},t;\vec{b}) = \alpha_0 \theta(n_p(\vec{b},\vec{s}) - n_c) \, \delta(t) \tag{8}$$

and taking the limit $\alpha_0 \to \infty$. The delta function $\delta(t)$ has to be included in order to recover Eq. (6), but it reflects our preconception that the energy density is highest at $t = 0$ and absorption therefore most likely.

There are various ways to introduce another time structure into the absorption term. In this paper we investigate the generalization

$$\alpha(\vec{s},\vec{p}_t,t;\vec{b}) = \alpha_0 \, \theta \left(n_p(\vec{b},\vec{s}) - n_c \right) \delta(t - t_A). \tag{9}$$

The idea of a threshold density is kept, but it acts at a later time t_A, which time is then determined from a comparison with the data. For $\alpha_0 \to \infty$ and a time $t_A^+ = t_A + \varepsilon$, one finds from the general solution eq. (3)

$$f^\psi(\vec{s},\vec{p}_t,t_A^+) = \theta \left(n_c - n_p(\vec{b},\vec{s}) \right) f_N^\psi(\vec{s} + \vec{v}_t t_A, \vec{p}_t), \tag{10}$$

which differs from the expression (6), by the motion in phase space of f_N^ψ during the time $0 \le t \le t_A$. For $t > t_A$ the momentum distribution derived from f^ψ does not change any more. The physical interpretation of t_A is a kind of mean value replacing a time structure of longer duration.

We calculate the suppression $\sigma^\psi(E_t,t_A)/\sigma^{DY}(E_t)$ and $\langle p_t^2 \rangle^\psi (E_t,t_A)$ with the distribution f^ψ from eq. (10). We use the parameters of ref. [8] where available, i.e. for J/ψ: $\sigma_{abs}^{J/\psi} = 6.4$ mb, $n_c = 3.75$ fm^{-2}. The parameters for the generation of $\langle p_t^2 \rangle$ by gluon rescattering are taken from a fit to the pA data [2]: $\langle p_t^2 \rangle_{NN} = 1.11$ (GeV/c)^2 and $a_{gN} = 0.081$ (GeV/c^2) fm^{-1}. We also account for the transverse energy fluctuations [8] which have been shown to be significant for the explanation of the sharp drop of J/ψ suppression in the domain of very large E_t values, by replacing n_p by $\frac{E_t}{\langle E_t \rangle} n_p$ where $\langle E_t \rangle$ is the mean transverse energy at given b. We then calculate $\sigma^{J/\psi}/\sigma^{DY}$ as a function of t_A. Since the critical density for J/ψ is quite high, the leakage affects only the very high momentum J/ψ's, and $\sigma^{J/\psi}/\sigma^{DY}$ increases only slightly as t_A increases (Fig. 1 upper left).

We turn to a discussion of $\langle p_t^2 \rangle^{J/\psi}(E_t)$. Fig. 1 (upper right) shows calculated curves for values of $t_A = 0$ to 4 fm/c. The dotted line ($t_A = 0$) is the result of the original threshold model with immediate anomalous suppression (and has been predicted in [12]). It fails badly at large values of E_t. Also no other curve with a given t_A describes the data for all values of E_t. We have to conclude that t_A depends on E_t: $t_A(E_t)$. The larger the values E_t the later anomalous suppression acts. From a comparison with data we have $t_A \lesssim 2$ fm/c for $E_t < 80$ GeV, $t_A \simeq 2.5$ fm/c for $80 \le E_t \le 100$ GeV, and $t_A \simeq 3.5$ fm/c for $E_t > 100$ GeV.

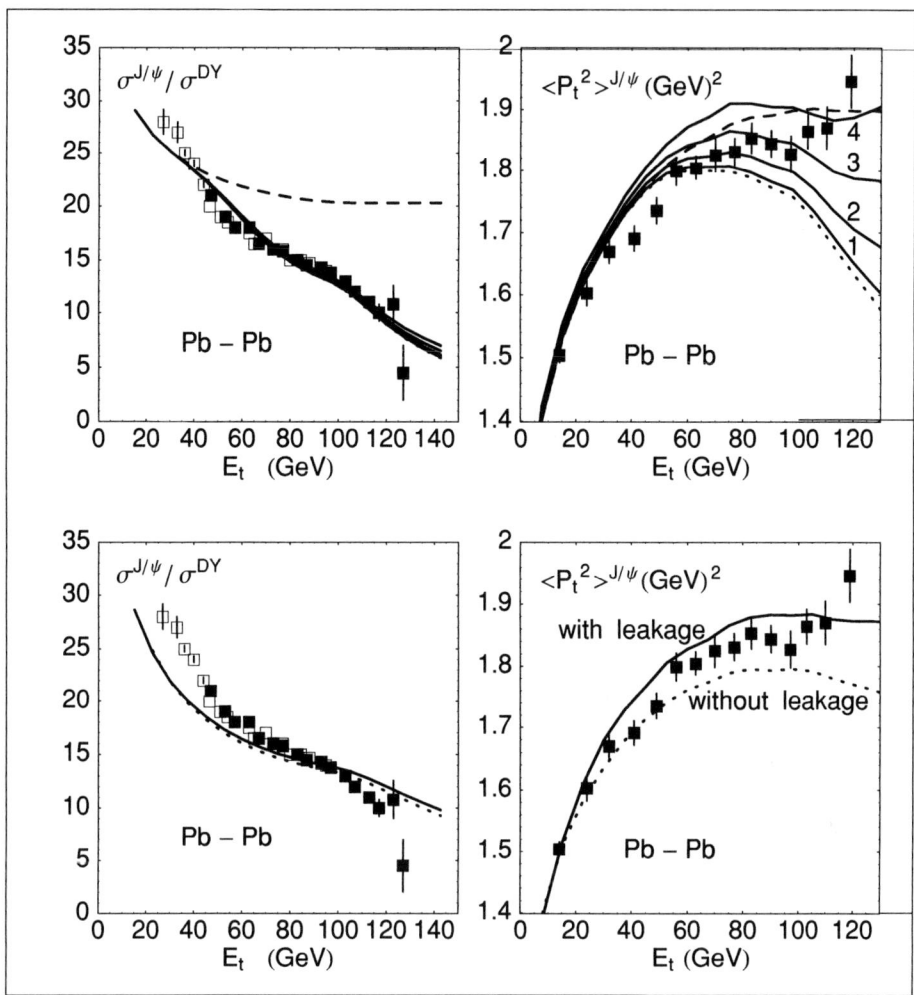

FIGURE 1. Nuclear suppression $\sigma^{J/\psi}/\sigma^{DY}$ and $\langle p_t^2 \rangle^{J/\psi}$ as a function of transverse energy E_t. Data are from [1] for $\sigma^{J/\psi}/\sigma^{DY}$ and from [2] for $\langle p_t^2 \rangle^{J/\psi}$. Upper two curves: Dashed curves show the result of normal suppression alone. The other curves include anomalous suppression within the QGP model eqs. (6) and (10), where anomalous suppression is assumed to act at time t_A. The curves are labled by the values $t_A = 1, 2, 3, 4$ fm/c. Also the curves in $\sigma^{J/\psi}/\sigma^{DY}$ carry these lables. Lower two figures: Dotted and solid lines are calculated in the comover model. Dotted line from ref. [15], which in our case corresponds to eq. (3) but setting $\vec{v}_t = 0$. Solid line full calculations including $\vec{v}_t$.

The results for the extracted values t_A, when anomalous suppression acts within the QGP model, are strange: Can we believe that the J/ψ is absorbed only at a mean time $t_A = 3\text{-}4$ fm/c after the two nuclei have separated? In the sence of the mean value, only one half of anomalous suppression acts for $t < t_A$ and one half for $t > t_A$.

Before we answer this questin we proceed to model II, the scenario of comovers:

Partons and/or hadrons which move with the ψ may destroy the ψ with a cross section σ_{co}^{ψ}. The comover density $n_{co}(t)$ depends on time, for which the Bjorken expansion scenario predicts t^{-1}. The absorptive term α in eq. (2) then takes the form

$$\alpha(\vec{s}, \vec{p}_t, t; \vec{b}) = \sigma_{co}^{\psi} \frac{n_{co}(\vec{b}, \vec{s})}{t} \theta \left(\frac{n_{co}(\vec{b}, \vec{s})}{n_f} t_0 - t \right) \theta(t - t_0), \qquad (11)$$

i.e. absorption by comovers starts at $t = t_0$ and ends at t_1, when the comover density $n_{co}(\vec{b}, \vec{s}) \cdot t_0 / t_1$ has reached a value n_f independent of $\vec{b}$ and $\vec{s}$. The comover approach contains a definite time structure for anomalous suppression and we have not changed it. In the choice of parameters we have followed [10, 11]: $n_{co}(\vec{b}, \vec{s}) = 1.5 n_p(\vec{b}, \vec{s})$ with the participant density from eq. (7), $t_0 = 1$ fm/c, $n_f = 1$ fm^{-2}, $\sigma_{abs}^{J/\psi} = 4.5$ mb and $\sigma_{co}^{J/\psi} = 1$ mb.

The calculated E_t dependence of the suppression shown in Fig. 1 (lower left) fits the data acceptably but less well than model I. For $\langle p_t^2 \rangle$, we compare the results of two calculations with the data. The solid curves represent the result of the full calculation, including leakage, i.e. expression f^{ψ} eq.(3). It fits the data reasonably well for the full range of transverse energies E_t. The dashed lines in Fig. 1 (lower right) show the calculation leaving out leakage. Formally this limit (which has been calculated already in [15]) is obtained from eq. (3) by setting $\vec{v}_t = 0$ in the exponent and in f_N^{ψ}. The case without leakage fits the data less well, through the difference between the two curves is not very big. We stress: the calculation of $\langle p_t^2 \rangle (E_t)$ in the comover model is a true prediction in the sense that no parameter is adjusted above those which are fitted to the suppression $S^{\psi}(E_t)$. We have also calculated the mean time $\langle t_A^{\psi} \rangle$ for comover action by studying the suppression $S^{\psi}(E_t)$ as a function of time and taking the mean of t with the weight $dS^{\psi}(E_t)/dt$ and find

$$\langle t_A^{J/\psi} \rangle = 3.5 \, \text{fm}/c, \qquad (12)$$

which values include the time t_0, eq. (11), between the end of normal suppression and the beginning of comover action. The values eq. (12) are found to be rather independent of E_t.

We summarize: we have described a method to derive information about the time structure of anomalous charmonium suppression from the data on $\langle p_t^2 \rangle (E_t)$ measured for $Pb - Pb$ collisions at SPS-energies, where anomalous suppression has been discovered. The time evolution of the ψ during anomalous suppression including leakage is described within a general transport equation. The formalism is applied to two models with rather contradictory underling physical assumptions, the QGP and the comover models with the following results:

(i) Calculations within the original models, where leakage is left out, do not describe the data for $\langle p_t^2 \rangle^{J/\psi} (E_t)$, the discrepancy being particularly strong for the QGP model at high values of E_t.

(ii) Including leakage into the comover model, without changing its structure nor its parameters leads to a good agreement with the data for $\langle p_t^2 \rangle (E_t)$ for J/ψ over the full range of values E_t.

(iii) The assumption in the threshold model that anomalous suppression acts instanta-neously at $t_A = 0$, i.e. right after normal suppression is not supported by experiment, but a mean value $t_A \approx 3.5$ fm fits the data for central events.

We conclude: According to the present study, anomalous suppression of charmonia is a process which takes a certain time, ≈ 2 fm/c in peripheral collisions and ≈ 4 fm/c in central collisions. Unless the partonic phase with its high density and rapid suppression has an unusual time structure[4, 16], the results of our paper favor the comover scenario with its rather weak but continuous suppression over a comparably long time.

Acknowledgements: We are grateful to our friends and colleagues A. Gal, J. Dolejšı, Yu. Ivanov, B.Z. Kopeliovich, H.J. Pirner and C.Volpe for help and valuable comments. One of the authors (P.Z.) thanks for the hospitality at the Institute of Theoretical Physics. This work has been supported by the grant 06HD954 of the German Federal Ministry of Science and Research and by the Chinese National Science Foundation.

REFERENCES

1. M.C. Abreu *et al.*, *Nucl. Phys. A* **610** 404c (1996).
2. NA50 Collaboration, M.C. Abreu *et al.*, *Phys. Lett. B* **499** 85 (2001).
3. F. Karsch, R. Petronzio, *Z. Phys. C* **37** 627 (1988);
   T. Matsui, H. Satz, *Phys. Lett. B* **178** 416 (1986);
   J.P. Blaizot, J.Y. Ollitrault, *Phys. Lett. B* **199** 627 (1987).
4. D. Kharzeev, M. Nardi, H. Satz, *Phys. Lett. B* **405** 14 (1997);
   E. Shuryak, D. Jeaney, *Phys. Lett. B* **430** 37 (1998).
5. J. Hüfner, P. Zhuang, *Phys. Lett. B* **515** 115 (2001).
6. R. Vogt, *Phys. Rep.* **310** 197 (1999).
7. C. Gerschel, J. Hüfner, *Annu. Rev. Nucl. Part. Sci.* **49** 255 (1999).
8. J.P. Blaizot, P.M. Dinh, J.Y. Ollitrault, *Phys. Rev. Lett.* **85** 4010 (2000);
   J.P. Blaizot, J.Y. Ollitrault, *Phys. Rev. Lett.* **77** 1703 (1996).
9. J. Hüfner, B.Z.Kopeliovich, *Phys. Rev. Lett.* **76** 192 (1996).
10. A. Capella, E.G. Feirreiro and A.B. Kaidalov, *Phys. Rev. Lett.* **85** 2080 (2000).
11. D. Kharzeev, C. Lovrenco, M. Nardi, H. Satz, *Z. Phys. C* **74** 307 (1997).
12. D. Kharzeev, M. Nardi, H. Satz, *Phys. Lett. B* **405** 14 (1997).
13. M.C. Abreu *et al.*, *Phys. Lett. B***477** 28 (2000).
14. M.C. Abreu *et al.*, *Nucl. Phys. A* **638** 261c (1998).
15. N. Armesto, A. Capella, E.G. Feirreiro, *Phys. Rev. C* **59** 395 (1999).
16. D. Rischke, M. Gyulassy, *Nucl. Phys. A* **597** 701 (1996).

Rare Events, J/ψ Production in Heavy Ion Collisions and Percolation

Jorge Dias de Deus

CENTRA and Department of Physics, IST
jdd@fisica.ist.utl.pt

Abstract. We discuss rare events in particle production, with an emphasis on the J/ψ suppression in heavy ion and nucleon collisions, and the role of two dimensional continuum percolation.

INTRODUCTION

We know that nuclei are made up of nucleons and that nucleons are made up of quarks. It is then natural to see nucleus-nucleus collisions as resulting from a superposition of nucleon-nucleon collsions and nucleon-nucleon collisions themselves as a consequence of the participation of quark and gluon degrees of freedom.

In the ideal situation we have a convolution equation which can formally be written as:

$$PHYSICS = ELEMENTARY\ COLLISION\ PHYSICS$$
$$\otimes \qquad\qquad (1)$$
$$SUM\ OVER\ ELEMENTARY\ COLLISIONS.$$

The classical example of this scheme is Glauber calculus, see [1] for a review and developments, where a set of rules is given on how to sum up the elementary contributions.

Sometime ago [2], it was shown that in good approximation the formal equation (1) can be reduced to the second term:

$$PHYSICS \cong SUM\ OVER\ ELEMENTARY\ COLLISIONS. \qquad (2)$$

What happens is that experimentally fluctuations in particle distributions, $P(n)$, and in transverse energy distributions, $P(E_T)$, are much larger in nucleus-nucleus and hadron-hadron collisions than in the more elementary e^+e^- collisions. Tipically, in multiparticle production, the ratio $(D/\bar{n})^2$, where D is the dispersion, $D^2 \equiv \overline{n^2} - \bar{n}^2$, and $\bar{n}$ the average multiplicity, is of the order of 0.09 in $e^+e^- \to q\bar{q} \to n$, while it is in the range $0.25 - 1$ in nucleus-nucleus and hadron-hadron collisions. This means that fluctuations in the elementary collision can be neglected, i.e., the approximation of

CP660, *Hadron Physics: Effective Theories of Low Energy QCD*, edited by A. H. Blin et al.
© 2003 American Institute of Physics 0-7354-0120-9/03/$20.00

assuming that one always produces the average number of collisions in the elementary collision is quite good. Equation (2) then follows from (1).

If in addition one assumes that the elementary collisions are all equivalent, i.e., producing particles with the same density in central rapidity and the same transverse momentum distribution, then the number ν of elementary collisions is a measure of the number of produced particles and of the transverse energy:

$$\nu \sim n,$$
$$\nu \sim E_T. \tag{3}$$

The validity of (3), in the sense that $n \sim E_T$, was recently confirmed by [3].

Please, note that our arguments do not apply to quantities that are strongly dependent on flavour and energy-momentum conservation (for instance: forward-backward production, quantum number flow). We concentrate here in quantities which are centrally produced: particles and transverse energy.

Finally, the additivity procedure implied in our arguments, if, on one hand, it is always relevant, on the other it is, in general, not sufficient. Medium effects, essentially non-linear effects, tend to distort results making Eq. (2) not valid. We then have

$$PHYSICS \cong SUM\ OVER\ ELEMENTARY\ COLLISIONS$$
$$\otimes \tag{4}$$
$$MEDIUM\ EFFECTS$$

The rest of this talk is organized as follows. In section II we introduce the notion of rare event. In section III we discuss the problem of the J/ψ suppression. In section IV we make a short introduction to percolation and its possible application to the J/ψ suppression.

RARE EVENTS

For reasons explained above, we shall focus here on production in central rapidity region (called, in the old times, pionization region). The production may be of charged particles, n, or transverse energy, E_T. In both cases the distributions reflect the distribution in the number of collisions, ν. We can be more precise. In the limit of validity of (2) we have

$$\frac{1}{\bar{n}} P\left(\frac{n}{\bar{n}}\right) = \frac{1}{\bar{E}_T} P\left(\frac{E_T}{\bar{E}_T}\right) = \frac{1}{\bar{\nu}} P\left(\frac{\nu}{\bar{\nu}}\right), \tag{5}$$

where $P(x)$ is a probability distribution, with, in the continuous approximation,

$$\int P(x)dx = \int xP(x)dx = 1, \qquad (6)$$

and $\bar{n}, \bar{E}_T, \bar{v}$ being average values, and

$$C_k \equiv \frac{\overline{n^k}}{\bar{n}^k} = \frac{\overline{E_T^k}}{\bar{E}_T^k} = \frac{\overline{v^k}}{\bar{v}^k}, \qquad k = 0,1,\dots. \qquad (7)$$

We shall consider two types of events in a AB collision: a) minimum bias events or unconstrained events (except for detector limitations) and b) triggered events (where, in addition, C occurs). In an event of type a) one measures a certain multiplicity n or transverse energy E_T, in an event of type b) one measures the same and, in addition, for instance, a di-muon pair, or a W or something else. It should be clear that events of type a) include events of type b).

We assume now that the limit (2) applies (no elementary fluctuations and no medium effects). If α_C is the probability of C to occur in an elementary collision, $N(v)$ the number of events of type a) for v elementary collisions, and $N_C(v)$ the number of events with C occurring, we have

$$N_C(v) = \sum_{s=1}^{v} \binom{v}{s} \alpha_C^s (1-\alpha_C)^{v-s} N(v). \qquad (8)$$

Note that (8) is easily obtained having in mind the identity $N(v) \equiv (\alpha_C + (1-\alpha_C))^v N(v)$ and the binomial expansion.

An event of type b) is denominated as rare if α_C is small, such that in (8) one can, in good approximation, keep just the linear term in α_C:

$$N_C(v) \cong \alpha_C v N(v) \qquad (9)$$

As

$$\sum_v N(v) = N,$$

$$\sum_v v N(v) = \bar{v} N, etc., \qquad (10)$$

we can construct the probability distributions, $P(v)$ and $P_C(v)$, such that [4]

$$P_C(v) = \frac{v}{\bar{v}} P(v). \qquad (11)$$

268

This is a remarkable result as it is universal, independent of the nature of C (to the extent that $\alpha_C \ll 1$). In general, it says that $P_C(v) < P(v)$, for $v < \bar{v}$, and $P_C(v) > P(v)$, for $v > \bar{v}$. It also follows that $\bar{v}_C = C_2 \bar{v}$ or, as $C_2 \geq 1$, $\bar{v}_C \geq \bar{v}$.

Eq. (11) was quantitatively checked in several situations, and similarly for Eq. (9):

i)   Associated E_T distributions to Drell-Yan dimuons, compared to minimum bias distributions in S-U (NA 38) [5] and Pb-Pb (NA50) collisions at the CERN SPS, [6].

ii)   Associated multiplicity distributions associated to $W^{\pm}$ and Z° production, compared to minimum bias distribution in $p\bar{p}$ collisions at the Fermi Lab. Tevatron [4].

In a different context, an equation similar to (11) was qualitatively tested in comparing associated distributions triggered by a sub-threshold pion to minimum bias distributions [7].

THE J/ψ PROBLEM

Production of $J/\psi\,(c\bar{c})$ can be considered as a rare event because the production cross-section and multiplicity are very small. However, clearly, Eq. (9) and (11) do not work:

$$N_{J/\psi}(v) \neq \alpha_{J/\psi} v N(v) \tag{12}$$

and

$$P_{J/\psi}(v) \neq \frac{v}{\bar{v}} P(v) \tag{13}$$

For instance, the ratio

$$R(E_T) \equiv \frac{N^{J/\psi}(E_T)}{N^{DY}(E_T)} \tag{14}$$

of the E_T associated distribution in the case of J/ψ and Drell-Yan, if Eq. (9) was true in both cases, should be constant, independent of E_T. Experimentally it is not constant, it decreases with E_T [8]. On the other hand, with Eq. (11), when comparing the J/ψ associated production distribution to the minimum bias distribution, in the case of S-U, agreement does not exist [4].

There are two proposed explanations for the decrease of $R(E_T)$, (14), as E_T increases: absorption and quark-gluon plasma formation. In the case of absorption the created J/ψ is subsequentely destroyed by interactions with the medium [9]. It is a problem similar to the attenuation of a beam moving through some length of matter. In practice, this means that the effective number of collisions is smaller. In (9) and (11) we make the change

$$v \to v^\varepsilon , \varepsilon < 1 . \tag{15}$$

Note that, comparing to the standard absorption formula, $R \sim \exp(-AL)$, L being the length, we have $L \sim (1-\varepsilon)\ln E_T$.

The other possibility is quark-gluon plasma formation at some critical $E_T, E_T^{crit.}$, with destruction of J/ψ by Debye screening [10] in the region of the phase transition:

$$\alpha_{J/\psi} \to \alpha_{J/\psi} , E_T < E_T^{crit.}$$

$$\tag{16}$$

$$\alpha_{J/\psi} \to 0 , E_T > E_T^{crit.} .$$

Finite size effects make the transition not as sharp as in (16).

We have then 3 possibilities. There is no interaction with the medium and $R(E_T) = const.$; there is absorption and $R(E_T) \sim E_T^{-(1-\varepsilon)}$; there occurs Debye screening and $R(E_T)$ drops abruptly at some value $E_T^{crit.}$.

The essential difference between absorption and Debye screening is in the curvature of $R(E_T)$: positive in the case of absorption and negative in the case of quark-gluon plasma formation. Experimentally, in most cases, the curve is compatible with absorption (see, for instance, $R(E_T)$ in S-U, NA38 [5]). However, for denser Pb-Pb collisions there is the possibility in the curve $R(E_T)$ of the presence of one or more drops, with change in curvature [8].

THE CASE FOR PERCOLATION

In the Dual Parton Model [11] the interaction of hadrons and nuclei can be seen as occurring via the formation of strings, carrying the quantum numbers at the ends. There are long strings in rapidity, associated to Valence parton interactions, in a number proportional to the number of participating nucleons, N, and short and more central strings, related to Sea parton interactions, in a number roughly proportional to the number of collisions v, a number growing with energy. As this number increases the strings start to overlap (fuse) in the impact parameter, generating a situation similar to two dimensional continuum percolation [12]. When percolation occurs we move from a situation of a colour insulator (nucleons, hadrons) to a situation of colour conduction (quark gluon plasma).

The relevant parameter in percolation is the percolation parameter η, which measures the transverse density. In the present context

$$\eta \equiv \left(\frac{r_0}{R}\right)^2 2k\nu , \qquad (17)$$

where $r_0 \left(r_0 \cong 0.2\,fm\right)$ is the transverse radius of the string, $R\left(R \cong 1.14 N^{1/3}\right)$ the radius of the interaction region, ν the number of collisions and $2k$ the number of strings (strings are produced in pairs) produced in an elementary collision.. The quantity η is an increasing function of the energy: as energy increases more Sea strings are produced. The parameter k controls this increase. Numerical studies show that th phase threshold occurs for $\eta_c \cong 1.12 - 1.15$ [12,13], increasing for $\eta_c \cong 1.5$ in the case, more realistic in nucleus-nucleus collisions, of non-uniform distributions [14].

In [15], see also [16], a model to describe the J/ψ suppression based on absorption and percolation, as mentioned in section III, was developed. Better and higher energy data are required to check the validity of this approach.

More recently, in [17], it was discussed the possibility of J/ψ suppression, by quark-gluon plasma formation, in pp collisions. In this case, in (17), $\nu = 1$ and $r_0/R \cong 1/5$. A strong suppression is predicted, in the Tevatron/LHC energy region.

REFERENCES

1. Braun, V. M., and Shabelski, Yu., *Int. J. of Mod. Phys.* **A3**, 2417 (1988).
2. Dias de Deus, J., Pajares, C., and Salgado, C.A.., *Phys. Letters* **B407**, 335 (1997).
3. Bazilevsky, A., (Phenix Collaboration), nucl-ex/0209025 (2002).
4. Dias de Deus, J., Pajares, C., and Salgado, C.A.., *Phys. Letters* **B409**, 474 (1997) and **B408,** 417 (1997).
5. Charlot, C., (NA38), in *Proceedings of the XXV Rencontres de Moriond*, Ed. Frontières (1994).
6. Cicaló, C., (NA50 Collaboration), in Quark Matter 99, Torino, Italy (1999).
7. Dias de Deus, J., Peña, M. T., and Seixas, J. C., *Phys. Rev.* **C64**, 044603 (2001).
8. Abreu, M. C., et al, (NA50 Collaboration), *Phys. Letters* **B477**, 28 (2000); *Phys. Lett.* **B410**, 337 (1997); Baglin, C., et al, *Phys. Lett.* **B220**, 471 (1989); **B251**, 465 (1990); **B255**, 459 (1991).
9. Capella, A., Merino, C., Pajares, C., Ramalho, A. V., and Tran Thanh Van, J., *Phys. Lett.* **B206**, 354 (1988); Gavin, S., and Vogt, R., *Phys. Rev. Lett.* **78**, 1006 (1997); Capella, A., Kaidalov, A., Kouidir Akil, A., and Gerschel, C., *Phys. Lett.* **B393**, 431 (1997).
10. Matsui, T., and Satz, H., *Phys. Lett.* **B178**, 416 (1986).
11. Capella, A., Sukhatme, U. P., Tan, C. I., and Tran Thanh Van, J., *Phys. Rep.* **236**, 225 (1994).
12. Isichenko, M. B., *Rev. Mod. Phys.* **64**, 961 (1992).
13. Pine, G. E., and Seager, C. H., *Phys. Rev.* **B10**, 1421 (1974).
14. Rodrigues, A., Ugoccioni, R., and Dias de Deus, J., *Phys. Lett.* **B458**, 402 (1999).
15. Dias de Deus, J., Ugoccioni, R., and Rodrigues, A., *Euro. Phys. J.* **C16**, 537 (2000).
16. Ugoccioni, R., and Dias de Deus, J., *Nucl. Phys.* **B92**, 83 (2001).
17. Dias de Deus, J., and Rodrigues, A., APS/1-HEP (2002).

J/ψ dissociation and hadron formfactors

Yu.L. Kalinovsky[*], D. Blaschke[†] and G. Burau[†]

[*]Laboratory of Information Technologies, JINR Dubna, 141980 Dubna, Russia
[†]Fachbereich Physik, Universität Rostock, D-18051 Rostock, Germany

Abstract. We summarize the results of the $SU(4)$ chiral meson Lagrangian approach to the cross section for J/ψ breakup by pion impact. The major weakness of this approach is the arbitrariness in the choice of hadronic form factors. We evaluate the dependence of the cross section on the masses of the final D-meson states and compare the result to a parametrization that has been employed for the study of in-medium effects on this quantity.

INTRODUCTION

The J/ψ meson plays a key role in the experimental search for the quark-gluon plasma (QGP) in heavy-ion collision experiments where an anomalous suppression of its production cross section relative to the Drell-Yan continuum as a function of the centrality of the collision has been found by the CERN-NA50 collaboration [1]. An effect like this has been predicted to signal QGP formation [2] as a consequence of the screening of color charges in a plasma in close analogy to the Mott effect (metal-insulator transition) in dense electronic systems [3]. However, a necessary condition to explain J/ψ suppression in the static screening model is that a sufficiently large fraction of $c\bar{c}$ pairs after their creation have to traverse regions of QGP where the temperature (resp. parton density) has to exceed the Mott temperature $T_{J/\psi}^{\text{Mott}} \sim 1.2 - 1.3T_c$ [4, 5] for a sufficiently long time interval $\tau > \tau_{\text{f}}$, where $T_c \sim 170$ MeV is the critical phase transition temperature and $\tau_{\text{f}} \sim 0.3$ fm/c is the J/ψ formation time. Within an alternative scenario [6], J/ψ suppression does not require temperatures well above the deconfinement one but can occur already at T_c due to impact collisions by quarks from the thermal medium. An important ingredient for this scenario is the lowering of the reaction threshold for string-flip processes which lead to open-charm meson formation and thus to J/ψ suppression. This process has an analogue in the hadronic world, where e.g. $J/\psi + \pi \rightarrow D^* + \bar{D} + h.c.$ could occur provided the reaction threshold of $\Delta E \sim 640$ MeV can be overcome by pion impact. It has been shown recently [7] that this process and its in-medium modification can play a key role in the understanding of anomalous J/ψ suppression as a deconfinement signal. Since at the deconfinement transition the D- mesons enter the continuum of unbound (but strongly correlated) quark- antiquark states (Mott- effect), the relevant threshold for charmonium breakup is lowered and the reaction rate for the process gets critically enhanced. Thus a process which is negligible in the vacuum may give rise to additional (anomalous) J/ψ suppression when conditions of the chiral/ deconfinement transition and D- meson Mott effect are reached in a heavy-ion collision but the dissociation of the J/ψ itself still needs impact to overcome the threshold which is still present

CP660, *Hadron Physics: Effective Theories of Low Energy QCD*, edited by A. H. Blin et al.
© 2003 American Institute of Physics 0-7354-0120-9/03/$20.00

but dramatically reduced.

For this alternative scenario as outlined in [7] to work the J/ψ breakup cross section by pion impact is required and its dependence on the masses of the final state D- mesons has to be calculated. Both, nonrelativistic potential models [12, 15] and chiral Lagrangian models [9, 10, 11] have been employed to determine the cross section in the vacuum. The results of the latter models appear to be strongly dependent on the choice of formfactors for the meson-meson vertices. This is considered as a basic flaw of these approaches which could only be overcome when a more fundamental approach, e.g. from a quark model, can determine these input quantities of the chiral Lagrangian approach.

In the present paper we would like to reduce the uncertainties of the chiral Lagrangian approach by constraining the formfactor from comparison with results of a nonrelativistic approach which makes use of meson wave functions [15]. Finally, we will obtain a result for the off-shell J/ψ breakup cross section which can be compared to the fit formula used in [7]. This quantity is required for the calculation of the in-medium modification of the J/ψ breakup due to the Mott-effect for mesonic states at the deconfinement/chiral restoration transition which has been suggested [7, 8] as an explanation of the anomalous J/ψ suppression effect observed in heavy-ion collisions at the CERN-SPS [1].

EFFECTIVE CHIRAL LAGRANGIAN

We start from QCD at low energy. The effective chiral Lagrangian for pseudoscalar (Goldstone) mesons can be written as

$$L_0 = \frac{F_\pi^2}{8} \text{tr} \left[\partial_\mu U(x) \partial_\mu U^+(x) \right] , \tag{1}$$

with $F_\pi = 132$ MeV being the weak pion decay constant, and $U(x) = \exp\left[2i\varphi(x)/F_\pi\right]$. The usual multiplet of pseudoscalar mesons is $\varphi = \varphi^a \lambda^a/\sqrt{2}$, λ^a are Gell - Mann matrices. Notice that $U(x)$ transforms in a so-called non-linear representation of the $SU(N_f)_L \times SU(N_f)_R$ group. To introduce vector and axial-vector mesons we follow the procedure which is connected with a replacement

$$L_0 \longrightarrow L = \frac{F_\pi^2}{8} \text{tr} \left[D_\mu U D_\mu U^+ \right] , \tag{2}$$

given by

$$D_\mu = \partial_\mu U - igA_\mu^L U + igU A_\mu^R . \tag{3}$$

The left - and right- handed spin-1 fields, A_μ^L and A_μ^R, are combinations of vector and axial-vector meson fields

$$A_\mu^L = \frac{1}{2} \left(V_\mu + A_\mu \right) ,$$

$$A_\mu^R = \frac{1}{2} \left(V_\mu - A_\mu \right) . \tag{4}$$

The coupling of these mesons to pseudoscalars is introduced as the gauge coupling: g is the gauge coupling constant and can be determined from the $\rho \longrightarrow \pi\pi$ decay: $g_{\rho\pi\pi} = 8.6$. Therefore, the Lagrangian involving spin-1 and spin-0 mesons takes the form

$$
\begin{aligned}
L(\varphi,V,A) &= \frac{1}{8}F_\pi^2\text{tr}\left[D_\mu U(D_\mu U)^+\right] + \frac{1}{8}F_\pi^2\text{tr}\left[M(U + U^+ - 2)\right] \\
&\quad - \frac{1}{2}\text{tr}\left[(F_{\mu\nu}^L)^2 + (F_{\mu\nu}^R)^2\right] + m_0^2\text{tr}\left[(A_\mu^L)^2 + (A_\mu^R)^2\right] \\
&\quad - i\xi\text{tr}\left[(D_\mu U)(D_\nu U)^+ F_{\mu\nu}^L + (D_\mu U)^+(D_\nu U)F_{\mu\nu}^R\right] \\
&\quad + \gamma\text{tr}\left[F_{\mu\nu}^L U F_{\mu\nu}^R U^+\right] .
\end{aligned}
\tag{5}
$$

The second term is proportional to M(mass matrix) and describes the "soft" breaking of the chiral $SU(N_f)_L \times SU(N_f)_R$ symmetry. The corresponding field strength tensors are given by

$$
F_{\mu\nu}^{L,R} = \partial_\mu A_\nu^{L,R} - \partial_\nu A_\mu^{L,R} - ig\left[A_\mu^{L,R}, A_\nu^{L,R}\right] .
\tag{6}
$$

The third and fourth terms in (5) correspond to the free Lagrangian of the spin-1 particles. At this level of the chiral symmetry all spin-1 mesons have the same "bare" mass, m_0. The last terms are so-called non-minimal terms since it contains of higher order in derivatives.

To obtain the Lagrangian with the "hidden" chiral symmetry (vector mesons as dynamic gauge bosons) we choose a gauge where left- and right-handed fields in the Lagrangian will be identical to the vector field V_μ: $A_\mu^{L'} = A_\mu^{R'} = V_\mu$ and $A_\mu' = 0$. This can be done by a gauge transformation which conserves the $SU(N_f)_L \times SU(N_f)_R$ symmetry

$$
\begin{aligned}
A_\mu^L &= A_\mu^R = V_\mu , \\
U &\longrightarrow U_L U U_R^+ , \\
A_\mu^L &\longrightarrow U_L A_\mu^L U_L^+ + \frac{i}{g} U_L \partial_\mu U_R^+ , \\
A_\mu^L &\longrightarrow U_R A_\mu^L U_R^+ + \frac{i}{g} U_R \partial_\mu U_R^+ ,
\end{aligned}
\tag{7}
$$

with the specific choice $U_L = U^{\frac{1}{2}}$ and $U_R = U^{-\frac{1}{2}}$, so that pseudoscalar mesons are gauge parameters. Now we can rewrite the Lagrangian (5) as a sum of three Lagrangians

$$
L_0 = \frac{F_\pi^2}{8}\text{tr}\left(D_\mu U D_\mu U^+\right) ,
\tag{8}
$$

$$
L_1 = -\frac{1}{2}\text{tr}\left((F_{\mu\nu}^L)^2 + (F_{\mu\nu}^R)^2\right) + \gamma\text{tr}\left(F_{\mu\nu}^L U F_{\mu\nu}^R U^+\right) ,
\tag{9}
$$

$$L_2 = m_0^2 \text{tr}\left((A_\mu^L)^2 + (A_\mu^R)^2\right) + B\,\text{tr}\left(A_\mu^L U A_\mu^R U^+\right)$$
$$+ C\,\text{tr}\left(A_\mu^L A^{R\mu} + A_\mu^R A^{L\mu}\right) . \tag{10}$$

Note that we have added two gauge invariant terms to the Lagrangian (17). The second term (with the coefficient B) in (17) plays an important role in the description of the width of the $\rho \to \pi\pi$ decay, and the third term (with the coefficient C) maintains the gauge invariance of the $J/\psi + \pi \to D^* + \bar{D}$ decay. Applying the substitutions (14) to the Lagrangian (5), we obtain

$$L_0 \to L'_0 = 0 ,$$

$$L_1 \to L'_1 = (\gamma - 1)\,\text{tr}\left(F_{\mu\nu}^V F^{\mu\nu V}\right) ,$$

$$L_2 \to L'_2 = \frac{m_V^2}{2}\text{tr}\left(V_\mu^2\right) - i\frac{g_{V\varphi\varphi}}{2}\text{tr}\left(V_\mu \left(\varphi \overset{\leftrightarrow}{\partial}_\mu \varphi\right)\right)$$
$$+ \frac{8C}{F_\pi^2}\text{tr}\left((V_\mu\varphi)^2 - V_\mu^2\varphi^2\right) + L(\varphi) , \tag{11}$$

by $m_V^2 = 2(B + 2m_0^2 + 2C)$, $g_{V\varphi\varphi} = 2(B - 2C + 2m_0^2)/(gF_\pi^2)$, where the vector mass and the vector-pseudoscalar-pseudoscalar coupling are defined.

J/ψ ABSORPTION CROSS SECTIONS

The above effective Lagrangian allows us to study the following processes for J/ψ absorption by π and ρ mesons:

$$J/\psi + \pi \to D^* + \bar{D}, \ J/\psi + \pi \to D + \bar{D}^*, \tag{12}$$
$$J/\psi + \rho \to D + \bar{D}, \ J/\psi + \rho \to D^* + \bar{D}^* . \tag{13}$$

The corresponding diagrams for this process, except the process $J/\psi + \pi \longrightarrow D + \bar{D}^*$, which has the same cross section as the process $J/\psi + \pi \longrightarrow D^* + \bar{D}$, are shown in Fig. 1.

The full amplitude for the first process $J/\psi + \pi \longrightarrow D^* + \bar{D}$, without isospin factors and before summing and averaging over external spins, is given by

$$M_1 \equiv M_1^{\mu\nu}\varepsilon_{1\mu}\varepsilon_{3\nu} = \left(\sum_{i=a,b,c} M_{1i}^{\mu\nu}\right)\varepsilon_{1\mu}\varepsilon_{3\nu} , \tag{14}$$

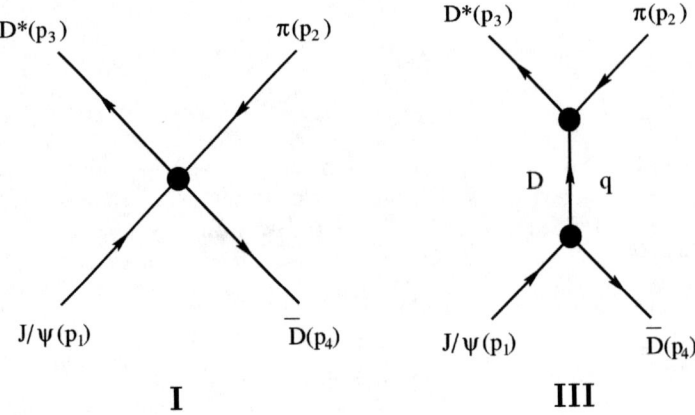

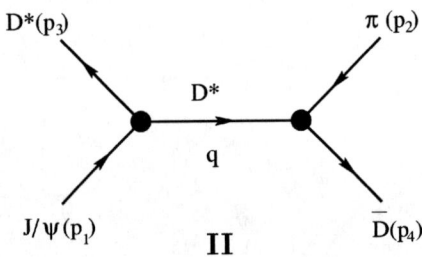

FIGURE 1. Diagrams for J/ψ breakup by pion impact: $J/\psi + \pi \to D^* + \bar{D}$; I - contact term, II+III - D-meson exchange processes.

with

$$M_{1a}^{\mu\nu} = -g_{\pi DD^*}g_{J/\psi DD}(-2p_2 + p_3)^{\nu}\left(\frac{1}{u - m_D^2}\right)(p_2 - p_3 + p_4)^{\mu},$$

$$
\begin{aligned}
M_{1b}^{\mu\nu} &= g_{\pi DD^*}g_{J/\psi D^*D^*}(-p_2 - p_4)^{\alpha}\left(\frac{1}{t - m_{D^*}^2}\right) \\
&\times \left[g^{\alpha\beta} - \frac{(p_2 - p_4)^{\alpha}(p_2 - p_4)^{\beta}}{m_{D^*}^2}\right] \\
&\times \left[(-p_1 - p_3)^{\beta}g^{\mu\nu} + (-p_2 + p_1 + p_4)^{\nu}g^{\beta\mu} + (p_2 + p_3 - p_4)^{\mu}g^{\beta\nu}\right],
\end{aligned}
$$

$$M_{1c}^{\mu\nu} = -g_{J/\psi\pi DD^*}\, g^{\mu\nu}. \tag{15}$$

Similarly, the full amplitude for the second process $J/\psi + \rho \to D + \bar{D}$ is given by

$$M_2 \equiv M_2^{\mu\nu} \varepsilon_{1\mu} \varepsilon_{2\nu} = \left(\sum_{i=a,b,c} M_{2i}^{\mu\nu} \right) \varepsilon_{1\mu} \varepsilon_{2\nu} \tag{16}$$

with

$$M_{2a}^{\mu\nu} = -g_{\rho DD} g_{J/\psi DD} (p_2 - 2p_3)^\mu \left(\frac{1}{u - m_D^2} \right) (p_2 - p_3 + p_4)^\nu ,$$

$$M_{2b}^{\mu\nu} = -g_{\rho DD} g_{J/\psi DD} (-p_2 + 2p_4)^\mu \left(\frac{1}{t - m_D^2} \right) (-p_2 - p_3 + p_4)^\nu ,$$

$$M_{2c}^{\mu\nu} = g_{J/\psi\rho DD} \, g^{\mu\nu} . \tag{17}$$

For the third process $J/\psi + \rho \to D^* + \bar{D}^*$, the full amplitude is given by

$$M_3 \equiv M_3^{\mu\nu\lambda\omega} \varepsilon_{1\mu} \varepsilon_{2\nu} \varepsilon_{3\lambda} \varepsilon_{4\omega} = \left(\sum_{i=a,b,c} M_{3i}^{\mu\nu\lambda\omega} \right) \varepsilon_{1\mu} \varepsilon_{2\nu} \varepsilon_{3\lambda} \varepsilon_{4\omega} , \tag{18}$$

$$
\begin{aligned}
M_{3a}^{\mu\nu\lambda\omega} = & \; g_{\rho D^* D^*} g_{J/\psi D^* D^*} \left[(-p_2 - p_3)^\alpha g^{\mu\lambda} + 2p_2^\lambda g^{\alpha\mu} + 2p_3^\mu g^{\alpha\lambda} \right] \\
& \times \left(\frac{1}{u - m_{D^*}^2} \right) \left[g^{\alpha\beta} - \frac{(p_2 - p_3)^\alpha (p_2 - p_3)^\beta}{m_{D^*}^2} \right] \\
& \times \left[-2p_1^\omega g^{\beta\nu} + (p_1 + p_4)^\beta g^{\nu\omega} - 2p_4^\nu g^{\beta\omega} \right] ,
\end{aligned}
$$

$$
\begin{aligned}
M_{3b}^{\mu\nu\lambda\omega} = & \; g_{\rho D^* D^*} g_{J/\psi D^* D^*} \left[-2p_2^\omega g^{\alpha\mu} + (p_2 + p_4)^\alpha g^{\mu\omega} - 2p_4^\mu g^{\alpha\omega} \right] \\
& \times \left(\frac{1}{t - m_{D^*}^2} \right) \left[g^{\alpha\beta} - \frac{(p_2 - p_4)^\alpha (p_2 - p_4)^\beta}{m_{D^*}^2} \right] \\
& \times \left[(-p_1 - p_3)^\beta g^{\nu\lambda} + 2p_1^\lambda g^{\beta\nu} + 2p_3^\nu g^{\beta\lambda} \right] ,
\end{aligned}
$$

$$M_{3c}^{\mu\nu\lambda\omega} = g_{J/\psi\rho D^* D^*} (g^{\mu\lambda} g^{\nu\omega} + g^{\mu\omega} g^{\nu\lambda} - 2g^{\mu\nu} g^{\lambda\omega}) . \tag{19}$$

In the above, p_j denotes the momentum of particle j. We choose the convention that particle 1 and 2 represent initial-state mesons while particles 3 and 4 represent final-state mesons on the left and right sides of the diagrams shown in Fig. 1, respectively. The indices μ, ν, λ and ω denote the polarization components of external particles while the indices α and β denote those of the exchanged mesons.

After averaging (summing) over initial (final) spins and including isospin factors, the cross sections for the the three processes are given by

$$\frac{d\sigma_1}{dt} = \frac{1}{96\pi s p_{i,c.m.}^2} M_1^{\mu\nu} M_1^{*\mu'\nu'} \left(g^{\mu\mu'} - \frac{p_1^\mu p_1^{\mu'}}{m_1^2}\right) \left(g^{\nu\nu'} - \frac{p_3^\nu p_3^{\nu'}}{m_3^2}\right), \qquad (20)$$

$$\frac{d\sigma_2}{dt} = \frac{1}{288\pi s p_{i,c.m.}^2} M_2^{\mu\nu} M_2^{*\mu'\nu'} \left(g^{\mu\mu'} - \frac{p_1^\mu p_1^{\mu'}}{m_1^2}\right) \left(g^{\nu\nu'} - \frac{p_2^\nu p_2^{\nu'}}{m_2^2}\right), \qquad (21)$$

$$\frac{d\sigma_3}{dt} = \frac{1}{288\pi s p_{i,c.m.}^2} M_3^{\mu\nu\lambda\omega} M_3^{*\mu'\nu'\lambda'\omega'} \left(g^{\mu\mu'} - \frac{p_1^\mu p_1^{\mu'}}{m_1^2}\right) \left(g^{\nu\nu'} - \frac{p_2^\nu p_2^{\nu'}}{m_2^2}\right)$$
$$\times \left(g^{\lambda\lambda'} - \frac{p_3^\lambda p_3^{\lambda'}}{m_3^2}\right) \left(g^{\omega\omega'} - \frac{p_4^\omega p_4^{\omega'}}{m_4^2}\right), \qquad (22)$$

with $s = (p_1 + p_2)^2$, and

$$p_{i,c.m.}^2 = \frac{[s - (m_1 + m_2)^2][s - (m_1 - m_2)^2]}{4s}, \qquad (23)$$

is the squared momentum of initial-state mesons in the center-of-momentum (c.m.) frame. The definition of $p_{f,c.m.}$ for the final-state mesons is analogous with the replacement $(m_1, m_2) \to (m_3, m_4)$.

HADRONIC FORMFACTORS

The chiral Lagrangian aproach for J/ψ breakup by light meson impact makes the assumption that mesons and meson-meson interaction vertices are local (four-momentum independent) objects. This neglect of the finite extension of mesons as quark-antiquark bound states has dramatic consequences: it leads to a monotonously rising behaviour of the cross sections for the corresponding processes, see the dashed lines in Fig. 2. This result, however, cannot be correct away from the rection threshold where the tails of the mesonic wave functions determine the high-energy behaviour of the quark exchange (in the nonrelativistic formulation of [12, 15]) or quark loop (in the relativistic formulation [16]) diagrams describing the micrscopic processes underlying the J/ψ breakup by meson impact. As long as the mesonic wave functions describe quark-antiquark bound states which have a finite extension in coordinate- and momentum space, the J/ψ breakup cross section is expected to be decreasing above the eaction threshold and asymptotically small at high c.m. energies. This result of the quark model approaches to meson-meson interactions [12, 15, 16] can be mimicked within chiral meson Lagrangian approaches by the use of formfactors at the interaction vertices [10, 11]. We will follow here the

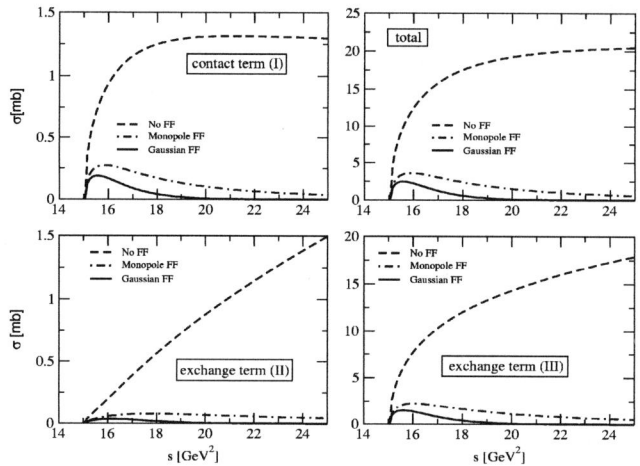

FIGURE 2. Upper right panel: total cross section for J/ψ breakup by pion impact without formfactor (dashed line), with monopole type formfactor (dash-dotted line) and with Gaussian formfactor (solid line) as a function of the squared c.m. energy of initial - state mesons. The partial contributions from the diagrams I, II, and III of Fig. 1 shown in the other panels.

definitions of Ref. [10], where the formfactor of the four-point vertices of Fig.1, i.e. of the box diagram (I) as well as of the meson exchange diagrams (II, III) is taken as the product of the triangle diagram formfactors

$$F_4^i(\mathbf{q}^2) = \left[F_3(\mathbf{q}^2) \right]^2 \quad , \ i = I, II, III \ , \tag{24}$$

with the squared three-momentum $\mathbf{q}^2$ given by the average value of the squared three-momentum transfers in the t and u channels

$$\mathbf{q}^2 = \frac{1}{2} \left[(\mathbf{p_1} - \mathbf{p_3})^2 + (\mathbf{p_1} - \mathbf{p_4})^2 \right]_{\text{c.m.}} = p_{i,\text{c.m.}}^2 + p_{f,\text{c.m.}}^2 \ . \tag{25}$$

For the triangle diagrams, we use formfactors with a momentum dependence in the monopole form (M)

$$F_3^M(\mathbf{q}^2) = \frac{\Lambda^2}{\Lambda^2 + \mathbf{q}^2} \ , \tag{26}$$

and in the Gaussian (G) form

$$F_3^G(\mathbf{q}^2) = \exp(-\mathbf{q}^2/\Lambda^2) \ . \tag{27}$$

At this point we have to add the comment that this choice, however, is obviously not supported by the underlying quark substructure diagrams that can provide a justification for the use of formfactors: While the triangle diagram is of third order in the wave functions so that the meson exchange diagrams are suppressed at large momentum transfer by six wave functions, the box diagram appears already at fourth order thus being less suppressed than suggested by the ansatz (24) of Ref. [10]. For the cross

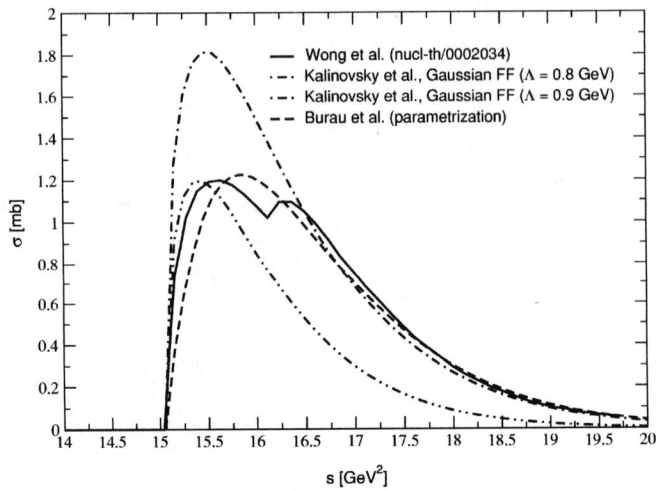

FIGURE 3. Total J/ψ breakup cross section with Gaussian formfactor ($\Lambda = 0.9$ GeV - dot-dashed line, $\Lambda = 0.8$ GeV - dot-dot-dashed line) compared to the nonrelativistic quark exchange model (Wong et al. [15] - solid line) and its parametrization by Burau et al. [8] (dashed line).

section of the three diagrams including the effect of hadronic formfactors, we multiply the bare expressions with the formfactors given above. The results are depicted in Fig. 2. In the last Section, we want to discuss the results and their possible implications for phenomenological applications.

RESULTS AND DISCUSSION

The J/ψ breakup cross section by π and ρ meson impact has been formulated within a chiral $U(4)$ Lagrangian approach. Numerical results have been obtained for the pion impact processes with the result that the D-meson exchange in the t-channel is the dominant subprocess contributing to the J/ψ breakup. The use of formfactors at the meson-meson vertices is mandatory since otherwise the high-energy asymptotics of the processes with hadronic final states will be overestimated, see Fig. 2. From a comparison with results of a nonrelativistic potential model calculation, we can choose the shape of the formfactor to be Gaussian and fix the range $\Lambda = 0.9$ GeV from the asymptotic high energy behaviour, see Fig. 3. Within our semi-quantitative discussion, we do not attempt a high accuracy description of the nonrelativistic result which accounts for another final state D-meson pair, see [12, 15].

Finally, we want to present an explore the influence of a variation of the final state D-meson masses on the effective J/ψ breakup cross section. Our motivation for considering mesonic states to be off their mass-shell is their compositeness which can become apparent in a high-temperature (and density) environment at the deconfinement/chiral restoration transition, when these states change their character qualitatively being reso-

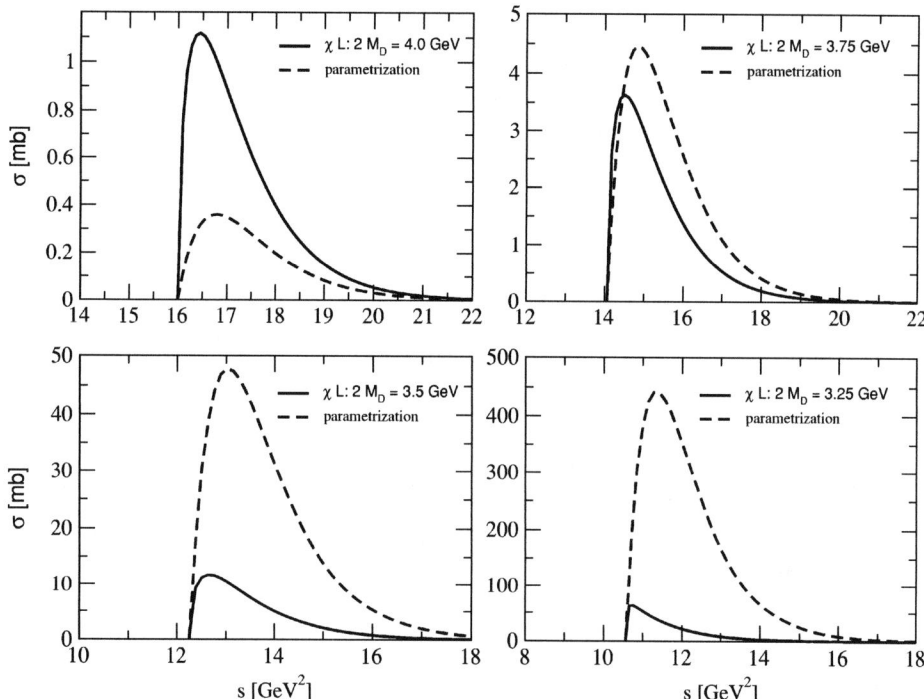

FIGURE 4. Total J/ψ breakup cross section in the chiral Lagrangian model with Gaussian formfactor ($\Lambda = 0.9$ GeV - solid line) compared to the parametrization of the nonrelativistic quark exchange model [15] by Burau et al. [8] (dashed line). The four panels illustrate the differences between both models when the final state masses $M_{D_1} = M_{D_2} = M_D$ are varied.

nant quark- antiquark scattering states in the quark plasma rather than on-shell mesonic bound states.

The consequence of this Mott-transition from bound to resonant states for the J/ψ breakup has been explored by Burau et al. [8, 7], see also these proceedings, using a fit formula for the D-mass dependence of the breakup cross section which shows a strong enhancement when the process becomes subthreshold ($M_D < M_D^{\text{vac}}$). This behaviour is qualitatively approved within the present chiral U(4) Lagrangian + formfactor model although the subthreshold enhancement is more moderate, see Fig. 4. A more consistent description should include a quark model derivation of the formfactors for the meson-meson vertices and their possible medium dependence. Such an investigation is in progress.

ACKNOWLEDGEMENT

Yu.K. acknowledges support from the Deutsche Forschungsgemeinschaft under grant no. RUS 17/129/02. D.B. and Yu.K. acknowledge discussions with J. Hüfner and M. Nielsen as well as the hospitality at the University of Coimbra.

REFERENCES

1. M.C. Abreu et al. (NA50 Collab.), Phys. Lett. **B 477**, 28 (2000).
2. T. Matsui, H. Satz, Phys. Lett. **B 178**, 416 (1986).
3. R. Redmer, Phys. Rep. **282**, 35 (1997).
4. F. Karsch, M.T. Mehr, H. Satz, Z. Phys. **C 37**, 617 (1988).
5. G. Röpke, D. Blaschke, H. Schulz, Phys. Lett. **B 202**, 479 (1988).
6. G. Röpke, D. Blaschke, H. Schulz, Phys. Rev. **D 38**, 3589 (1988).
7. D. Blaschke, G. Burau, Yu. Kalinovsky, *Mott dissociation of D-mesons at the chiral phase transition and anomalous J/ψ suppression*, in: Progress in Heavy Quark Physics 5, Dubna (2000); [nucl-th/0006071].
8. G. Burau, D. Blaschke, Yu. Kalinovsky, *Mott effect at the chiral phase transition and anomalous J/ψ suppression*, Phys. Lett. **B 506**, 297 (2001); [nucl-th/0012030].
9. S.G. Matinyan, B. Müller, Phys. Rev. **C 58**, 2994 (1998).
10. Z. Lin, C.M. Ko, Phys. Rev. **C 62**, 034903 (2000).
11. K. Haglin, C. Gale, *Hadronic Interactions of the J/ψ*, [nucl-th/0010017].
12. K. Martins, D. Blaschke, E. Quack, Phys. Rev. **C 51**, 2723 (1995).
13. P. Braun-Munzinger, K. Redlich, Nucl. Phys. **A 661**, 546 (1999).
14. C.M. Ko, X.-N. Wang, B. Zhang, X.F. Zhang, Phys. Lett. **B 444**, 237 (1998).
15. C.-Y. Wong, E. Swanson, T. Barnes, Phys. Rev. **C 62**, 045201 (2000).
16. D. Blaschke, G. Burau, M. Ivanov, Yu. Kalinovsky, P.C. Tandy, in: *Progress in Nonequilibrium Green Functions*, Ed. M. Bonitz, World Scientific, Singapore, 2000, p. 392; [hep-ph/0002047].
17. J.-C. Itzykson, P.-B. Zuber, *Quantum Field Theory*, McGraw-Hill, New York, 1980.
18. F.E. Close, *An Introduction to Quarks and Partons*, Academic Press, London, 1979.

Electroproduction of Charmonia
off Protons and Nuclei

Yu.P. Ivanov[*†], B.Z. Kopeliovich[**‡], A.V. Tarasov[†] and J. Hüfner[*]

*Institut für Theoretische Physik der Universität, 69120 Heidelberg, Germany
†Joint Institute for Nuclear Research, Dubna, 141980 Moscow Region, Russia
**Institut für Theoretische Physik der Universität, 93040 Regensburg, Germany
‡Max-Planck Institut für Kernphysik, Postfach 103980, 69029 Heidelberg, Germany

Abstract. Elastic virtual photoproduction of charmonia on nucleons is calculated in a parameter free way with the light-cone dipole formalism and the same input: factorization in impact parameters, light-cone wave functions for the photons and the charmonia, and the universal phenomenological dipole cross section which is fitted to other data. The charmonium wave functions are calculated with four known realistic potentials, and two models for the dipole cross section are tested. Very good agreement with data for the cross section of charmonium electroproduction is found in a wide range of s and Q^2. Using the ingredients from those calculations we calculate also exclusive electroproduction of charmonia off nuclei. Here new effects become important, (i) color filtering of the $c\bar{c}$ pair on its trajectory through nuclear matter, (ii) dependence on the finite lifetime of the $c\bar{c}$ fluctuation (coherence length) and (iii) gluon shadowing in a nucleus compared to the one in a nucleon. Total coherent and incoherent cross sections for C, Cu and Pb as functions of s are presented. The results can be tested with future electron-nucleus colliders or in the peripheral collisions of relativistic heavy ions.

INTRODUCTION

In contrast to hadro-production of charmonia, where the mechanism is still debated, electro(photo)production of charmonia seems better understood: the $c\bar{c}$ fluctuation of the incoming real or virtual photon interacts with the target (proton or nucleus) via the dipole cross section $\sigma_{q\bar{q}}$ and the result is projected on the wave function of the observed hadron [1]. The aim of this paper is not to propose a conceptually new scheme, but to calculate within this approach as accurately as possible and without any free parameters. Wherever there is room for arbitrariness, like form for the color dipole cross section and for for charmonium wave function, we use and compare other author's proposals, which have been tested on different data.

In the light-cone (LC) dipole approach the virtual photoproduction of charmonia (here Ψ stands for J/ψ or ψ') looks as shown in Fig. 1 [1]. The corresponding expression for the forward amplitude reads

$$\mathcal{M}_{\gamma p}(s, Q^2) = \sum_{\mu, \bar{\mu}} \int_0^1 d\alpha \int d^2\vec{r}_T \, \Phi_\Psi^{*(\mu,\bar{\mu})}(\alpha, \vec{r}_T) \, \sigma_{q\bar{q}}(r_T, s) \, \Phi_\gamma^{(\mu,\bar{\mu})}(\alpha, \vec{r}_T, Q^2). \quad (1)$$

Here the summation runs over spin indexes μ, $\bar{\mu}$ of the c and $\bar{c}$ quarks, Q^2 is the photon

CP660, Hadron Physics: Effective Theories of Low Energy QCD, edited by A. H. Blin et al.

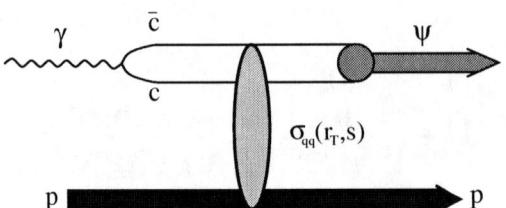

FIGURE 1. Schematic representation of the amplitude for the reaction $\gamma p \to \Psi p$ in the rest frame of the proton. The $c\bar{c}$ fluctuation of the photon with transverse separation r_T and c.m. energy $\sqrt{s}$ interacts with the target proton via the dipole cross section $\sigma_{q\bar{q}}(r_T, s)$ and produces a Ψ.

virtuality, $\Phi_\gamma(\alpha, r_T, Q^2)$ is the LC distribution function of the photon for a $c\bar{c}$ fluctuation of separation r_T and relative fraction α of the photon LC momentum carried by c or $\bar{c}$. Correspondingly, $\Phi_\Psi(\alpha, \vec{r}_T)$ is the LC wave function of J/ψ or ψ'. The dipole cross section $\sigma_{q\bar{q}}(r_T, s)$ mediates the transition.

This paper is organized as follows. First we review the expressions of the factorized LC approach to electroproduction of heavy quarkonia (more details one can find in [2]) and then compare with available experimental data for J/ψ production. The calculations which are parameter free demonstrate good agreement with data.

After it we analyse exclusive electroproduction of charmonia off nuclei $\gamma A \to \Psi X$, where $X = A$ (coherent) or $X = A^*$ (incoherent, where A^* is an exited state of the A-nucleon system). In these processes new phenomena are to be expected. (i) Color filtering, i.e. inelastic interactions of the $c\bar{c}$ pair on its way through the nucleus is expected to lead to a suppression of Ψ production relative to $A\sigma_{\gamma p \to \Psi p}$. (ii) Production of a $c\bar{c}$ pair in a nucleus and its absorption are also determined by the values of the coherence length (lifetime of the $c\bar{c}$ fluctuation),

$$l_c = \frac{2\nu}{Q^2 + M_{c\bar{c}}^2} \approx \frac{2\nu}{Q^2 + M_{J/\psi}^2}, \tag{2}$$

where ν is the energy of the virtual photon in the rest frame of the nucleus. (iii) Since the dipole cross section $\sigma_{q\bar{q}}$ also depends on the gluon distribution in the target (p of A), nuclear shadowing of the gluon distribution is expected to reduce $\sigma_{q\bar{q}}$ in a nuclear reaction relative to the one on the proton.

The predictions in this paper may be used for planning of future experiments for electron-nucleus collisions at high energies like in the eRHIC project. Another possibility to observe photoproduction off nuclei is heavy ion relativistic collisions (see, for example, review [4]). In the last section we present our results for J/ψ production in such processes.

LIGHT-CONE DIPOLE APPROACH

The LC variable describing longitudinal motion of the quarks is the fraction $\alpha = p_c^+ / p_\gamma^+$ of the photon LC momentum $p_\gamma^+ = E_\gamma + p_\gamma$ carried by the quark or antiquark. α

is Lorentz-boost invariant. In the nonrelativistic approximation (assuming no relative motion of c and $\bar{c}$) $\alpha = 1/2$ (e.g. [1]), otherwise one should integrate over α (see Eq. (1)).

Photon wave function

For transversely (T) and longitudinally (L) polarized photons the perturbative photon-quark distribution function in Eq. (1) reads [5, 6],

$$\Phi_{T,L}^{(\mu,\bar{\mu})}(\alpha,\vec{r}_T,Q^2) = \frac{\sqrt{N_c\,\alpha_{em}}}{2\pi} Z_c \chi_c^{\mu\dagger} \widehat{O}_{T,L} \tilde{\chi}_{\bar{c}}^{\bar{\mu}} K_0(\varepsilon r_T)\,, \qquad (3)$$

where $\tilde{\chi}_{\bar{c}} = i\sigma_y\chi_{\bar{c}}^*$, χ and $\bar{\chi}$ are the spinors of the c-quark and antiquark respectively; $Z_c = 2/3$. $K_0(\varepsilon r_T)$ is the modified Bessel function with $\varepsilon^2 = \alpha(1-\alpha)Q^2 + m_c^2$. The operators $\widehat{O}_{T,L}$ have the form:

$$\widehat{O}_T = m_c\,\vec{\sigma}\cdot\vec{e}_\gamma + i(1-2\alpha)(\vec{\sigma}\cdot\vec{n})(\vec{e}_\gamma\cdot\vec{\nabla}_{r_T}) + (\vec{n}\times\vec{e}_\gamma)\cdot\vec{\nabla}_{r_T}\,, \qquad (4)$$

$$\widehat{O}_L = 2Q\alpha(1-\alpha)\,\vec{\sigma}\cdot\vec{n}\,, \qquad (5)$$

where $\vec{n} = \vec{p}/p$ is a unit vector parallel to the photon momentum and $\vec{e}$ is the polarization vector of the photon. Effects of the non-perturbative interaction within the $q\bar{q}$ fluctuation are negligible for the heavy charmed quarks.

Charmonium wave function

The charmonium wave function is well defined in its rest frame where one can rely on the Schrödinger equation. As soon as the rest frame wave function is known, one may be tempted to apply the Lorentz transformation to the $c\bar{c}$ pair as it would be a classical system and boost it to the infinite momentum frame. However, quantum effects are important and in the infinite momentum frame a series of different Fock states emerges from the Lorentz boost. Therefore the lowest $|c\bar{c}\rangle$ component in the infinite momentum frame does not represent the $|c\bar{c}\rangle$ in the rest frame. We rely on the widely used procedure [11] for the generation of the LC wave functions of charmonia.

In the rest frame the spatial part of the $c\bar{c}$ pair wave function satisfying the Schrödinger equation

$$\left(-\frac{\Delta}{m_c} + V(r)\right)\Psi_{nlm}(\vec{r}) = E_{nl}\,\Psi_{nlm}(\vec{r}) \qquad (6)$$

is represented in the form

$$\Psi(\vec{r}) = \Psi_{nl}(r)\cdot Y_{lm}(\theta,\varphi)\,, \qquad (7)$$

where $\vec{r}$ is 3-dimensional $c\bar{c}$ separation, $\Psi_{nl}(r)$ and $Y_{lm}(\theta,\varphi)$ are the radial and orbital parts of the wave function. The following four potentials $V(r)$ have been used

- "COR": Cornell potential [12],

$$V(r) = -\frac{k}{r} + \frac{r}{a^2} \tag{8}$$

with $k = 0.52$, $a = 2.34\,\mathrm{GeV}^{-1}$ and $m_c = 1.84\,\mathrm{GeV}$.
- "BT": Potential suggested by Buchmüller and Tye [13] with $m_c = 1.48\,\mathrm{GeV}$. It has a similar structure as the Cornell potential: linear string potential at large separations and Coulomb shape at short distances with some refinements, however.
- "LOG": Logarithmic potential [14]

$$V(r) = -0.6635\,\mathrm{GeV} + (0.733\,\mathrm{GeV})\log(r \cdot 1\,\mathrm{GeV}) \tag{9}$$

with $m_c = 1.5\,\mathrm{GeV}$.
- "POW": Power-law potential [15]

$$V(r) = -8.064\,\mathrm{GeV} + (6.898\,\mathrm{GeV})(r \cdot 1\,\mathrm{GeV})^{0.1} \tag{10}$$

with $m_c = 1.8\,\mathrm{GeV}$.

The shapes of the four potentials differ from each other only at large r ($\geq 1\,\mathrm{fm}$) and very small r ($\leq 0.05\,\mathrm{fm}$) separations. Note, however, that COR and POW use $m_c \approx 1.8\,\mathrm{GeV}$, while BT and LOG use $m_c \approx 1.5\,\mathrm{GeV}$ for the mass of the charmed quark. This difference will have significant consequences.

For the ground state $1S$ all the potentials provide a very similar behavior for the radial part $\Psi_{nl}(r)$ at $r > 0.3\,\mathrm{fm}$, while for small r the predictions differ by up to 30%. The peculiar property of the $2S$ state wave function is the node at $r \approx 0.4\,\mathrm{fm}$ which causes strong cancellations in the matrix elements Eq. (1) and as a result, a suppression of photoproduction of ψ' relative to J/ψ [1, 16].

The lowest Fock component $|c\bar{c}\rangle$ in the infinite momentum frame is not related by simple Lorentz boost to the wave function of charmonium in the rest frame. This makes the problem of building the LC wave function for the lowest $|c\bar{c}\rangle$ component difficult, no unambiguous solution is yet known. There are only recipes in the literature, a simple one widely used [11], is the following. One applies a Fourier transformation from coordinate to momentum space to the known spatial part of the non-relativistic wave function (7), $\Psi(\vec{r}) \Rightarrow \Psi(\vec{p})$, which can be written as a function of the effective mass of the $c\bar{c}$, $M^2 = 4(p^2 + m_c^2)$, expressed in terms of LC variables

$$M^2(\alpha, p_T) = \frac{p_T^2 + m_c^2}{\alpha(1-\alpha)} . \tag{11}$$

In order to change integration variable p_L to the LC variable α one relates them via M, namely $p_L = (\alpha - 1/2)M(p_T, \alpha)$. In this way the $c\bar{c}$ wave function acquires a kinematical factor

$$\Psi(\vec{p}) \Rightarrow \sqrt{2}\frac{(p^2 + m_c^2)^{3/4}}{(p_T^2 + m_c^2)^{1/2}} \cdot \Psi(\alpha, \vec{p}_T) \equiv \Phi_\psi(\alpha, \vec{p}_T) . \tag{12}$$

This procedure was used in [17] and the result is applied to calculation of the amplitudes (1). The result was discouraging, since the ψ' to J/ψ ratio of the electroproduction cross sections are far too low in comparison with data. However, the oversimplified dipole cross section $\sigma_{q\bar{q}}(r_T) \propto r_T^2$ was used, and what is even more essential, the important ingredient of Lorentz transformations, the Melosh spin rotation, was left out.

The 2-dimensional spinors χ_c and $\chi_{\bar{c}}$ describing c and $\bar{c}$ respectively in the infinite momentum frame are known to be related via the Melosh rotation [11, 18] to the spinors $\bar{\chi}_c$ and $\bar{\chi}_{\bar{c}}$ in the rest frame: $\overline{\chi}_{\mathbf{c}} = \hat{\mathbf{R}}(\alpha, \vec{\mathbf{p}}_T) \chi_{\mathbf{c}}$ and $\overline{\chi}_{\bar{c}} = \hat{\mathbf{R}}(1 - \alpha, -\vec{\mathbf{p}}_T) \chi_{\bar{c}}$, where the matrix $R(\alpha, \vec{p}_T)$ has the form:

$$\hat{R}(\alpha, \vec{p}_T) = \frac{m_c + \alpha M - i [\vec{\sigma} \times \vec{n}] \vec{p}_T}{\sqrt{(m_c + \alpha M)^2 + p_T^2}} . \tag{13}$$

Since the potentials we use contain no spin-orbit term, the $c\bar{c}$ pair is in S-wave. In this case spatial and spin dependences in the wave function factorize and we arrive at the following LC wave function of the $c\bar{c}$ in the infinite momentum frame

$$\Phi_\psi^{(\mu,\bar{\mu})}(\alpha, \vec{p}_T) = U^{(\mu,\bar{\mu})}(\alpha, \vec{p}_T) \cdot \Phi_\psi(\alpha, \vec{p}_T) , \tag{14}$$

where

$$U^{(\mu,\bar{\mu})}(\alpha, \vec{p}_T) = \chi_c^{\mu\dagger} \hat{R}^\dagger(\alpha, \vec{p}_T) \vec{\sigma} \cdot \vec{e}_\psi \sigma_y \hat{R}^*(1 - \alpha, -\vec{p}_T) \sigma_y^{-1} \tilde{\chi}_{\bar{c}}^{\bar{\mu}} . \tag{15}$$

Now we can determine the LC wave function in the mixed longitudinal momentum - transverse coordinate representation:

$$\Phi_\psi^{(\mu,\bar{\mu})}(\alpha, \vec{r}_T) = \frac{1}{2\pi} \int d^2\vec{p}_T \, e^{-i\vec{p}_T \vec{r}_T} \Phi_\psi^{(\mu,\bar{\mu})}(\alpha, \vec{p}_T) . \tag{16}$$

Phenomenological dipole cross section

The color dipole cross section $\sigma_{q\bar{q}}(r_T, s)$ is poorly known from first principles. It is expected to vanish $\propto r_T^2 \ln r_T$ at small $r_T \to 0$ due to color screening [7] and to level off at large separations. We use a phenomenological form which interpolates between the two limiting cases of small and large separations. Few parameterizations are available in the literature, we choose two of them which are simple, but quite successful in describing data and denote them by the initials of the authors as "GBW" [8] and "KST" [9].

$$\text{"GBW":} \quad \sigma_{q\bar{q}}(r_T, x) = 23.03 \left[1 - e^{-r_T^2/r_0^2(x)} \right] \text{mb} , \tag{17}$$

$$r_0(x) = 0.4 \left(\frac{x}{x_0} \right)^{0.144} \text{fm} ,$$

where $x_0 = 3.04 \cdot 10^{-4}$. The proton structure function calculated with this parameterization fits well all available data at small x and in wide range of Q^2 [8]. However, it obviously fails describing the hadronic total cross sections, since it never exceeds the value

23.03 mb. The x-dependence guarantees Bjorken scaling for DIS at high Q^2, however, Bjorken x is not a well defined quantity in the soft limit. Instead we use the prescription of [10], $x = (M_\psi^2 + Q^2)/s$, where M_ψ is the charmonium mass.

This problem with limited dipole cross section as well as the difficulty with the definition of x have been fixed in [9]. The dipole cross section is treated as a function of the c.m. energy $\sqrt{s}$, rather than x, since $\sqrt{s}$ is more appropriate for hadronic processes. A similarly simple form for the dipole cross section is used

$$\text{"KST"}: \qquad \sigma_{\bar{q}q}(r_T, s) = \sigma_0(s)\left[1 - e^{-r_T^2/r_0^2(s)}\right]. \qquad (18)$$

The values and energy dependence of hadronic cross sections are reproduced with the following expressions

$$\sigma_0(s) = 23.6 \left(\frac{s}{s_0}\right)^{0.08} \left(1 + \frac{3}{8}\frac{r_0^2(s)}{\langle r_{ch}^2\rangle}\right) \text{ mb}, \qquad (19)$$

$$r_0(s) = 0.88 \left(\frac{s}{s_0}\right)^{-0.14} \text{ fm}. \qquad (20)$$

The energy dependent radius $r_0(s)$ is fitted to data for the proton structure function $F_2^p(x, Q^2)$, $s_0 = 1000 \, \text{GeV}^2$ and the mean square of the pion charge radius $\langle r_{ch}^2\rangle = 0.44 \, \text{fm}^2$. The improvement at large separations leads to a somewhat worse description of the proton structure function at large Q^2. Apparently, the cross section dependent on energy, rather than x, cannot provide Bjorken scaling. Indeed, parameterization (18) is successful only up to $Q^2 \approx 10 \, \text{GeV}^2$.

In fact, the cases we are interested in, charmonium production and interaction, are just in between the regions where either of these parameterization is successful. Therefore, we suppose that the difference between predictions using Eq. (17) and (18) is a measure of the theoretical uncertainty which fortunately turns out to be rather small.

ELECTROPRODUCTION OFF PROTONS

Having Eq. (1) and the expressions from the previous section (LC wave functions and dipole cross section), we can calculate cross sections for the virtual photoproduction $\gamma p \to \Psi p$

$$\sigma_{\gamma p \to \Psi p}(s, Q^2) = \frac{|\widetilde{\mathcal{M}}_T(s, Q^2)|^2 + \varepsilon|\widetilde{\mathcal{M}}_L(s, Q^2)|^2}{16 \pi B}, \qquad (21)$$

where ε is the photon polarization (for H1 data $\langle\varepsilon\rangle = 0.99$); B is the slope parameter in reaction $\gamma^* p \to \psi p$. We use the experimental value [20] $B = 4.73 \, \text{GeV}^{-2}$. $\widetilde{\mathcal{M}}_{T,L}$ includes also the correction for the real part of the amplitude:

$$\widetilde{\mathcal{M}}_{T,L}(s, Q^2) = \mathcal{M}_{T,L}(s, Q^2)\left(1 - i\frac{\pi}{2}\frac{\partial \ln \mathcal{M}_{T,L}(s, Q^2)}{\partial \ln s}\right), \qquad (22)$$

where we apply the well known derivative analyticity relation between the real and imaginary parts of the forward elastic amplitude [19]. The correction from the real part is not small since the cross section of charmonium electroproduction is a rather steep function of energy.

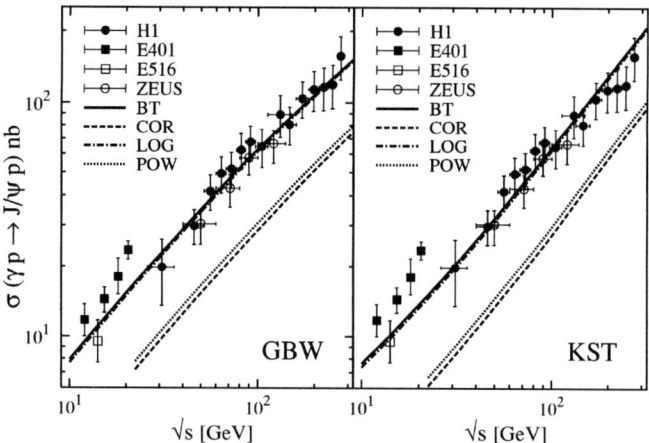

FIGURE 2. Integrated cross section for elastic photoproduction with real photons ($Q^2 = 0$) calculated with GBW and KST dipole cross sections and for four potentials to generate J/ψ wave functions. Experimental data points from the H1 [20], E401 [21], E516 [22] and ZEUS [23] experiments.

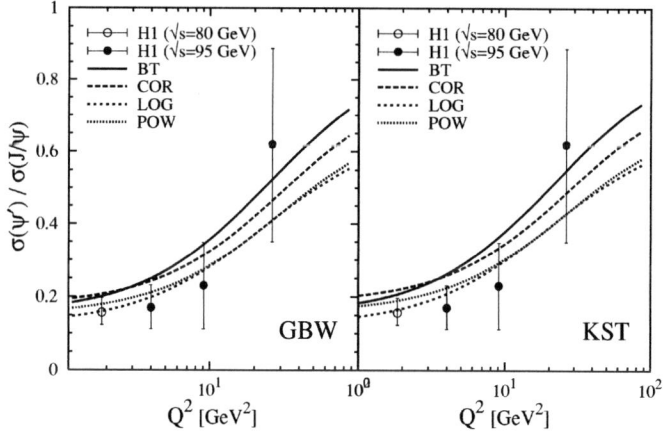

FIGURE 3. The ratio of ψ' to J/ψ virtual photoproduction cross sections as a function of the photon virtuality Q^2 at energy $\sqrt{s} = 90$ GeV for four potentials and with GBW and KST parameterizations for the dipole cross section. Experimental data points from the H1 experiment [24].

We present here only results for energy dependence of the J/ψ integrated cross section (Fig. 2) and for Q^2 dependence of the ratio ψ' to J/ψ photoproduction (Fig. 3). More results are presented in [2]. All calculations are performed with GBW and KST parameterizations for the dipole cross section and for wave functions of the J/ψ calculated from BT, LOG, COR and POW potentials. One can see that there are no major differences for the results using the GBW and KST parameterizations. The BT and LOG potentials describe the data very well, while the potentials COR and POW underestimate them by a factor of two. The different behavior has been traced to the following origin: BT and LOG use $m_c \approx 1.5\,\text{GeV}$, but COR and POW $m_c \approx 1.8\,\text{GeV}$. While the bound state wave functions of J/ψ are little affected by this difference, the photon wave function Eq. (3) depends sensitively on m_c.

It turns out that the effects of spin rotation have a gross impact on the ratio $R = \sigma(\psi')/\sigma(J/\psi)$. This effects add 30-40% to the J/ψ electroproduction cross section. But they have a much more dramatic impact on ψ' increasing the electroproduction cross section by a factor 2-3. This spin effects explain the large values of the ratio R observed experimentally. Our results for R are about twice as large as evaluated in [25] and even more than in [17].

ELECTROPRODUCTION OFF NUCLEI

Exclusive charmonium production off nuclei, $\gamma A \rightarrow \Psi X$ is called coherent, when the nucleus remains intact, i.e. $X = A$, or incoherent, when X is an excited nuclear state which contains nucleons and nuclear fragments but no other hadrons. The cross sections depend on the polarization ε of the virtual photon (in all figures below we will imply $\varepsilon = 1$),

$$\sigma_{\gamma A}(s, Q^2) = \sigma_T(s, Q^2) + \varepsilon\, \sigma_L(s, Q^2)\,, \tag{23}$$

where the indexes T, L correspond to transversely or longitudinally polarized photons, respectively. At high energies the coherence length Eq. (2) may substantially exceed the nuclear radius ($l_c \gg R_A$). In this case the transverse size of the $c\bar{c}$ wave packet is "frozen" by Lorentz time dilation, i.e. it does not fluctuate during propagation through the nucleus, and the expressions for the incoherent (inc) and coherent (coh) cross sections are simple [1]:

$$\sigma_{T,L}^{coh}(s, Q^2) = \int d^2b \left| \widetilde{\mathcal{M}}_{T,L}^{coh}(s, Q^2, b) \right|^2\,, \tag{24}$$

$$\sigma_{T,L}^{inc}(s, Q^2) = \int d^2b \left| \widetilde{\mathcal{M}}_{T,L}^{inc}(s, Q^2, b) \right|^2 \frac{T_A(b)}{16\pi B(s)}\,, \tag{25}$$

where $T_A(b) = \int_{-\infty}^{\infty} dz\, \rho_A(b, z)$ is the nuclear thickness function given by the integral of the nuclear density along the trajectory at a given impact parameter b and expressions for $\widetilde{\mathcal{M}}(s, Q^2, b)$ are given by Eqs. (22) and (1) with replacement

$$\sigma_{q\bar{q}}(r_T, s) \Rightarrow \begin{cases} 1 - \exp\left[-\sigma_{q\bar{q}}(r_T, s)\, T_A(b)/2\right], & coh. \\ \sigma_{q\bar{q}}(r_T, s)\exp\left[-\sigma_{q\bar{q}}(r_T, s)\, T_A(b)/2\right], & inc. \end{cases} \tag{26}$$

But for charmonium production off nuclei the "frozen" approximation is not enough and additional phenomena should be taken into account: effects of finite coherence length and gluon shadowing.

Finite coherence length

The "frozen" approximation (26) is valid only for $l_c \gg R_A$ and can be used only at $\sqrt{s} > 20 \div 30\,\text{GeV}$. The low-energy part should be corrected for the effects related to the finiteness of l_c. A strictly quantum-mechanical treatment of a fluctuating $q\bar{q}$ pair propagating through an absorptive medium based on the LC Green function approach has been suggested recently in [26]. However, an analytical solution for the LC Green function is known only for the simplest form of the dipole cross section $\sigma_{q\bar{q}}(r_T) \propto r_T^2$. With a realistic form of $\sigma_{q\bar{q}}(r_T)$ it is possible only to solve this problem numerically, what is still a challenge. Here we use the approximation suggested in [27] to evaluate the corrections arising from the finiteness of l_c by multiplying the cross sections for coherent and incoherent production evaluated for $l_c \to \infty$ by a kind of formfactor F^{coh} and F^{inc} respectively:

$$\sigma_{\gamma A}(s, Q^2) \Rightarrow \sigma_{\gamma A}(s, Q^2) \cdot F\left(s, l_c(s, Q^2)\right), \tag{27}$$

where

$$F^{coh}(s, l_c) = \int d^2 b \left| \int_{-\infty}^{\infty} dz\, \rho_A(b, z)\, F_1(s, b, z)\, e^{iz/l_c} \right|^2 / (...)|_{l_c=\infty}, \tag{28}$$

$$F^{inc}(s, l_c) = \int d^2 b \int_{-\infty}^{\infty} dz\, \rho_A(b, z) \left| F_1(s, b, z) - F_2(s, b, z, l_c) \right|^2 / (...)|_{l_c=\infty}, \tag{29}$$

$$F_1(s, b, z) = \exp\left(-\frac{1}{2} \sigma_{\Psi N}(s) \int_z^{\infty} dz'\, \rho_A(b, z') \right), \tag{30}$$

$$F_2(s, b, z, l_c) = \frac{1}{2} \sigma_{\Psi N}(s) \int_{-\infty}^{z} dz'\, \rho_A(b, z')\, F_1(b, z')\, e^{-i(z-z')/l_c}. \tag{31}$$

For the charmonium nucleon total cross section $\sigma_{\Psi N}(s)$ we use our previous results [2].

Gluon shadowing

The gluon density in nuclei at small Bjorken x is expected to be suppressed compared to a free nucleon due to interferences. This phenomenon, called gluon shadowing, renormalizes the dipole cross section,

$$\sigma_{q\bar{q}}(r_T, x) \Rightarrow \sigma_{q\bar{q}}(r_T, x)\, R_G(x, Q^2, b). \tag{32}$$

where the factor $R_G(x, Q^2, b)$ is the ratio of the gluon density at x and Q^2 in a nucleon of a nucleus to the gluon density in a free nucleon. No data are available so far which could provide information about gluon shadowing. Currently it can be evaluated only theoretically. To calculate function R_G we use the approach developed in [9] and applied for charmonia production in [3].

Numerical results

Combining the results above (i.e. including finite coherence length and gluon shadowing) we obtain the results for charmonia (J/ψ and ψ') electroproduction off nuclei. As it is common practice we express nuclear cross sections in the form of the ratio

$$R_\Psi(s,Q^2) = \frac{\sigma_{\gamma A}(s,Q^2)}{A\,\sigma_{\gamma p}(s,Q^2)}, \tag{33}$$

where the numerator stands for the expression Eq. (23) (with Eq. (24) for coherent and Eq. (25) for incoherent scattering) and the denominator is given by Eq. (21). We present here only s dependences (plots for Q^2 dependences and momentum transfer $\vec{k}_T$ distributions are given in [3]) for coherent (Fig. 4) and incoherent (Fig. 5) production of charmonia with the GBW [8] parameterization for the dipole cross section. KST [9] parameterization gives quite close results (they differ at most 10% at high energies). It is not a surprise that the ratios for coherent production exceed one: in the absence of $c\bar{c}$ attenuation the forward coherent production would be proportional to A^2, while integrated over momentum transfer it behaves as $A^{4/3}$. It is a result of our definition Eq. (33) that R_Ψ^{coh} exceeds one.

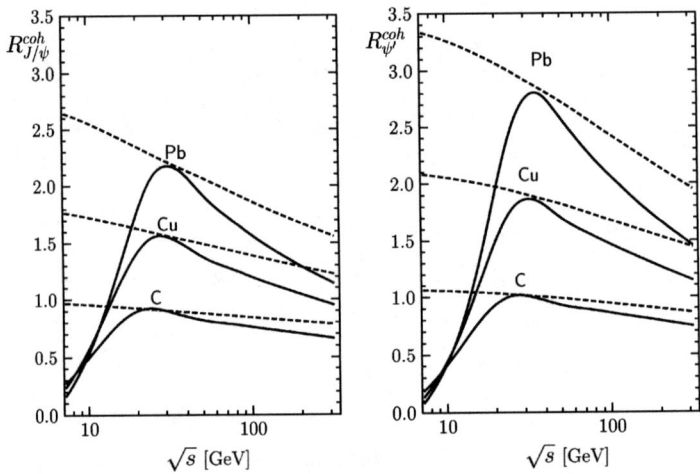

FIGURE 4. Ratios $R_{J/\psi}^{coh}$ and $R_{\psi'}^{coh}$ for coherent production on carbon, copper and lead as a function of $\sqrt{s}$ and at $Q^2 = 0$ calculated with GBW parameterization of $\sigma_{q\bar{q}}$. Solid curves display the modifications due to the finite coherence length l_c and gluon shadowing corrections while the dashed lines are without ("frozen" approximation).

We see that in comparison with the "frozen" approximation, l_c corrections noticeable change ratios at low energies (especially for ψ'). For coherent production in the limit of low energies $l_c \to 0$ the strongly oscillating exponential phase factor in (28) makes the integral very small and thus $F^{coh} \approx 0$. Then the cross section rises with l_c unless it saturates at $l_c \gg R_A$ when the phase factor becomes constant. Apparently, this transition region is shifted to higher energies for larger nuclear radius. For incoherent production

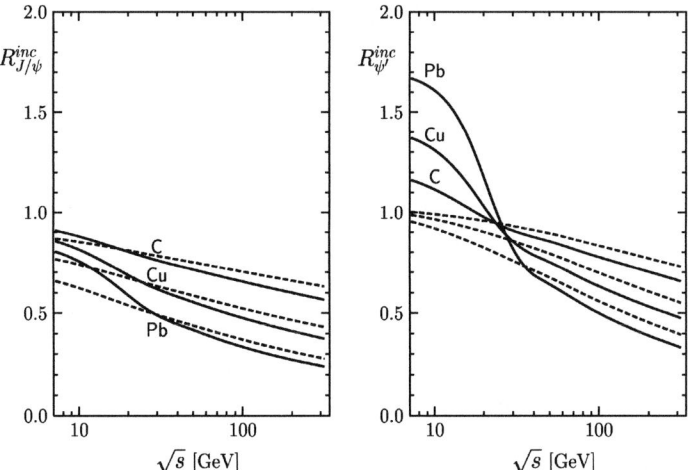

FIGURE 5. Ratios R_{ψ}^{inc} for J/ψ and ψ' incoherent production on nuclei as function of $\sqrt{s}$ and at $Q^2 = 0$. The meaning of the different lines is the same as in Fig. 4.

the observed nontrivial energy dependence is easy to interpret. At low energies $l_c \ll R_A$ and the photon propagates without any attenuation inside the nucleus where it develops for a short time t_c a $c\bar{c}$ fluctuation which momentarily interacts to get on mass shell. The produced $c\bar{c}$ pair attenuates along the path whose length is a half of the nuclear thickness on the average. On the other hand, at high energies when $l_c \gg R_A$ the $c\bar{c}$ fluctuation develops long before its interaction with the nucleus. As a result, it propagates through the whole nucleus and the mean path is twice as long as at low energies. This is why the nuclear transparency drops when going from the regimes of short to long l_c.

At high energies gluon shadowing becomes important. We see that the onset of gluon shadowing happens at a c.m. energy of few tens of GeV. Remarkably, the onset of shadowing is delayed with rising nuclear radius. Nuclear suppression of J/ψ production becomes stronger with energy. This is an obvious consequence of the energy dependence of $\sigma_{q\bar{q}}(r_T, s)$, which rises with energy. For ψ' the suppression is rather similar as for J/ψ. In particular we do not see any considerable nuclear enhancement of ψ' which has been found earlier [1, 28], where the oversimplified form of the dipole cross section, $\sigma_{q\bar{q}}(r_T) \propto r_T^2$ and the oscillator form of the wave function had been used.

HEAVY ION PERIPHERAL COLLISIONS

The large charge Z of heavy nuclei gives rise to strong electromagnetic fields: the photon field of one nucleus can produce a photo-nuclear interaction in the other. The cross section of the charmonia photoproduction by the induced photons reads

$$k\frac{d\sigma}{dk} = \int d^2b \int d^2b' \, n(k, \vec{b}' - \vec{b}) \, \sigma_A(b, s), \qquad (34)$$

where k is the photon momentum. Photon flux induced by projectile nucleus with Lorenz factor γ is

$$n(k,\vec{b}) = \frac{\alpha_{em}Z^2k^2}{\pi^2\gamma^2}K_1^2\left(\frac{bk}{\gamma}\right).$$ (35)

Cross sections $\sigma_A(b,s)$ for coherent and incoherent production are

$$\sigma_A^{coh}(s,b) = \left|\widetilde{\mathcal{M}}_A^{coh}(s,b)\right|^2,$$ (36)

$$\sigma_A^{inc}(s,b) = \left|\widetilde{\mathcal{M}}_A^{inc}(s,b)\right|^2 \frac{T(b)}{16\pi B(s)},$$ (37)

where expressions $\widetilde{\mathcal{M}}(s,b)$ correspond to $\mathcal{M}$ in Eqs. (24, 25) at $Q^2 = 0$. Our predictions for coherent J/ψ production at RHIC and LHC energies are presented on Fig. 6. We see

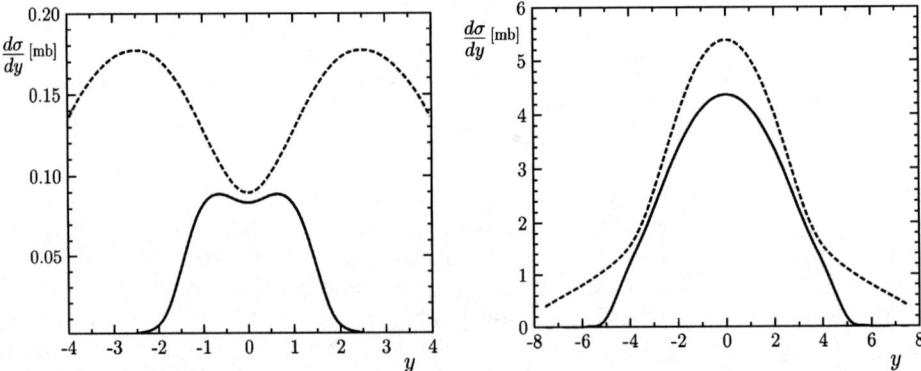

FIGURE 6. Rapidity distribution for coherent J/ψ production in heavy ion peripheral gold-gold collisions at RHIC (left) and lead-lead collisions at LHC (right) calculated with the GBW parameterization of $\sigma_{q\bar{q}}$. Solid curves display the modifications due to gluon shadowing and finite coherence length l_c while the dashed lines are without ("frozen" approximation).

that l_c corrections modify the distribution at the edges (positive and negative y) while suppression around $y = 0$ is provided mostly by gluon shadowing (especially for LHC, where energies should be much higher).

CONCLUSION

In this paper we use the LC dipole approach for description of charmonium electroproduction off protons and nuclei. We have no free parameters and our calculations are in good agreement with existing experimental data. Our predictions can be tested in future experiments at high energies with electron-nuclear colliders (eRHIC) and in peripheral heavy ion collisions (RHIC and LHC).

Acknowledgment: The authors gratefully acknowledge the partial support by a grant from the Gesellschaft für Schwerionenforschung Darmstadt (grant no. GSI-OR-SCH) and by the Federal Ministry BMBF (grant no. 06 HD 954).

REFERENCES

1. Kopeliovich, B.Z., and Zakharov, B.G., *Phys. Rev. D* **44**, 3466 (1991).
2. Huefner, J., *et al.*, *Phys. Rev. D* **62**, 094022 (2000); hep-ph/0007111.
3. Huefner, J., *et al.*, *Phys. Rev. C* **66**, 024903 (2002); hep-ph/0202216.
4. Baur, G., Hencken, K., and Trautmann, D., *Prog. Part. Nucl. Phys.* **42**, 357 (1999); nucl-th/9810078.
5. Kogut, J.B., and Soper, D.E., *Phys. Rev. D***1** 2901 (1970).
6. Bjorken, J.M., Kogut, J.B., and Soper, D.E., *Phys. Rev. D***3** 1382 (1971).
7. Zamolodchikov, A.B., Kopeliovich, B.Z., and Lapidus, L.I., *JETP Lett.* **33** 595 (1981).
8. Golec-Biernat, K., and Wüsthoff, M., *Phys. Rev. D***53** 014017 (1999); hep-ph/9903358.
9. Kopeliovich, B.Z., Schäfer, A., and Tarasov, A.V., *Phys. Rev. D***62** 054022 (2000); hep-ph/9908245.
10. Ryskin, M.G., Roberts, R.G., Martin, A.D., and Levin, E.M., *Z. Phys. C***76** 231 (1997).
11. Terent'ev, M.V., *Sov. J. Nucl. Phys.* **24** 106 (1976).
12. Eichten, E., *et al.*, Phys. Rev. D **17** 3090 (1978); *Phys. Rev. D* **21** 203 (1980).
13. Buchmüller, W., and Tye, S.-H.H., *Phys. Rev. D* **24** 132 (1981).
14. Quigg, C., and Rosner, J.L., *Phys. Lett. B* **71** 153 (1977).
15. Martin, A., *Phys. Lett. B* **93** 338 (1980).
16. Benhar, O., *et al.*, *Phys. Rev. Lett.* **69** 1156 (1992).
17. Hoyer, P., and Peigné, S., *Phys. Rev. D* **61** 031501 (2000).
18. Melosh, H.J., *Phys. Rev. D* **9** 1095 (1974); Jaus, W., *Phys. Rev. D* **41** 3394 (1990).
19. Bronzan, J.B., Kane, G.L., and Sukhatme, U.P., *Phys. Lett. B* **49** 272 (1974).
20. H1 Collaboration, Adloff, C., *et al.*, *Phys. Lett. B* **483** 23-35 (2000); hep-ex/0003020.
21. E401 Collaboration, Binkley, M., *et al.*, *Phys. Rev. Lett.* **48** 73 (1982).
22. E516 Collaboration, Denby, B.H., *et al.*, *Phys. Rev. Lett.* **52** 795 (1984).
23. ZEUS Collaboration, Breitweg, J., *et al.*, *Z. Phys. C* **75** 215 (1997).
24. H1 Collaboration, Adloff, C., *et al.*, *Eur. Phys. J. C* **10** 373 (1999); hep-ex/9903008.
25. Suzuki, K., *et al*. *Phys. Rev. D* **62** 031501 (2000); hep-ph/0005250.
26. Kopeliovich, B.Z., *et al.*, *Phys. Rev. C* **65**, 035201 (2002); hep-ph/0107227.
27. Hüfner, J., Kopeliovich, B.Z., and Zamolodchikov, Al.B., *Z. Phys. A* **357**, 113 (1997).
28. Kopeliovich, B.Z., *et al.*, *Phys. Lett. B* **324**, 469 (1994).

Form Factors and Cross Section for J/ψ Dissociation

M.Nielsen*, F.S. Navarra*, R.D. Matheus*, R. Rodrigues da Silva* and
M.E. Bracco†

*Instituto de Física, USP
C.P. 66318, 05315-970 São Paulo, SP, Brazil
†Instituto de Física, UERJ
Rua São Francisco Xavier 524, Maracanã, 20559-900, Rio de Janeiro, RJ, Brazil

Abstract. We use the QCD sum rules to evaluate the $D^*D\pi$, $DD\rho$ and DDJ/ψ form factors as well as the $J/\psi\,\pi \to$ open charm cross sections. The form factors are evaluated as a function of the momentum Q^2 of the off-shell meson and the cross sections as a function of $\sqrt{s}$.

INTRODUCTION

Matter at very high energy density in thermal equilibrium is believed to form a Quark Gluon Plasma (QGP). The central goal of the heavy-ion program at the RHIC at Brookhaven National Laboratory is to identify new forms of matter like the QGP. Various signatures of the QGP creation have been proposed, and the suppression of J/ψ mesons [1] is among the most promising ones. Anomalous J/ψ suppression has been already observed in Pb+Pb collisions at $158A$ GeV [2]. Many different mechanisms have been proposed to explain the phenomenon. It has been suggested that the suppression was due to a new phase, the QGP [3], a change of the equation of state due to the QCD phase transition [4], the melting of charmonia in QGP [5], or a percolation deconfinement [6]. It has also been suggested that the anomalous suppression comes from the absorption by various comovers produced in the Pb+Pb collisions [7, 8, 9, 10, 11, 12]. A definitive understanding on the nature of J/ψ suppression in Pb+Pb collisions is still lacking.

The absorption of J/ψ by comovers depends on the J/ψ dissociation cross sections in collision with hadrons. The J/ψ dissociation cross sections in collision with hadrons have been considered previously in several theoretical studies, but the predicted cross sections show great variation at low energies, largely due to different assumptions regarding the dominant scattering mechanism. Kharzeev and Satz [13] employed the parton model and perturbative QCD "short-distance" approach of Bhanot and Peskin [14] and found remarkably small low-energy cross sections for collisions of J/ψ with light hadrons. There are also calculations based on quark forces to obtain the underlying scattering amplitudes from explicit nonrelativistic quark model wavefunctions of the initial and final mesons.The wave function and the interaction are then used to evaluate the cross section for the quark-interchange process [15].

CP660, Hadron Physics: Effective Theories of Low Energy QCD, edited by A. H. Blin et al.
© 2003 American Institute of Physics 0-7354-0120-9/03/$20.00

Matinyan and Müller [16], Haglin [17], Lin and Ko [18], Song and Lee [19], Ivanov et all. [20], and Navarra et all. [21] recently reported results for these dissociation cross sections in meson exchange models. These references all use effective meson Lagrangians, but differ in the interaction terms included in the Lagrangian. Results depend on a choice of meson coupling schemes and assumed form factors which are not phenomenologically known.

Here we report the use of the QCD sum rule (QCDSR) method [22], based on the three-point function, to evaluate the $D^*D\pi$, $DD\rho$ and DDJ/ψ hadronic form factors and coupling constants, need in the calculation of the $J/\psi - \pi(\rho)$ cross section using meson exchange models. We also report the use of a QCD sum rule four-point function to directly evaluate the $J/\psi - \pi$ cross section.

THE QCDSR CALCULATION

The fundamental assumption of the QCDSR approach is the principle of duality. Specifically one assumes that there exist a momentum interval over which a correlation function may be equivalentely described at both, the quark level and at the hadron level. Therefore, the underlying procedure of the QCDSR technique is the following: on one hand we calculate the correlation function at the quark level in terms of quark and gluon fields. On the other hand, the correlation function is calculated at the hadron level introducing hadron characteristics as masses, decay constants, form factors, etc. At the quark level the complex structure of the QCD vacuum leads us to employ the Wilson's operator product expansion (OPE). The calculation of the correlation function at the hadron level (or phenomenological side) proceeds by writing a dispersion relation for each one of the invariant structures of the correlator.

As usual, our goal is to make a match between the two representations of the correlation function. To improve the matching a Borel transform [22] is performed in both sides of the sum rule. The Borel transform is defined by

$$\tilde{f}(M^2) = \lim_{Q^2, n \to \infty} \frac{(Q^2)^{n+1}}{n!} \left(-\frac{d}{dQ^2}\right)^n f(Q^2),$$

(1)

where $M^2 \equiv Q^2/n$ is the Borel mass. The Borel transform suppresses higher dimension operators in the OPE side by a factorial factor, and changes a power-law suppression to an exponential suppression of the higher mass states in the phenomenological side.

The basic idea, supported by ample successful applications, is that taking into account only the first few terms in the OPE, complemented by rather simple assumptions for the higher mass contributions to the dispersion relation, will already provide a good estimate to the amplitudes of interest.

Let us start with the three-point function associated with a $HH'M$ vertex, where H and H' are the external mesons and M is the exchanged meson, which is given by:

$$\Gamma_M(p, p') = \int d^4x \, d^4y \, \langle 0|T\{j_{H'}(x) j_M(y) j_H^\dagger(0)\}|0\rangle \, e^{ip' \cdot x} e^{-i(p'-p) \cdot y},$$

(2)

where j_i is the interpolating field of the hadron i, written in terms of the quark content and with the same quantum numbers of these hadrons. In the case of the $D^*D\pi$ vertex with an off-shell pion, for instance, one has $j_M = j_\pi = i\bar{u}\gamma_5 d$, $j_H^\dagger = j_\mu^{D^*} = \bar{d}\gamma_\mu c$ and $j_{H'} = j_D = i\bar{c}\gamma_5 u$, which are the interpolating fields for π, D^* and D respectively.

The QCD side, or theoretical side, of the vertex function is evaluated by performing Wilson's operator product expansion (OPE) of the operator in Eq. (2). Writing Γ_M in terms of the invariant amplitudes, we can write a double dispersion relation for each one of the invariant amplitudes, over the virtualities p^2 and p'^2 holding q^2 fixed:

$$\Gamma(p^2, p'^2, q^2) = -\frac{1}{4\pi^2} \int_{m_Q^2}^{s_0} ds \int_{m_{Q'}^2}^{u_0} du \frac{\rho(s,u,q^2)}{(s-p^2)(u-p'^2)} , \qquad (3)$$

where $\rho(s,u,q^2)$ equals the double discontinuity of the amplitude $\Gamma(p^2, p'^2, q^2)$ and $m_Q(m_{Q'})$ is the mass of the heaviest quark, $Q(Q')$ in the meson $H(H')$.

The phenomenological side of the vertex function, $\Gamma_M(p,p')$, is obtained by the consideration of H and H' state contribution to the matrix element in Eq. (2):

$$\Gamma_M^{phen}(p,p') = \lambda_H \lambda_{H'} \lambda_M g_{HH'M}^{(M)}(q^2) \Delta_{H'}(p) \Delta_H(p') \Delta_M(q) , \qquad (4)$$

where Δ_i is the propagator of the meson i, $q = p - p'$, and $\lambda_H = \langle 0|j_H|H \rangle$ defines the couplings of the currents with the respective hadronic states, which is proportional to the meson decay constant. The form factor we want to determine, for an off-shell meson M, are represented by $g_{HH'M}^{(M)}(q^2)$.

The contribution of higher resonances and continuum in Eq. (4) will be taken into account as usual in the standard form of ref. [23], through the continuun thresholds s_0 and u_0, for the H and H' mesons respectively, introduced in Eq. (3).

The expressions for the sum rules related with the vertices $D^*D\pi$, $DD\rho$ and DDJ/ψ are given in refs. [24, 25, 26].

To evaluate the $J/\psi - \pi$ cross section using the QCDSR we consider the four-point function for the general process $J/\psi\,\pi \to M_3 M_4$:

$$\begin{aligned}
\Pi_{\mu 34} &= \int d^4x\, d^4y\, d^4z\, e^{-ip_1 \cdot x}\, e^{ip_3 \cdot y}\, e^{ip_4 \cdot z} \\
&\times \langle 0|T\{j_\pi(x) j_3(y) j_\mu^\psi(0) j_4(z)\}|0 \rangle ,
\end{aligned} \qquad (5)$$

with the currents given by $j_\pi = \bar{d}i\gamma_5 u$, $j_\mu^\psi = \bar{c}\gamma_\mu c$, $j_3 = \bar{u}\Gamma_3 c$ and $j_4 = \bar{c}\Gamma_4 d$. p_1, p_2, p_3 and p_4 are the four-momenta of the mesons π, J/ψ, M_3 and M_4 respectively, with $p_1 + p_2 = p_3 + p_4$. Γ_3 and Γ_4 denote specific gamma matrices corresponding to the process envolving the mesons M_3 and M_4. For instance, for the process $J/\psi\,\pi \to \bar{D}D^*$ we will have $\Gamma_3 = \gamma_v$ and $\Gamma_4 = i\gamma_5$.

The phenomenological side of the correlation function, $\Pi_{\mu 34}$, is obtained by the consideration of J/ψ, π, M_3 and M_4 state contribution to the matrix element in Eq. (5). The hadronic amplitudes are defined by the matrix element:

$$\begin{aligned}
i\mathcal{M} &= \langle \psi(p_2,\mu)| M_3(-p_3,v) M_4(-p_4,\rho)\,\pi(p_1) \rangle \\
&= i\,\mathcal{M}_{\mu 34}(p_1,p_2,p_3,p_4)\, \varepsilon_2^\mu f_3^{*v} f_4^{*\rho} ,
\end{aligned} \qquad (6)$$

where $f_i^{*\alpha} = \varepsilon_i^{*\alpha}$ for the D^* meson and $f_i^{*\alpha} = 1$ for the D meson.

The phenomenological side of the sum rule can be written as (for the part of the hadronic amplitude that will contribute to the cross section) [27]:

$$\Pi_{\mu34}^{phen} = -\frac{m_\pi^2 f_\pi}{m_u + m_d} \frac{m_\psi f_\psi \lambda_3 \lambda_4 \mathcal{M}_{\mu34}}{(p_1^2 - m_\pi^2)(p_2^2 - m_\psi^2)(p_3^2 - m_3^2)(p_4^2 - m_4^2)} + \text{h. r.}. \tag{7}$$

where h. r. means higher resonances, and where λ_i is related with the conrresponding meson decay constant: $\lambda_D = -\langle D|j_D|0\rangle = -m_D^2 f_D/m_c$ and $\lambda_{D^*} = \langle 0|j_\alpha|D^*\rangle = m_{D^*} f_{D^*}$.

We note that one has $1/p_1^2$ pole in Eq. (7) in the limit of a vanishing pion mass. Following Reinders, Rubinstein, and Yazaki [28], and others [29, 30, 24], we can write a sum rule at $p_1^2 = 0$ and single out the leading terms in the operator product expansion (OPE) of Eq. (5) that match the $1/p_1^2$ term. The perturbative diagram does not contribute with $1/p_1^2$ and, up to dimension four, only the diagrams proportional to the quark condensate contribute. After collecting the $1/p_1^2$ terms on the theoretical side and taking the limit $p_{1\mu} \to 0$ in the residue of the pion pole, one obtains for the contribution of the quark condensate [27]:

$$\Pi_{\mu\nu}^{<\bar{q}q>} = -\frac{2m_c\langle\bar{q}q\rangle}{p_1^2} \frac{p_{1\nu}(p_{1\mu} + p_{2\mu} - 2p_{3\mu}) - p_{1\mu}p_{2\nu}}{(p_3^2 - m_c^2)(p_4^2 - m_c^2)}, \tag{8}$$

for the $J/\psi\,\pi \to \bar{D}\,D^*$ process.

$$\Pi_\mu^{<\bar{q}q>} = -\frac{2\langle\bar{q}q\rangle}{p_1^2} \frac{\varepsilon_{\mu\alpha\beta\sigma} p_1^\alpha p_3^\beta p_4^\sigma}{(p_3^2 - m_c^2)(p_4^2 - m_c^2)}, \tag{9}$$

for the $J/\psi\,\pi \to \bar{D}\,D$ process, and

$$\Pi_{\mu\nu\alpha}^{<\bar{q}q>} = -\frac{2\langle\bar{q}q\rangle}{p_1^2(p_3^2 - m_c^2)(p_4^2 - m_c^2)} \left[(m_c^2 + p_3 \cdot p_4)\varepsilon_{\alpha\mu\nu\beta}\, p_1^\beta + E_{\mu\nu\alpha} \right], \tag{10}$$

for the $J/\psi\,\pi \to \bar{D}^*\,D^*$ process, where

$$\begin{aligned}
E_{\mu\nu\alpha} &= p_1^\beta p_3^\lambda p_4^\gamma(-\varepsilon_{\nu\beta\lambda\gamma}g_{\alpha\mu} + \varepsilon_{\mu\beta\lambda\gamma}g_{\alpha\nu} - \varepsilon_{\alpha\beta\lambda\gamma}g_{\mu\nu}) \\
&+ \varepsilon_{\mu\nu\beta\lambda}(p_1^\beta p_4^\lambda p_{3\alpha} - p_3^\beta p_4^\lambda p_{1\alpha}) + \varepsilon_{\alpha\nu\beta\lambda}(p_1^\beta p_3^\lambda p_{1\mu} \\
&+ p_3^\beta p_4^\lambda p_{1\mu} - p_1^\beta p_4^\lambda p_{3\mu} - p_1^\beta p_3^\lambda p_{4\mu}) + \varepsilon_{\alpha\mu\beta\lambda}(-p_1^\beta p_3^\lambda p_{1\nu} \\
&- p_3^\beta p_4^\lambda p_{1\nu} + p_1^\beta p_4^\lambda p_{1\nu} + p_1^\beta p_3^\lambda p_{4\mu}).
\end{aligned} \tag{11}$$

Comparing Eqs. (8), (9) and (10) with Eq. (7) the structures defining $\mathcal{M}_{\mu34}$ in Eq. (7) are easily identified. Therefore, defining

$$\mathcal{M}_{\mu\nu} = \Lambda_{DD^*} \left(p_{1\mu}p_{1\nu} - p_{1\mu}p_{2\nu} - 2p_{1\nu}p_{3\mu} \right), \tag{12}$$

$$\mathcal{M}_\mu = \Lambda_{DD}\, \varepsilon_{\mu\alpha\beta\sigma} p_1^\alpha p_3^\beta p_4^\sigma \,, \tag{13}$$

and

$$\mathcal{M}_{\mu\nu\alpha} = \Lambda_{D^*D^*}\, E_{\mu\nu\alpha} \,. \tag{14}$$

we can write sum rules for Λ_{DD^*}, Λ_{DD} and $\Lambda_{D^*D^*}$ in any of the structures appearing in Eqs. (12), (13) and (14). To improve the matching between the phenomenological and theoretical sides we make a single Borel transformation to all the external momenta (except p_1^2) taken to be equal: $-p_2^2 = -p_3^2 = -p_4^2 = P^2 \to M^2$. The problem of doing a single Borel transformation is the fact that terms associated with the pole-continuum transitions are not suppressed [31]. In ref. [31] it was explicitly shown that the pole-continuum transition has a different behavior as a function of the Borel mass as compared with the double pole contribution (triple pole contribution in our case) and continuum contribution: it grows with M^2 as compared with the contribution of the fundamental states. Therefore, the pole-continuum contribution can be taken into account through the introduction of a parameter A in the phenomenological side of the sum rule [24, 30, 31].

Thus, neglecting m_π^2 in the denominator of Eq. (7) and doing a single Borel transform we get

$$\frac{\Lambda_{DD^*} + A_{DD^*} M^2}{m_{D^*}^2 - m_\psi^2} \left[\frac{e^{-m_D^2/M^2} - e^{-m_\psi^2/M^2}}{m_\psi^2 - m_D^2} - (\psi \to D^*) \right]$$
$$= -2m_c \langle \bar{q}q \rangle \frac{e^{-m_c^2/M^2}}{M^2} \frac{m_c(m_u + m_d)}{m_\pi^2 m_D^2 m_{D^*} m_\psi f_\pi f_D f_{D^*} f_\psi} \,, \tag{15}$$

for the $J/\psi\, \pi \to \bar{D}\, D^*$ process, and

$$\frac{\Lambda_{MM} + A_{MM} M^2}{m_M^2 - m_\psi^2} f_M(M^2) = C_M \frac{m_u + m_d}{m_\pi^2 m_M^2 m_\psi f_\pi f_M^2 f_\psi}$$
$$\times 2\langle \bar{q}q \rangle \frac{e^{-m_c^2/M^2}}{M^2} \,, \tag{16}$$

where the subscript M stands for the D or D^* mesons, with $C_D = \frac{m_c^2}{m_D^2}$, $C_{D^*} = 1$ and

$$f_M(M^2) = \frac{e^{-m_M^2/M^2}}{M^2} - \frac{e^{-m_M^2/M^2} - e^{-m_\psi^2/M^2}}{m_\psi^2 - m_M^2} \,. \tag{17}$$

In Eqs. (15) and (16) we have transferred to the theoretical side the couplings of the currents with the mesons, and have introduced, in the phenomenological side, the parameter A_{MM} to account for possible nondiagonal transitions.

RESULTS FOR THE FORM FACTORS

The parameter values used in all calculations are $m_u + m_d = 14$ MeV, $m_c = 1.5$ GeV, $m_\pi = 140$ MeV, $m_D = 1.87$ GeV, $m_{D^*} = 2.01$ GeV, $m_\rho = 0.77$ GeV, $m_\psi = 3.1$ GeV,

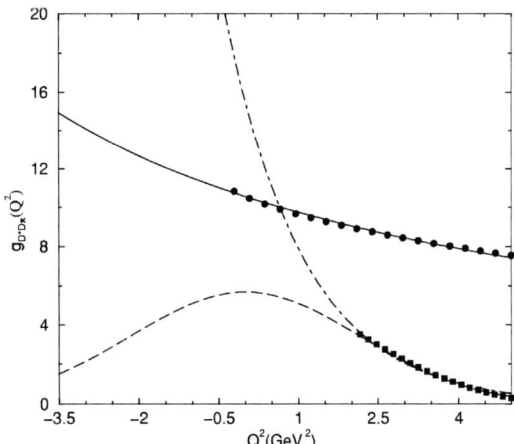

FIGURE 1. Momentum dependence of the $D^*D\pi$ form factor. The solid, dashed and dot-dashed lines give the parametrization of the QCDSR results through Eq. (22) for the circles, and Eqs. (19) and (20) for the squares.

$f_D = 160\,\text{MeV}$, $f_\pi = 131.5\,\text{MeV}$, $g_\rho = 5.45$, $\langle \bar{q}q \rangle = -(0.23)^3\,\text{GeV}^3$, $\langle g^2G^2 \rangle = 0.5\,\text{GeV}^4$. We parametrize the continuum thresholds as

$$s_0 = (m_H + \Delta_s)^2, \qquad u_0 = (m_{H'} + \Delta_u)^2 . \tag{18}$$

We first discuss the $D^*D\pi$ form factor with an off-shell pion. In ref. [24] we have analized the behavior of the perturbative and the gluon condensate contribution to the form factor $g_{D^*D\pi}(Q^2 = -q^2)$ as a function of the Borel mass M^2. We got that the gluon condensate, despite being small, helps the stability of the curve, as a function of M^2, providing a rather stable plateau for $M^2 \geq 3\,\text{GeV}^2$.

Fixing $M^2 = 3.5\,\text{GeV}^2$ we show, in Fig. 1, the momentum dependence of the form factor (squares) in the interval $2 \leq Q^2 \leq 5\,\text{GeV}$. Since the present approach cannot be used at $Q^2 = 0$, to extract the $g_{D^*D\pi}$ coupling, defined as the value of the form factor at the pole of the off-shell meson, from the form factor we need to extrapolate the curve to $Q^2 = 0$ (in the approximation $m_\pi^2 = 0$).

As can be seen in Fig. 1, the Q^2 dependence of the form factor, represented by the squares, can be well described by a gaussian form (dashed line)

$$g_{D^*D\pi}^{(\pi)}(Q^2) = 5.7\,e^{-Q^4/9.17} , \tag{19}$$

and can also be well reproduced by the exponential parametrization (dot-dashed line)

$$g_{D^*D\pi}^{(\pi)}(Q^2) = 15.5\,e^{-Q^2/1.48} . \tag{20}$$

Off course, the two parametrizations in Eqs. (19) and (20) lead to very different values for the $D^*D\pi$ coupling constant:

$$g_{D^*D\pi} = \begin{cases} 5.7 & \text{with the gaussian parametrization} \\ 15.5 & \text{with the exponential parametrization} \end{cases} \tag{21}$$

To solve this problem we analyse the $D^*D\pi$ form factor for an off-shell D meson. In Fig. 1 we also show, through the circles, the momentum dependence of the $D^*D\pi$ form factor for an off-shell D meson, in the interval $-0.5 \leq Q^2 \leq 5\,\text{GeV}$, where we expect the sum rules to be valid (since in this case the cut in the t channel starts at $t \sim m_c^2$ and thus the Euclidian region stretches up to that threshold). From this figure we can see that the Q^2 dependence of the form factor represented by the circles can be well reproduced by the monopole parametrization (solid line)

$$g_{D^*D\pi}^{(D)}(Q^2) = \frac{126.1}{Q^2 + 11.95} . \tag{22}$$

From the parametrization in Eq. (22) we can also extract the $D^*D\pi$ coupling constant, which now is defined as the value of the form factor at the D pole ($Q^2 = -m_D^2$). We get

$$g_{D^*D\pi} = 14.9 , \tag{23}$$

in an excelent agreement with the exponential parametrization of $g_{D^*D\pi}^{(\pi)}(Q^2)$. Therefore, we conclude that the gaussian parametrization, despite describing very well the behaviour of the QCDSR results for $g_{D^*D\pi}^{(\pi)}(Q^2)$ in the region $2 \leq Q^2 \leq 5\,\text{GeV}$, can not be used to do the extrapolation at $Q^2 = -m_\pi^2$. By imposing that the two form factors: $g_{D^*D\pi}^{(\pi)}(Q^2)$ and $g_{D^*D\pi}^{(D)}(Q^2)$ should lead to the same coupling constant, $g_{D^*D\pi}$, we find that the exponential parametrization of $g_{D^*D\pi}^{(\pi)}(Q^2)$ is the correct parametrization of the QCDSR results. It is very interesting to notice that the obtained coupling constant, besides being still smaller than the experimental value [32]: $g_{D^*D\pi} = 17.9 \pm 0.3 \pm 1.9$, is much closer to it than other QCDSR evaluations [24].

There is another important information that we can extract from the parametrization of the QCDSR results which is the value of the cut-off. Defining the coupling constant as the value of the form factor at $Q^2 = -m_M^2$, where m_M is the mass of the off-shell meson, the monopole and the exponential parametrizations of the form factor can be written as (neglecting m_π^2):

$$g_{D^*D\pi}^{(D)}(Q^2) = g_{D^*D\pi} \frac{\Lambda_D^2 - m_D^2}{Q^2 + \Lambda_D^2} , \tag{24}$$

$$g_{D^*D\pi}^{(\pi)}(Q^2) = g_{D^*D\pi} e^{-\frac{Q^2}{\Lambda_\pi^2}} , \tag{25}$$

and from Eqs. (22) and (20) we get

$$\Lambda_D = 3.5\,\text{GeV}, \quad \Lambda_\pi = 1.2\,\text{GeV}. \tag{26}$$

Therefore, the form factor is harder if the off-shell meson is heavy, implying that the size of the vertex depends on the exchanged meson. This means that a heavy meson will see the vertex as pointlike, whereas a light meson will see its extension.

The same kind of behaviour is also obtained in the $DD\rho$ vertex [25]. Fixing $M'^2 = 4.7\,\text{GeV}^2$ (which corresponds to $M^2 = 0.8\,\text{GeV}^2$ for the case of off-shell D) we show, in Fig. 2, the momentum dependence of the form factor (circles for $g_{DD\rho}^{(D)}(Q^2)$ and squares for $g_{DD\rho}^{(\rho)}(Q^2)$) in the interval $0.1 \leq Q^2 \leq 5\,\text{GeV}$. In Fig. 2 we also show that the Q^2 dependence of the form factor represented by the circles can be well reproduced by the monopole parametrization (solid line)

$$g_{DD\rho}^{(D)}(Q^2) = \frac{37.5}{Q^2 + 12.12}, \tag{27}$$

and that the form factor represented by the squares can be well reproduced by the exponential parametrization (solid line)

$$g_{DD\rho}^{(\rho)}(Q^2) = 2.53e^{-\frac{Q^2}{0.98}}. \tag{28}$$

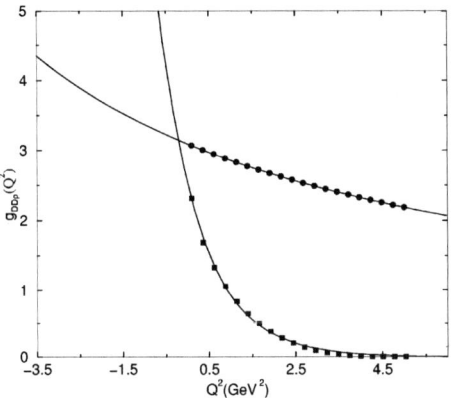

FIGURE 2. Momentum dependence of the $DD\rho$ form factor for $\Delta_s = \Delta_u = 0.5\,\text{GeV}$. The solid lines give the parametrization of the QCDSR results through Eq. (27) for the circles, and Eq. (28) for the squares.

The coupling constants and cut-offs resulting of the parametrizations in Eqs. (27) and (28), for the case of a $DD\rho$ vertex, are given in Table 1.

TABLE 1. Values of the coupling constants and cut-offs in the $DD\rho$ vertex.

	$g_{DD\rho}^{(M)}(Q^2 = -m_M^2)$	$\Lambda_M\,(\text{GeV})$
D off-shell	4.4	3.5
ρ off-shell	4.6	1.0

We see that there is a very good agreement betwen the two values of the coupling constant, extracted from the QCDSR results.

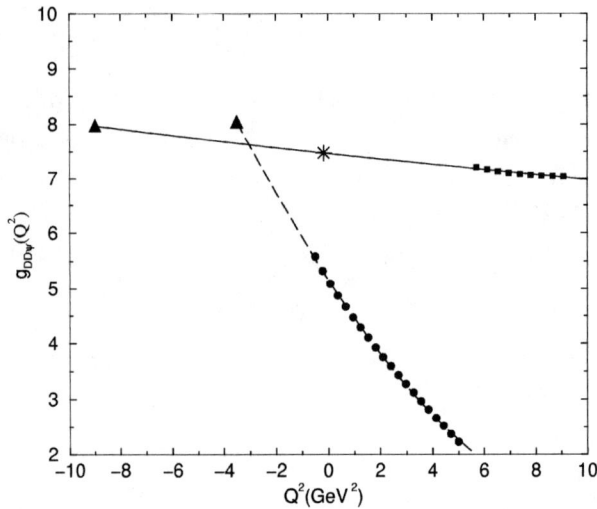

FIGURE 3. Momentum dependence of the $DDJ\psi$ form factor. Circles and squares represent our numerical calculations for the D and J/ψ off-shell respectively. The dashed and solid lines give the parametrization of the QCDSR results through Eq. (29) for the circles and Eq. (30) for the squares. The triangles give the form factors at the poles of the particles (which we indentify with the coupling constant). The star shows the form factor at $Q^2 = 0$.

In the case of the $J/\psi DD$ vertex, fixing $M^2 = 11$ GeV^2 we calculate the momentum dependence of the form factor $g^{(D)}_{DDJ/\psi}(Q^2)$ in the interval $-0.5 \leq Q^2 \leq 5.0 GeV^2$ [26]. Our numerical calculations can be well reproduced by the gaussian parametrization:

$$g^{(D)}_{DDJ/\psi}(Q^2) = 16.4\, e^{-[\frac{(Q^2+16.2)^2}{228}]}\,. \tag{29}$$

Along the same lines we choose $M^2 = 8$ GeV^2 and calculate the momentum dependence of the form factor $g^{(J/\psi)}_{DDJ/\psi}(Q^2)$. The numerical results can be fitted by the monopole parametrization:

$$g^{(J/\psi)}_{DDJ/\psi}(Q^2) = \frac{1069.76}{Q^2 + 143.18}\,. \tag{30}$$

All our numerical results and parametrizations are shown in Fig. 3. The circles and squares correspond to the numerical results for the D and J/ψ off-shell respectively. These points are fitted by the dashed (Eq. (29)) and solid lines (Eq. (30)) respectively and extrapolated to the meson poles which are represented by the triangles, which give the numbers for the coupling constant.

The coupling constants and cut-offs resulting of the parametrizations in Eqs. (30) and (29), for the case of a $J/\psi DD$ vertex, are given in Table 2.

Again we obtain a very good agreement between the two values of the coupling constant, extracted from the QCDSR results.

TABLE 2. Values of the coupling constants and cut-offs in the $J/\psi DD$ vertex.

	$g^{(M)}_{DDJ/\psi}(Q^2 = -m_M^2)$	Λ_M (GeV)
D off-shell	8.1	3.9
J/ψ off-shell	8.0	12.0

When neutral vector mesons are off-shell, it is also possible to define the coupling constant as the value of the form factor at $Q^2 = 0$. In Table 3 we give the coupling constants resulting of the parametrizations in Eqs. (30) and (28) evaluated at $Q^2 = 0$ The

TABLE 3. Values of the coupling constants evaluated at $Q^2 = 0$.

$g^{(J/\psi)}_{DDJ/\psi}(Q^2 = 0)$	$g^{(\rho)}_{DD\rho}(Q^2 = 0)$
7.5	2.5

star in Fig. 3 indicates the J/ψ off-shell form factor taken at $Q^2 = 0$. It is very interesting to notice that the values used in refs. [16, 18, 19] for $g_{DD\rho}$ and $g_{DDJ/\psi}$, obtained in the framework of vector meson dominance, are: $g_{DDJ/\psi} = 7.7$ and $g_{DD\rho} = 2.52$, in complete agreement with our value extracted from the form factor normalized at $Q^2 = 0$.

RESULTS FOR THE $J/\psi - \pi$ CROSS SECTION

The QCD sum rule results for $\Lambda_{DD^*} + A_{DD^*}M^2$, $\Lambda_{DD} + A_{DD}M^2$ and $\Lambda_{D^*D^*} + A_{D^*D^*}M^2$ in Eqs. (15) and (16), as a function of M^2 follow a straight line in the Borel region $8 \leq M^2 \leq 16$ GeV^2 [27]. The value of the amplitude Λ is obtained by the extrapolation of the line to $M^2 = 0$ [30, 24, 31]. Fitting the QCD sum rule results to a straight line we get

$$\Lambda_{DD^*} \simeq 17.71\,\text{GeV}^{-2}, \quad \Lambda_{DD} \simeq 12.25\,\text{GeV}^{-1}, \quad \Lambda_{D^*D^*} \simeq 11.39\,\text{GeV}^{-3}. \quad (31)$$

As expected, in our approach Λ is just a number and all dependence of $\mathcal{M}_{\mu 34}$ (Eq . (12),(13) and (14)) on particle momenta is contained in the Dirac structure. This is a consequence of our low energy approximation.

Having the QCD sum rule results for the amplitude of the three processes $J/\psi\,\pi \to \bar{D}\,D^*$, $\bar{D}\,D$, $\bar{D}^*\,D^*$, we can evaluate the differential cross section.

We show, in Fig. 4, the cross section for the $J/\psi\,\pi$ dissociation. It is important to keep in mind that, since our sum rule was derived in the limit $p_1 \to 0$, we can not extend our results to large values of $\sqrt{s}$.

Our first conclusion is that our results show that, for values of $\sqrt{s}$ far from the $J/\psi\,\pi \to \bar{D}^*\,D^*$ threshold, $\sigma_{J/\psi\pi \to \bar{D}^*D^*} \geq \sigma_{J/\psi\pi \to \bar{D}D^* + D\bar{D}^*} \geq \sigma_{J/\psi\pi \to \bar{D}D}$, in agreement with the model calculations discussed, for example, in ref. [21] but in disagreement with the results obtained with the nonrelativistic quark model of ref. [15], which show that

305

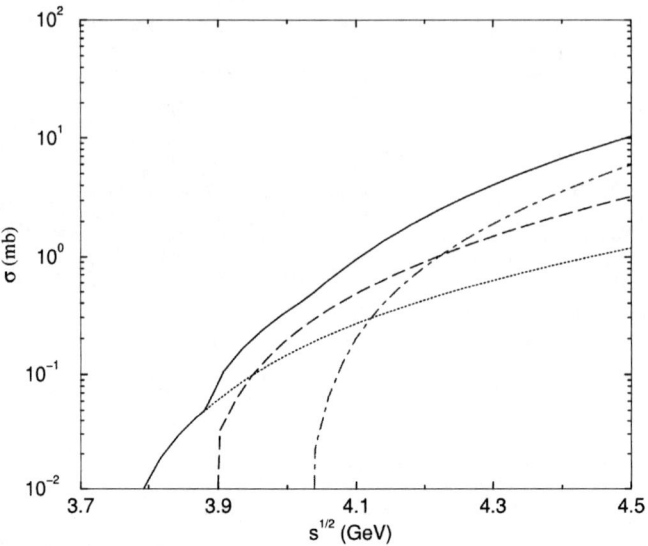

FIGURE 4. Total cross sections of the processes $J/\psi\,\pi \to \bar{D}\,D^* + D\,\bar{D}^*$ (dashed line), $\bar{D}\,D$ (dotted line) and $\bar{D}^*\,D^*$ (dot-dashed line). The solid line gives the total $J/\psi\,\pi$ dissociation cross section.

the state $\bar{D}^*D$ has a larger production cross section than $\bar{D}^*D^*$. Furthermore, our curves indicate that the cross section grows monotonically with the c.m.s. energy but not as fast, near the thresholds, as it does in the calculations in refs. [16, 19, 21]. Again, this behavior is in opposition to [15], where a peak just after the threshold followed by continuous decrease in the cross section was found.

At higher energies, due to our low energy approximation, our approach gradually looses validity. In the fiducial region, close to threshold, $4.1 \le \sqrt{s} \le 4.3$ GeV, we find $2.5 \le \sigma \le 4.0$ mb and these values are much smaller than those obtained with the effective Lagrangians without form factors in the hadronic vertices, but agree in order of magnitude with the quark model calculations of [15].

CONCLUSIONS

We have used the method of QCD sum rules to compute form factors and coupling constants in $DD\rho$, $D^*D\pi$ and DDJ/ψ vertices. Our results for the couplings show once more that this method is robust, yelding numbers which are approximately the same regardless of which particle we choose to be off-shell and depending weakly on the choice of the continuum theshold. As for the form factors, we obtain a harder (softer) form factor when the off-shell particle is heavier (lighter). Therefore we can say that heavy mesons "see" smaller sizes in the light mesons, but the latter can not resolve small structures in the former.

We have also used the QCD sum rule approach to evaluate the hadronic amplitude of the $J/\psi\,\pi$ dissociation. From the hadronic amplitude we have evaluated the $J/\psi\,\pi \to$ charmed mesons dissociation cross section, and have obtained $2.5 \leq \sigma \leq 4.0\,\mathrm{mb}$ at $4.1 \leq \sqrt{s} \leq 4.3\,\mathrm{GeV}$. In view of the uncertainties discussed above these numbers should be taken as upper limits.

It is interesting to remember that Bhanot and Peskin [14] have also used the OPE in the short distance limit to study the charmonium hadron cross section. This work was latter enlarged and updated by Kharzeev et al. [13] and also by Oh et al. [33]. In these papers the crucial assumption was made that the charmonium is very small and resolves the partonic structure of the light hadron. In our approach we do not use this assumption, and we obtain larger values for the cross section. This seems to indicate that size effects are important, and that the J/ψ cannot be considered as a nearly point like object.

ACKNOWLEDGMENTS

Part of the work reported in this manuscript was done in collaboration with A. Lozéa, M. Chiapparini, G. Krein and R. Marques de Carvalho. This work has been supported by FAPESP and CNPq - Brazil.

REFERENCES

1. T. Matsui and H. Satz, *Phys. Lett.* **B178**, 416 (1986).
2. NA50 Collaboration (M.C. Abreu et al.), *Phys. Lett.* **B450**, 456 (1999); *Phys. Lett.* **B477**, 28 (2000).
3. J.-P. Blaizot and J.-Y. Ollitrault, *Phys. Rev. Lett.* **77**, 1703 (1996); C. Y. Wong, *Nucl. Phys.* **A610**, 434c (1996); **A630**, 487c (1998); D. Kharzeev, C. Lourenco, M. Nardi, and H. Satz, *Z. Phys.* **C74**, 307 (1997).
4. E. Shuryak and D. Teaney, *Phys. Lett.* **B430**, 37 (1998).
5. R. Vogt, *Phys. Lett.* **B430**, 15 (1998).
6. M. Nardi and H. Satz, *Phys. Lett.* **B442**, 14 (1998).
7. S. Gavin and R. Vogt, *Nucl. Phys.* **A610**, 442c (1996); W. Cassing and C. M. Ko, *Phys. Lett.* **B396**, 39 (1997); W. Cassing and E. L. Bratkovskaya, *Nucl. Phys.* **A623**, 570 (1997).
8. N. Armesto and A. Capella, *Phys. Lett.* **B430**, 23 (1998).
9. J. D. de Deus and J. Seixas, *Phys. Lett.* **B430**, 363 (1998).
10. J. Hüfner and B. Z. Kopeliovich, *Phys. Lett.* **B445**, 223 (1998).
11. J. Geiss, C. Greiner, E. L. Bratkovskaya, W. Cassing, and U. Mosel, *Phys. Lett.* **B447**, 31 (1999).
12. D. E. Kahana and S. H. Kahana, *Phys. Rev.* **C59**, 1651 (1999).
13. D. Kharzeev and H. Satz, *Phys. Lett.* **B334**, 155 (1994).
14. G. Bhanot and M.E. Peskin, *Nucl. Phys.* **B156**, 391 (1979); M.E. Peskin, *Nucl. Phys.* **B156**, 365 (1979).
15. C.-Y. Wong, E. S. Swanson and T. Barnes, *Phys. Rev.* **C62**, 045201 (2000), and references therein.
16. S.G. Matinyan and B. Müller, *Phys. Rev.* **C58** (1998) 2994.
17. K.L. Haglin, *Phys. Rev.* **C61** (2000) 031902; K.L. Haglin and C. Gale, *Phys. Rev.* **C63** (2001) 065201.
18. Z. Lin and C.M. Ko, *Phys. Rev.* **C62** (2000) 034903.
19. Yongseok Oh, Taesoo Song and Su Houng Lee, *Phys. Rev.* **C63** (2001) 034901.
20. V.V. Ivanov, Yu.L. Kalinovsky, D.B. Blaschke and G.R.G. Burau, hep-ph/0112354.
21. F.S. Navarra, M. Nielsen and M.R. Robilotta, *Phys. Rev.* **C64** (2001) 021901(R).
22. M.A. Shifman, A.I. Vainshtein and V.I. Zakharov, *Nucl. Phys.* **B120**, 316 (1977).
23. B.L. Ioffe and A.V. Smilga, *Nucl. Phys.* **B232**, 109 (1984).

24. F.S. Navarra et al., *Phys. Lett.* **B489**, 319 (2000); F.S. Navarra, M. Nielsen and M.E. Bracco, *Phys. Rev.* **D65**, 037502 (2002).
25. M.E. Bracco et al., *Phys. Lett.* **B521**, 1 (2001).
26. R.D. Matheus et al., *Phys. Lett.* **B541**, 265 (2002).
27. F.S. Navarra, M. Nielsen, R.S. Marques de Carvalho and G. Krein, *Phys. Lett.* **B529**, 87.
28. L.J. Reinders, H. Rubinstein and S. Yazaki, *Phys. Rep.* **127**, 1 (1985).
29. S. Choe, M.K. Cheoun and S.H. Lee, *Phys. Rev.* **C53**, 1363 (1996).
30. M.E. Bracco, F.S. Navarra and M. Nielsen, *Phys. Lett.* **B454**, 346 (1999).
31. B.L. Ioffe and A.V. Smilga, *Nucl. Phys.* **B232** 109 (1984).
32. S. Ahmed et al., (CLEO Collaboration), hep-ex/0108013.
33. Y. Oh, S. Kim and S.H. Lee, hep-ph/0111132.

QUARK MODELS

Meson Form Factors in the Quark-Level LσM

M. D. Scadron[*], F. Kleefeld[†], G. Rupp[†] and E. van Beveren[**]

[*]*Physics Department, University of Arizona, Tucson, AZ 85721, USA*
[†]*Centro de Física das Interacções Fundamentais, Instituto Superior Técnico, Lisbon, Portugal*
[**]*Departamento de Física, Universidade de Coimbra, Coimbra, Portugal*

Abstract. We study meson (π,K) form factors in general, and specialize at a later stage to a specific scheme, namely the quark-level linear σ model (LσM). We compute a variety of electromagnetic and weak observables of light mesons, including pion and kaon form factors and charge radii, charged-pion polarizabilities, semileptonic weak K_{l3} decay, semileptonic weak radiative pion and kaon form factors, radiative decays of vector mesons, and nonleptonic weak $K_{2\pi}$ decay. We find good to very good agreement with experiment of all these predicted observables.

1. LσM AND CHIRAL GOLDBERGER–TREIMAN RELATIONS

The NJL [1] and $Z = 0$ compositeness [2] relations

$$m_\sigma = 2\hat{m} \quad , \quad g = \frac{2\pi}{\sqrt{N_c}} , \tag{1}$$

with $\hat{m} \sim M_N/3$ and meson-quark coupling $g = 2\pi/\sqrt{3} = 3.628$, follow by nonperturbatively solving [3] the strong-interaction Nambu-type gap equations $\delta f_\pi = f_\pi$ and $\delta \hat{m} = m$ (where f_π is the pion decay constant and $\hat{m}$ is the nonstrange constituent quark mass) in quark-loop order, regularization schemes. For a more detailed description of the quark-level LσM, see Ref. [4], or the Appendices of Ref. [5]. Here, we collect instead powerful results of the LσM on meson form factors and related data for strong, e.m., and weak interactions.

This chiral LσM is based on the quark-level pion and kaon Goldberger–Treiman relations (GTRs)

$$f_\pi g = \hat{m} = \frac{1}{2}(m_u + m_d) \quad , \quad f_K g = \frac{1}{2}(m_s + \hat{m}) , \tag{2}$$

for $f_\pi \approx 93$ MeV ($f_\pi \approx 90$ MeV in the chiral limit (CL) [6]), $f_K/f_\pi \approx 1.22$, and $m_s \approx 1.44\hat{m}$ (from Eq. (2)). We begin in Sec. 2 by studying meson vector form factors and their measured charge radii. In Sec. 3 we survey charged-pion polarizabilities for $\gamma\gamma \to \pi\pi$, and compare the results with the LσM predictions. In Sec. 4 we study the semileptonic weak K_{l3} decays and the form factor $f_+(k^2)$ evaluated at $k^2 = 0$. Then in Sec. 5 we examine the radiative semileptonic weak form factors for $\pi^+ \to e^+\nu\gamma$ and $K^+ \to e^+\nu\gamma$ decays, with the observed pion second axial-vector form factor implying a pion charge radius $r_\pi \sim 0.6$ fm, also found in Sec. 2 from data [7] and from the theoretical LσM. In Sec. 6 we return to the LσM and its link with vector-meson

CP660, Hadron Physics: Effective Theories of Low Energy QCD, edited by A. H. Blin et al.

dominance (VMD). Finally, in Sec. 7 we begin by studying the $\Delta I = 1/2$ rule for two-pion decays of the kaon in connection with the σ as the pion's chiral partner, and end by showing that the mass of the now experimentally confirmed scalar κ meson is consistent with the observed $K \to 2\pi$ decay rate. We summarize our results and draw our conclusions in Sec. 8.

2. MESON VECTOR FORM FACTORS AND CHARGE RADII

The charged-pion and kaon e.m. vector currents are defined as

$$
\begin{aligned}
\langle \pi^+(q')|V_{em}^\mu(0)|\pi^+(q)\rangle &= F_\pi(k^2)\,(q'+q)^\mu , \\
\langle K^+(q')|V_{em}^\mu(0)|K^+(q)\rangle &= F_K(k^2)\,(q'+q)^\mu ,
\end{aligned} \tag{3}
$$

with $k_\mu = q'_\mu - q_\mu$. The former pion form factor $F_\pi(k^2)$ can be — perturbatively — characterized by the (constituent) quark udu and dud loop graphs of Fig. 1a, while the charged-kaon form factor $F_K(k^2)$ is in a similar manner determined by the usu and sus

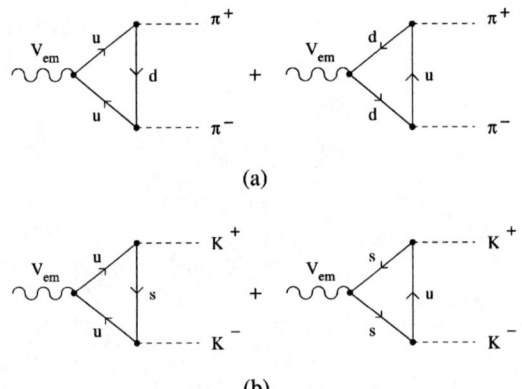

(a)

(b)

FIGURE 1. VPP quark triangle graphs.

loop graphs depicted in Fig. 1b. Even if each of the diagrams in Fig. 1 appears to be linearly divergent by naive power counting, gauge invariance enforces every single quark triangle (QT) to be merely logarithmically divergent. After evaluation of spin traces, the form factors in Eq. (3) can be brought to the form (with color number $N_c = 3$)

$$
F_\pi(k^2)_{\mathrm{QT}} = -4ig^2 N_c \left(+\frac{2}{3}I(k^2, m_u^2, m_d^2, m_\pi^2) + \frac{1}{3}I(k^2, m_d^2, m_u^2, m_\pi^2) \right) , \tag{4}
$$

$$
F_K(k^2)_{\mathrm{QT}} = -4ig^2 N_c \left(+\frac{2}{3}I(k^2, m_u^2, m_s^2, m_K^2) + \frac{1}{3}I(k^2, m_s^2, m_u^2, m_K^2) \right) . \tag{5}
$$

The integral $I(k^2, m_q^2, m_Q^2, M^2)$ is defined in Ref. [5].

The perturbative QT expressions (4) and (5) in the CL (i.e. $M \to 0$) should be compared to a CL non-perturbative LσM result [3, 8]

$$F_\pi(k^2)^{\text{CL}}_{\text{L}\sigma\text{M}} = -4ig^2N_c \int_0^1 dx \int d^{-4}p \left[p^2 - \hat{m}^2 + x(1-x)k^2\right]^{-2}, \quad (6)$$

$$F_K(k^2)^{\text{CL}}_{\text{L}\sigma\text{M}} = -4ig^2N_c \int_0^1 dx \int d^{-4}p \left[p^2 - m^2_{us} + x(1-x)k^2\right]^{-2}, \quad (7)$$

where $m_{us} = (m_s + \hat{m})/2$ and where $d^{-4}p = d^4p\,(2\pi)^{-4}$. The logarithmic divergence of these expressions has been guaranteed through a rerouting procedure [8, 9]. When $k^2 = 0$, these form factors become automatically normalized to unity, i.e.,

$$F_\pi(0)^{\text{CL}}_{\text{L}\sigma\text{M}} = -4ig^2N_c \int d^{-4}p \left[p^2 - \hat{m}^2\right]^{-2} = 1, \quad (8)$$

$$F_K(0)^{\text{CL}}_{\text{L}\sigma\text{M}} = -4ig^2N_c \int d^{-4}p \left[p^2 - m^2_{us}\right]^{-2} = 1, \quad (9)$$

due to the GTRs in Eq. (2), and the definition of the pion and kaon decay constants $\left\langle 0|A^\mu_3|\pi^0\right\rangle = if_\pi q^\mu$, $\left\langle 0|A^\mu_{4-i5}|K^+\right\rangle = i\sqrt{2}f_K q^\mu$, with $f_\pi \approx 93$ MeV and $f_K/f_\pi \approx 1.22$ [3, 9].

In contrast, the perturbative QT results yield in the CL

$$F_{\pi^+}(0)^{\text{CL}}_{\text{QT}} = -4ig^2N_c \int d^{-4}p \left[p^2 - \hat{m}^2\right]^{-2}, \quad (10)$$

$$F_{K^+}(0)^{\text{CL}}_{\text{QT}} = -4ig^2N_c \left\{ \int d^{-4}p \left[\left(p^2 - \hat{m}^2\right)\left(p^2 - m^2_s\right)\right]^{-1} \right.$$

$$\left. - \frac{i\pi^2}{(2\pi)^4}\frac{1}{2(m_s + \hat{m})^2}\left(m^2_s + \hat{m}^2 - \frac{2m^2_s\hat{m}^2}{m^2_s - \hat{m}^2}\ln\left|\frac{m^2_s}{\hat{m}^2}\right|\right)\right\}, \quad (11)$$

being — up to an finite constant correction term in the case of the kaon — normalized by the logarithmically divergent gap equations (LDGEs) [10]

$$1 = -4ig^2N_c \int d^{-4}p \left[p^2 - \hat{m}^2\right]^{-2}, \quad (12)$$

$$1 = -4ig^2N_c \int d^{-4}p \left[\left(p^2 - \hat{m}^2\right)\left(p^2 - m^2_s\right)\right]^{-1}. \quad (13)$$

To proceed, given Eqs. (6) and (7), the meson charge radii are computed in the LσM as

$$\langle r^2_{\pi^+}\rangle^{\text{CL}}_{\text{L}\sigma\text{M}} = 6\left.\frac{dF_\pi(k^2)}{dk^2}\right|_{k^2=0} = \frac{8iN_c}{(2\pi)^4}g^2\left(\frac{-i\pi^2}{2\hat{m}^2}\right) = \frac{N_c}{4\pi^2 f^2_\pi} \approx (0.61\,\text{fm})^2 \quad (14)$$

and (the obvious $SU(3)$ extension)

$$\langle r^2_{K^+}\rangle^{\text{CL}}_{\text{L}\sigma\text{M}} = 6\left.\frac{dF_K(k^2)}{dk^2}\right|_{k^2=0} = \frac{8iN_c}{(2\pi)^4}g^2\left(\frac{-i\pi^2}{2m^2_{us}}\right) = \frac{N_c}{4\pi^2 f^2_K} \approx (0.49\,\text{fm})^2. \quad (15)$$

313

Here we have evaluated the charge radii in the CL [3, 6, 11], with $f_\pi^{CL} \approx 90$ MeV, $f_K^{CL} \approx 110$ MeV.

At this point we may return to the perturbative QT results (4) and (5), from which we derive in the CL

$$\langle r_{\pi+}^2 \rangle_{QT}^{CL} = \frac{g^2 N_c}{4\pi^2 \hat{m}^2} \overset{!}{=} \langle r_{\pi+}^2 \rangle_{L\sigma M}^{CL} , \tag{16}$$

$$\langle r_{K+}^2 \rangle_{QT}^{CL} = \frac{g^2 N_c}{4\pi^2 \hat{m}^2} \left(1 - \frac{5}{6}\delta + \frac{3}{5}\delta^2 - \frac{4}{9}\delta^3 + \frac{22}{63}\delta^4 - \frac{2}{7}\delta^5 + \ldots \right) , \tag{17}$$

with $m_s = (1+\delta)\hat{m}$, i.e., $\delta = (m_s/\hat{m}) - 1 \approx 0.44$. The coefficients of the presented Taylor expansion in the SU(3)-breaking parameter δ coincide with the ones given in Ref. [12], while our full result is also in agreement with the expressions originally derived by Tarrach [13]. Taking into account the first three terms of this expansion, we may estimate the ratio r_K/r_π to be

$$\frac{\langle r_{K+}^2 \rangle}{\langle r_{\pi+}^2 \rangle} \approx 1 - \frac{5}{6}\delta + \frac{3}{5}\delta^2 \approx 0.750 \quad \text{or} \quad \frac{\langle r_{K+} \rangle}{\langle r_{\pi+} \rangle} \approx 0.866 . \tag{18}$$

Here we note that the observed pion charge radius is [7]

$$r_\pi = (0.642 \pm 0.002) \text{ fm} , \tag{19}$$

and the analogue charged-kaon charge radius is [14]

$$r_K = (0.560 \pm 0.031) \text{ fm} . \tag{20}$$

If we take the experimental value $r_{\pi+} \approx 0.64$ fm from Eq. (19), the latter ratio (18) implies $< r_{K+} > \approx 0.556$ fm, which is compatible with Eq. (20).

In summary, the more detailed perturbative results of Eqs. (4), (5), (10), (11), (16), and (17) are compatible with the simpler non-perturbative (SU(3)-symmetry) scheme of Eqs. (6)–(9), (14), and (15) above. Thus, no further renormalization needs be considered in either case. Note, too, that these detailed or simple field-theory versions of the charged-pion form factor can be recovered in an even simpler fashion by using a once-subtracted dispersion relation for the pion charge radius, yielding in the CL

$$r_\pi^2 = \frac{6}{\pi} \int_0^\infty \frac{dq^2 \, \Im F_\pi(q^2)}{(q^2)^2} = \frac{N_c}{4\pi^2 (f_\pi^{CL})^2} = \frac{1}{\hat{m}^2} , \tag{21}$$

where we use [3] the GTRs Eq. (2), along with $g = 2\pi/\sqrt{N_c}$ from Eq. (1). This suggests that the tightly bound "fused" $\bar{q}q$ pion charge radius in the CL is

$$r_\pi^{CL} = \frac{1}{\hat{m}} = \frac{197.3 \text{ MeV fm}}{325 \text{ MeV}} \approx 0.61 \text{ fm} , \tag{22}$$

with $\hat{m}_{CL} \approx 325$ MeV $\sim M_N/3$, as expected from the GTR $m_{CL} = f_\pi^{CL} g \approx 90$ MeV $\times 3.628 \approx 325$ MeV.

314

3. CHARGED-PION POLARIZABILITIES FOR $\gamma\gamma \to \pi\pi$

Analyzing Crystal-Ball [15], MARK-II [16] and CELLO [17] $\gamma\gamma \to \pi\pi$ data, Kaloshin *et al.* [18, 19, 20] obtain for the charged electric pion polarizability

$$\alpha_{\pi^+} = (2.75 \pm 0.50) \times 10^{-4}\,\text{fm}^3 \,. \tag{23}$$

The LσM prediction follows from the model-independent value [21], given by

$$\alpha_{\pi^+} = \frac{\alpha}{8\pi^2 m_\pi f_\pi^2}\, \gamma \,, \tag{24}$$

where $\gamma \equiv F_A(0)/F_V(0)$. In Sec. 5 we find $\gamma = 2/3$. Hence, the LσM result

$$\alpha_{\pi^+}^{\text{L}\sigma\text{M}} = \frac{\alpha}{12\pi^2 m_\pi f_\pi^2} \approx 3.9 \times 10^{-4}\,\text{fm}^3 \tag{25}$$

is reasonably near the data shown in Eq. (23) above. It is, moreover, in good agreement with e.g. the prediction $3.6 \times 10^{-4}\,\text{fm}^3$ of a quark confinement model that also yields good results for heavy-meson semileptonic form factors [22].

Another consistency check is the detailed quark-plus-meson-loop analysis of Ref. [23]:

$$\alpha_{\pi^+}^{\text{L}\sigma\text{M}} = \frac{\alpha}{8\pi^2 m_\pi f_\pi^2} - \frac{\alpha}{24\pi^2 m_\pi f_\pi^2} = \frac{\alpha}{12\pi^2 m_\pi f_\pi^2} \,, \tag{26}$$

requiring $\gamma^{\text{L}\sigma\text{M}} = 2/3$ from Eq. (24).

Finally we comment on low-energy $\gamma\gamma \to 2\pi^0$ scattering, where there is no pole term, and the neutral polarizabilities α_{π^0}, β_{π^0} are much smaller than α_{π^+}, β_{π^+}. In Ref. [24] it was shown that a $\gamma\gamma \to 2\pi^0$ cross section of ~ 10 nb (generated by a $\sigma(700)$ meson pole) reasonably anticipated the later 1990 Crystal-Ball data [15] in the 0.3–0.7 GeV range.

4. SEMILEPTONIC WEAK $K_{\ell 3}$ DECAY AND FORM-FACTOR SCALE $f_+(0)$

The semileptonic weak $K^+ \to \pi^0 e^+ \nu$ ($K_{\ell 3}$) decay width is measured as [14]

$$\Gamma(K^+ \to \pi^0 e^+ \nu) = \frac{\hbar}{\tau_{K^+}}(4.87 \pm 0.06)\% = (25.88 \pm 0.32) \times 10^{-16}\,\text{MeV} \,. \tag{27}$$

Taking a q^2 form-factor dependence $f_+(q^2) = f_+(0)[1 + \lambda_+ q^2/m_\pi^2]$, the standard $V-A$ (vector here) weak current predicts a $K_{\ell 3}$ decay width ($y = m_{\pi^0}^2/m_{K^+}^2$, $m_e = m_\nu = 0$; see also Ref. [25])

$$
\begin{aligned}
\Gamma(K^+ \to \pi^0 e^+ \nu) &= \\
&= \frac{G_F^2 |V_{us}|^2 m_{K^+}^5}{2\pi^3\,768}\, f_+^2(0)\left(0.5792 + 0.1600\,\frac{m_{K^+}^2}{m_{\pi^0}^2}\lambda_+ + 0.01770\,\frac{m_{K^+}^4}{m_{\pi^0}^4}\lambda_+^2\right) \\
&= f_+^2(0)\,(25.90 \pm 0.07) \times 10^{-16}\,\text{MeV} \,,
\end{aligned}
\tag{28}
$$

where $G_F = 11.6639 \times 10^{-6} \text{ GeV}^{-2}$, $V_{us} = 0.2196 \pm 0.0026$, and $\lambda_+ = 0.0278 \pm 0.0019$ [14]. If we neglect here the term quadratic in λ_+, as e.g. done in Ref. [25], the leading factor in Eq. (28) becomes 25.80 instead of 25.90. Moreover, accounting for a nonvanishing electron mass yields a totally negligible correction of the order of 0.001%. In any case, comparison with the data in Eq. (27) clearly shows that the form-factor scale $f_+(0)$ must be near unity. However, electroweak radiative corrections to $\Gamma(K^+ \rightarrow \pi^0 e^+ \nu)$ are *not* negligible on the scale of the experimental errors in V_{us} and λ_+, giving rise to an enhancement of $|V_{us}|$ by more than 2% [26], suggesting that $f_+(0)$ should be a trifle less than unity.

As a matter of fact, the nonrenormalization theorem [27] *requires* the form factor $f_+(q^2)$ to be close to unity when $q^2 = 0$. Furthermore, in the infinite-momentum frame (IMF), tadpole graphs are suppressed and so [28]

$$1 - f_+^2(0) = \mathcal{O}(\delta^2) \approx 6\% \tag{29}$$

is second order in $SU(3)$-symmetry breaking. Of similar order are, for example, $(m_\pi/m_K)^2 = 7.7\%$, and $(1 - f_K/f_\pi)^2 = 5\%$, for $f_K/f_\pi = 1.22$.

Next we follow the (constituent) quark-model triangle graph of Fig. 2, with

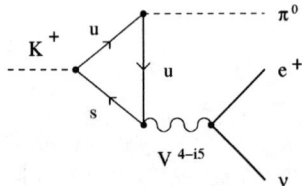

FIGURE 2. Quark-loop contribution to $K^+ \rightarrow \pi^0 e^+ \nu$.

$$\sqrt{2} \left\langle \pi^0 | V_\mu^{4-i5} | K^+ \right\rangle = f_+(t)(p_K + p_\pi)_\mu + f_-(t)(p_K - p_\pi)_\mu . \tag{30}$$

Note that, for this process, the f_- form factor can be disposed of, since it is weighted by $m_e \ll m_K$ [25], giving rise to a m_e^2/m_K^2 suppression of the corresponding contributions to $\Gamma(K^+ \rightarrow \pi^0 e^+ \nu)$. To test $SU(2)$-symmetry breaking in $K_{\ell 3}$ decays as in Eqs. (28) and (30) above, we note the present data consistency [14] of $\lambda_+(K_{e3}^+) = 0.0278 \pm 0.0019$, $\lambda_+(K_{e3}^0) = 0.0291 \pm 0.0018$, $\lambda_+(K_{\mu 3}^+) = 0.033 \pm 0.010$, and $\lambda_+(K_{\mu 3}^0) = 0.033 \pm 0.005$. Then, expanding in the $SU(3)$-breaking parameter $\delta = (m_s/\hat{m}) - 1$ (as already used to obtain Eq. (18)) and working in the soft-pion CL, the Feynman graph of Fig. 2 predicts [29] (recall the value of the meson-quark coupling $g \approx 3.628$ in Eq. (1))

$$f_+(0) = 1 - \frac{g^2 \delta^2}{8\pi^2} \approx 0.968 . \tag{31}$$

This value slightly below unity is not only in agreement with the nonrenormalization theorem Eq. (29), as $1 - f_+^2(0) = 1 - (0.968)^2 = 6.3\%$, but also quantitatively compatible with Eqs. (27) and (28), if we account for the mentioned radiative corrections contributing with about -2% to $f_+(0)$, and the experimental errors in V_{us} and λ_+.

5. SEMILEPTONIC WEAK RADIATIVE FORM FACTORS FOR $\pi^+ \to E^+ \nu\gamma$ AND $K^+ \to E^+ \nu\gamma$

From Ref. [14], the $\pi^+ \to e^+ \nu\gamma$ and $K^+ \to e^+ \nu\gamma$ matrix elements are

$$M_V = \frac{-eG_F V_{qq'}}{\sqrt{2}\,m_P} \varepsilon^\mu \ell^\nu F_V^P \varepsilon_{\mu\nu\sigma\tau} k^\sigma q^\tau , \tag{32}$$

$$M_A = \frac{-ieG_F V_{qq'}}{\sqrt{2}\,m_P} \varepsilon^\mu \ell^\nu \{F_A^P [(s-t)g_{\mu\nu} - q_\mu k_\nu] + R^P t\, g_{\mu\nu}\} , \tag{33}$$

where $V_{qq'}$ is the corresponding Cabibbo-Kobayashi-Maskawa (CKM) mixing-matrix element, ε^μ is the photon polarization vector, ℓ^ν is the lepton-neutrino current, q and k are the meson and photon four-momenta, respectively, with $s = q \cdot k$, $t = k^2$, and P stands for π or K. The weak vector (pion) form factor F_V^π in Eq. (32) and the second axial vector form factor R^π in Eq. (33) are model independent [30], with F_V^π determined only by conserved vector currents (CVC), and R^π related via the pion charge radius ($r_\pi = 0.642 \pm 0.002$ fm) to partially conserved (pion) axial currents (PCAC). Specifically, F_V^π was long ago determined by CVC [30], viz.

$$F_V^\pi(0) = \frac{\sqrt{2}\,m_{\pi^+}}{8\pi^2 f_\pi} \approx 0.027 , \tag{34}$$

reasonably close to data [14] 0.017 ± 0.008. Furthermore, PCAC predicts (PCAC is manifest in the LσM [31, 32])

$$R^\pi = \frac{1}{3} m_{\pi^+} f_{\pi^+} r_{\pi^+}^2 = 0.064 \pm 0.001 , \tag{35}$$

where $f_{\pi^+} = 130.7 \pm 0.1$ MeV [14] and we use $r_{\pi^+} = 0.642 \pm 0.002$ fm. Then Eq. (35) is near data [33] $R^\pi = 0.059^{+0.009}_{-0.008}$.

To apply the LσM theory, we consider the quark-plus-meson-loop graphs of Fig. 3. Then the ratio $\gamma = F_A(0)/F_V(0)$ is predicted as [34]

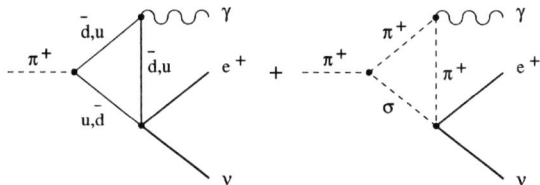

FIGURE 3. Quark- and meson-loop contribution to $\pi^+ \to \gamma e^+ \nu$.

$$\gamma^{\mathrm{L}\sigma\mathrm{M}} = 1 - \frac{1}{3} = \frac{2}{3} , \tag{36}$$

with

$$F_A^\pi(0) = \sqrt{2}\,m_\pi[(8\pi^2 f_\pi)^{-1} - (24\pi^2 f_\pi)^{-1}] = \sqrt{2}\,\frac{m_\pi}{12\pi^2 f_\pi} \approx 0.0179 . \tag{37}$$

Thus, the form-factor ratio of Eq. (37) divided by Eq. (34) gives $\gamma^{L\sigma M} = 0.0179/0.027 \approx 0.66$, compatible with Eq. (36) and with data [14]:

$$\gamma_{\text{data}} = \frac{0.0116 \pm 0.0016}{0.017 \pm 0.008} = 0.68 \pm 0.33 . \tag{38}$$

With hindsight, this ratio $\gamma^{L\sigma M} = 2/3$ is near the original current-algebra (CA) estimate 0.6 found in Ref. [35], and exactly the same γ found in Eq. (24) from the LσM Eq. (26).

Extending the above LσM picture to $SU(3)$ symmetry, we first assume a scalar nonet pattern below 1 GeV (e.g. $f_0(600)$, $\kappa(800)$, $f_0(980)$, $a_0(980)$) as found from a kinematic IMF scheme [36], or from a dynamical coupled-channel unitarized model [37]. Then the $K^+ \to e^+ \nu \gamma$ quark-plus-meson LσM form-factor loop amplitudes predict [38] at $k^2 = 0$

$$|F_V^K(0) + F_A^K(0)|_{L\sigma M} \approx 0.109 + 0.044 = 0.153 , \tag{39}$$

close to the $K^+ \to e^+ \nu \gamma$ data [14]

$$|F_V^K(0) + F_A^K(0)|_{\text{data}} = 0.148 \pm 0.010 . \tag{40}$$

An $SU(3)$ LσM theory is reasonably detailed [10] due to resonances below 1 GeV, but the LσM kaon form-factor sum in Eq. (39) is easily tested via the data in Eq. (40). The same is true for the pion form-factor values in Eqs. (34–38), partly based on the measured pion charge radius [7] $r_\pi = 0.642 \pm 0.002$ fm.

6. VMD: LσM VIA VPP AND VPV OR PVV LOOPS

We first confirm the (crucial) value of the pion charge radius [7] $r_\pi = 0.642 \pm 0.002$ fm via Sakurai's vector-meson-dominance (VMD) prediction [39]

$$r_\pi = \frac{\sqrt{6}}{m_\rho} \approx 0.63 \,\text{fm} . \tag{41}$$

Recall that the tightly bound $\bar{q}q$ chiral pion in Eq. (21), with constituent quark mass $\hat{m} \approx 325$ MeV (near $\hat{m} \approx M_N/3$), has CL charge radius $r_\pi^{CL} = 1/\hat{m} \approx 0.61$ fm. So the close agreement between Eqs. (41) and (21) means we must take the VMD scheme along with the LσM as the basis of our chiral theory.

The ρ^0 form factor predicts, from $udu + dud$ quarks loops in the CL (see Fig. 4),

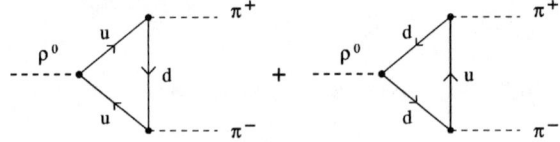

FIGURE 4. Vector-mesonic VPP quark triangle graphs.

$$g_{\rho\pi\pi} = -i4N_c g^2 g_\rho \int d^4 p \, (p^2 - \hat{m}^2)^{-2} = g_\rho , \tag{42}$$

by virtue of the LDGE Eq. (12) [9]. Then, folding in the mesonic π-σ-π loop changes the VMD prediction (42) only slightly to [11]

$$g_{\rho\pi\pi} = g_\rho + \frac{1}{6} g_{\rho\pi\pi} = \frac{6}{5} g_\rho \,, \tag{43}$$

compatible with the observed couplings $g_{\rho\pi\pi} \approx 6.04$ and $g_\rho \approx 5.01$, since (for $p_{CM} = 358$ MeV)

$$\Gamma_{\rho\pi\pi} = \frac{p_{CM}^3 g_{\rho\pi\pi}^2}{6\pi m_\rho^2} = 149.2 \pm 0.7 \,\text{MeV} \implies g_{\rho\pi\pi} \approx 6.04 \tag{44}$$

$$\Gamma_{\rho ee} = \frac{e^4 m_\rho}{12\pi g_\rho^2} = 6.85 \pm 0.11 \,\text{keV} \implies g_\rho \approx 5.01 \,, \tag{45}$$

with $e \approx 0.3028$ (i.e., $\alpha \approx 1/137$). Also, the quark-loop VPV or PVV (see Fig. 5) amplitudes are [40], using $\Gamma_{VPV} = p^3 |F_{VPV}|^2/12\pi$,

$$|F(\rho \to \pi\gamma)| = \frac{eg_\rho}{8\pi^2 f_\pi} \approx 0.207 \,\text{GeV}^{-1} \,, \quad |F(\omega \to \pi\gamma)| = \frac{eg_\omega}{8\pi^2 f_\pi} \approx 0.704 \,\text{GeV}^{-1} \,,$$

$$|F(\pi^0 \to 2\gamma)| = \frac{\alpha}{\pi f_\pi} = \frac{e^2}{4\pi^2 f_\pi} \approx 0.025 \,\text{GeV}^{-1} \,, \tag{46}$$

for $g_\rho \approx 5.01$ and $g_\omega \approx 17.06$, very close to the data 0.222 ± 0.012 GeV^{-1} [14], 0.698 ± 0.014 GeV^{-1} [41], 0.0252 ± 0.0009 GeV^{-1} [14], respectively. Equivalently, VMD predicts at tree level $|F_{\rho\pi\gamma}| e/g_\rho = |F_{\omega\pi\gamma}| e/g_\omega = |F_{\pi^0\gamma\gamma}|/2$, then compatible with the LσM quark loops in Eq. (46).

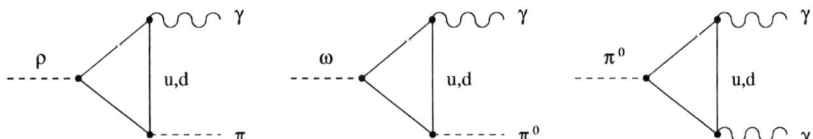

FIGURE 5. PVV quark triangle graphs for $\rho \to \pi\gamma$, $\omega \to \gamma\pi^0$, and $\pi^0 \to \gamma\gamma$.

7. NONLEPTONIC WEAK $K_{2\pi}$ $\Delta I = 1/2$ RULE AND σ, κ MESONS

The well-known [14] $\Delta I = 1/2$ rule $\Gamma(K_S \to \pi^+\pi^-)/\Gamma(K^+ \to \pi^+\pi^0) \approx 450$ for nonleptonic weak $K_{2\pi}$ decays suggests [42] that the parity-violating (PV) amplitude $\langle 2\pi | H_w^{pv} | K_S \rangle$ could be dominated by the $\Delta I = 1/2$ weak transition $\langle \sigma | H_w^{pv} | K_S \rangle$. The σ-pole graph of Fig. 6, with LσM coupling $\langle 2\pi | \sigma \rangle = m_\sigma^2/2f_\pi$ for m_σ near m_K and $\Gamma_\sigma \sim m_\sigma$, predicts [43]

$$|\langle 2\pi | H_w^{pv} | K_S \rangle| = \left| \frac{2\langle 2\pi | \sigma \rangle \langle \sigma | H_w^{pv} | K_S \rangle}{m_K^2 - m_\sigma^2 + im_\sigma \Gamma_\sigma} \right| \approx \frac{1}{f_\pi} |\langle \sigma | H_w^{pv} | K_S \rangle| \,. \tag{47}$$

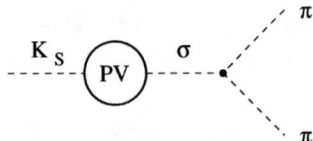

FIGURE 6. Parity-violating two-pion decay of K_S dominated by σ pole.

But pion PCAC (manifest in the LσM) requires, using the weak chiral commutator $[Q_5 + Q, H_w] = 0$,

$$|\langle 2\pi | H_w^{pv} | K_S \rangle| \rightarrow \frac{1}{f_\pi} |\langle \pi | [Q_5^\pi, H_w] | K_S \rangle| \approx \frac{1}{f_\pi} |\langle \pi^0 | H_w^{pc} | K_L \rangle| \, , \qquad (48)$$

with both pions being consistently reduced in Ref. [44]. To reconfirm Eq. (48), one considers the $\Delta I = 1/2$ weak tadpole graph, giving

$$|\langle 2\pi | H_w^{pv} | K_S \rangle| = \frac{|\langle 0 | H_w | K_S \rangle \langle K_S 2\pi | K_S \rangle|}{m_K^2} \, , \qquad (49)$$

and then one invokes the Weinberg-Osborn [45] strong chiral coupling $|\langle K_S 2\pi | K_S \rangle| = m_{K_S}^2 / 2 f_\pi^2$, together with the usual PCAC relation $|\langle 0 | H_w^{pv} | K_S \rangle| = |2 f_\pi \langle \pi^0 | H_w^{pv} | K_L \rangle|$, to recover Eq. (48) [46].

In either case, equating Eq. (48) to Eq. (47) leads to

$$|\langle \sigma | H_w^{pv} | K_S \rangle| \approx |\langle \pi^0 | H_w^{pc} | K_L \rangle| \, , \qquad (50)$$

suggesting that the π and σ mesons are "chiral partners", at least for nonleptonic weak interactions. But of course, Secs. 1–6 above also show that the π and the σ are chiral partners for strong, e.m., and semileptonic weak interactions, as well. To compare this chiral-partner $K \rightarrow \pi$ transition with $K_{2\pi}$ data, we return to the PCAC equation (48) to write, for $f_\pi \approx 93$ MeV,

$$|\langle 2\pi | H_w^{pv} | K_S \rangle| \approx \frac{1}{f_\pi} |\langle \pi^0 | H_w^{pc} | K_L \rangle| \approx 38 \times 10^{-8} \, \text{GeV} \, , \qquad (51)$$

midway between the observed $K_S \rightarrow \pi^+ \pi^-$ and $K_S \rightarrow \pi^0 \pi^0$ amplitudes

$$\left| M_{K_S \rightarrow \pi\pi}^{+-} \right|_{\text{PDG}} = m_{K_S} \left[\frac{8\pi \Gamma_{+-}^{K_S}}{q} \right]^{\frac{1}{2}} = (39.1 \pm 0.1) \times 10^{-8} \, \text{GeV} \, , \qquad (52)$$

$$\left| M_{K_S \rightarrow \pi\pi}^{00} \right|_{\text{PDG}} = m_{K_S} \left[\frac{16\pi \Gamma_{00}^{K_S}}{q} \right]^{\frac{1}{2}} = (37.1 \pm 0.2) \times 10^{-8} \, \text{GeV} \, , \qquad (53)$$

320

suggesting $|\langle \pi^0 | H_w^{pc} | K_L \rangle| \approx 3.58 \times 10^{-8}$ GeV^2. In fact, when one statistically averages *eleven* first-order weak data sets for $K_S \rightarrow 2\pi$, $K \rightarrow 3\pi$, $K_L \rightarrow 2\gamma$, $K_L \rightarrow \mu^+\mu^-$, $K^+ \rightarrow \pi^+ e^+ e^-$, $K^+ \rightarrow \pi^+\mu^+\mu^-$, and $\Omega^- \rightarrow \Xi^0\pi^-$, one finds [47]

$$|\langle \pi^0 | H_w^{pc} | K_L \rangle| = |\langle \pi^+ | H_w^{pc} | K^+ \rangle| = (3.59 \pm 0.05) \times 10^{-8} \, \text{GeV}^2 . \quad (54)$$

To induce theoretically at the quark level the $\Delta I = 1/2$ $s \rightarrow d$ single-quark-line (SQL) transition scale β_w in a model-independent manner, one considers the second-order weak (see Fig. 7) $K_L - K_S$ mass difference Δm_{LS} diagonalized to [48]

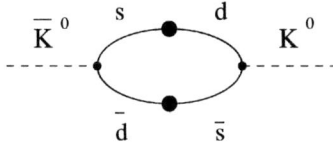

FIGURE 7. $\bar{K}^0 \leftrightarrow K^0$ SQL graph. Each dot represents the SQL weak scale β_w.

$$2\beta_w^2 = \frac{\Delta m_{LS}}{m_K} = (0.70126 \pm 0.00121) \times 10^{-14} \implies |\beta_w| \approx (5.9214 \pm 0.0051) \times 10^{-8} .$$
$$(55)$$

Then using Eq. (54), one predicts from the soft-meson theorem, or from Cronin's chiral Lagrangian [49],

$$|\langle \pi^0 | H_w^{pc} | K_L \rangle| = 2\beta_w m_{K_L}^2 \frac{f_K}{f_\pi} = (3.5785 \pm 0.0031) \times 10^{-8} \, \text{GeV}^2 , \quad (56)$$

given $f_K / f_\pi \approx 1.22$. This SQL scale β_w in Eq. (55) and the $K \rightarrow \pi$ weak amplitude in Eq. (56) (or in Eq. (54)), correspond to a "truly weak" interaction, which Weinberg [50] shows cannot be transformed away in the electroweak standard model.

To test the latter weak scale (56) (or the similar data averages (54), we first re-express the neutral chiral-partner relation (extended to the κ transition [43]) as

$$\langle \pi^0 | H_w^{pc} | K^0 \rangle = \langle \sigma | H_w^{pv} | K^0 \rangle = \langle \pi^0 | H_w^{pv} | \kappa^0 \rangle = \frac{3.58 \times 10^{-8} \, \text{GeV}^2}{\sqrt{2}} = 2.53 \times 10^{-8} \, \text{GeV}^2 .$$
$$(57)$$

We fix this $\kappa^0 \rightarrow \pi^0$ weak transition (57) to the weak PV K^0 tadpole graph of Fig. 8, via the $K^0 \rightarrow$ vacuum PCAC scale, as

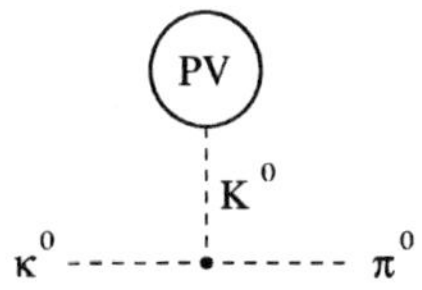

FIGURE 8. Parity-violating weak K^0 tadpole graph for $\kappa^0 \rightarrow \pi^0$ transition.

$$|\langle 0 | H_w^{pv} | K^0 \rangle| = \frac{2f_\pi^2}{1 - m_\pi^2/m_K^2} |\langle 2\pi^0 | H_w^{pv} | K^0 \rangle| = 0.51 \times 10^{-8} \, \text{GeV}^3 , \quad (58)$$

using $|\langle 2\pi^0|H_w^{pv}|K^0\rangle| = 26.26 \times 10^{-8}$ GeV from data, while eliminating the 4% $\Delta I = 3/2$ component (see Ref. [50], third paper). Then Fig. 8 predicts the amplitude magnitude

$$|\langle \pi^0|H_w^{pv}|\kappa^0\rangle| = \frac{|\langle 0|H_w^{pv}|K^0\rangle|}{m_{K^0}^2} g_{\kappa^0 K^0 \pi^0} \approx 2.53 \times 10^{-8}\,\text{GeV}^2\,, \qquad (59)$$

scaled to Eq. (57) above, *provided* one uses the LσM coupling, for $f_\pi = 92.4$ MeV [14],

$$|g_{\kappa^0 K^0 \pi^0}| = \frac{m_\kappa^2 - m_K^2}{4f_\pi} = 1.229\,\text{GeV}\,, \qquad (60)$$

corresponding to a κ mass of 838 MeV. This value is not too distant from our earlier $m_\kappa = 730$–800 MeV predictions [36, 37], and the very recent E791 observed mass $m_\kappa \approx 800$ MeV [51]. Moreover, the $SU(3)$ analogue to Eq. (60), i.e., $|g_{\sigma\pi\pi}| = (m_\sigma^2 - m_\pi^2)/2f_\pi$ suggests $m_\sigma = 687$ MeV, reasonably near the predicted CL-LσM value [3, 52] $m_\sigma = 650$ MeV.

8. SUMMARY AND CONCLUSIONS

In Sec. 1 we reviewed the solution of the LσM at the quark-loop level. In Sec. 2 we used $SU(2)$, $SU(3)$ Goldberger–Treiman quark relations to normalize the π and K form factors to unity at $k^2 = 0$, after which we differentiated these form factors to predict the LσM charge radii, both being compatible with data. Next in Sec. 3 we briefly reviewed e.m. charged-pion polarizabilities for $\gamma\gamma \to \pi\pi$, and compared them with LσM predictions. In Sec. 4 we used quark loops to match the observed form factor $f_+(0)$. In Sec. 5 we showed that the LσM form factors F_V^π, R^π, $F_V^K + F_A^K$, and the ratio F_A^π/F_V^π are all in agreement with the measured values. Then in Sec. 6 we compared tree-level VMD with LσM VPP and PVV quark loops. Both theories agree well with data. Finally, in Sec. 7 we successfully extended this LσM picture to nonleptonic weak decays, in particular to the $\Delta I = 1/2$-dominated $K_{2\pi}$ decays and inferred $\sigma(687)$ and $\kappa(838)$ masses.

Next, we discuss low-energy QCD. While an exact match via the LσM is not possible, QCD at the 1-GeV scale generates a dynamical quark mass [53] $m_{\text{dyn}} = [4\pi\alpha_s \langle -\bar{\psi}\psi\rangle_{1\,\text{GeV}}/3]^{1/3} \approx 320$ MeV, near the LσM quark mass $\hat{m} = 2\pi f_\pi/\sqrt{3} \approx 325$ MeV in the CL. Such approximate agreement also holds for the quark condensate as well. Moreover, the frozen coupling strength in QCD at infrared freeze-out [54] is $\alpha_s = \pi/4$, with $\alpha_s^{\text{eff}} = (4/3)\alpha_s = \pi/3$. This exactly matches the LσM strength $\alpha_{\text{LσM}} = g^2/4\pi = \pi/3$, with $g = 2\pi/\sqrt{3}$. Also, QCD with $\alpha_s(m_\sigma) = \pi/4$ leads to [55] $m_\sigma^2/m_{\text{dyn}}^2 = \pi/\alpha_s(m_\sigma) \approx 4$, simulating the NJL–LσM value $m_\sigma^2/\hat{m}^2 = 4$ in the CL. Lastly, the chiral restoration temperature T_c computed in $N_f = 2$ lattice simulations gives [56] $T_c = 173 \pm 8$ MeV, close to the LσM value [57] $T_c = 2f_\pi \approx 180$ MeV in the CL.

To conclude, we mention a very recent large-N_c renormalization-group-flow analysis of the quark-level LσM [58], using the Schwinger proper-time regularization, which finds (for $f_\pi = 93$ MeV) $\lambda = 23.6$, $g = 3.44$, $m_q = 320$ MeV, and $m_\sigma = 650$ MeV, strikingly close to our above theoretical values $\lambda = 8\pi^2/3 = 26.3$, $g = 2\pi/\sqrt{3} \approx 3.628$,

$m_q = 325$ MeV, and $m_\sigma = 650$ MeV, respectively. Therefore, our present results, as well as our recent findings in Ref. [59], appear to confirm the assumption of the authors of Ref. [58]: *"We assume the linear σ model to be a valid description of Nature below scales of 1.5 GeV."*

ACKNOWLEDGMENTS

The authors are indebted to A. E. Kaloshin for valuable information on pion polarizabilities. This work was partly supported by the *Fundação para a Ciência e a Tecnologia* (FCT) of the *Ministério do Ensino Superior, Ciência e Tecnologia* of Portugal, under Grant no. PRAXIS XXI/BPD/20186/99 and under contract number CERN/P/-FIS/43697/2001.

REFERENCES

1. Y. Nambu and G. Jona-Lasinio, Phys. Rev. **122**, 345 (1961).
2. A. Salam, Nuovo Cim. **25**, 224 (1962); S. Weinberg, Phys. Rev. **130**, 776 (1963); M. D. Scadron, Phys. Rev. D **57**, 5307 (1998) [hep-ph/9712425].
3. R. Delbourgo and M. D. Scadron, Mod. Phys. Lett. A **10**, 251 (1995) [hep-ph/9910242];
4. M. D. Scadron, Acta Phys. Polon. B **32**, 4093 (2001); Kyoto sigma workshop, June 2000, hep-ph/0007184.
5. M. D. Scadron, F. Kleefeld, G. Rupp and E. van Beveren, arXiv hep-ph/0211275.
6. M. D. Scadron, Rept. Prog. Phys. **44**, 213 (1981); S. A. Coon and M. D. Scadron, Phys. Rev. C **23**, 1150 (1981).
   These authors show that a once-subtracted dispersion relation requires $1 - f_\pi^{CL}/f_\pi = m_\pi^2/8\pi^2 f_\pi^2 \approx 0.03$, so the observed $f_\pi \approx 93$ MeV corresponds to $f_\pi^{CL} \approx 90$ MeV.
7. A statistical oscillator scheme of A. F. Grashin and M. V. Lepeshkin, Phys. Lett. B **146**, 11 (1984), accurately fitting both nucleon and pion form factors over a wide range $0 < q^2 < 5$ GeV^2, then refines the early E. B. Dally *et al.*, Phys. Rev. Lett. **48**, 375 (1982) pion-charge-radius data from $r_\pi = 0.663 \pm 0.023$ fm to 0.633 ± 0.008 fm, and the new PDG result $r_\pi = 0.672 \pm 0.008$ fm [14] to 0.642 ± 0.002 fm. Also note that the PDG of 2002 did not fold in the above result of Grashin and Lepeshkin.
8. N. Paver and M. D. Scadron, Nuovo Cim. A **78**, 159 (1983); also see E. Ruiz Arriola, hep-ph/0210007.
9. T. Hakioglu and M. D. Scadron, Phys. Rev. D **43**, 2439 (1991).
10. R. Delbourgo and M. D. Scadron, Int. J. Mod. Phys. A**13**, 657 (1998) [hep-ph/9807504];
11. A. Bramon, Riazuddin, and M. D. Scadron, J. Phys. G **24**, 1 (1998) [hep-ph/9709274].
12. C. Ayala and A. Bramon, Europhys. Lett. **4**, 777 (1987).
13. R. Tarrach, Z. Phys. C **2**, 221 (1979); S. B. Gerasimov, Sov. J. Nucl. Phys. **29**, 259 (1979) (Erratumibid. **32**, 156 (1980)) [Yad. Fiz. **29**, 513 (1979)]. Also see V. Bernard, B. Hiller, and W. Weise, Phys. Lett. B **205**, 16 (1988).
14. K. Hagiwara *et al.* [Particle Data Group Collaboration], Phys. Rev. D **66**, 010001 (2002).
15. H. Marsiske *et al.* [Crystal Ball Collaboration], Phys. Rev. D **41**, 3324 (1990).
16. J. Boyer *et al.* [MARK-II Collaboration], Phys. Rev. D **42**, 1350 (1990).
17. H. J. Behrend *et al.* [CELLO Collaboration], Z. Phys. C **56**, 381 (1992).
18. A. E. Kaloshin and V. V. Serebryakov, Z. Phys. C **64**, 689 (1994) [hep-ph/9306224].
19. A. E. Kaloshin, V. M. Persikov, and V. V. Serebryakov, Phys. Atom. Nucl. **57**, 2207 (1994) [Yad. Fiz. **57N12**, 2298 (1994)] [hep-ph/9402220].
20. A. E. Kaloshin, V. M. Persikov, and V. V. Serebryakov, hep-ph/9504261.
21. M. V. Terentev, Sov. J. Nucl. Phys. **16**, 87 (1973) [Yad. Fiz. **16**, 162 (1972)].

22. M. A. Ivanov and T. Mizutani, Phys. Rev. D **45**, 1580 (1992); M. A. Ivanov, P. Santorelli, and N. Tancredi, Eur. Phys. J. A **9**, 109 (2000) [hep-ph/9905209].
23. A. I. Lvov, Sov. J. Nucl. Phys. **34**, 289 (1981) [Yad. Fiz. **34**, 522 (1981)].
24. A. E. Kaloshin and V. V. Serebryakov, Z. Phys. C **32**, 279 (1986).
25. M. D. Scadron, *Advanced Quantum Theory*, Springer-Verlag (1991), see Eq. (13.48); R. E. Marshak, Riazuddin, and C. P. Ryan, *Theory of Weak Interactions in Particle Physics*, Wiley-Interscience, NY (1969), see Eq. (5.112).
26. W. J. Marciano and A. Sirlin, Phys. Rev. Lett. **71**, 3629 (1993); V. Cirigliano, M. Knecht, H. Neufeld, H. Rupertsberger, and P. Talavera, Eur. Phys. J. C **23**, 121 (2002) [hep-ph/0110153].
27. M. Ademollo and R. Gatto, Phys. Rev. Lett. **13**, 264 (1964).
28. S. Fubini and G. Furlan, Physics **4**, 229 (1965).
29. N. Paver and M. D. Scadron, Phys. Rev. D **30**, 1988 (1984).
30. V. G. Vaks and B. L. Ioffe, Nuovo Cim. **10**, 342 (1958).
31. M. Gell-Mann and M. Lévy, Nuovo Cim. **16**, 705 (1960).
32. V. de Alfaro, S. Fubini, G. Furlan, and C. Rossetti, in *Currents in Hadron Physics*, North-Holland Publ., Amsterdam, Chap. 5 (1973).
33. S. Egli *et al.* [SINDRUM Collaboration], Phys. Lett. B **222**, 533 (1989).
34. A. Bramon and M. D. Scadron, Europhys. Lett. **19**, 663 (1992).
35. T. Das, V. S. Mathur, and S. Okubo, Phys. Rev. Lett. **19**, 859 (1967).
36. M. D. Scadron, Phys. Rev. **D26**, 239 (1982); N. Paver and M. D. Scadron, Nuovo Cim. A **79**, 57 (1984), see Eq. 32b.
37. E. van Beveren, T. A. Rijken, K. Metzger, C. Dullemond, G. Rupp, and J. E. Ribeiro, Z. Phys. **C30**, 615 (1986).
38. R. E. Karlsen, M. D. Scadron, and A. Bramon, Mod. Phys. Lett. A **8**, 97 (1993).
39. J. J. Sakurai, Annals Phys. **11**, 1 (1960).
40. R. Delbourgo, D. S. Liu, and M. D. Scadron, Int. J. Mod. Phys. A **14**, 4331 (1999) [hep-ph/9905501].
41. M. Benayoun, S. I. Eidelman, and V. N. Ivanchenko, Z. Phys. C **72**, 221 (1996).
42. T. Morozumi, C. S. Lim, and A. I. Sanda, Phys. Rev. Lett. **65**, 404 (1990).
43. R. E. Karlsen and M. D. Scadron, Mod. Phys. Lett. A **6**, 543 (1991);
44. R. E. Karlsen and M. D. Scadron, Phys. Rev. D **45**, 4108 (1992).
45. S. Weinberg, Phys. Rev. Lett. **17**, 616 (1966); H. Osborn, Nucl. Phys. B **15**, 501 (1970).
46. A. D. Polosa, N. A. Tornqvist, M. D. Scadron, and V. Elias, Mod. Phys. Lett. A **17**, 569 (2002) [hep-ph/0005106]; Also see: E. van Beveren, F. Kleefeld, G. Rupp, and M. D. Scadron, Mod. Phys. Lett. A **17**, 1673 (2002) [hep-ph/0204139].
47. J. Lowe and M. D. Scadron, hep-ph/0208118.
48. M. D. Scadron and V. Elias, Mod. Phys. Lett. A **10**, 1159 (1995); S. R. Choudhury and M. D. Scadron, Phys. Rev. D **53**, 2421 (1996), see Eq. (24).
49. J. A. Cronin, Phys. Rev. **161**, 1483 (1967), see Eq. (48).
50. S. Weinberg, Phys. Rev. D **8**, 605 (1973); Phys. Rev. Lett. **31**, 494 (1973). Also see B. H. McKellar and M. D. Scadron, Phys. Rev. D **27**, 157 (1983).
51. E. M. Aitala *et al.* [E791 Collaboration], Phys. Rev. Lett. **89**, 121801 (2002) [hep-ex/0204018].
52. See e.g. M. D. Scadron, Eur. Phys. J. C **6**, 141 (1999) [hep-ph/9710317].
53. See e.g.: V. Elias and M. D. Scadron, Phys. Rev. D **30**, 647 (1984); L. R. Babukhadia, V. Elias, and M. D. Scadron, J. Phys. G **23**, 1065 (1997) [hep-ph/9708431].
54. A. C. Mattingly and P. M. Stevenson, Phys. Rev. Lett. **69**, 1320 (1992) [hep-ph/9207228].
55. V. Elias and M. D. Scadron, Phys. Rev. Lett. **53**, 1129 (1984).
56. F. Karsch, E. Laermann, and A. Peikert, Nucl. Phys. B **605**, 579 (2001) [hep-lat/0012023].
57. D. Bailin, J. Cleymans, and M. D. Scadron, Phys. Rev. D **31**, 164 (1985); J. Cleymans, A. Kocic, and M. D. Scadron, *ibid* **39**, 323 (1989); N. Bilic, J. Cleymans, and M. D. Scadron, Int. J. Mod. Phys. A **10**, 1169 (1995) [hep-ph/9402201]; M. D. Scadron and P. Zenczykowski, Hadronic Journal, in press (2002) [hep-th/0106154].
58. J. Meyer, K. Schwenzer, H. J. Pirner, and A. Deandrea, Phys. Lett. B **526**, 79 (2002) [hep-ph/0110279], see Table 1; H. J. Pirner, hep-ph/0209221.
59. Frieder Kleefeld, Eef van Beveren, George Rupp, and Michael D. Scadron, Phys. Rev. D **66**, 034007 (2002) [hep-ph/0109158].

Consistent relativistic Quantum Theory for systems/particles described by non-Hermitian Hamiltonians and Lagrangians

Frieder Kleefeld

Centro de Física das Interacções Fundamentais (CFIF), Instituto Superior Técnico,
Edifício Ciência, Piso 3, Av. Rovisco Pais, P-1049-001 LISBOA, Portugal
e-mail: kleefeld@cfif.ist.utl.pt

... to Manuela, Alexander and Sara ...

Abstract. A causal, non-Hermitian, renormalizable, local, unitary and Lorentz convariant formulation of Quantum Theory (QT) (= Quantum Mechanics (QM) and Quantum Field Theory (QFT)) is developed which is free of formalistic problems we face in the commonly used Hermitian formulation of QT. Side effects of the new formulation of QT are the derivation of a consistent (anti)causal neutrino Lagrangian, the enrichment of chiral symmetries, the removal of the Dirac sea, the separation of positive and negative energy states including a reformulation of the anti-particle concept and a critical analysis of the concept of probability currents. In a first step we apply the new formulation of QT to establish a relation between perturbative Quantum Chromodynamics (QCD) and the (axial)vector meson extended Quark-Level Linear Sigma Model (QLLSM) at high energies.

INTRODUCTION TO (ANTI)CAUSAL QUANTUM THEORY

It is well known — that the present Hermitian formulation of QT (see e.g. [1]) gives rise to a variety of formalistic — unresolved — problems and paradoxa. Without going into details[1] we want to stress that such conceptual shortcomings get more pronounced, if we try to apply the Hermitian QT (HQT) to "non-Hermitian" (e.g. absorptive, resonant, bifurcating, CP-violating, ...) problems. Simple consequences are e.g. the loss of locality, causality, stability of the effective action and the positivity of spectral functions. We will show that many of these problems can be understood and circumvented by treating QT in strict sense (anti)causal. In standard HQT — as we teach it to the students — causality is implemented by analytical continuation of Green's functions or propagators according to (anti)causal boundary conditions to be imposed. What we don't teach the students is that this analytical continuation (usually performed in generating functionals) not only spoils the unitarity of the HQT (due to an infinitesimally non-Hermitian effective action, i.e. a non-unitary S-matrix), yet also has severe negative implications on its consistency

---

[1] Some examples: problems in the probability concept of QT [2], the EPR paradoxon [3], the Klein paradoxon, Zitterbewegung, supercritical fields and the problem of gauge fixing [4], the problem of bifurcation or compositeness in QM (noted e.g. by T.D. Lee [5]), in QFT (noted e.g. by L.D. Landau (p. 18 in [6] and references therein)), the problem of regularization and anomalies (e.g. [7]).

CP660, *Hadron Physics: Effective Theories of Low Energy QCD*, edited by A. H. Blin et al.
© 2003 American Institute of Physics 0-7354-0120-9/03/$20.00

when afterwards demanding field operators or wavefunctions to stay Hermitian yielding a lot of paradoxa mentioned above[2]. E.g. in Refs. [9, 10] it is noted that that decay rates of the electroweak gauge Bosons cannot be described in a gauge-invariant manner, if they are not determined on the complex mass shell of the respective gauge Bosons.

What is to be understood by a complex mass shell? Let us consider for simplicity a neutral non-interacting Klein-Gordon (KG) field $\phi(x)$. In QFT the real mass m of $\phi(x)$ has to be analytically continued according to $m \to m - i\varepsilon$, yielding a *complex(valued)* mass $M = m - \frac{i}{2}\Gamma$ with $\Gamma \to \varepsilon = +0$. For resonances like the electroweak gauge Bosons the width Γ will be even finite. As the complex mass M is entering the analytically continued *causal* generating functional the respective KG field will obey the classical equation of motion $((i\partial)^2 - M^2)\phi(x) = 0$. A neutral spin 1/2 Fermion will correspondingly fulfil the respective causal Dirac equation $(i\slashed{\partial} - M)\psi(x) = 0$. As the imaginary part of M is finite the *causal* KG and Dirac equations have to be solved by a *Laplace-transform* and not by a simple *Fourier-transform* as taught in textbooks. Consequently e.g. the neutral KG field $\phi(x)$ will be *complex(valued)* and *not* real or Hermitian as stated in textbooks[3],[4]. More explicitly we obtain[5]:

$$
\phi(x) = \int \frac{d^4p}{(2\pi)^3} \delta(p^2 - M^2) a(p) e^{-ipx} \overset{!}{=} \sum_{\pm} \int \frac{d^3p}{(2\pi)^3 2\omega(\vec{p})} a(\pm p) e^{\mp ipx}\bigg|_{p^0 = \omega(p)}
$$

$$
\psi(x) = \sum_s \int \frac{d^3p}{(2\pi)^3 2\omega(\vec{p})} \left[b(p,s) u(p,s) e^{-ipx} + b(-p,s) u(-p,s) e^{ipx} \right]\bigg|_{p^0 = \omega(p)}
$$

The symbolic δ-distribution "$\delta(p^2 - M^2)$" should be understood in the sense of residuum calculus and Cauchy's theorem and finds its solid mathematical basis in a formalism worked out by N. Nakanishi [17, 18]. The definition of spinors (and polarization vectors) requires the introduction of a Lorentz boost for objects with *complex mass* M [6].

---

[2] There are also non-Hermiticities inserted in QT by hand, which may lead to serious problems, if one treats the field operators or wavefunctions furtheron Hermitian. Examples are complex mixing angles [8], resonance propagators [9], CP-violating phases [10, 11] in high energy, hadronic or solid-state physics, imaginary chemical potentials e.g. in QCD [12], optical potentials in nuclear physics, ...

[3] The mass shell condition in 4-momentum space $p^2 = M^2$ will therefore be *complex* yielding the singular points $p^0 = \pm\omega(\vec{p})$ (with $\omega(\vec{p}) := \sqrt{\vec{p}^2 + M^2}$ and $\omega(\vec{0}) := M$) in the complex p^0-plane. Hence in performing the Laplace-transform the solution of the causal KG (or Dirac) equation can be decomposed by $\phi(x) = \phi^{(+)}(x) + \phi^{(-)}(x)$ (or $\psi(x) = \psi^{(+)}(x) + \psi^{(-)}(x)$), while $\phi^{(\pm)}(x)$ (or $\psi^{(\pm)}(x)$) is associated with $p^0 = \pm\omega(\vec{p})$, respectively.

[4] Hermitian conjugation of the causal KG equation will yield what we call *anticausal* KG equation, i.e. $((i\partial)^2 - M^{+2})(\phi(x))^+ = 0$. Due to its properties we call the *anticausal* KG field a KG *hole field* [14]. As $M^+ \neq M$, it is clear that $(\phi(x))^+ \neq \phi(x)$ and $\phi^{(+)}(x) \neq (\phi^{(-)}(x))^+$. In QM *causal* and *anticausal* states are called *Gamow states* and *anti-Gamow states*, respectively (see e.g. Refs. [15, 16]).

[5] We introduce a Dirac spinor $u(p,s) \equiv v(-p,s)$) fulfilling the equation $(\slashed{p} - \sqrt{p^2})u(p,s) = 0$ or, equivalently, the *transposed* equation $\overline{u^c}(p,s)(-\slashed{p} - \sqrt{p^2}) = 0$. Important identities: $u^c(p,s) = u(-p^*,s) = v(p^*,s)$ and $\text{sgn}[\text{Re}(p^0)] \sum_s u(p,s)\overline{v^c}(p,s) = \slashed{p} + \sqrt{p^2}$ (for $\text{Re}[p^0] \neq 0$). Furthermore we may define: $a(\vec{p}) := a(p)|_{p^0 = \omega(p)}$, $c^+(\vec{p}) := a(-p)|_{p^0 = \omega(p)}$, $b(\vec{p},s) := b(p,s)|_{p^0 = \omega(p)}$, $d^+(\vec{p},s) := b(-p,s)|_{p^0 = \omega(p)}$.

[6] A Lorentz transformation (LT) Λ^μ_ν is defined by $\Lambda^\mu_\rho g_{\mu\nu} \Lambda^\nu_\sigma = g_{\rho\sigma}$ for any metric $g_{\mu\nu}$. Let n^μ be a timelike unit 4-vector ($n^2 = 1$) and ξ^μ an *arbitrary complex* 4-vector with $\xi^2 \neq 0$. The 4 independent LTs

Hermiticity is also broken in the causal time-dependent Schrödinger equation:

$$i\partial_t \psi(x) = \left(-\frac{1}{2M}\vec{\nabla}^2 + V_C(x)\right)\psi(x) = H_C\,\psi(x). \tag{1}$$

Due to the fact that M is complex and the *causal potential* $V_C(x)(\neq V_C^+(x) =: V_A(x))$ is essentially the Laplace-transform of a causal propagator we notice that the *causal Hamilton operator* H_C entering the causal Schrödinger equation is not Hermitian, i.e. $H_C \neq H_C^+ =: H_A$ (while H_A is called the *anticausal Hamilton operator*)[7]. As H_C is not Hermitian its right $(\psi(x))$ and left eigenfunctions $(\bar{\psi}(x))$ are *not* related by Hermitian conjugation[8]. $\bar{\psi}(x)$ may be called *adjoint causal* Schrödinger wavefunction fulfilling the *adjoint causal Schrödinger equation* $-i\partial_t\,\bar{\psi}(x) = \bar{\psi}(x)H_C$.

Pars pro toto we are now going to discuss further aspects in the context of the (anti)causal KG field. For convenience we define the quantity $\sqrt{Z} := |M|/M$ and the field $\varphi(x) := \phi(x)/\sqrt{NZ}$. The non-Hermitian *causal (anticausal)* Lagrangian $\mathscr{L}_C(x)$ $(\mathscr{L}_A(x))$ with $\mathscr{L}_C(x) = (\mathscr{L}_A(x))^+$ yielding the *causal (anticausal)* KG equation is:

$$\mathscr{L}_C(x) = \frac{1}{2N}\left((\partial\phi(x))^2 - M^2\phi(x)^2\right) = \frac{1}{2}\left(Z(\partial\varphi(x))^2 - |M|^2\varphi(x)^2\right). \tag{2}$$

The complex constant N is just the square of an overall normalization of the causal field $\phi(x)$[9]. Typically we choose N to be real and positive (e.g. $N=1$) for (anti)causal fields, while we call fields for which N is of opposite sign (e.g. $N=-1$) (anti)causal *ghost fields*. We see that — in the context of a HQT — the quantity $\sqrt{Z}$ gets — with respect to $\varphi(x)$ — the meaning of a traditional *wavefunction renormalization*. Due to its definition we know that $|Z|=1$. For $Z=\exp(i\varepsilon)$ (with $\varepsilon=+0$) the causal fields $\phi(x)$ and $\varphi(x)$ may be called *asymptotic stable*, as they can represent asymptotic stable systems[10]. If $\mathrm{Im}[\sqrt{Z}] > 0$ and non-infinitesimal, the width $\Gamma > 0$ will also be finite. Then the fields $\phi(x)$ and $\varphi(x)$ may be called *non-asymptotic* or *unstable*, as they represent *intermediate states* with a finite life time[11]. In the context of HQT it is the general belief (see e.g.

---

relating ξ^μ with its "restframe" (i.e. $\xi^\mu = \Lambda^\mu_{\ \nu}(\xi)\,n^\nu\sqrt{\xi^2}$ and $n_\nu\sqrt{\xi^2} = \xi_\mu\,\Lambda^\mu_{\ \nu}(\xi)$) are:

$$\Lambda^\mu_{\ \nu}(\xi) = \pm\left\{g^\mu_{\ \rho} - \frac{\sqrt{\xi^2}}{\sqrt{\xi^2}\mp\xi\cdot n}\left[n^\mu\mp\frac{\xi^\mu}{\sqrt{\xi^2}}\right]\left[n_\rho\mp\frac{\xi_\rho}{\sqrt{\xi^2}}\right]\right\}P^\rho_{\ \nu} \quad \text{and} \quad \Lambda^\mu_{\ \rho}(\xi)P^\rho_{\ \nu}$$

(with $P^\mu_{\ \nu} := 2n^\mu n_\nu - g^\mu_{\ \nu}$ = reflection matrix). Lorentz covariance of the Dirac equation requires $S^{-1}(\Lambda(p))\,\gamma^\mu\,S(\Lambda(p)) = \Lambda^\mu_{\ \nu}(p)\,\gamma^\nu$ for $u(p,s) = S(\Lambda(p))\,u(\sqrt{p^2}\,n,s)$. The metric $(+,-,-,-)$ yields

$$u(p,s)\big|_{\mathrm{Re}[p^0]>0} = \frac{\not{p}+\sqrt{p^2}}{\sqrt{2\sqrt{p^2}\,(p^0+\sqrt{p^2})}}\,u(\sqrt{p^2},\vec{0},s) \overset{p^0=\omega(\vec{p})}{\longrightarrow} \frac{\not{p}+M}{\sqrt{2M\,(\omega(\vec{p})+M)}}\,u(M,\vec{0},s).$$

Similarly construct $v(p,s)$, $u^c(p,s)$, $v^c(p,s)$, $\bar{u}(p,s)$, $\bar{v}(p,s)$, $\overline{u^c}(p,s)$ and $\overline{v^c}(p,s)$ for $\mathrm{Re}[p^0]>0$!

[7] For $M = m - i\varepsilon$ we may call H_C *quasi-Hermitian*, if $V_C(x)$ is in a similar sense quasi-Hermitian.
[8] Causal "bra's" ("out-states") are *not* Hermitian conjugate to causal "ket's" ("in-states") $(\langle\bar{\psi}| \neq |\psi\rangle^+)$!
[9] We may call $\sqrt{N}$ the *causal norm* of a state/field.
[10] Such fields appear in the context of the S-matrix as in- and out-states.
[11] This is rather consistent with the need of *complex wavefunction renormalizations* stressed in [9, 10].

Ref. [13]) that composite states are considered to be states without kinetic energy, i.e. with vanishing wavefunction renormalization, i.e. $Z \to 0$. In a (anti)causal QT *this is not possible* due to the constraint $|Z| = 1$. Nevertheless we may declare states/fields to be *composite*, when $\text{Re}[\sqrt{Z}] \to 0$. A transition from $\text{Im}[\sqrt{Z}] > 0$ to $\text{Im}[\sqrt{Z}] < 0$ will bring us from causal to anticausal fields. The Hermitian situation $\text{Im}[\sqrt{Z}] = 0$ describes Hermitian *acausal* fields, whose propagators are half causal and half anticausal (= principal value prescription). For historical reasons we will call such ("standing wave", i.e. non-propagating) states *shadow states* [19, 20][12]. Finally — in accordance with earlier discussions in the literature on HQT — we want to call here $\text{Re}[\sqrt{Z}]$ the *traditional norm* of a state/field. With the help of the preceding discussion we are now in the position to consider selected historical difficulties in the formalistic developement of a Hermitian *and* causal QT which persist upto now.

In a first step we want to address here the "bifurcation" or "compositeness problem". In 1954 T.D. Lee [5] (see also [22, 17, 18, 23, 24, 25]) observed in the *Hermitian* "Lee-model" that the renormalization of the model required a purely imaginary unrenormalized coupling constant to obtain a finite real renormalized coupling constant, and that beyond a certain critical value of the coupling constant, at which the wavefunction renormalization went to zero (i.e. $Z \to 0$), the *Hermitian* Hamiltonian developed *Hermitian conjugate complex pairs of eigenvalues* for pairs of *zero traditional norm eigenstates* to be considered as *metric partners* [13] forming a *biorthogonal basis* [32] [14]. The problem of the treatment of zero traditional norm states and negative causal norm states ("ghosts") had — since Dirac [33, 34] — induced a lot of theoretical activities which culminated in those times among other things in the concept of an indefinite metric and a metric operator (for details see e.g. Refs. [18, 25, 35, 32]). The application of such concepts has faced upto now enormous difficulties in the context of HQT, mainly because most calculations involved a (blind) *mixture* of purely Hermitian states ("shadow states") and zero traditional norm states ("(anti)causal states") and their interactions. As we will argue in the following *a consistent treatment within a (anti)causal QT requires the sole existence of causal and anticausal states (consisting of a special superposition of "shadow states") such that both classes states — causal and anticausal — don't interact !* As a consequence the shadow states interact in a very restricted way which is governed by the Lorentz covariance of the theory [19, 20]. Furthermore, as $|Z| = 1$, intermediate unstable fields cannot be easily integrated out from a (anti)causal Lagrangian. We also have to realize that the restriction to a "Hermitian" Lehmann-Källén spectral representation [36, 37] (in which spectral functions are real functions of real invarinant mass variables) is an inadmittable corset imposed by HQT, which does not allow QT to bifurcate[15].

---

[12] These shadow states shouldn't be confused with the shadow states addressed e.g. by T. Regge [21]!

[13] A similar singular phenomenon is known in perturbative QED as "Landau pole" [26]. As a nonperturbative phenomen bifurcation has been observed e.g. in QED in Dyson-Schwinger calculations at a critical value of $\alpha_{cr} > 0$ [27, 28, 29] (signalled by numerical instabilities and hugely enhanced calculation time or complex branchcuts), in variational methods at $\alpha_{cr} \leq 0$ [30, 39] and on the lattice [31].

[14] T.D. Lee surprised that a purely imaginary (unrenormalized) coupling constant enters a HQT remarked: "This difficulty may, however, be overcome by a modification of the present rules of quantum mechanics".

[15] This explains partially the technical and numerical difficulties we have in describing QED close to the Landau pole or QCD close to the chiral or (de)confinement phase transition(s).

In (anti)causal QT spectral functions are complex functions of complex invariant mass variables even for asymptotic states [38, 39].

Secondly we want to show an obvious inconsistency in the probability concept of traditional QT. The way students learn to derive probability currents in QT is to subtract classical equations of motion and its *Hermitian conjugate* yielding in (anti)causal QT the following continuity-like equations for neutral KG, Dirac and Schrödinger theory[16]:

$$\partial_\mu [\phi^+(x)\, \partial^\mu \phi(x) - (\partial^\mu \phi^+(x))\, \phi(x)] = 2i\varepsilon\, \phi^+(x)\, \phi(x) \tag{3}$$

$$\partial_\mu [i\,\overline{\psi}(x)\, \gamma^\mu\, \psi(x)] = -2i\varepsilon\, \overline{\psi}(x)\, \psi(x) \tag{4}$$

$$i\partial_t [\psi^+(x)\, \psi(x)] + \frac{1}{2m} \overrightarrow{\nabla} \cdot [\psi^+(x)\, \overrightarrow{\nabla}\, \psi(x) - (\overrightarrow{\nabla}\, \psi^+(x))\, \psi(x)] =$$

$$= \psi^+(x)[V_C(x) - V_C^+(x)]\, \psi(x) - \frac{i\varepsilon}{2m}[\psi^+(x)\, \overrightarrow{\nabla}^2\, \psi(x) + (\overrightarrow{\nabla}^2\, \psi^+(x))\, \psi(x)] \tag{5}$$

We see that *none* of the three currents will be conserved, if we use a complex mass M and a causal potential $V_C(x)$. This means for the Schrödinger theory that $|\psi(x)|^2$ *cannot* be considered to be the density of a conserved current (being a minimal requirement in the context of probability theory) and hence should *not* be interpreted as a probability density as originally done by M. Born [40]! Alternatively, if we perform the subtraction of the causal KG equation for $\phi^{(\pm)}(x)$ (Dirac equation for $\psi^{(\pm)}(x)$, Schrödinger equation for $\psi(x)$, respectively) and the *transposed* (or adjoint) equation for $\phi^{(\mp)}(x)$ ($\psi^{(\mp)c}(x)$ [17] and $\tilde{\psi}(x)$, respectively) we will obtain the following continuity equations:

$$\partial_\mu [\phi^{(\mp)}(x)\, \partial^\mu \phi^{(\pm)}(x) - (\partial^\mu \phi^{(\mp)}(x))\, \phi^{(\pm)}(x)] = 0 \tag{6}$$

$$\partial_\mu [i\, \overline{\psi^{(\mp)c}}(x)\, \gamma^\mu\, \psi^{(\pm)}(x)] = 0 \tag{7}$$

$$i\partial_t [\tilde{\psi}(x)\, \psi(x)] + \frac{1}{2M} \overrightarrow{\nabla} \cdot [\tilde{\psi}(x)\, \overrightarrow{\nabla}\, \psi(x) - (\overrightarrow{\nabla}\, \tilde{\psi}(x))\, \psi(x)] = 0 \tag{8}$$

The respective *non-vanishing*[18] causal currents are *conserved* even for *arbitrary* complex mass M and causal potential $V_C(x)$[19]. Within the (anti)causal framework charge, standard gauge invariance and the anti-particles are introduced according to the isospin concept, i.e. by considering N causal fields (i.e. $\phi_r(x)$ or $\psi_r(x)$ with $r = 1,\ldots,N$) of equal complex mass M. Unitarity of the theory is restored by considering the Hermitian Lagrangian $\mathcal{L}(x) = \mathcal{L}_C(x) + \mathcal{L}_A(x)$ (or Hamiltonian) containing causal *and* anticausal fields[20]. For the (anti)causal free KG [19, 18, 14] and Dirac [42, 43, 44] the-

---

[16] For simplicity we display here only the quasi-real case with $M = m - i\varepsilon$ and $\varepsilon = +0$.

[17] We adopt the standard notation $\overline{\psi^c}(x) := \psi^T(x)\, C$ (with $C = i\gamma^2 \gamma^0$) and $\overline{\psi}(x) := \psi^+(x)\, \gamma^0$.

[18] Even in the case of the neutral KG field this current is not vanishing!

[19] The respective isospin currents for neutral KG and Dirac fields vanish (due to a tricky cancellation of the underlying currents of Eqs. (6) and (7)), i.e. $\phi(x)\, \partial^\mu \phi(x) - (\partial^\mu \phi(x))\, \phi(x) = 0$ and $i\,\overline{\psi^c}(x)\, \gamma^\mu\, \psi(x) = 0$, if we treat $\phi(x)$ as a *commuting* field and $\psi(x)$ as a *anticommuting Grassmann* field!

[20] I.e. a (anti)causal prescription of QT yields a "doubling" of degrees of freedom with respect to the Hermitian (acausal) description. Obviously this "doubling" commonly performed at finite temperature T for consistency reasons [41] is for causality and unitarity reasons already present at $T = 0$.

ory and e.g. for the simple (anti)causal Harmonic oscillator in QM (e.g. [16]) we have $(\bar{M} := \gamma_0 M^+ \gamma_0)$:

$$\mathcal{L}(x) = \sum_r \frac{1}{2}\left((\partial \phi_r(x))^2 - M^2(\phi_r(x))^2 + (\partial \phi_r^+(x))^2 - M^{*2}(\phi_r^+(x))^2\right) \quad (9)$$

$$\mathcal{L}(x) = \sum_r \frac{1}{2}\left(\overline{\psi_r^c}(x)(\frac{1}{2}i\overleftrightarrow{\partial} - M)\psi_r(x) + \overline{\psi}_r(x)(\frac{1}{2}i\overleftrightarrow{\partial} - \bar{M})\psi_r^c(x)\right) \quad (10)$$

$$\langle z, z^*|H|z', z'^*\rangle = \left(-\frac{1}{2M}\frac{d^2}{dz^2} + \frac{1}{2}M\Omega^2 z^2 - \frac{1}{2M^*}\frac{d^2}{dz^{*2}} + \frac{1}{2}M^*\Omega^{*2}z^{*2}\right)\langle z, z^*|z', z'^*\rangle$$

For $N = 1$ Eq. (10) is describing the causal Lagrangian of a neutral Fermion (i.e. a Majorana–like neutrino [11])[21]. For $N = 2$ we may define the "charged" fields $\psi_\pm(x) := (\psi_1(x) \pm i\psi_2(x))/\sqrt{2}$ (representing e.g. a causal positron and electron[22]) and denote the respective minimally coupled Dirac-like Lagrangian of such a simply charged Fermion:

$$\mathcal{L}(x) = \overline{\psi_+^c}(x)(\frac{1}{2}i\overleftrightarrow{\partial} + g\,\slashed{A}(x) - M)\,\psi_-(x) + \overline{\psi}_-(x)(\frac{1}{2}i\overleftrightarrow{\partial} + g^*\gamma^\mu A_\mu^+(x) - \bar{M})\psi_+^c(x) \quad (11)$$

being invariant under the local gauge transformations $g\,\slashed{A}' = g\,\slashed{A} + [\slashed{\partial}, \Lambda(x)]$, $\psi'_-(x) = \exp(+i\Lambda(x))\,\psi_-(x)$, $\psi'_+(x) = \exp(-i(\Lambda(x))^T)\,\psi_+(x)$ [23]. The same holds for a non-Abelian gauge theory, i.e. for $A_\mu(x) = A_\mu^a(x)\lambda^a/2$ and $\Lambda(x) = \Lambda^a(x)\lambda^a/2$ [24]. Causal (anti)particles want to minimize their energy [14], anticausal (anti)holes want to maximize their energy. As long as the vacuum for (Fermionic or Bosonic) (anti) particles and (anti)holes is situated at zero energy and no interactions between (anti)particles and (anti)holes are permitted, positive and negative energy states are nicely separated. There will be no need for any Dirac sea[25], it will be not necessary to forbid the Bose-Einstein condensation of Bosons [33] and there will be no paradoxa like the Zitterbewegung or the Klein paradoxon. Finally it is of interest to study the Hermiticity content of (anti)causal fields. Hence we decompose the (anti)causal *neutral* KG fields $\phi^+(x)$, $\phi(x)$

---

[21] This can be understood from the Grassmann–nature of the Fermion fields as $(\overline{\psi^c}(x)\,\slashed{A}(x)\,\psi(x))^T = -\overline{\psi^c}(x)\,\slashed{A}(x)\,\psi(x) = 0$ for a "symmetric" Abelian/non-Abelian gauge field $A_\mu(x) = (A_\mu(x))^T$. In HQT the kinetic term of such a Majorana–like field would vanish, while in (anti)causal QT it survives!

[22] In a isospin–concept (causal) particles and antiparticles (described by $\psi_\pm(x)$) having positive energy and moving forward in time [14] possess same parity, while the parity of (anticausal) holes and antiholes (described by $\psi_\mp^c(x)$) bearing negative energy and moving backward in time [14] is opposite. This is perfectly possible without spoiling experimentally confirmed results of theoretical calculations in HQFT, e.g. meson or positronium spectra, as this type of internal parity is subject to a superselection rule [45].)

[23] Within such an isospin concept we obtain for $A^\mu(x) = 0$ the following continuity equations for $N = 2$: $\partial_\mu[i\overline{\psi_\mp^c}(x)\,\gamma^\mu\,\psi_\pm(x)] = 0$ (Dirac–field) and $\partial_\mu[\phi_\mp(x)\,\partial^\mu\phi_\pm(x) - (\partial^\mu\phi_\mp(x))\,\phi_\pm(x)] = 0$ (KG field).

[24] Recently Jun-Chen Su [46] showed that a renormalizable Lagrangian for a massive (non-)Abelian vector Boson originally derived by 't Hooft [47] within the context of a Higgs-mechanism persists renormalizable without relying on any Higgs mechanism. It is easy to extend such a Lagrangian consistently to complex mass vector fields implying that the whole standard renormalization concept of QFT on the basis of gauge invariance is working particularly well for (anti)causal fields of arbitrary complex mass. In this context we have to accept that vector Bosons consist of *three* physical degrees of freedom!

[25] Due to its infinite mass a Dirac sea would induce the immediate gravitational collapse of the universe!

and neutrino fields $\psi^c(x)$, $\psi(x)$ in shadow fields[26] yielding the following Lagrangians:

$$\mathscr{L}(x) = \frac{1}{2}\Big((\partial\phi_{(1)}(x))^2 - \mathrm{Re}[M^2](\phi_{(1)}(x))^2\Big) - \frac{1}{2}\Big((\partial\phi_{(2)}(x))^2 - \mathrm{Re}[M^2](\phi_{(2)}(x))^2\Big)$$
$$+ \ \mathrm{Im}[M^2]\,\phi_{(1)}(x)\,\phi_{(2)}(x) \tag{12}$$

$$\mathscr{L}(x) = \frac{1}{2}\overline{\psi}_{(1)}(x)\left(\frac{1}{2}i\,\overleftrightarrow{\partial} - \mathrm{Re}[M]\right)\psi_{(1)}(x) - \frac{1}{2}\overline{\psi}_{(2)}(x)\left(\frac{1}{2}i\,\overleftrightarrow{\partial} - \mathrm{Re}[M]\right)\psi_{(2)}(x)$$
$$+ \ \frac{1}{2}\,\mathrm{Im}[M]\,\Big(\overline{\psi}_{(2)}(x)\,\psi_{(1)}(x) + \overline{\psi}_{(1)}(x)\,\psi_{(2)}(x)\Big) \tag{13}$$

We observe that the diagonal (anti)causal Lorentz covariant Lagrangians Eqs. (9) and (12) are *non-diagonal* in the shadow fields and that the two (anti)causal fields consist of one shadow field with *positive* and one with *negative* norm[27]. Lagrangian Eq. (13) is particularly useful to study chiral symmetry in the context of (anti)causal QT. It is straight forward to show that a simultaneous chiral transformation in of the shadow fields according to[28] $\psi_{(1)}(x) \to \exp(i\gamma_5\,\alpha)\,\psi_{(1)}(x)$ and $\psi_{(2)}(x) \to \exp(-i\gamma_5\,\alpha)\,\psi_{(2)}(x)$ will yield the continuity-like equation[29] $\partial_\mu[\overline{\psi}(x)\,\gamma^\mu\,\gamma_5\psi(x)] \propto \mathrm{Re}[M]$, while the chiral rotation $\psi(x) \to \exp(i\gamma_5\,\alpha)\,\psi(x)$ and $\psi^c(x) \to \exp(i\gamma_5\,\alpha)\,\psi^c(x)$ in Eq. (10) for $N=1$ yields the standard continuity-like equation[30] $\partial_\mu[\overline{\psi^c}(x)\,\gamma^\mu\,\gamma_5\psi(x)] \propto M$. The first (broken) chiral symmetry related to the current $\overline{\psi}(x)\,\gamma^\mu\,\gamma_5\,\psi(x)$ mixes causal and anticausal fields, while the second (broken) chiral symmetry related to the current $\overline{\psi^c}(x)\,\gamma^\mu\,\gamma_5\,\psi(x)$ mixes only causal fields or anticausal fields. HQT can't distinguish between the two chiral currents and runs therefore into anomalies yielding different results depending on the choice of regularization [7]. The (anti)causal neutrino Lagrangian (Eq. (10) for $N=1$) can be studied, too, by decomposing the the neutrino fields into their chiral components [31], i.e.:

$$\mathscr{L}(x) =$$
$$= \ \frac{1}{2}\Big\{\overline{\psi^c_L}(x)\frac{i}{2}\overleftrightarrow{\partial}\psi_R(x) + \overline{\psi^c_R}(x)\frac{i}{2}\overleftrightarrow{\partial}\psi_L(x) - M\Big(\overline{\psi^c_R}(x)\,\psi_R(x) + \overline{\psi^c_L}(x)\,\psi_L(x)\Big)\Big\} + \mathrm{h.c.}$$
$$= \ \frac{1}{2}\Big\{\overline{\chi^c_+}(x)\frac{1}{2}\overleftrightarrow{\partial}\chi_+(x) - \overline{\chi^c_-}(x)\frac{1}{2}\overleftrightarrow{\partial}\chi_-(x) - M\Big(\overline{\chi^c_+}(x)\chi_-(x) + \overline{\chi^c_-}(x)\chi_+(x)\Big)\Big\} + \mathrm{h.c.}$$

Note that $\mathscr{L}(x)$ in terms of the new fields $\chi_\pm(x) := (\psi_R(x) \pm i\psi_L(x))/\sqrt{2}$ is invariant under $\chi_\pm(x) \to \exp(\pm i\gamma_5\,\alpha)\,\chi_\pm(x)$ even for *arbitrary complex Fermion mass M*! Further aspects (quantization, Wick's theorem, CPT, ...) of the presented (anti)causal QT are either discussed in Refs. [14, 42, 43, 44] or will be published elsewhere (e.g. Ref. [48]).

---

[26] This is done by $\phi(x) =: (\phi_{(1)}(x) + i\phi_{(2)}(x))/\sqrt{2}$, $\phi^+(x) =: (\phi_{(1)}(x) - i\phi_{(2)}(x))/\sqrt{2}$ and $\psi(x) =:$ $(\psi_{(1)}(x) + i\psi_{(2)}(x))/\sqrt{2}$, $\psi^c(x) =: (\psi_{(1)}(x) - i\psi_{(2)}(x))/\sqrt{2}$.

[27] N. Nakanishi [18, 19] was the first who investigated the KG Lagrangians Eqs. (9) and (12) in more detail. Unfortunately he lost — as we think — the interest in his "Complex-Ghost Relativistic Field Theory" due to singularities he faced by allowing interactions between causal and anti-causal fields!

[28] We mention that $\gamma_C := -iC$ behaves in many aspects like γ_5 — particularly in (anti)causal QT.

[29] Of course there exist also the respective Hermitian conjugate and transposed continuity-like equations!

[30] This type of chiral symmetry is broken even for a "massless", yet *causal* neutrino due to $M \to -i\varepsilon$!

[31] $\psi(x) = \psi_R(x) + \psi_L(x)$ with $\psi_R(x) := P_R\,\psi(x)$, $\psi_L(x) := P_L\,\psi(x)$ and $P_R := (1+\gamma_5)/2, P_L := (1-\gamma_5)/2$.

QCD AND THE QUARK-LEVEL LINEAR SIGMA MODEL

As an interesting application of the formalism described above we want to illustrate an attractive relation between the Lagrangian of QCD [49] and the unbroken Lagrangian of the QLLSM [50] including vector mesons[32]. We start with the (anti)causal QCD Lagrangian in R_α gauge [46, 47] for (anti)quarks $(q_\pm)$ and gluons (B) with respective complex masses M_q and M_B [33,34,35]:

$$\mathscr{L}(x) = \overline{q^c_+}(x) \left(\frac{i}{2} \overset{\leftrightarrow}{\partial} + g\, \slashed{B}(x) - M_q \right) q_-(x)$$

$$- \frac{1}{2} \operatorname{tr}_c\left[B_{\mu\nu}(x) B^{\mu\nu}(x) \right] + M_B^2 \operatorname{tr}_c\left[B_\mu(x) B^\mu(x) \right] - \frac{1}{\alpha} \operatorname{tr}_c\left[\left(\partial_\mu B^\mu(x) \right)^2 \right]$$

$$- 2 \operatorname{tr}_c\left[\left(\partial_\mu \bar{C}(x) \right) \left(\frac{\alpha M_B^2}{\partial^2} + 1 \right) \partial^\mu C(x) + i g \left\{ \left(\partial_\mu \bar{C}(x) \right), C(x) \right\} B^\mu(x) \right] + \text{h.c.}$$

For later convenience we insert unit matrices in spin and flavour space, i.e.:

$$\mathscr{L}(x) = \overline{q^c_+}(x) \left(\frac{i}{2} \overset{\leftrightarrow}{\partial} + g\, \slashed{B}(x) \, [1]_F - M_q \right) q_-(x)$$

$$+ \frac{1}{4N_F} \operatorname{tr}\left[-\frac{1}{2} B_{\mu\nu}(x) [1]_S [1]_F B^{\mu\nu}(x) [1]_S [1]_F \right.$$

$$+ M_B^2 B_\mu(x) [1]_S [1]_F B^\mu(x) [1]_S [1]_F - \frac{1}{\alpha} \left(\partial_\mu B^\mu(x) [1]_S [1]_F \right)^2$$

$$- 2 \left(\left(\partial_\mu \bar{C}(x) [1]_S [1]_F \right) \left(\frac{\alpha M_B^2}{\partial^2} + 1 \right) \partial^\mu C(x) [1]_S [1]_F \right.$$

$$\left. + i g \left\{ \left(\partial_\mu \bar{C}(x) [1]_S [1]_F \right), C(x) [1]_S [1]_F \right\} B^\mu(x) [1]_S [1]_F \right) \right] + \text{h.c.}, \quad (14)$$

while "tr" indicates the trace in colour, spin and flavour space, i.e. "$\operatorname{tr}_{c,S,F}$". For the following arguments it seems imperative to use Feynman gauge ($\alpha \to 1$) for which the gluon ($iD^{ab}_{\mu\nu}(k)$) and ghost ($i\Delta^{ab}(k)$) propagators take their simplest form, i.e. $iD^{ab}_{\mu\nu}(k) \to -i\delta^{ab} g_{\mu\nu}/(k^2 - M_B^2)$ and $i\Delta^{ab}(k) \to -i\delta^{ab}/(k^2 - M_B^2)$. We know that at high energies — in the perturbative regime of QCD — quark–quark scattering is predominantly determined by one-gluon exchange (OGE) whose spin structure is — in Feynman gauge — described by $[\gamma_\mu]^{(1)} [\gamma^\mu]^{(2)}$ being subject to the Fierz identity [51]

$$[\gamma_\mu]_{ij}[\gamma^\mu]_{k\ell} = [1]_{i\ell}[1]_{kj} + [i\gamma_5]_{i\ell}[i\gamma_5]_{kj} - \frac{1}{2}\left([\gamma_\mu]_{i\ell}[\gamma^\mu]_{kj} + [\gamma_\mu\gamma_5]_{i\ell}[\gamma^\mu\gamma_5]_{kj} \right).$$

---

[32] It has to be stated that the presented "derivation" is hardly possible in a HQT.

[33] For "massless" quarks and gluons one has to perform the replacement $M \to -i\varepsilon$.

[34] In the following we will use the indices c, S, F for "colour", "spin" and "flavour", respectively. E.g. $[1]_c$, $[1]_S$ and $[1]_F$ are the unit matrices in colour (c), spin (S) and flavour (F) space.

[35] Be reminded that $D_\mu := \partial_\mu - ig B_\mu(x)$, $B_\mu(x) := B^a_\mu(x) [\lambda^a/2]_c$, $B_{\mu\nu}(x) := B^a_{\mu\nu}(x) [\lambda^a/2]_c = \frac{i}{g}[D_\mu, D_\nu]$, $C(x) := C^a(x) [\lambda^a/2]_c$ and $\bar{C}(x) := \bar{C}^a(x) [\lambda^a/2]_c$, while "$\operatorname{tr}_c$" indicates a trace and $[\lambda^a]_c$ $(a = 1, \ldots, 8)$ represent Gell-Mann matrices in colour (c) space.

Inspection of these Fierz identities suggests that a t-channel one-gluon OGE can be replaced by the simultaneous u-channel exchange ($j \leftrightarrow \ell$) of scalar (S), pseudo-scalar (P), vector (V) and axial-vector (Y) fields, while the u-channel OGE may be replaced by the respective simultaneous t-channel exchange of S, P, V and Y fields. Of course, this idea can't be considered isolated from the respective dynamics in colour and flavour space [36]. The relevant Fierz identity in colour space is:

$$[\lambda^a/2]_{ij}\,[\lambda^a/2]_{k\ell} = \frac{1}{2}\left(1 - \frac{1}{N_c^2}\right)[1]_{i\ell}\,[1]_{kj} - \frac{1}{N_c}\,[\lambda^a/2]_{i\ell}\,[\lambda^a/2]_{kj}\,,$$

while in flavour space we have $[1]_{ij}\,[1]_{k\ell} = \frac{1}{N_F}\,[1]_{i\ell}\,[1]_{kj} + 2\,[\lambda^a/2]_{i\ell}\,[\lambda^a/2]_{kj}$. The Fierz identity in flavour space seems to support the idea that the exchanged S, P, V and Y particles underlying the t-channel OGE are e.g. flavour nonets (for $N_F = 3$), while the Fierz identity in colour space indicates that in the large N_c limit ($N_c \to \infty$) these S, P, V and Y fields may be considered to be colour singlets. Hence one might tend to call these exchanged S, P, V and Y fields either "mesons" or "glue balls". Defining $[\lambda^0/2]_F := [1]_F/\sqrt{2N_F}$ and keeping in mind that in quark-quark scattering always t- and u-channel OGE graphs contribute both — due to the Pauli exclusion principle with opposite sign — we may claim the validity of the following Fierz replacement in the product of spin, flavour and colour space:

$$\left([\gamma_\mu]^{(1)}\,[\gamma^\mu]^{(2)}\right)\left([\lambda^a/2]_c^{(1)}\,[\lambda^a/2]_c^{(2)}\right)\left([1]_F^{(1)}\,[1]_F^{(2)}\right)$$

$$\longrightarrow\; -\left([1]^{(1)}[1]^{(2)} + [i\gamma_5]^{(1)}[i\gamma_5]^{(2)} - \frac{1}{2}\left\{[\gamma_\mu]^{(1)}[\gamma^\mu]^{(2)} + [\gamma_\mu\gamma_5]^{(1)}[\gamma^\mu\gamma_5]^{(2)}\right\}\right)$$

$$\left(\frac{1}{2}\left(1 - \frac{1}{N_c^2}\right)[1]_c^{(1)}[1]_c^{(2)} - \frac{1}{N_c}[\lambda^a/2]_c^{(1)}[\lambda^a/2]_c^{(2)}\right)$$

$$\left(\frac{1}{N_F}[1]_F^{(1)}[1]_F^{(2)} + 2\,[\lambda^a/2]_F^{(1)}[\lambda^a/2]_F^{(2)}\right)$$

$$\xrightarrow{N_c \to \infty}\; -\left([1]^{(1)}[1]^{(2)} + [i\gamma_5]^{(1)}[i\gamma_5]^{(2)} - \frac{1}{2}\left\{[\gamma_\mu]^{(1)}[\gamma^\mu]^{(2)} + [\gamma_\mu\gamma_5]^{(1)}[\gamma^\mu\gamma_5]^{(2)}\right\}\right)$$

$$\left([1]_c^{(1)}[1]_c^{(2)}\right)\left([\lambda^0/2]_F^{(1)}[\lambda^0/2]_F^{(2)} + [\lambda^a/2]_F^{(1)}[\lambda^a/2]_F^{(2)}\right). \tag{15}$$

After defining $N_F \times N_F$ "meson" [37] field matrices in flavour space ($\Sigma := \sum_{a=0}^{N_F^2-1}$), i.e. $S(x) := \sqrt{2}\,\Sigma\,\sigma_a(x)\,[\lambda^a/2]_F$ ("scalar"), $P(x) := \sqrt{2}\,\Sigma\,\eta_a(x)\,[\lambda^a/2]_F$ ("pseudo–scalar"), $V^\mu(x) := \sqrt{2}\,\Sigma\,\omega_a^\mu(x)\,[\lambda^a/2]_F$ ("vector") and $Y^\mu(x) := \sqrt{2}\,\Sigma\,v_a^\mu(x)\,[\lambda^a/2]_F$ ("axial–vector") we may translate Fierz replacement Eq. (15) for $N_c \to \infty$ into a respective replacement prescription for the OGE propagator in terms of meson propagators:

$$[\gamma^\mu]^{(1)}\,[\gamma^\nu]^{(2)}\,\langle 0|T\left[B_\mu^{(1)}(x)\,B_\nu^{(2)}(y)\right]|0\rangle\,\left([1]_F^{(1)}\,[1]_F^{(2)}\right)\;\xrightarrow{N_c \to \infty}$$

---

[36] This statement applies to QCD. In a similiar (feasible) consideration for QED in the perturbative (low energy) regime the Fierz identities in colour and flavour space are absent. The results are expected to be similar to the observations in [52].

[37] In the following we will use for simplicity the notatation "meson" rather than "glue ball".

$$\frac{1}{2} \left([1]^{(1)}[1]^{(2)} Z_s^{-1} \langle 0| T\left[S^{(1)}(x)\, S^{(2)}(y)\right] |0\rangle \right.$$

$$+ [i\gamma_5]^{(1)}[i\gamma_5]^{(2)} Z_p^{-1} \langle 0| T\left[P^{(1)}(x)\, P^{(2)}(y)\right] |0\rangle$$

$$+ \frac{1}{2}\left\{ [\gamma^\mu]^{(1)}[\gamma^\nu]^{(2)} Z_v^{-1} \langle 0| T\left[V_\mu^{(1)}(x)\, V_\nu^{(2)}(y)\right] |0\rangle \right.$$

$$\left. \left. + [\gamma^\mu \gamma_5]^{(1)}[\gamma^\nu \gamma_5]^{(2)} Z_y^{-1} \langle 0| T\left[Y_\mu^{(1)}(x)\, Y_\nu^{(2)}(y)\right] |0\rangle \right\} \right) \left([1]_c^{(1)}[1]_c^{(2)}\right). \quad (16)$$

The renormalization constants Z_s, Z_p, Z_v and Z_y will be determined later. Treating each element of the meson field matrices $S(x)$, $P(x)$, $V^\mu(x)$ and $Y^\mu(x)$ as an independent field the propagator replacement Eq. (16) will be achieved by replacing in the quark-gluon vertex of the QCD-Lagrangian ($g\, \overline{q_+^c}(x)\, \not{B}(x)\,[1]_F\, q_-(x)$) the gluon field according to[38]:

$$\not{B}(x)\,[1]_F \overset{N_c \to \infty}{\longrightarrow} \frac{1}{\sqrt{2}}\, \gamma_\mu \left(\frac{1}{4}\left(\frac{c_s}{\sqrt{Z_s}}\, \gamma^\mu\, S(x) + \frac{c_p}{\sqrt{Z_p}}\, i\gamma^\mu\, \gamma_5 P(x)\right) \right.$$

$$\left. + \frac{1}{\sqrt{2}}\left(\frac{c_v}{\sqrt{Z_v}}\, V^\mu(x)\,[1]_s + \frac{c_y}{\sqrt{Z_y}}\, Y^\mu(x)\, \gamma_5\right)\right)[1]_c \quad (17)$$

with $c_s, c_p, c_v, c_y \in \{+1, -1\}$. Inspection of Eq. (17) and $\not{B}(x) = \gamma_\mu B^\mu(x)$ suggests the following large N_c replacement of the gluon fields in terms of S, P, V and Y mesons[39]:

$$B^\mu(x)\,[1]_s[1]_F \overset{N_c \to \infty}{\longrightarrow} \frac{1}{\sqrt{2}}\left(\frac{1}{4}\left(\frac{c_s}{\sqrt{Z_s}}\, \gamma^\mu\, S(x) + \frac{c_p}{\sqrt{Z_p}}\, i\gamma^\mu\, \gamma_5 P(x)\right)\right.$$

$$\left. + \frac{1}{\sqrt{2}}\left(\frac{c_v}{\sqrt{Z_v}}\, V^\mu(x)\,[1]_s + \frac{c_y}{\sqrt{Z_y}}\, Y^\mu(x)\, \gamma_5\right)\right)[1]_c. \quad (18)$$

Using Eq. (18) it is also straight forward to replace the gluon field-strength tensor $B_{\mu\nu}(x)$ keeping in mind that $B_{\mu\nu}(x)\,[1]_s\,[1]_F = \frac{i}{g}[D_\mu\,[1]_s\,[1]_F$, $D_\nu\,[1]_s\,[1]_F] = [\partial_\mu, B_\nu(x)\,[1]_s[1]_F] - [\partial_\nu, B_\mu(x)\,[1]_s[1]_F] - ig[B_\mu(x)\,[1]_s[1]_F, B_\nu(x)\,[1]_s[1]_F]$. In order to guarantee the renormalizability of the resulting Lagrangian the Grassmann ghosts $C(x)$ and $\bar{C}(x)$ may by decomposed for the V and Y fields into $C_v(x)\,c_v/\sqrt{Z_v}$, $\bar{C}_v(x)\,c_v/\sqrt{Z_v}$, $C_y(x)\,c_y/\sqrt{Z_y}$ and $\bar{C}_y(x)\,c_y/\sqrt{Z_y}$ with $C_v(x) = \sum C_v^a(x)\,[\lambda^a/2]_F$, $C_y(x) = \sum C_y^a(x)\,[\lambda^a/2]_F$, $\bar{C}_v(x) = \sum \bar{C}_v^a(x)\,[\lambda^a/2]_F$, $\bar{C}_y(x) = \sum \bar{C}_y^a(x)\,[\lambda^a/2]_F$ and $\sum = \sum_{a=0}^{N_F^2 - 1}$. The scalar and pseudo-scalar fields renormalize each other in the sense of the LSM, if they form a chiral circle. After application of the replacements Eq. (17) and (18) to the QCD Lagrangian Eq. (14) we may choose Z_s, Z_p, Z_v and Z_y such that the kinetic terms of the new S, P, V and Y fields take the standard form. This is achieved by:

$$\frac{c_\ell}{\sqrt{Z_\ell}} = \frac{4is_\ell}{\sqrt{3 + \frac{1}{\alpha}}} \sqrt{\frac{N_F}{N_c}} \quad (\ell \in \{s, p\}), \qquad \frac{c_{\ell'}}{\sqrt{Z_{\ell'}}} = \sqrt{2}\, s_{\ell'} \sqrt{\frac{N_F}{N_c}} \quad (\ell' \in \{v, y\}), \quad (19)$$

---

[38] Remember that $\gamma_\mu \gamma^\mu = 4\,[1]_s$.

[39] It should be noted that the suggested replacement is not completely unique. Nevertheless it is strongly suggested by the requirement of renormalizability of the resulting Lagrangian and the structure of Lagrangians obtained in the context of extended Nambu-Jona-Lasinio models.

with $s_s, s_p, s_v, s_y \in \{+1, -1\}$. As an overall result of our replacements we obtain in Feynman gauge ($\alpha \to 1$) the following renormalizable Lagrangian of a QLLSM including (axial)vector-mesons, which is expected to describe the properties of QCD at high energies[40,41]:

$$\mathscr{L}(x) \xrightarrow{N_c \to \infty}$$

$$\to\ \overline{q^c_+}(x)\left(\frac{i}{2}\overset{\leftrightarrow}{\partial} - M_q\right) q_-(x)$$

$$+\ \overline{q^c_+}(x)\, g\sqrt{\frac{N_F}{N_c}}\left(\sqrt{2}\, i\left(s_s S(x) + s_p\, i\gamma_5 P(x)\right) + \frac{1}{\sqrt{2}}\left(s_v\, \slashed{V}(x) + s_y\, \slashed{Y}(x)\,\gamma_5\right)\right) q_-(x)$$

$$+\ \frac{1}{8}\,\mathrm{tr}_F\left[\left(\partial^\mu S(x)\right)\left(\partial_\mu S(x)\right)\right] - \frac{1}{2}M_B^2\, \mathrm{tr}_F\left[\left(S(x)\right)^2\right]$$

$$+\ \frac{1}{8}\,\mathrm{tr}_F\left[\left(\partial^\mu P(x)\right)\left(\partial_\mu P(x)\right)\right] - \frac{1}{2}M_B^2\, \mathrm{tr}_F\left[\left(P(x)\right)^2\right]$$

$$+\ \frac{3}{8}\,\mathrm{tr}_F\left[\left(\partial^\mu S(x) - ig\,\sqrt{\frac{N_F}{2N_c}}\left(s_v\,[V^\mu(x),\, S(x)] - is_s s_p s_y\,\{Y^\mu(x),\, P(x)\}\right)\right)^2\right]$$

$$+\ \frac{3}{8}\,\mathrm{tr}_F\left[\left(\partial^\mu P(x) - ig\,\sqrt{\frac{N_F}{2N_c}}\left(s_v\,[V^\mu(x),\, P(x)] + is_s s_p s_y\,\{Y^\mu(x),\, S(x)\}\right)\right)^2\right]$$

$$-\ \frac{1}{8}\,\mathrm{tr}_F\left[\left(V_+^{\mu\nu}(x)\right)^2\right] + \frac{1}{2}M_B^2\, \mathrm{tr}_F\left[\left(V^\mu(x)\right)^2\right] - \frac{1}{2}\,\mathrm{tr}_F\left[\left(\partial^\mu V_\mu(x)\right)^2\right]$$

$$-\ \frac{1}{8}\,\mathrm{tr}_F\left[\left(V_-^{\mu\nu}(x)\right)^2\right] + \frac{1}{2}M_B^2\, \mathrm{tr}_F\left[\left(Y^\mu(x)\right)^2\right] - \frac{1}{2}\,\mathrm{tr}_F\left[\left(\partial^\mu Y_\mu(x)\right)^2\right]$$

$$+\ \frac{3}{8}\left(-ig\,\sqrt{\frac{N_F}{N_c}}\right)^2 \mathrm{tr}_F\left[\left(\left(S(x) + iP(x)\right)\left(S(x) - iP(x)\right)\right)^2\right]$$

$$-\ \mathrm{tr}_F\left[\left(\partial_\mu \bar{C}_v(x)\right)\left(\frac{M_B^2}{\partial^2} + 1\right)\partial^\mu C_v(x) + ig\sqrt{\frac{N_F}{2N_c}}\left\{\left(\partial_\mu \bar{C}_v(x)\right),\, C_v(x)\right\}V^\mu(x)\right]$$

$$-\ \mathrm{tr}_F\left[\left(\partial_\mu \bar{C}_y(x)\right)\left(\frac{M_B^2}{\partial^2} + 1\right)\partial^\mu C_y(x) + ig\sqrt{\frac{N_F}{2N_c}}\left\{\left(\partial_\mu \bar{C}_y(x)\right),\, C_y(x)\right\}Y^\mu(x)\right]$$

$$+\ \text{h.c.} \tag{20}$$

---

[40] This statement should be understood in the sense of conclusions drawn in Ref. [53] by N.D. Hari Dass and V. Soni. In their paper they discuss "non-Abelian gauge theories with fermions and scalars that nevertheless possess asymptotic freedom". They claim that "even the possibility that" such a theory (being hardly distinguishable from the QLLSM including vector-mesons) "is indistinguishable from QCD could not be ruled out". Their analysis shows that "for deep inelastic scattering the leading behavior of this theory in the ultraviolet is identical to that of QCD".

[41] We want to mention that for the QCD anomalous term we find the replacement $\mathrm{tr}_c[B^{\mu\nu}(x)\,\tilde{B}_{\mu\nu}(x)] \xrightarrow{N_c \to \infty}$ $\frac{1}{4}\,\mathrm{tr}_F[V_+^{\mu\nu}(x)\,\tilde{V}_{+\mu\nu}(x)] + \frac{1}{4}\,\mathrm{tr}_F[V_-^{\mu\nu}(x)\,\tilde{V}_{-\mu\nu}(x)]$ (affecting only (axial–)vector mesons). In order to obtain anomalous structures like the 't Hooft determinant (see e.g. Ref. [54]) involving also (pseudo-)scalars we suggest to apply our replacement strategy to vanishing terms like e.g. $\mathrm{tr}_c[B^{\mu\nu}(x)\, B_{\mu\nu}(x)\, \gamma_5]$.

with $V^{\mu\nu}_{\pm}(x) := [\mathscr{D}^{\mu}_{\pm}, \mathscr{D}^{\nu}_{\pm}]/\left(-ig\,\sqrt{N_F/(2N_c)}\,\right)$ being defined by the covariant derivatives $\mathscr{D}^{\mu}_{\pm} := \partial^{\mu} - ig\,\sqrt{N_F/(2N_c)}\,(s_v V^{\mu}(x) \pm s_y Y^{\mu}(x))$. In spite of limited space a discussion of Eq. (20) is imperative. Apart from slight, yet significant differences the Lagrangian is very similar to the Lagrangian of the "traditional" QLLSM (see e.g. Refs. [50, 55, 56] and references therein). The V and Y mesons couple to the S and P mesons in a similar manner as in an 1-loop effective action of the Gauged LSM in Ref. [51] or the Extended Chiral Quark Model in Ref. [57] obtained by a tedious Bosonization. The individual interaction terms in Eq. (20) follow strictly the weighting rules of the $1/N_c$ expansion (see [58] and Chapter 8 in [49] including references). A characteristic new property of Eq. (20) is that the couplings of the scalar and pseudo-scalar fields to (anti)quarks contain an extra factor i (= imaginary unit)[42]. It appears as a surprise that the quasi-Hermitian QCD Lagrangian has been translated at high energies into a seemingly non-Hermitian Lagrangian with the same properties. Non-Hermitian QT allows now to relate consistently the high-energy Lagrangian to a low-energy Lagrangian by performing a Higgs-mechanism (HM) with a complex-valued scalar condensate $\langle S \rangle$ yielding at low energies complex selfenergies of mesons and (anti)quarks and complex coupling constants without getting in conflict with unitarity, causality, renormalizability and Lorentz covariance. E.g. a real-valued scalar condensate $\langle S \rangle$ would yield — due to the additional phasefactor in the coupling of scalars to (anti)quarks — purely imaginary selfenergy for (anti)quarks[43]. There is also the scenario of (close to) purely imaginary scalar condensates $\langle S \rangle$ yielding on one hand quasi-real (anti)quark selfenergies, on the other hand "vacuum replica" [59] or "vacuum instability" like structures [60] in the "composite field-space" [44],[45]. A further advantage of the derived QLLSM Lagrangian is that one gets a very clear understanding how the electromagnetic interaction couples to quarks and mesons. Photons couple in the QCD Lagrangian only to (anti)quarks, not to gluons, which yields that they also don't couple at high energies to the new S, P, V and Y fields. A complex Renormalization Group Transformation or — equivalently — a complex-valued HM relating Eq. (20) to the low energy version of the QLLSM Lagrangian may mix the photon fields with the vector-meson fields such that the photons couple solely to vector-mesons and not to quarks[46]. The discussed implementation of QED in QCD makes clear why one faces double-counting ambiguities [62] at low energies, whenever one tries to couple photons to both, (anti)quarks *and* mesons.

---

[42] This complex phase being related to the choice of Z_s and Z_p in Eq. (19) maps the negative spatial components of the $(+,-,-,-)$ metric into its positive time-like component yielding negative relative signs between the mass or kinetic terms of (axial-)vector fields relative to (pseudo-)scalar fields. In the language of Ref. [52] it would map a "wrong sign mass term" into a right sign mass term or vice-versa.

[43] Confinement would therefore just be a notation for highly unstable (or composite), i.e. very short lived, (anti)quarks decaying at low energies very fast into some asymptotic colourless meson states with quasi-real selfenergies

[44] In the discussion of the "Composite-Field Effective Potential" [52, 60] the role of the scalar fields is represented by so called "composite fields". These considerations find their origin in a stability and convergence analysis of the parameter space of the LSM which may be found in Ref. [61].

[45] We warmly like to thank J.E.F.T. Ribeiro for illustrating — in private communication — properties of "Vacuum Replicas" [59] and drawing our attention to related work of P. Rembiesa (e.g. Ref. [60])!

[46] This scenario is commonly called Vector-Meson Dominance (VMD).

ACKNOWLEDGMENTS

We particularly want to thank George Rupp, Eef van Beveren and Jun-Chen Su for very valuable remarks. A comprehensive form of the mainly unpublished results has been presented for the first time in a seminar (5.2.2002) and lecture course (21.3.-11.4.2002) in the CFIF, Lisbon. This work has been supported by the *Fundação para a Ciência e a Tecnologia* (FCT) of the *Ministério da Ciência e da Tecnologia (e do Ensinio Superior)* of Portugal, under Grant no. PRAXIS XXI/BPD/20186/99 and SFRH/BDP/9480/2002.

REFERENCES

1. H. Araki, *Mathematical theory of quantum fields*, ©1999 by Oxford University Press; Lecture Notes by F. Strocci, *General properties of QFT* (Lecture Notes in Physics — Vol. 51), ©1993 by World Scientific; R. Haag, *Local Quantum Physics* (2nd ed.), ©1992, 1996 by Springer-Verlag; R. Jost, *The general theory of quantized fields*, Providence, R.I.: Am. Phys. Soc., 1965; R.F. Streater, A.S. Wightman, *PCT, spin and statistics, and all that*, Benjamin, New York, 1964; N.N. Bogoliubov, D.V. Shirkov, *Introduction to the Theory of Quantized Fields*, ©1980 by John-Wiley.
2. W. Mückenheim et al., *Phys. Rep.* **133** (1986) 337.
3. A. Afriat, F. Selleri, *The Einstein, Podolski, and Rosen Paradox in Atomic, Nuclear and Particle Physics*, ©1999 by Plenum Press, New York.
4. W. Greiner, B. Müller, J. Rafelski, *Quantum Electrodynamics of Strong Fields*, ©1985 by Springer-Verlag Berlin Heidelberg; W. Greiner, *Relativistic Quantum Mechanics — Wave Equations*, ©1997 by Springer-Verlag Berlin Heidelberg; S.D. Joglekar, hep-th/0106264.
5. T.D. Lee, *Phys. Rev.* **95** (1954) 1329.
6. E.S. Fradkin, "Method of Green's Functions in Quantum Field Theory and in Quantum Statistics", in *Quantum Field Theory and Hydrodynamics*, edited by D.V. Skobel'tsyn, Proc. of the P.N. Lebedev Physics Inst. of the Academy of Sciences of the USSR, Moscow, Vol. 29, ©1967 by Consultants Bureau, Plenum Publishing Corp., New York (Lib. of Congress Card. Card No. 66-12629), pp. 1–132.
7. J. Zinn-Justin, "The Regularization Problem and Anomalies in Quantum Field Theory", in Proc. XVIII Lisbon Autumn School on *Topology of Strongly Correlated Systems* (October 8–13, 2000, CFIF, Lisbon, Portugal), edited by P. Bicudo, J.E. Ribeiro, P. Sacramento, V. Seixas, V. Vieira, ©2001 by World Scientific Publishing Co. Pte. Ltd. (ISBN 981-02-4572-6), pp. 141–183.
8. M. Benayoun, H.B. O'Connell, *Eur. Phys. J.* C **22** (2001) 503.
9. B.A. Kniehl, A. Sirlin, *Phys. Lett.* B **530** (2002) 129.
10. D. Espriu, J. Manzano, P. Talavera, *Phys. Rev.* D **66** (2002) 076002.
11. A. de Gouvêa, B. Kayser, R.N. Mohapatra, hep-ph/0211394.
12. M. D'Elia, M.-P. Lombardo, hep-lat/0209146, hep-lat/0205022.
13. N.D. Hari Dass, V. Soni, hep-th/0204177.
14. F. Kleefeld, E. van Beveren, G. Rupp, *Nucl. Phys.* A **694** (2001) 470.
15. H. Kaldass, A. Bohm, S. Wickramasekara, *Int. J. Mod. Phys.* A **17** (2002) 3749; R. de la Madrid, quant-ph/0201091; R.C. Fuller, *Phys. Rev.* **188** (1969) 1649; A.O. Barut, "Dispersion Relations and Resonance Scattering", in *Lectures in Theoretical Physics* (Vol. IV), delivered at Summer Inst. Theor. Phys., Univ. Colorado, Boulder, 1961, edited by W.E. Brittin et al., ©1962 by John Wiley & Sons (Lib. of Congress Cat. Card No. 59-13034), pp. 460–523.
16. N. Moiseyev, *Phys. Rep.* **302** (1998) 211.
17. N. Nakanishi, *Prog. Theor. Phys.* **19** (1958) 607.
18. N. Nakanishi, *Prog. Theor. Phys. Suppl.* **51** (1972) 1.
19. N. Nakanishi, *Phys. Rev.* D **5** (1972) 1968.
20. H.P. Stapp, *Il Nuovo Cimento* **23** A (1974) 357.
21. T. Regge, *Il Nuovo Cimento (Serie X)* **18** (1960) 947.
22. W. Heisenberg, "Einführung in die Theorie der Elementarteilchen", lecture given at University of Munich in 1961, lecture notes prepared by H. Rechenberg, K. Lagally (ICTP Trieste library number

539.12 H473), see p. 95 ff.

23. D. Hittner, *Il Nuovo Cimento* **15 A** (1973) 401.
24. F. Casagrande, L.A. Lugiato, *Il Nuovo Cimento* **17 A** (1973) 464.
25. I. Rabuffo, G. Vitiello, *Il Nuovo Cimento* **44 A** (1978) 401.
26. R. Tarrach, hep-th/9502020.
27. S. Ahlig, R. Alkofer, *Annals Phys.* **275** (1999) 113.
28. V. Šauli, hep-ph/0211221, hep-ph/0209046.
29. P. Maris, *Phys. Rev.* **D 52** (1995) 6087; *Phys. Rev.* **D 50** (1994) 4189.
30. A.W. Schreiber, R. Rosenfelder, C. Alexandrou, hep-th/0007182; C. Alexandrou, R. Rosenfelder, A.W. Schreiber, *Phys. Rev.* **D 62** (2000) 085009.
31. Helen R. Quinn, privat communication (May 7, 2002).
32. A. Mostafazadeh, math-ph/0209018; G. Scolarici, L. Solombrino, quant-ph/0211161.
33. P.A.M. Dirac, *Proc. Roy. Soc. (London)* **A 180** (1942) 1.
34. P.A.M. Dirac, *Comm. Dublin Inst. Adv. Studies, A.*, No. 1 (1943).
35. W. Pauli, *Rev. Mod. Phys.* **15** (1943) 175; S.N. Gupta, *Proc. Phys. Soc.* **63** (1950) 681 and *Proc. Phys. Soc.* **64** (1951) 850; K. Bleuler, *Helv. Phys. Acta* **23** (1950) 567; K. Bleuler, W. Heitler, *Prog. Theor. Phys.* **5** (1950) 600; J.M. Jauch, R. Rohrlich, *The theory of photons and electrons* (2nd corr. printing 1980), published by Springer-Verlag, ©1955 by J.M. Jauch, R. Rohrlich, see p. 103 ff; E.C.G. Sudarshan, "Lectures in theoretical physics", lecture given at Madras Univ. in 1970/71, lect. notes prepared by M. Seetharaman, K. Eswaran (available in the ICTP Trieste library).
36. H. Lehmann, *Il Nuovo Cimento* **11** (1954) 342; G. Källén, *Helv. Phys. Acta* **25** (1952) 417.
37. J. Schwinger, *Annals Phys.* **9** (1960) 169.
38. T. Spitzenberg, K. Schwenzer, H.-J. Pirner, *Phys. Rev.* **D 65** (2002) 074017; M.A. Stephanov, *Nucl. Phys. B (Proc. Suppl.)* **53** (1997) 469.
39. V.E. Markushin, R. Rosenfelder, A.W. Schreiber, *Il Nuovo Cimento* **117 B** (2002) 75.
40. A. Pais, "Max Born and the statistical interpretation of QM", KEK library preprint 198301341.
41. H.A. Weldon, *Nucl. Phys.* **B 534** (1998) 467.
42. F. Kleefeld, *Doctoral Thesis* (University of Erlangen-Nürnberg, 1999).
43. F. Kleefeld, *Acta Physica Polonica* **B30** (1999) 981.
44. F. Kleefeld, "Consistent Effective Description of Nucleonic Resonances in an Unitary Relativistic Field-Theoretic Way", in *Relativistic Nuclear Physics and Quantum Chromodynamics* (2 volumes), Proc. XIV Int. Seminar on High Energy Physics Problems (ISHEPP 98), 17.–22.8.1998, JINR, Dubna, Russia, edited by A.M. Baldin, V.V. Burov, ©2000 by Joint Institute for Nuclear Research, Dubna (ISBN 5-85165-570-4 (Vol. I)), pp. 69-77 in Vol. I (nucl-th/9811032).
45. G.C. Wick, A.S. Wightman, E.P. Wigner, *Phys. Rev.* **88** (1952) 101.
46. Jun-Chen Su, hep-th/9805195, hep-th/9805192, hep-th/9805193, hep-th/9805194.
47. G. 't Hooft, *Nucl. Phys.* **B 35** (1971) 167.
48. F. Kleefeld, "Does it make any sense to talk about a Δ-isobar?", in Proc. 2002 CFIF Fall Workshop on *Nuclear Dynamics: from Quarks to Nuclei*, 31.10.–2.11.2002, Lisbon, Portugal, nucl-th/0212008.
49. *Handbook of QCD* (3 volumes), edited by M. Shifman, ©2001 by World Scientific Publishing.
50. R. Delbourgo, M.D. Scadron, *Mod. Phys. Lett.* **A 10** (1995) 251; M. Gell-Mann, M. Lévy, *Il Nuovo Cimento* **16** (1960) 705; M. Lévy, *Il Nuovo Cimento* **A 52** (1967) 23.
51. R. Alkofer, H. Reinhardt, *Chiral Quark Dynamics* (Lecture Notes in Physics; New Series m: Monographs; Vol. 33 (= m 33)), ©1995 by Springer-Verlag Berlin Heidelberg.
52. R.W. Haymaker, J. Perez-Mercader, *Phys. Rev.* **D 27** (1983) 1353 and *Phys. Lett.* **106 B** (1981) 201.
53. N.D. Hari Dass, V. Soni, *Phys. Rev.* **D 65** (2002) 095005.
54. G. 't Hooft, *Phys. Rep.* **142** (1986) 357.
55. M.D. Scadron, F. Kleefeld, G. Rupp, E. van Beveren, hep-ph/0211275.
56. E. van Beveren, F. Kleefeld, G. Rupp, M.D. Scadron, *Mod. Phys. Lett.* **A 17** (2002) 1673.
57. A.A. Andrianov, D. Espriu, *J. HEP* **10** (1999) 022; J. Bijnens, *Phys. Rep.* **265** (1996) 369.
58. G. 't Hooft, *Nucl. Phys.* **B 72** (1974) 461, **B 75** (1974) 461; E. Witten, *Nucl. Phys.* **B 160** (1979) 57.
59. P.J.A. Bicudo, J.E.F.T. Ribeiro, A.V. Nefediev, *Phys. Rev.* **D 65** (2002) 085026.
60. P. Rembiesa, *Phys. Rev.* **D 38** (1988) 1916.
61. S. Okubo, V.S. Mathur, *Phys. Rev.* **D 1** (1970) 2046; P. Carruthers, R.W. Haymaker, *Phys. Rev.* **D 4** (1971) 1808 and *Phys. Rev.* **D 6** (1972) 1528; J.J. Brehm, *Phys. Rev.* **D 9** (1974) 1818.
62. A. Bramon, Riazuddin, M.D. Scadron, *J. Phys.* **G 24** (1998) 1.

Pion Structure Function and Violation of the Momentum Sum Rule

J.P. Lansberg[1], F. Bissey, J.R. Cudell, J. Cugnon, M. Jaminon and
P. Stassart

Université de Liège, Département de Physique B5, Sart Tilman,B-4000 LIEGE 1, Belgium

Abstract.
    We present a method to evaluate the pion structure functions from a box diagram
calculation. Pion and constituent quark fields are coupled through the simplest pseudoscalar
coupling. The $\gamma^\star \pi \to q\bar{q}$ cross-section is evaluated and related to the structure functions.
We then show that the introduction of non-perturbative effects, related to the pion size
and preserving gauge invariance, provides us with a straighforward relation with the quark
distribution. It is predicted that higher-twist terms become negligible for Q^2 larger than
about 2 GeV^2 and that quarks in the pion have a momentum fraction smaller than in the
proton. We enlarge the discussion concerning this violation of the momentum sum rule,
emphasizing that the sum rule is recovered in the chiral limit and also when the finite size
condition is not imposed.

1. INTRODUCTION

The usual way to describe quarks inside hadrons relies on the structure functions.
Quantum Chromodynamics (QCD), being the theory of the interactions between
quarks and gluons, is supposed to predict the structure functions, but in fact, these
quantities result from non-perturbative effects and hence are not easily modeled.
QCD is only able to predict their evolution as the virtuality of the probing photon
(Q^2) changes. It is our purpose here to build a simple model for these quantities.

    Phenomenological quark models, which possess some non-perturbative aspects
and which are rather successful in reproducing low-energy properties of hadrons, are
expected to help us understand the connection between deep-inelastic scattering
(DIS) data and non-pertubative inputs. For very simple systems such as pions
and other low-mass mesons, there exist some effective models (*e.g.* the Nambu-
Jona-Lasinio (NJL) model) that incorporate, in some simplified way, special QCD
features such as spontaneous chiral symmetry breaking and anomalies.

    However, investigations of the structure functions along these lines have given
rather different results [2, 3, 4]. This situation originates from the fact that the
NJL model needs to be regularized and that different regularizations yield different
results.

    We present here an alternative way to evaluate of the pion structure function

---

[1] email: JPH.Lansberg@ulg.ac.be

CP660, *Hadron Physics: Effective Theories of Low Energy QCD*, edited by A. H. Blin et al.
© 2003 American Institute of Physics 0-7354-0120-9/03/$20.00

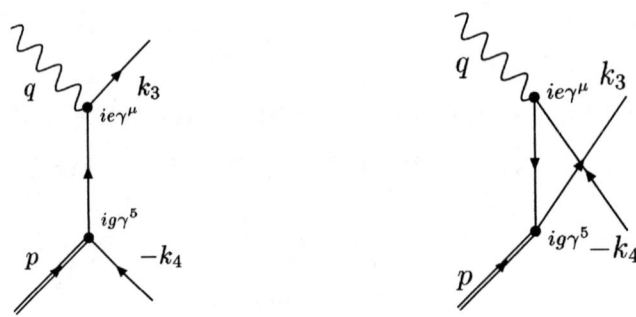

FIGURE 1. Feynman diagrams for the process $\gamma^\star \pi_0 \to q\bar{q}$.

based on Ref. [1], where the $q\bar{q}\pi$ vertex is represented by the simplest pseudoscalar coupling, *i.e.* $ig\gamma_5$ and the pion size is mimicked by the introduction of a gauge preserving cut-off.

2. CROSS-SECTION CALCULATION FOR $\gamma^\star \pi_0 \to Q\bar{Q}$ BY INTEGRATION OVER T

In this section, we calculate the cross-section for $\gamma^\star \pi_0 \to q\bar{q}$ which can be related to the form factors W_1 and W_2 as we shall see later. Another way to achieve this goal consists in calculating the imaginary part of the forward amplitude for $\gamma^\star \pi_0 \to \gamma^\star \pi_0$.

At the leading order in the loop expansion, we have the two diagrams shown in Fig. 1 which, when squared, give two types of contributions. The first one (corresponding to box-diagrams for the $\gamma^\star \pi_0 \to \gamma^\star \pi_0$ amplitude) enables us to calculate the structure functions straightforwardly. The second involves crossed quark lines and is related to processes where the two photons (in the $\gamma^\star \pi_0 \to \gamma^\star \pi_0$ scattering) are connected to different quark lines. This is incompatible with a probabilistic interpretation in terms of structure functions. As we will conclude from the calculations, contributions of this kind are fortunately higher-twists, *i.e.* they scale as $\frac{1}{Q^2}$, when we introduce the cut-off.

Their introduction is nevertheless mandatory, as this ensures gauge invariance (see Ref. [1]). This is required for a proper evaluation of the cross-section for the $\gamma^\star \pi_0 \to q\bar{q}$ process, which is a physical process, observable and thus gauge invariant.

2.1. Kinematics of the Process

For the sake of convenience, we work with cartesian coordinates in the CM (center of momentum) frame. The four-momenta of the four particles are given by:

$$q = (E(= q_0), \vec{q}), p = (E_\pi, -\vec{q}), k_3 = (k_{3,0}, \vec{k}_3), k_4 = (k_{4,0}, -\vec{k}_3). \tag{1}$$

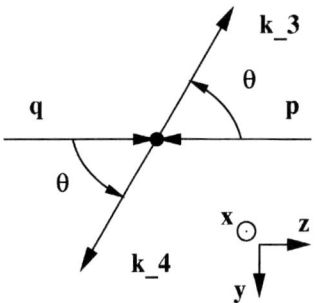

FIGURE 2. Kinematics and definition of the angle θ in the CM frame.

Let us now introduce the Mandelstam variables:

$$s = (p+q)^2 = (k_3+k_4)^2 = E_{CM}^2 = (E+E_\pi)^2,$$
$$t = (k_3-q)^2 = (k_4-p)^2, u = (k_3-p)^2 = (k_4-q)^2. \tag{2}$$

We then define the square momenta:

$$q^2 = -Q^2, p^2 = m_\pi^2, k_3^2 = m^2, k_4^2 = m^2 \tag{3}$$

The variables of Eq. (1) can be written in terms of these quantities and of Mandelstam variables. One has:

$$k_{3,0} = k_{4,0} = \frac{\sqrt{s}}{2} \quad , \quad |\vec{k_3}| = |\vec{k_4}| = \sqrt{\frac{s}{4} - m^2}, \tag{4}$$

$$E = \frac{s - m_\pi^2 - Q^2}{2\sqrt{s}}, \tag{5}$$

and

$$|\vec{p}|^2 = |\vec{q}|^2 = E^2 + Q^2 = \frac{(s - m_\pi^2 - Q^2)^2 + 4sQ^2}{4s}. \tag{6}$$

The angle θ between the vectors $\vec{k_3}$ and $\vec{q}$ (see Fig. 2) is related to t by

$$t = m^2 + m_\pi^2 - s + \sqrt{s}\frac{s - m_\pi^2 - Q^2}{2\sqrt{s}} + 2\sqrt{\frac{s}{4} - m^2}\frac{\sqrt{(s - m_\pi^2 - Q^2)^2 + 4sQ^2}}{2\sqrt{s}}\cos\theta$$

Eventually we obtain:

$$\cos\theta = \frac{t - m^2 - \frac{m_\pi^2}{2} + \frac{s}{2} + \frac{Q^2}{2}}{2\sqrt{\frac{s}{4} - m^2}\frac{\sqrt{(s-m_\pi^2-Q^2)^2+4sQ^2}}{2\sqrt{s}}} = \frac{t - m^2 - \frac{m_\pi^2}{2} + \frac{s}{2} + \frac{Q^2}{2}}{2\sqrt{\frac{s}{4} - m^2}\sqrt{E^2 + Q^2}}. \tag{7}$$

The physical values of t are lying between boundaries, corresponding to $\cos\theta = \pm 1$:

$$t_{\max\atop\min} = m^2 + \frac{m_\pi^2}{2} - \frac{s}{2} - \frac{Q^2}{2} \pm 2\sqrt{\frac{s}{4} - m^2}\frac{\sqrt{(s - m_\pi^2 - Q^2)^2 + 4sQ^2}}{2\sqrt{s}} \qquad (8)$$

Finally, assuming the incident direction along the z axis and the scattering in the $y - z$ plane, we get:

$$q = (\frac{s - m_\pi^2 - Q^2}{2\sqrt{s}}, 0, 0, \sqrt{\frac{(s - m_\pi^2 - Q^2)^2 + 4sQ^2}{4s}}) = (E, 0, 0, \sqrt{E^2 + Q^2}),$$

$$p = (\sqrt{m_\pi^2 + \frac{(s - m_\pi^2 - Q^2)^2 + 4sQ^2}{4s}}, 0, 0, -\sqrt{\frac{(s - m_\pi^2 - Q^2)^2 + 4sQ^2}{4s}})$$

$$k_3 = (\frac{\sqrt{s}}{2}, 0, -\sqrt{\frac{s}{4} - m^2}\sin(\theta), \sqrt{\frac{s}{4} - m^2}\cos\theta),$$

$$k_4 = (\frac{\sqrt{s}}{2}, 0, \sqrt{\frac{s}{4} - m^2}\sin(\theta), -\sqrt{\frac{s}{4} - m^2}\cos\theta). \qquad (9)$$

These quantities will be explicitly used in the determination of scalar products involving the polarisation vectors.

2.2. Amplitude Calculation

Applying Feynman rules for the processes shown in Fig. 1, we obtain the analytical expression of the square amplitude in a straightforward way. For a given polarisation of the virtual photon, $i = L, T_1, T_2$, and with well-known conventions[2] we have for the squared of the first diagram shown in Fig. 1,

$$|\mathcal{M}_i|^2 = 3g^2(e_u^2 + e_d^2)\text{Tr}\left((-\rlap{/}k_4 + m)\gamma^5\frac{\rlap{/}k_3 - \rlap{/}q + m}{(k_3 - q)^2 - m^2}\gamma^{\mu'}\right.$$
$$\left.(\rlap{/}k_3 + m)\gamma^\mu\frac{\rlap{/}k_3 - \rlap{/}q + m}{(k_3 - q)^2 - m^2}\gamma^5\right)\varepsilon_{i\mu}\varepsilon_{i\mu'}^\star. \qquad (10)$$

The expression for the square of the second diagram as well as for the interference terms have similar expressions.

We first proceed to the calculation of the expressions of the scalar products of the polarisation vectors with q, p, k_3 and k_4 for the three possible polarisation states.

We shall use the following definition of the longitudinal polarisation vector,

$$\varepsilon_{L,\mu} = \frac{1}{\sqrt{Q^2}}(q_z, 0, 0, E). \qquad (11)$$

---

[2] The isospin/charge factor $(e_u^2 + e_d^2)$ corresponds to the following choice of the isospin matrix:

$$\pi^- : \begin{pmatrix} 0 & 0 \\ \sqrt{2} & 0 \end{pmatrix} - \pi^0 : \begin{pmatrix} 1 & 0 \\ 0 & -1 \end{pmatrix} - \pi^+ : \begin{pmatrix} 0 & \sqrt{2} \\ 0 & 0 \end{pmatrix} - \gamma : \begin{pmatrix} e_u & 0 \\ 0 & e_d \end{pmatrix}$$

One has

$$\varepsilon_L \cdot q = 0,$$

$$\varepsilon_L \cdot p = \frac{1}{\sqrt{Q^2}} \sqrt{s} \sqrt{E^2 + Q^2},$$

$$\varepsilon_L \cdot k_3 = \frac{1}{\sqrt{Q^2}} \left(\frac{\sqrt{s}\sqrt{E^2+Q^2}}{2} - \frac{E(t - m^2 - \frac{m_\pi^2}{2} + \frac{s}{2} + \frac{Q^2}{2})}{2\sqrt{E^2+Q^2}} \right). \tag{12}$$

For the linear transverse polarisation vectors, we use the following definition:

$$\varepsilon_{T_1,\mu} = (0,1,0,0), \quad \varepsilon_{T_2,\mu} = (0,0,1,0). \tag{13}$$

This and Eq. (9) lead to:

$$\varepsilon_{T_i} \cdot q = 0 = \varepsilon_{T_i} \cdot p, \quad i = 1,2,$$

$$\varepsilon_{T_1} \cdot k_3 = 0,$$

$$\varepsilon_{T_2} \cdot k_3 = \sqrt{\frac{s}{4} - m^2} \sqrt{1 - \left(\frac{t - m^2 - \frac{m_\pi^2}{2} + \frac{s}{2} + \frac{Q^2}{2}}{2\sqrt{\frac{s}{4} - m^2}\sqrt{E^2 + Q^2}} \right)^2}. \tag{14}$$

2.3. Polarised Cross-Sections

The total cross-section is given by the following relation [5], with $i = L, T_1, T_2$:

$$\frac{d\sigma_i}{dt} = \frac{1}{16\pi\lambda} |\mathcal{M}_i|^2 \tag{15}$$

with[3] $\lambda = \lambda(s, -Q^2, m_\pi^2)$. The latter directly links the amplitude –which is obtained, in our case, with REDUCE– and the cross-section differential upon t. Integrating over t from t_{min} to t_{max} (see Eq. (8)) gives the total cross-sections for given polarisations:

$$\sigma_i = \frac{1}{16\pi\lambda} 3g^2(e_u^2 + e_d^2) \int dt \, T^{\mu\mu'} \varepsilon_{i,\mu} \varepsilon_{i\mu'}^\star, \tag{16}$$

with $T^{\mu\mu'} \equiv \text{Tr}\left((-\not{k}_4 + m)\gamma^5 \frac{\not{k}_3 - \not{q} + m}{(k_3-q)^2 - m^2} \gamma^{\mu'} (\not{k}_3 + m)\gamma^\mu \frac{\not{k}_3 - \not{q} + m}{(k_3-q)^2 - m^2} \gamma^5 \right)$.

It is also of interest to consider

$$\sigma_{contr.} = \frac{1}{16\pi\lambda} 3g^2(e_u^2 + e_d^2) \int dt \, T^{\mu\mu'} \left(g_{\mu\mu'} + \frac{q_\mu q_{\mu'}}{Q^2} \right) \tag{17}$$

since it does not involve any polarisation vectors and its calculation is somewhat easier. However, it should not confused with $\sigma_L + \sigma_{T_1} + \sigma_{T_2}$ (see later), which could be measured in some limiting experimental conditions.

---

[3] The function $\lambda(x,y,z)$ is defined as $\lambda(x,y,z) \equiv x^2 + y^2 + z^2 - 2xy - 2xz - 2yz$

2.4. The Cut-Off

2.4.1. Cut-off and Higher-twists

We introduce a cut-off in our calculations: instead of integrating over t from $t_{\min}$ to $t_{\max}$, we integrate from[4] $Max\{t_{\min}, \Lambda_t\}$ to $t_{\max}$. As shown in Ref. [1], this is equivalent to setting a cut-off for the square of the relative momentum of the quarks inside the pion, which can be considered as resulting from the finite size of the pion.

Irrespectively of the factors arising in the extraction of the structure functions (see section (3)), we can simply convince ourselves that in the presence of a cut-off the contribution of the crossed diagrams (the first photon hits a quark and the second an antiquark) are higher-twists, by plotting the ratio of these two contributions as a function of Q^2. The suppression is clear on Fig. 3 for whatever (finite) value of the cut-off; the curves are even indistinguishable in this case.

As we already said, in order to get a physical interpretation of our results –in terms of structure functions–, we need the crossed diagram contributions to be suppressed. The importance of a cut-off, which is not a regulator in the studied process, is thus manifest.

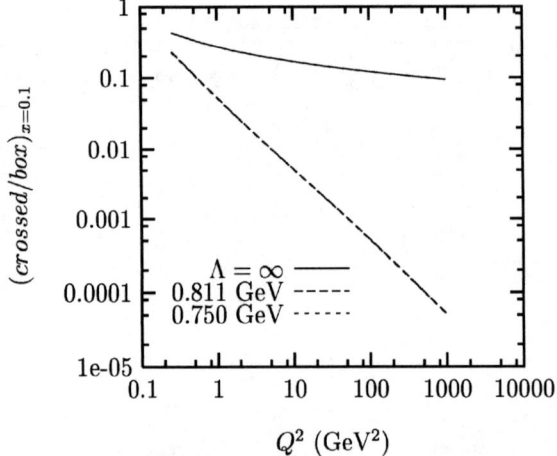

FIGURE 3. Ratio of the direct contributions to the crossed contributions to the cross-section for $\gamma^* \pi^0 \to q\bar{q}$ at $x = 0.1$ and for three values of the cut-off Λ.

---

[4] Recall that $t < 0$.

2.4.2. Cut-off and Gauge Invariance

In most cases, the presence of a cut-off implies the breakdown of gauge invariance. This is the main reason why one commonly uses other regularization methods such as Pauli-Villars or dimensional regularization. In the context of an effective theory, a simple momentum cut-off has the critical advantage that the regularization parameter (which is kept to a finite value) can be interpreted as a physical one. In the case of dimensional regularization, the signification of fractional dimensional space is not at all clear. As for Pauli-Villars regularization, the introduction of wrong-sign fermionic contribution is not intuitive.

Nevertheless, in our model, as the intermediate states in $\gamma^\star \pi^0 \to \gamma^\star \pi^0$ are on-shell, the cut-off on the internal loop momentum[5] acts on a physical observable, that is t of the process $\gamma^\star \pi^0 \to q\bar{q}$. The conclusion is thus staightforward: this cut-off does not break gauge invariance.

This procedure would have been fully justified if we had only one pion vertex, and thus one condition to implement. But we have two vertices with two different expressions for the relative momenta. To keep gauge invariance, the cut-off on t must be on the overall cross-section as the physical observable is the full cross-section (the square of the sum of the two diagrams) at a given t.

To keep the constraint coming from the pion size, we should have imposed two conditions on the cross-section to be integrated over t. However, this would have given zero. As a makeshift we imposed one condition or the other.

Even if the physical interpretation is not so obvious, we thus kept gauge invariance, information from the pion size, suppressed the crossed diagrams and hence obtained an interpretation in terms of structure functions.

3. $\sigma_{\gamma^\star \pi_0 \to Q\bar{Q}}$, W_1, W_2, F_1 AND F_2

At this stage we have calculated the polarised cross-sections and we should be able to link them to the structure functions for neutral pion. Before doing this, we shall first recall some mandatory formulae derived from the analysis of deep inelastic proton scattering.

We introduce the usual form of the hadronic tensor

$$W_{\mu\nu}(p,q) = (-g_{\mu\nu} - \frac{q_\mu q_\nu}{Q^2})W_1(Q^2,\nu) + [p_\mu + (\frac{p \cdot q}{Q^2})q_\mu][p_\nu + (\frac{p \cdot q}{Q^2})q_\nu]\frac{W_2(Q^2,\nu)}{M^2}, \quad (18)$$

with M, the hadron mass, and $W_1(Q^2,\nu)$ and $W_2(Q^2,\nu)$ the structure functions.

In the same spirit as for the definition of the cross-section of $\gamma^\star p \to X$ [6], we define the polarised cross-sections for $\gamma^\star \pi_0 \to X$ by [6]:

---

[5] which corresponds, up to some constants, to a cut-off on the relative mometum between the two quarks at the $\pi^0 q\bar{q}$ vertex.

[6] This definition is independent of the hadron spin because, on the one hand, we have to sum the contributions coming from the different polarisations and on the other hand, we have a spin

345

$$\sigma_i(\gamma^\star \pi_0 \to X) = \frac{4\pi^2\alpha}{K}\varepsilon_i^{\star\mu}\varepsilon_i^\nu W_{\mu\nu}, \tag{19}$$

with the flux factor $K = \frac{1}{2}\frac{\sqrt{\lambda(s,-Q^2,m_\pi^2)}}{m_\pi}$.

Defining the transverse cross-section through

$$\sigma_T \equiv \frac{1}{2}(\sigma_{T_1} + \sigma_{T_2}), \tag{20}$$

we can easily show that:

$$\begin{aligned}
\sigma_T &= \frac{4\pi^2\alpha}{K}W_1, \\
\sigma_L &= \frac{4\pi^2\alpha}{K}\left[(1+\frac{\nu}{2xm_\pi^2})W_2 - W_1\right].
\end{aligned} \tag{21}$$

These last two formulae establish the link between the functions W_1 and W_2 and previously calculated polarised cross-sections for $\gamma^\star\pi_0 \to q\bar{q}$ (Eq. (16))[7].

---

factor arising in the relation between $W_{\mu\nu}$ and the hadronic current.

[7] Besides we would like to stress that one has to be cautious if one wants to consider non-polarised processes. Usually, a simple contraction of the amplitude for a given helicity with the expression,

$$-g^{\mu\nu} + \frac{q^\mu q^\nu}{M^2} \tag{22}$$

seems to be the most convenient way to proceed, instead of considering separately the different polarisations. This is in most cases licit because

$$\sum_{i=L,T_1,T_2} \varepsilon_i^{\star\mu}\varepsilon_i^\nu = -g^{\mu\nu} + \frac{q^\mu q^\nu}{M^2}, \tag{23}$$

for massive particles, and thus the –strict– sum over polarisations gives the *correct* tensor.

For spacelike virtual particles, nevertheless, we have the following normalisations

$$\begin{aligned}
\varepsilon_T \cdot \varepsilon_T &= -1, \\
\varepsilon_L \cdot \varepsilon_L &= +1.
\end{aligned} \tag{24}$$

Therefore, the following relation –which indeed reduces to Eq. (23) for massive particles for which Eq. (24) does not hold–,

$$\sum_i \frac{\varepsilon_i^{\star\mu}\varepsilon_i^\nu}{\varepsilon_i \cdot \varepsilon_i} = g^{\mu\nu} + \frac{q^\mu q^\nu}{Q^2} \tag{25}$$

leads, in this case, to

$$\sigma_{contr.} = \frac{4\pi^2\alpha}{K}\left(g^{\mu\nu} + \frac{q^\mu q^\nu}{Q^2}\right)W_{\mu\nu} = \sigma_L - 2\sigma_T \tag{26}$$

but not to $\sigma_L + 2\sigma_T$!

With all these tools, we can now extract the functions W_1 and W_2. Inverting for W_1 and W_2, we get

$$W_1 = \frac{\sqrt{\lambda}}{8\pi^2 m_\pi \alpha} \sigma_T \tag{27}$$

and

$$W_2 = \frac{\sqrt{\lambda}}{8\pi^2 m_\pi \alpha} \frac{2x m_\pi^2}{\nu + 2x m_\pi^2} (\sigma_L + \sigma_T). \tag{28}$$

We then get F_1 and F_2 through the usual scaling relations:

$$F_1(x) = M W_1(Q^2, \nu) \ , F_2(x) = \frac{\nu}{M} W_2(Q^2, \nu) \ , x = \frac{Q^2}{2p \cdot q} = \frac{Q^2}{2\nu}, \tag{29}$$

still with M for the hadron mass.

4. SUM RULES, PLOTS AND RESULTS

4.1. Number of Particles Sum Rule

The structure functions can now be related to the (valence) quark distributions:

$$F_1 = \frac{4}{18}(u_v(x) + \bar{u}_v(x)) + \frac{1}{18}(d_v(x) + \bar{d}_v(x)), \tag{30}$$

$$F_2 = 2x F_1. \tag{31}$$

We stress that the last relation, known as the Callan-Gross relation, comes out of our calculation. This does indicate that our approximations have been done consistently[8].

So far, we have only described the model. In order to make predictions, we need to fix its parameters, namely Λ, m and g. The latter can be thought of as the normalisation of the quark wave function, and we determine it by imposing that there are only two constituent quarks in the pion. In our model, the valence quark distributions are equal

$$u_v(x) = \bar{u}_v(x) = d_v(x) = \bar{d}_v(x) \equiv v(x). \tag{32}$$

The condition $\int_0^1 v(x)dx = 1/2$ then yields

$$\int_0^1 F_1(x)dx = \frac{5}{18}. \tag{33}$$

---

[8] For charged pions, the additional diagrams implying a direct coupling of the virtual photon to the pion are suppressed by a factor $1/s$ and the leading-twist results are the same, except for the charge coefficients entering Eq. (30) .

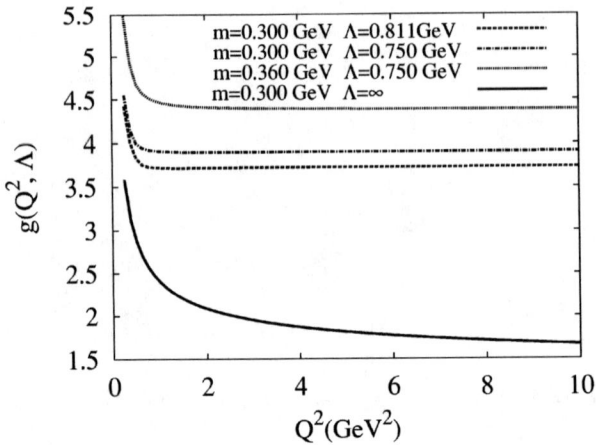

FIGURE 4. Values of the coupling constant $g(Q^2, \Lambda)$ which fulfill the sum rule (Eq. (33)), for the values of the parameters indicated at the top.

As F_1 is a function, not only of m and Λ, but also of Q^2, this gives us a coupling constant that evolves with Q^2. The resulting values of g are shown in Fig. 4 (right): for a finite cut-off Λ, the cross-section at fixed g would grow with energy, until the pion reaches its maximum allowed size, in which case the cross section would remain constant. If we impose relation Eq. (33), this means that $g(Q^2)$ will first decrease until the cut-off makes it reach a plateau for $Q^2 \gg \Lambda^2$ (in practice, the plateau is reached around $Q^2 \approx 2\Lambda^2$). The plateau value depends on m/Λ (and m_π/Λ).

4.2. Momentum Sum Rule

To constrain further our parameters, we can try to use the momentum sum rule

$$2 \langle x \rangle = 4 \int_0^1 xv(x)dx = \frac{18}{5} \int_0^1 F_2(x)dx. \qquad (34)$$

In the parton model, this integral should be equal to 1 as $Q^2 \to \infty$, as we do not have gluons in the model.

Analyzing Fig. 5, it is clear that – at least for a finite value of Λ– the momentum sum rule is violated. For instance the momentum carried by the quarks at large Q^2 is 0.6 for $m = 300$ MeV and $\Lambda = 750$ MeV.

Nevertheless, there are two limits of our model that fulfill the condition $2 \langle x \rangle = 1$ automatically and for which the quarks carry the entire momentum of the pion. First of all, if we do not impose that the pion has a finite size, then asymptotically we get $2 \langle x \rangle = 1$. In the limit $m_\pi = 0$, we obtain the simple expression

$$2 \langle x \rangle \, (\Lambda = \infty, m_\pi = 0) = \frac{4\ln(2\nu/m^2) - 3}{4\ln(2\nu/m^2) - 1}. \qquad (35)$$

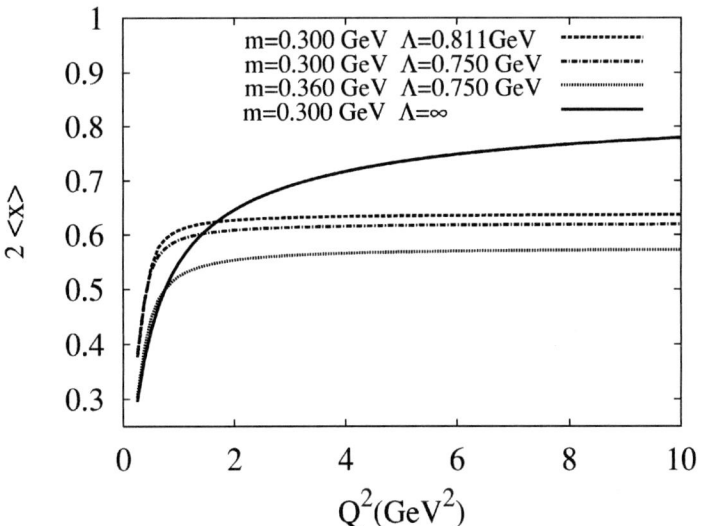

FIGURE 5. Momentum fraction of the quarks inside the neutral pion (Eq. 34), as a function of Q^2, for the values of the parameters indicated at the top of the figure.

This means that in that regime, for sufficiently high Q^2, the quarks behave as free particles, and the usual derivation based on the OPE holds [8].

The second case where this holds confirms this interpretation: if we impose $m \leq m_\pi/2$, the expressions we have given develop an infrared divergence, which corresponds to the case where both quarks emerging from the pion are on-shell and free. This divergence can be re-absorbed into the normalisation Eq. (33) of g, and the sum rule $2\langle x \rangle = 1$ is again automatically satisfied at large Q^2.

However, in the physical pion case, it makes more sense to consider that one of the quarks remains off-shell: the introduction of a cut-off changes the sum rule value, as the fields can never be considered as free. Hence, because of the Goldstone nature of the pion, one expects that the momentum sum rule will take a smaller value than in the case of other hadrons. This may explain why fits that assume the same momentum fraction for valence quarks in protons and pions [9] do not seem to leave any room for sea quarks [10].

We come back to the results for $2\langle x \rangle$ shown in Fig. 5. Again, the curves show a plateau at sufficiently large Q^2, with a value depending on m/Λ and m_π/Λ. It is easy to show that this value is always smaller than 1 as it can be guessed from Fig. 6.

Following the discussion in [1], we choose the following conservative values for the remaining parameters of our model: $m = 300$ MeV or 360 MeV and $\Lambda \simeq 800$ MeV.

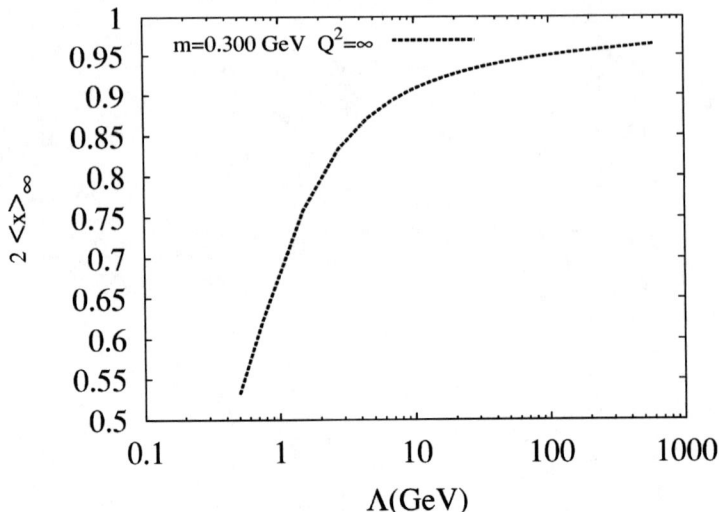

FIGURE 6. Asymptotic value (for large Q^2) of the momentum fraction of the quarks inside the neutral pion (Eq. 34), as a function of the cut-off Λ.

4.3. Plots of F_2

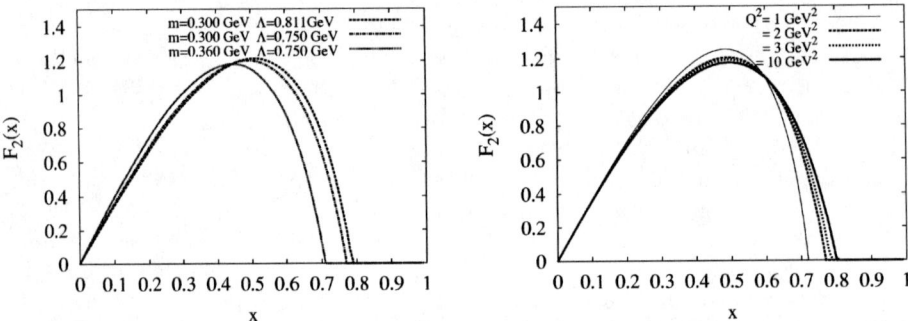

FIGURE 7. Structure function F_2 for the neutral pion. On the left, $Q^2=2$ GeV^2 and the values of the parameters m and Λ are as indicated. On the right, $m=0.3$ GeV and $\Lambda=0.75$ GeV and the values of Q^2 are as indicated.

Let us finally examine the properties of the distribution $v(x)$, or equivalently, of the function F_2. Some of our results are summarized in Fig. 7, for $Q^2 = 2$ GeV^2. The most striking feature is the vanishing of this function for x larger than some value x_{max}. This is again due to the fact that, because the quarks are not free in this model, the actual value of their mass does matter. It can clearly be seen that this effect originates from kinematical cuts (see Ref. [1]).

As a result, putting the cut quarks on-shell will be easier for small than for large x. Therefore, the x-distribution (F_1) is expected to be enhanced on the low x side, leading to a momentum fraction smaller than unity. The vanishing at large x is not obtained in similar works, in particular in the one of Ref. [2]. In this reference, the Bjorken limit is taken first and the kinematical constraint is not applied. This procedure seems questionable for evaluating cross-sections at finite Q^2.

5. CONCLUSION

We have discussed the simplest model allowing to relate $\gamma^\star \pi \to q\bar{q}$ to the pion quark distributions. The model consists in calculating the simplest diagrams for a γ^5 coupling between pion and constituent quark fields.

We show that the crossed diagrams, which are to be included in order to guarantee gauge invariance, preclude the existence of the relation between the the cross-section and the quark distribution. The introduction of a cut-off on t allows such a relation without breaking gauge invariance. The cut-off on t being equivalent to a cut-off on the square of the relative momentum of the quarks inside pion, this effect can be traced back to the finite size of the pion. For reasonable value of the cut-off, the crossed diagrams become negligible as soon as Q^2 is larger than $2\,\text{GeV}^2$ allowing quark distributions to be meaningful even for relatively small Q^2.

One of the most important outcomes of our calculation is the deviation from the momentum sum rule. However, we recover this sum rule $(2\langle x \rangle = 1)$ in two limiting cases, which correspond to free quarks. The first one is the limit $\Lambda \to \infty$. The second one is for $m < \frac{m_\pi}{2}$. Then, an infrared divergence appears, which can be reabsorbed into the normalisation of g and the momentum sum rule is again recovered for large Q^2. Hence, the violation of the sum rule can tentatively be attributed to the Goldstone boson nature of the pion.

ACKNOWLEDGMENTS

This work has been performed in the frame of ESOP Collaboration and has benefited from the financial support of the EU (contract N° HPRN-CT-2000-00130).

REFERENCES

1. F. Bissey, J. R. Cudell, J. Cugnon, M. Jaminon, J. P. Lansberg and P. Stassart, Phys. Lett. B **547**, 210 (2002) [arXiv:hep-ph/0207107].
2. T. Shigetani, K. Suzuki and H. Toki, Phys. Lett. B **308**, 383 (1993) [arXiv:hep-ph/9402286].
3. R. M. Davidson and E. Ruiz Arriola, Phys. Lett. B **348**, 163 (1995) ; H. Weigel, E. Ruiz Arriola and L. P. Gamberg, Nucl. Phys. B **560**, 383 (1999) [arXiv:hep-ph/9905329].
4. R. M. Davidson and E. Ruiz Arriola, Acta Phys. Polon. B **33**, 1791 (2002) [arXiv:hep-ph/0110291].
5. V.D. Barger and R.J.N. Philips, *Collider Physics*, Addison-Wesley, Menlo Park, 1987.
6. I.J.R. Aitchison and A.J.G. Hey, *Gauge Theories in Particles Physics* (first edition), Institute of Physics Publishing, Bristol, 1982, p. 95.
7. C. G. Callan and D. J. Gross, *Phys. Rev. Lett.* **22**, 156 (1969).
8. C. Itzykson and J. B. Zuber, *Quantum Field Theory*, New York, U.S.A.: McGraw-Hill (1980) 705 P. (International Series In Pure and Applied Physics).
9. M. Glück, E. Reya and I. Schienbein, Eur. Phys. J. C **10**, 313 (1999) [arXiv:hep-ph/9903288].
10. M. Klasen, J. Phys. G **28**, 1091 (2002) [arXiv:hep-ph/0107011].

The light scalar mesons within quark models

E. van Beveren*, G. Rupp†, N. Petropoulos* and F. Kleefeld†

*Centro de Física Teórica, Departamento de Física, Universidade, P3004-516 Coimbra, Portugal
†Centro de Física das Interacções Fundamentais, Instituto Superior Técnico, Edifício Ciência,
P1049-001 Lisboa Codex, Portugal

Abstract. Low-energy meson-meson scattering data are a powerful testing ground for quark models. Here, we describe the behaviour at threshold of S-wave scattering-matrix singularities.
The majority of the full scattering-matrix mesonic poles stem from an underlying confinement spectrum. However, the light scalar mesons $K_0^*(830)$, $a_0(980)$, $f_0(400-1200)$, and $f_0(980)$ do not, but instead originate in $^3P_0$-barrier semi-bound states. We show that the behaviour of the corresponding poles is identical at threshold.
In passing, the light-meson sector is given a firm basis.

Introduction. It is generally understood that, once the meson sector of strong interactions is fully and consistently described, not too many complications are expected upon including the baryons as well. However, as things stand, important steps have yet to be taken towards a simple quantitative theory for the description of mesons and their interactions [1]. Preferably, this should be a one-parameter theory which, in the limit of free quarks and gluons and some non-abelian interaction, approaches QCD. Here, we pay attention to the unification of all flavours.

For a lowest-order approximation of strong interactions, confinement models may be constructed. Their usefulness can be measured by the models' achievements when adjusting their parameters to experiment[2]. Observed spectra may be interpreted in terms of quark-antiquark or more complicated systems [3]. But in a full theory with quarks and mesons, one can study strong interactions through meson-meson scattering. Consequently, one needs more than confinement only. For further refinements of strong-interaction models, one should compare the models' predictions directly to experimental scattering cross sections and phase shifts when available [4, 5, 6, 7, 8, 9, 10, 11].

Here we will discuss the eternally disputed [12, 13] low-lying nonet of S-wave poles in meson-meson scattering cross sections, within a four-parameter model [7]. It should thereby be noted that the model's parameters are fitted to the $J^P = 1^-$ $c\bar{c}$ and $b\bar{b}$ spectra, as well as to P-wave meson-meson scattering data [6].

First, let us outline the motivation for our work. For the interaction in the vicinity of a resonance in meson-meson scattering, one may consider quark-exchange or quark-pair-creation processes, giving rise to an intermediate $q\bar{q}$ system. When the intermediate $q\bar{q}$ system is close enough to a genuine bound state of confinement, then the system will resonate, resulting in a resonance in meson-meson scattering. Another picture for the same phenomenon is to consider self-energy contributions from virtual meson loops. Either picture describes the same physical situation, namely a mesonic resonance or bound state [14], but in a rather different way. Our aim is to merge both pictures in one

CP660, *Hadron Physics: Effective Theories of Low Energy QCD*, edited by A. H. Blin et al.
© 2003 American Institute of Physics 0-7354-0120-9/03/$20.00

model.

Flavour-independent confinement. Let us assume that the spectrum of mesonic quark-antiquark systems can be described by flavour-independent harmonic-oscillator confinement. Then, for each pair of flavours, an infinite set of mesons exists with all possible spin, angular, and radial excitations. But, unfortunately, for most flavour pairs only a few angular and even fewer radial recurrencies are known [15]. When we do not distinguish up and down, but just refer to non-strange (n) quarks and, moreover, ignore the existence of top quarks, then we dispose of four different flavours: n, s, c, and b. These can be combined into ten different flavour pairs, each of which may come in two different spin states: 0 or 1. This gives rise to, in principle, twenty different meson spectra. With some 150 known mesons, this means 7.5 angular plus radial excitations on average, per flavour pair. This is much less than e.g. the known excitations of the positronium spectrum. No wonder that it requires some imagination to guess economic strategies for the description of mesons.

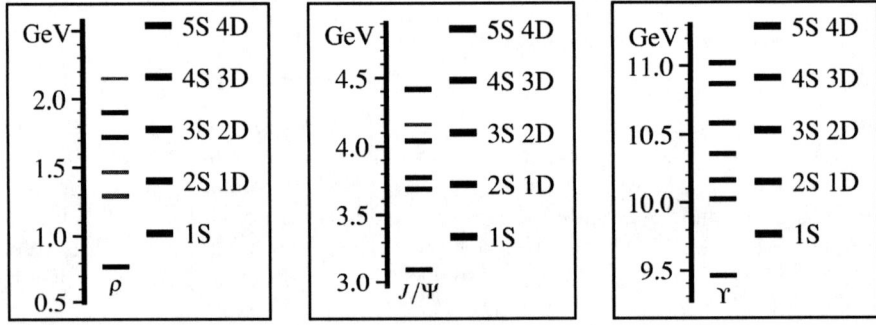

FIGURE 1. Non-strange, charmonium, and bottomonium $J^{PC} = 1^{--}$ states compared to the corresponding states from a harmonic-oscillator spectrum. The level spacing for the oscillator equals 0.38 GeV.

As one may verify from the latest Review of Particle Physics [15], vector states are more numerous than any other type of mesonic resonances, hence better known. Consequently, in order to structure a model we begin with the vector mesons, carrying quantum numbers $J^P = 1^-$. In figure (1) we compare the measured $n\bar{n}$, $c\bar{c}$, and $b\bar{b}$ vector states with the possible states of the harmonic oscillator. Most of the data are taken from Ref. [15]. The $\rho(1290)$ signal has been reported in Refs. [16, 17, 18, 19, 20]. The $\Upsilon(1D)$ is our interpretation of the D state which has been observed in Ref. [21].

The charmonium vector states, shown in figure (1), bear many similarities with the two-particle harmonic oscillator: a ground state in S wave, and higher radial excitations that are *almost* degenerate with the D-wave states. Also, except for the ground state, the level spacings are roughly equal. In Ref. [22] the mechanism is discussed which turns the oscillator spectrum into the charmonium spectrum.

For the ρ and Υ vector states, also shown in figure (1), we see a very similar pattern: the S-D splittings are slightly larger, while the $\rho(770)$ ground state of the ρ spectrum and the $\Upsilon(1S)$ ground state of the Υ spectrum also come out far below the corresponding oscillator ground states. From figure (1) one may moreover conclude that there is not much reason to separate the light-quark sector from the heavy quarks. Below, we discuss the mechanism which turns the oscillator states into the ρ and Υ resonances [6].

What we learn from the above comparison is that confinement is flavour independent. Hence, confinement should be described by flavour-independent dynamics. A non-relativistic Schrödinger equation with flavour-mass-dependent harmonic-oscillator [6] is just a perfect example of such dynamics. It casts quark confinement in the form of a one-parameter model. This parameter is the oscillator frequency ω, which comes out at about 0.19 GeV for the data.

Mechanism for more structure. At this point we dispose of a beautiful one-parameter model for mesons, which has a particularly simple spectrum for each of the flavour and spin excitations, given by

$$M\left(f,\bar{f};\ell,n\right)) = \omega\left(2n+\ell+\frac{2}{3}\right) + m_f + m_{\bar{f}} \ . \tag{1}$$

Here, f and $\bar{f}$ represent the flavours of respectively the quark and the antiquark, m_f and $m_{\bar{f}}$ their respective masses, ℓ and n their relative angular momentum and radial excitation.

Let us study some details of formula (1) in the following. The vector-meson states have unit total angular momentum, $J = 1$, and unit $q\bar{q}$ total spin, $s = 1$. Hence, since the parity of vector-meson states equals $P = -1$, their orbital angular momentum can be $\ell = 0$ (S wave) or $\ell = 2$ (D wave). From formula (1) we then understand that, within harmonic-oscillator confinement, the vector-meson states with $(n, \ell = 2)$ are degenerate with the vector-meson states with $(n+1, \ell = 0)$, as shown in figure (1). For other flavour and spin excitations similar results emerge. One obtains a very regular, equally spaced spectrum of quark-antiquark states. However, by comparing the experimental meson spectrum to the theoretical harmonic oscillator spectrum (1), we readily see that our simple model by far does not agree with the data.

In order to cure this disagreement, we may study other potentials, which generate spectra that agree better with the data, and even modify, whenever necessary, their parameters for different flavour sectors. But here we prefer to stick to our one-parameter model for all flavours.

When we ignore the electroweak interactions, then the mesons of our model are permanently stable quark-antiquark systems. For the great majority of mesons, this picture badly conflicts with observation. Strong interactions are not just confinement, but also hadronic decay, and elastic and inelastic scattering of hadrons. Our one-parameter model just describes confinement. All other strong phenomena must still be included.

Hadronic decay is quantitatively well described by the phenomenon of quark-pair creation [23], i.e., the creation of a valence quark and a valence antiquark out of the vacuum. Quark pairs are supposed to be created and annihilated all the time inside the realm of a hadron. Only once in a while such pair develops into a valence quark pair, which allows the hadron to decay into other hadrons. We will represent the probability for that process to occur by one parameter, λ, which, in the spirit of flavour symmetry, we assume to be constant for all flavours. In practice, we only consider the creation of non-strange and strange valence quark pairs, since the thresholds for charm and bottom are much higher [24], far above the few mesonic resonances which we want to describe.

Both λ and the oscillator frequency ω should be related to the fundamental parameter α of QCD. However, having no knowledge about such relation, we accept them as free parameters. Nevertheless, they are chosen independent of the flavour pair which constitutes the meson under consideration. This way we guarantee flavour independence of our model's results, as dictated by QCD, and reconfirmed by experiment [25].

We suppose that quark-pair creation takes place in the interior of a hadron, neither at very large (confinement), nor at very small (asymptotic freedom) interquark distances. In the coordinate representation, these considerations get translated into a potential V_t of the form as depicted in figure (2). This transition potential enables the communication between a meson and its decay products.

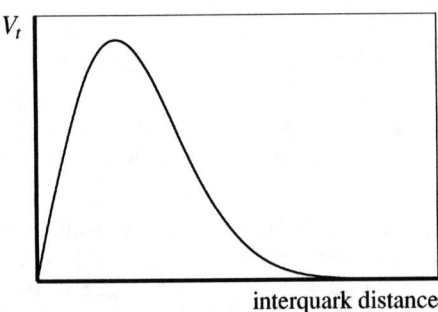

interquark distance

FIGURE 2. Form of the transition potential V_t in the coordinate representation

The principle of flavour independence of strong interactions demands that the transition potential cannot be a function of the pure interquark distance r, but has to be scaled by the reduced mass μ of the flavours of the system, *i.e.*

$$V_t = V_t(\sqrt{\mu}\, r) \quad . \tag{2}$$

The model's results are then to lowest order flavour independent. Differences, which also can be observed in the data, stem from the higher orders and kinematics.

We have chosen the form of V_t such that it depends on two, again flavour-independent, parameters. This completes the description of the model, which, besides the four constituent quark masses, has four model parameters.

Non-exotic meson-meson scattering. The model under discussion is taylor-made for describing non-exotic meson-meson scattering. In particular, for processes of the form

$$\text{meson } A + \text{meson } B \longrightarrow q\bar{q} \longrightarrow \text{meson } C + \text{meson } D \quad , \tag{3}$$

where it is understood that all flavours are such that the $q\bar{q}$ system couples to the AB two-meson system by light-quark-pair annihilation and to the CD two-meson system by light-quark-pair creation. Processes with more than two mesons in the final state are supposed to come from subsequent decays of the mesons in the initial CD two-meson final state, and hence to be of less importance for the properties we want to study.

In the interaction region, where our model applies, we have a finite probability to find an AB two-meson system, or a CD two-meson system, or a $q\bar{q}$ mesonic system.

The latter is usually far from its confinement eigenstates and thus highly unstable. The CD pair represents any of the unlimited number of possible two-meson final states. In Ref. [26] it is described how the two-meson final states can be limited to a finite number of possibilities. Additionally, we only take the lowest-lying pseudoscalar, π, K, η, or η', and vector mesons ρ, K^*, ω, or ϕ, as final-state particles. Nevertheless, the number of final-state channels comes out mostly somewhere between ten and twenty. However, in Ref. [27] it is shown that, when one takes relatively small errors of up to some 50 MeV for granted, then one may even obtain part of the results for low energies by just including the lowest-lying, or the most important, final-state channel.

The above-discussed parameter λ represents the overall three-meson-vertex coupling of the two-meson channels to the $q\bar{q}$ channel of expression (3). Relative couplings for the various three-meson vertices that may occur can be determined by a technique described in Ref. [26]. In some cases one also has various possible $q\bar{q}$ channels, as the isosinglets, $n\bar{n}$, $s\bar{s}$, ... mix with one another.

The probability to find a certain two-particle state in the interaction region can be determined from its wave function. Consequently, the model can be formulated in terms of a multichannel system of coupled-channel equations for the various wave functions. Details may be found in Ref. [6]. By a technique which has been described in Ref. [28], one can analytically solve the coupled-channel equations and determine the scattering matrix $S(\sqrt{s})$ as a function of the total centre-of-mass energy. Subsequently one may study elastic cross sections, or phase shifts, but also inelasticities and wave functions. Moreover, the solutions of the coupled-channel equations may be analytically continued to complex values of $\sqrt{s}$, which allows to study the pole structure of the scattering matrix in the complex-energy plane.

Bound states and resonances. Each of the two-meson final-state channels has a minimum value for the energy at which the two mesons can be formed, the threshold energy, which is given by the sum of the two meson masses. When the total energy $\sqrt{s}$ of the coupled-channel system is above the threshold of a particular channel, one says that the channel is open. It means that scattering is possible for that channel. Nevertheless, it is legitimate to study the solutions of the set of coupled-channel equations for energies below the thresholds, i.e., when channels are closed. In particular, below the lowest threshold, where all channels are closed and no scattering is possible, one may obtain the analytic continuation in $\sqrt{s}$ of the scattering matrix.

Singularities in $S(\sqrt{s})$ for real values of $\sqrt{s}$ below the lowest threshold, represent permanently bound states, when calculated in the correct Riemann sheet (i.e., with positive imaginary momenta k in all channels). The wave functions corresponding to those singularities have for all channels contributions that rapidly vanish at large interparticle distances. They represent the stable mesons, like the J/Ψ, which cannot decay strongly. For ground states of pseudoscalar and vector mesons, one finds the poles shifted to mass values which are far below the corresponding oscillator ground states. The sizes of the shifts are roughly proportional to λ^2, and may have values of several hundreds of MeV. The higher excitations of pseudoscalar and vector mesons shift much less. For this reason, one may start from harmonic-oscillator confinement and yet end up with realistic meson spectra [22]. We achieve this for all mesons with one fixed value for each of the four model parameters [6].

357

The fact that all channels contribute to the wave function of stable mesons implies that stable mesons, too, have two-meson components, not just $q\bar{q}$. This observation has important consequences for electromagnetic and weak transitions of stable mesons [29].

Above the lowest threshold no bound states can exsist. Accordingly, one does not find singularities in the scattering matrix on the real $\sqrt{s}$ axis. The poles which are found in that region of the complex $\sqrt{s}$ plane have real and imaginary parts. When a pole is encountered in the lower half of the complex $\sqrt{s}$ plane, and moreover in the right Rieman sheet and not too far from the real axis, then one may observe a nearby enhancement in the elastic cross sections of all open channels. These structures represent the resonances that show up in meson-meson scattering, as e.g. the ρ meson in $\pi\pi$ scattering, and the K^* meson in $K\pi$ scattering.

In the quark-exchange picture, we obtain a resonance in the particular partial-wave meson-meson-scattering cross section which matches the quantum numbers of the intermediate $q\bar{q}$ system of process (3). Such a phenomenon may be described by scattering phase shifts of the Breit-Wigner [30] form

$$\cotg\left(\delta_\ell(s)\right) \approx \frac{E_R - \sqrt{s}}{\Gamma_R/2} , \tag{4}$$

where E_R and Γ_R represent the central invariant meson-meson mass and the resonance width, respectively.

However, formula (4) is a good approximation for the scattering cross section only when the resonance shape is not very much distorted and the width of the resonance is small. Moreover, the intermediate state in such a process is essentially a constituent $q\bar{q}$ configuration that is part of a confinement spectrum (also referred to as bare or intrinsic state), and hence may resonate in one of the eigenstates. This implies that the colliding mesons scatter off the whole $q\bar{q}$ confinement spectrum of radial, and possibly also angular excitations, not just off one single state [31]. Consequently, a full expression for the phase shifts of formula (4) should contain all possible eigenstates of such a spectrum as long as quantum numbers are respected. Let us denote the eigenvalues of the relevant part of the spectrum by E_n ($n = 0, 1, 2, \ldots$), and the corresponding eigenstates by $\mathscr{F}_n$. Then, following the procedure outlined in Ref. [27], we may write for the partial-wave phase shifts the more general expression

$$\cotg\left(\delta(s)\right) = \left[I(s) \sum_{n=0}^{\infty} \frac{|\mathscr{F}_n|^2}{\sqrt{s} - E_n}\right]^{-1} \left[R(s) \sum_{n=0}^{\infty} \frac{|\mathscr{F}_n|^2}{\sqrt{s} - E_n} - 1\right] . \tag{5}$$

In $R(s)$ and $I(s)$ we have absorbed the kinematical factors and details of two-meson scattering, and moreover the three-meson vertices. The details of formula (5) can be found in Ref. [27].

For an approximate description of a specific resonance, and in the rather hypothetical case that the three-meson vertices have small coupling constants, one may single out, from the sum over all confinement states, one particular state (say number N), the eigenvalue of which is nearest to the invariant meson-meson mass close to the resonance. Then, for total invariant meson-meson masses $\sqrt{s}$ in the vicinity of E_N, one finds the approximation

$$\cot(\delta(s)) \approx \frac{\left[E_N + R(s)\,|\mathscr{F}_N|^2\right] - \sqrt{s}}{I(s)\,|\mathscr{F}_N|^2} . \tag{6}$$

Formula (6) is indeed of the form (4), with the central resonance position and width given by

$$E_R \approx E_N + R(s)\,|\mathscr{F}_N|^2 \quad \text{and} \quad \Gamma_R \approx 2I(s)\,|\mathscr{F}_N|^2 . \tag{7}$$

In experiment one observes the influence of the nearest bound state of the confinement spectrum, as in classical resonating systems. Nevertheless, formula (6) is only a good approximation when the three-meson couplings are small. Since the coupling of the meson-meson system to quark exchange is strong, the influence of the higher- and lower-lying excitations is not negligible.

In data analyses one usually represents a resonance in the elastic cross section by a Breit-Wigner structure, associated with a singularity in the complex $\sqrt{s}$ plane. In our model resonances are not well represented by Breit-Wigner structures, due to non-perturbative effects. A good example is the $\rho'(1250)$. This resonance is omitted from the meson tables of the Review of Particle Physics. Nevertheless, it is reported in data analyses as a clear signal [16, 17, 20]. In our model it comes out as a very tiny structure around 1.26 GeV in coupled $\pi\pi$, $K\bar{K}$, $\eta_n\rho$, $\pi\omega$, KK^*, $\rho\rho$, K^*K^* scattering in the tail of the $\rho(770)$ resonance [6]. Now, since in the data analyses quoted in [15] one uses Breit-Wigner structures, it is no surprise that for this tiny effect, which moreover has a rather large width, no evidence is found. At the position of the $\rho'(1470)$, which in Ref. [15] is claimed to be the first radial ρ excitation, we find a D resonance, which has the same quantum numbers as the S state. Naturally, we may know which resonance is dominantly S and which is dominantly D, since we can determine the wave functions in our model. From cross sections alone one cannot easily distinguish [32] between the two.

In the other hypothetical limit, namely of very large couplings, we obtain for the phase shift the expression

$$\cot(\delta(s)) \approx \frac{R(s)}{I(s)} , \tag{8}$$

which describes scattering off an infinitely hard cavity.

The physical values of the couplings come out somewhere in between the two hypothetical cases. Most resonances and bound states can be classified as stemming from a specific confinement state [33, 34]. However, some structures in the scattering cross section stem from the cavity which is formed by quark exchange or pair creation [27]. The most notable of such states are the low-lying resonances observed in S-wave pseudoscalar-pseudoscalar scattering [35, 36, 37, 38].

From the above discussion one may conclude that, to lowest order, the mass of a meson follows from the quark-antiquark confinement spectrum. It is, however, well-known that higher-order contributions to the meson propagator, in particular those from meson loops, cannot be neglected. Virtual meson loops give a correction to the meson mass, whereas decay channels also contribute to the strong width of the meson. One obtains for the propagator of a meson the form

$$\Pi(s) = \frac{1}{s - \left(M_{\text{confinement}} + \Sigma \, \Delta M_{\text{meson loops}}\right)^2} \, , \tag{9}$$

where ΔM develops complex values when open decay channels are involved.

For the full mass of a meson, all possible meson-meson loops have to be considered. A model for meson-meson scattering must therefore include all possible inelastic channels as well. Although in principle this could be done, in practice it is not feasible, unless a scheme exists dealing with all vertices and their relative intensities. In Ref. [26] relative couplings have been determined in the harmonic-oscillator approximation, assuming $^3P_0$ quark exchange. However, further kinematical factors must be worked out and included.

In its present form, the model is far from perfect. However, it allows for many predictions and for a good classification of the mesonic resonances. It moreover serves well as an interface between QCD quenched-lattice calculations and experiment. Quenched calculations describe confinement. As we have shown, the confinement spectrum and wave functions are very different from the hadronic reality for mesons. Applying the model, we may indicate what differences must be anticipated. For example, on the lattice one finds the lowest scalar-meson states at 1.3–1.5 GeV, exactly as we find for the pure harmonic oscillator. But the model yields a further nonet of scalar resonances well below 1.0 GeV. Hence, it is neither a failure of lattice QCD, nor of experiment! The light scalar resonances just do not form part of the confinement spectrum.

The spectrum. The complete model consists of an expression for the K matrix, similar to formula (5), but extended to many meson-meson scattering channels, several constituent quark-antiquark channels, and more complicated transition potentials [6, 7], which at the same time and with the same set of four parameters reproduces bound states, partial-wave scattering quantities, and the electromagnetic transitions of $c\bar{c}$ and $b\bar{b}$ systems [29].

The K matrix can be analytically continued below the various thresholds, even the lowest one, with no need of redefining any of the functions involved, in order to study the singularities of the corresponding scattering matrix. Below the lowest threshold, these poles show up on the real $\sqrt{s}$ axis, and can be interpreted as the bound states of the coupled system, to be identified with the stable mesons. For the light flavours one finds this way a nonet of light pseudoscalars, i.e., the pion, Kaon, eta, and eta$'$. For the heavy flavours, the lowest-lying model poles can be identified with the $D(1870)$, $D_s(1970)$, $\eta_c(1S)$, $J/\psi(1S)$, $\psi(3686)$, $B(5280)$, $B_s(5380)$, $\Upsilon(1S)$, $\Upsilon(2S)$, and $\Upsilon(3S)$.

Above the lowest threshold, the model's partial-wave cross sections and phase shifts for all included meson-meson channels can be calculated and compared to experiment, as well as the inelastic transitions. Out of the many singularities of the scattering matrix in a rather complex set of Riemann sheets, some come out with negative imaginary part in the $\sqrt{s}$ plane, and moreover close enough to the physical real axis so as to be noticed in the partial-wave phase shifts and cross sections. These can be identified with the known resonances, like the ρ pole in $\pi\pi$ scattering, or the K^* pole in $K\pi$ scattering. However, there may always be a pole in a nearby Riemann sheet just around the corner of one of the thresholds, which can be noticed in the partial-wave cross section. The study of poles is an interesting subject by itself [39, 40].

Scattering-matrix poles. In the hypothetical case of very small couplings for the three-meson vertices, we obtain poles in the scattering matrix that are close to the eigenvalues of the confinement spectrum. Let us denote by M_1 and M_2 the meson masses, and by ΔE the difference between the complex-energy pole of the scattering matrix and the energy eigenvalue, E_N, of the nearby state of the confinement spectrum. Using formula (7), we obtain

$$\Delta E \approx \{R(s) - iI(s)\}\,|\mathscr{F}_N|^2 \ . \tag{10}$$

We may distinguish two different cases:

(1) $E_N > M_1 + M_2$ (above threshold),
(2) $E_N < M_1 + M_2$ (below threshold).

When the nearby state of the confinement spectrum is in the scattering continuum, then ΔE has a **negative** imaginary part and a real part, since both $R(s)$ and $I(s)$ of formula (10) are real, and $I(s)$ is moreover positive. The resonance singularity of the scattering matrix is in the lower half of the complex-energy plane (second Riemann sheet), as is depicted on the right-hand side of threshold in Fig. (3).

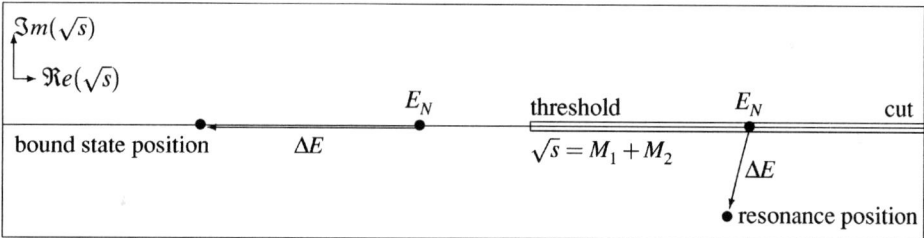

FIGURE 3. When the confinement state on the real $\sqrt{s}$ axis is below the lowest scattering threshold, then the bound-state singularity comes out on the real $\sqrt{s}$ axis. On the other hand, when the confinement state on the real $\sqrt{s}$ axis is in the scattering continuum, then for small coupling (perturbative regime) the resonance pole moves into the lower half of the complex $\sqrt{s}$ plane.

When the nearby state of the confinement spectrum is below the scattering threshold, then ΔE has only a real part, since $I(s)$ turns purely imaginary below threshold, whereas $R(s)$ remains real. The bound-state singularity of the scattering matrix corresponding to this situation remains on the real axis of the complex-energy plane, as is depicted on the left-hand side of threshold in Fig. (3).

Threshold behaviour. Near the lowest threshold, as a function of the overall coupling constant, S-wave poles behave very differently from P- and higher-wave poles. This can easily be understood from the effective-range expansion [41] at the pole position. There, the cotangent of the phase shift equals i. Hence, for S waves the next-to-lowest-order term in the expansion equals ik (k represents the linear momentum related to s and the lowest threshold). For higher waves, on the other hand, the next-to-lowest-order term in the effective-range expansion is proportional to k^2.

Poles for P and higher waves behave in the complex k plane as indicated in Fig. (4b). The two k-plane poles meet at threshold ($k = 0$). When the coupling constant of the model is increased, the poles move along the imaginary k axis. One pole moves towards negative imaginary k, corresponding to a virtual bound state below threshold on the real $\sqrt{s}$ axis, but in the wrong Riemann sheet. The other pole moves towards positive imaginary k, corresponding to a real bound state.

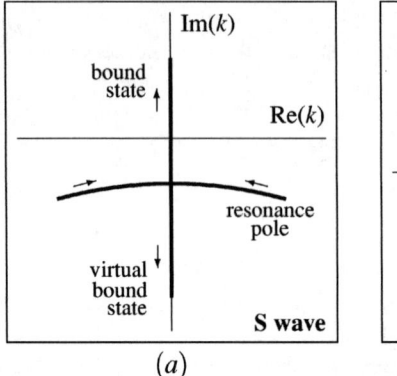 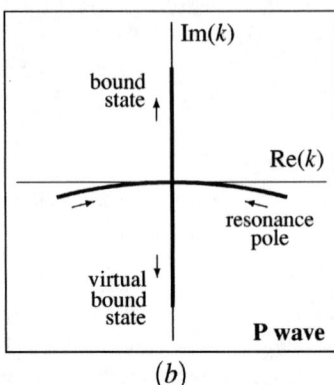

(a)   (b)

FIGURE 4. Variation of the positions of scattering-matrix poles as a function of hypothetical variations in the three-meson-vertex coupling, for S waves (a), and for P and higher waves (b). The arrows indicate increasing coupling constant.

For S-wave poles, the behaviour is shown in Fig. (4a). The two k-plane poles meet on the negative imaginary k axis. When the coupling constant of the model is slightly increased, both poles continue on the negative imaginary k axis, corresponding to two virtual bound states below threshold on the real $\sqrt{s}$ axis. Upon further increasing the coupling constant of the model, one pole moves towards increasing negative imaginary k, thereby remaining a virtual bound state for all values of the coupling constant. The other pole moves towards positive imaginary k, eventually crossing threshold ($k = 0$), thereby turning into a real bound state of the system of coupled meson-meson scattering channels. Hence, for a small range of hypothetical values of the coupling constant, there are two virtual bound states, one of which is very close to threshold. Such a pole certainly has a noticeable influence on the scattering cross section.

The low-lying nonet of S-wave poles. The nonet of low-lying S-wave poles behave as described above, with respect to variations of the model's overall coupling constant. However, they do not stem from the confinement spectrum, but rather from the cavity. For small values of the coupling, such poles disappear into the continuum, i.e., they move towards negative imaginary infinity [27], and not towards an eigenstate of the confinement spectrum as in Fig. (3).

In Fig. (5) we study the hypothetical pole positions of the $K_0^*(730)$ pole in $K\pi$ S-wave scattering. The physical value of the coupling constant equals 0.75, which is not shown in Fig. (5). A figure for smaller values of the coupling constants can be found in Ref. [27]. The physical pole in $K\pi$ isodoublet S-wave scattering, related to experiment [37, 36], comes out at $727 - i263$ MeV in Ref. [7]. Here we concentrate on the threshold behaviour of the hypothetical pole movements in the complex k and $\sqrt{s}$ planes. Until they meet on the axis, which is for a value of the coupling constant slightly larger than 1.24, we have only depicted the right-hand branch.

In the left-hand picture of Fig. (5) we observe how the poles arrive on the imaginary k axis, and then continue to move along that axis. One of the poles moves upwards, initially describing a virtual bound state, and crossing the real k axis for a value of the coupling constant slightly larger than 1.30. The other pole moves downwards, remaining a virtual bound state for further increasing values of the coupling constant.

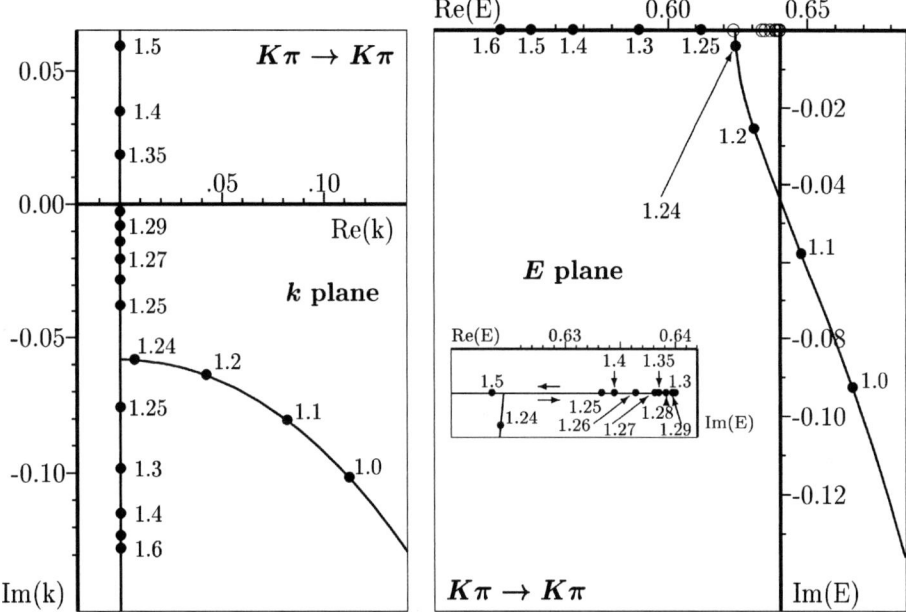

FIGURE 5. Hypothetical movement of the $K_0^*(730)$ pole in $K\pi$ S-wave scattering as a function of the coupling constant. The two branches on the imaginary k axis are discussed in the text. In the $E = \sqrt{s}$ plane these two branches come out on the real axis below threshold. The poles of the upper branch are shown as open circles in the main figure, and as closed circles in the inset. Units are in GeVs.

In the right-hand picture of Fig. (5) the same pole has been depicted in the complex $\sqrt{s}$ plane. Here, the situation is more confusing, since the pole positions are in the same interval of energies. The pole corresponding to the one moving downwards along the imaginary k axis moves to the left on the real $\sqrt{s}$ axis. Its positions as a function of the coupling constant are indicated by solid circles. The pole moving upwards along the imaginary k axis initially moves towards threshold and then turns back, following the former pole, but in a different Riemann sheet. The positions of the latter pole are indicated by open circles. In the inset we try to better clarify its motion. Notice that, since we took 0.14 GeV and 0.50 GeV for the pion and the Kaon mass, respectively, we end up with a threshold at 0.64 GeV.

It is interesting to notice that in a recent work of Boglione and Pennington [42] also a zero-width state is found below the $K\pi$ threshold in S-wave scattering. Here, we obtain such a state for *unphysical* values of the coupling.

In Fig. (6) we have depicted the movement of the $a_0(980)$ pole in S wave $I = 1$ KK scattering (threshold at 1.0 GeV) on the upwards-going branch. One observes a very similar behaviour as in the case of $K\pi$ scattering, but with two important differences, to be described next.

363

The $K_0^*(730)$ poles meet on the real $\sqrt{s}$ axis only 16 MeV below threshold (see Fig. 5), and for a value of the coupling constant which is well above the physical value of 0.75, whereas the $a_0(980)$ poles meet 238 MeV below threshold, when the coupling constant only equals 0.51. At the physical value of the coupling constant, the $a_0(980)$ pole is a real bound state some 9 MeV below threshold.

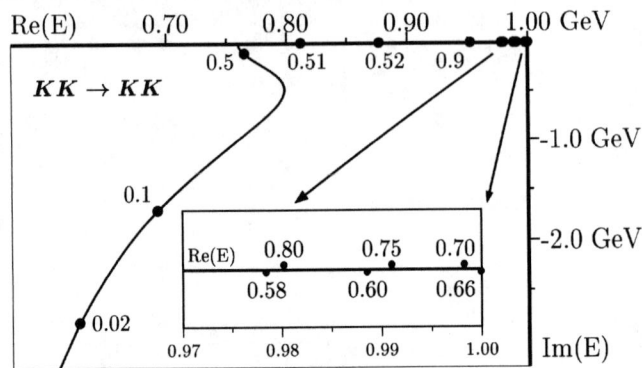

FIGURE 6. Pole movement as a function of the coupling constant for KK $I = 1$ S-wave scattering. Some values of the coupling constant are indicated in the figure. The six filled circles at the right end of the real axis correspond, from left to right, to the values 0.58, 0.80, 0.60, 0.75, 0.70, and 0.66 for the model's coupling constant. This situation is magnified in the inset.

But there is yet another difference. Whereas the $K\pi$ channel represents the lowest possible scattering threshold for the $K_0^*(730)$ system, KK is not the lowest channel for the $a_0(980)$. In a more complete description, at least all pseudoscalar meson-meson loops should be taken into account. One of these is the $\eta\pi$ channel, which has a threshold well below KK. Consequently, upon including the $\eta\pi$ channel in the model, the pole cannot remain on the real $\sqrt{s}$ axis, but has to acquire an imaginary part, in a similar way as shown in Fig. (3). In Ref. [7] we obtained a resonance-like structure in the $\eta\pi$ cross section, representing the physical $a_0(980)$. The corresponding pole came out at $962 - i28$ MeV.

For the $f_0(980)$ system the situation is very similar to that of the $a_0(980)$. Assuming a pure $s\bar{s}$ quark content [43], we obtain for the variation of the corresponding pole in KK $I = 0$ S-wave scattering a picture almost equal to the one shown in Fig. (6). However, only in lowest order the KK channel could be considered the lowest threshold for the $f_0(980)$ system. In reality, $s\bar{s}$ also couples to the non-strange quark-antiquark isosinglet through KK, and hence to $\pi\pi$ [8]. This coupling is nevertheless very weak, which implies that the resulting pole does not move far away from the KK bound state. In Ref. [7] we obtained a resonance-like structure in the $\pi\pi$ cross section representing the physical $f_0(980)$. The corresponding pole came out at $994 - i20$ MeV.

At lower energies, we found for the same cross section a pole which is the equivalent of the $K_0^*(730)$ system, but now in $\pi\pi$ isoscalar S-wave scattering. This pole at $470 - i208$ MeV may be associated with the σ meson, since it has the same quantum numbers, and lies in the ballpark of predicted pole positions in models of the σ (for a complete overview of σ poles, see Ref. [22]).

We do not find any other relevant poles in the energy region up to 1.0 GeV.

Conclusions. We have shown that the poles of the $a_0(980)$ and $f_0(980)$ belong to a nonet of scattering-matrix poles. The lower-lying isoscalar pole and the isodoublet poles in the complex-energy plane have real parts of 0.47 GeV and 0.73 GeV, respectively, and imaginary parts of 0.21 GeV resp. 0.26 GeV. Whether these poles represent real physical resonances [44] is not so relevant here. Important is that the $a_0(980)$ and $f_0(980)$ are well classified within a nonet of scattering-matrix poles with very specific characteristics, different from those of the poles stemming from confinement, like the confinement-ground-state nonet of scalar mesons $f_0(1370)$, $a_0(1450)$, $K_0^*(1430)$, and $f_0(1500)$. The latter poles vary as a function of the coupling constant exactly the way indicated in figure (3). For vanishing coupling constant they end up on the real $\sqrt{s}$ axis at the positions of the various ground-state eigenvalues of the confinement spectrum, which are the light-flavour $^3P_0$ states at 1.3 to 1.5 GeV [45, 46].

The low-lying S-wave poles related to the cross sections in the $f_0(470)$, $K_0^*(730)$, $f_0(980)$, and $a_0(980)$ regions move to negative imaginary infinity in the $\sqrt{s}$ plane for decreasing values of the coupling. The pole positions for the physical value of the coupling are well explained by their threshold behaviour. Whether or not these poles have large imaginary parts, leading to large widths and strong resonance distortion, depends in a subtle way on the thresholds and couplings of the various relevant scattering channels [47].

As to the nature of the light scalar mesons, which has recently been discussed in Refs. [12, 13, 48, 49], we can only remark that in a many-coupled-channel model each of the channels contributes to the states under the resonance, not just one specific channel.

We have moreover shown that the **principle of flavour independence for the strong interactions** has far-reaching consequences for the construction of hadron models. Applied as a guiding principle for building a largely non-relativistic many-coupled-channel Schrödinger equation for meson-meson scattering, it results in a surprisingly complete model for mesons and mesonic resonances. Its description of the light scalar resonances leaves no doubt on where these mesons are, nor on how to classify them.

ACKNOWLEDGMENTS

This work has been partly supported by the *Fundação para a Ciência e a Tecnologia* of the *Ministério da Ciência e da Tecnologia* of Portugal, under contract numbers POCTI/-35304/FIS/2000, CERN/FIS/43697/2001, PRAXIS XXI/BPD/20186/99, and SFRH/-BPD/9480/2002.

REFERENCES

1. J. Schechter, eConf **C010815** (2002) 76 [arXiv:hep-ph/0112205].
2. S. Godfrey and N. Isgur, Phys. Rev. D **32** (1985) 189.
3. J. Vijande, F. Fernández, A. Valcarce, B. Silvestre-Brac, Contributed to Meson 2002: 7th International Workshop on Meson Production, Properties and Interaction, Cracow, Poland, 24-28 May 2002, arXiv:hep-ph/0206263.
4. E. Eichten, *In *Cargese 1975, Proceedings, Weak and Electromagnetic Interactions At High Energies, Part A*, New York 1976, 305-328.*

5. E. Eichten, K. Gottfried, T. Kinoshita, K. D. Lane and T. M. Yan, Phys. Rev. D **21** (1980) 203.
6. E. van Beveren, G. Rupp, T. A. Rijken, and C. Dullemond, Phys. Rev. **D27** (1983) 1527.
7. E. van Beveren, T. A. Rijken, K. Metzger, C. Dullemond, G. Rupp and J. E. Ribeiro, Z. Phys. **C30** (1986) 615.
8. R. N. Cahn and P. V. Landshoff, Nucl. Phys. B **266** (1986) 451.
9. Nils A. Törnqvist and Matts Roos, Phys. Rev. Lett. **76** (1996) 1575 [arXiv:hep-ph/9511210].
10. Frank E. Close, Int. J. Mod. Phys. A **17** (2002) 3239 [arXiv:hep-ph/0110081].
11. Agnieszka Furman and Leonard Leśniak, Phys. Lett. B **538** (2002) 266 [arXiv:hep-ph/0203255].
12. Frank E. Close and Nils A. Törnqvist, J. Phys. G **28** (2002) R249 [arXiv:hep-ph/0204205].
13. Nils A. Törnqvist, arXiv:hep-ph/0201171.
14. Frank E. Close and Nathan Isgur, Phys. Lett. B **509** (2001) 81 [arXiv:hep-ph/0102067].
15. K. Hagiwara *et al.* [Particle Data Group Collaboration], *Review Of Particle Physics*, Phys. Rev. D **66** (2002) 010001.
16. D. Aston *et al.*, SLAC-PUB-5606 (1994).
17. D. Aston *et al.*, Nucl. Phys. Proc. Suppl. **21** (1991) 105.
18. A. B. Govorkov, JINR-P2-86-682 (1986).
19. S. Bartalucci *et al.*, Nuovo Cim. A **49** (1979) 207.
20. J. Ballam *et al.*, Nucl. Phys. B **76** (1974) 375.
21. S. E. Csorna *et al.* [CLEO Collaboration], arXiv:hep-ex/0207060.
22. Eef van Beveren and George Rupp, Proceedings of the Workshop on Recent Developments in Particle and Nuclear Physics, April 30, 2001, Coimbra (Portugal) ISBN 972-95630-3-9, pages 1-16, [arXiv:hep-ph/0201006].
23. G. Zweig, CERN Reports TH-401 and TH-412, see also *Developments in the Quark Theory of Hadrons*, Vol. 1*, 22-101 (1981) editted by D. B. Lichtenberg and S. P. Rosen.
24. N. Brambilla, Y. Sumino and A. Vairo, Phys. Rev. D **65** (2002) 034001 [arXiv:hep-ph/0108084].
25. K. Abe *et al.* [SLD Collaboration], Phys. Rev. D **59** (1999) 012002 [arXiv:hep-ex/9805023].
26. E. van Beveren, Z. Phys. **C21** (1984) 291.
27. Eef van Beveren and George Rupp, Eur. Phys. J. **C22**, 493 (2001) [arXiv:hep-ex/0106077].
28. E. van Beveren, C. Dullemond, T. A. Rijken, and G. Rupp, Lect. Notes Phys. **211** (1984) 182.
29. A. G. Verschuren, C. Dullemond, and E. van Beveren, Phys. Rev. **D44** (1991) 2803.
30. G. Breit and E. Wigner, Phys. Rev. **49** (1936) 519.
31. Tullio Regge, Nuovo Cim. **14** (1959) 951.
32. Stephen Godfrey, Gabriel Karl and Patrick J. O'Donnell, Z. Phys. C **31** (1986) 77.
33. A. V. Anisovich and A. V. Sarantsev, Phys. Lett. **B413** (1997) 137 [arXiv:hep-ph/9705401].
34. V. V. Anisovich and A. V. Sarantsev, arXiv:hep-ph/0204328.
35. Carla Göbel, on behalf of the E791 Collaboration, Proceedings of Heavy Quarks at Fixed Target (HQ2K), Rio de Janeiro, October 2000, 373-384 [hep-ex/0012009]
36. Carla Göbel, on behalf of the E791 Collaboration, AIP Conf. Proc. **619** (2002) 63 [arXiv:hep-ex/0110052].
37. E. M. Aitala *et al.* [E791 Collaboration], Phys. Rev. Lett. **89** (2002) 121801 [arXiv:hep-ex/0204018].
38. Eef van Beveren and George Rupp, AIP Conf. Proc. **619** (2002) 209 [arXiv:hep-ph/0110156].
39. Matthias Jamin, José Antonio Oller, and Antonio Pich, Nucl. Phys. **B587**, 331 (2000) [hep-ph/0006045].
40. J. A. Oller, E. Oset and J. R. Peláez, Phys. Rev. **D59** (1999) 074001 [Erratum-ibid. **D60** (1999) 099906] [arXiv:hep-ph/9804209].
41. V. De Alfaro and T. Regge, *Potential Scattering*, North Holland (Amsterdam, 1965).
42. M. Boglione and M. R. Pennington, Phys. Rev. D **65** (2002) 114010 [arXiv:hep-ph/0203149].
43. F. De Fazio and M. R. Pennington, Phys. Lett. B **521** (2001) 15 [arXiv:hep-ph/0104289].
44. S. N. Cherry and M. R. Pennington, Nucl. Phys. A **688** (2001) 823 [arXiv:hep-ph/0005208].
45. W. Lee and D. Weingarten, Phys. Rev. **D61**, 014015 (2000) [hep-lat/9910008]; D. Weingarten, private communication.
46. W. Lee and D. Weingarten, Nucl. Phys. Proc. Suppl. **53**, 236 (1997) [hep-lat/9608071].
47. Kim Maltman, Phys. Lett. **B462** (1999) 14 [arXiv:hep-ph/9906267].
48. Kim Maltman, Nucl. Phys. **A675** (2000) 209c.
49. C. M. Shakin, Phys. Rev. D **65** (2002) 114011.

Relativistic Unitarized Quark/Meson Model in Momentum Space

George Rupp[*] and Eef van Beveren[†]

[*]Centro de Física das Interacções Fundamentais, Instituto Superior Técnico, Edifício Ciência,
P1049-001 Lisboa Codex, Portugal
[†]Centro de Física Teórica, Departamento de Física, Universidade, P3004-516 Coimbra, Portugal

Abstract. An outline is given how to formulate a relativistic unitarized constituent quark model of mesons in momentum space, employing harmonic quark confinement. As a first step, the momentum-space harmonic-oscillator potential is solved in a relativistically covariant, three-dimensional quasipotential framework for scalar particles, using the spline technique. Then, an illustrative toy model with the same dynamical equations but now one $q\bar{q}$ and one meson-meson channel, coupled to one another through quark exchange describing the $^3P_0$ mechanism, is solved in closed form on a spline basis. Conclusions are presented on how to generalize the latter to a realistic multichannel quark/meson model.

1. INTRODUCTION

Mesonic $q\bar{q}$ states are the simplest color-singlet configurations in QCD. However, physical mesons appear to be much more complicated objects, in view of the quite disparate spectra of especially the light scalars and pseudoscalars, and the large variety of states ranging from OZI-stable quarkonia to extremely broad, often badly established resonances. Because of these complexities, theoretical approaches usually focus on parts of the meson data only. Thus, one has quarkonium potential models (see e.g. Ref. [1]), relativistic constituent quark models (see e.g. Ref. [2]), chiral quark models (see e.g. Ref. [3]), heavy-quark effective theory [4], the linear σ model (LσM) [5], Nambu–Jona-Lasinio models [6], (unitarized) chiral perturbation theory [7]), meson-exchange models (see e.g. Ref. [8]), and so on. While it lies outside the scope of this short paper to describe in detail the respective merits and drawbacks of all these approaches, on can generally state that a good reproduction of either spectra or scattering data is achieved, often at the cost of a non-negligible number of free fit parameters, but without accomplishing a unified description of mesonic spectra and meson-meson scattering. In our view, however, spectra and scattering of mesons are inexorably intertwined concepts, considering that most mesons are resonances, decaying strongly into final states of two or more lighter mesons.

Therefore, we believe one should treat the mechanisms responsible for quark confinement and for mesonic decay on an equal footing. This we have realized already in an essentially non-relativistic (NR) framework, employing the coupled-channel Schrödinger formalism, with several confined $q\bar{q}$ channels and many free two-meson channels. For a detailed discussion of the configuration-space model, we refer to a separate contribu-

CP660, *Hadron Physics: Effective Theories of Low Energy QCD*, edited by A. H. Blin et al.
© 2003 American Institute of Physics 0-7354-0120-9/03/$20.00

tion [9] to this workshop, so let us just emphasize here some important non-perturbative features. From the spectroscopic point of view, meson-loop effects on the $q\bar{q}$ spectra are included to all (ladder) orders, even those resulting from closed meson-meson channels. Conversely, the strong influence of s-channel $q\bar{q}$ states on non-exotic meson-meson scattering is taken into account. Furthermore, bound states, resonances, and scattering observables all follow directly from an analytic S-matrix derived in closed form, thus dispensing with any kind of perturbative methods, which could be highly untrustworthy for the often very large couplings involved. As a matter of fact, perhaps the most remarkable conquest of the model is its parameter-free prediction of the light scalar mesons [10], including the $\kappa(800)$ (K_0^*) [11], which has very recently been experimentally confirmed [12]. Precisely for the light scalars, perturbation theory manifestly breaks down [9, 13].

Other coupled-channel, also called unitarized, approaches are due to Eichten *et al.* (Ref. [1], first paper), Bicudo & Ribeiro (Ref. [3], fourth paper), and Törnqvist with collaborators (Ref. [14] and references therein). The conclusions of the latter two models are in qualitative agreement with ours, in the sense that large shifts are found owing to unitarization, even for (bare) states below the lowest decay threshold. However, in the case of the scalar mesons, important differences between our model and especially the one of Ref. [14] come to light, as the $\kappa(800)$ is not found in the latter reference (see Ref. [11] for a detailed comparison).

Notwithstanding its successes, our coordinate-space model has, of course, a number of shortcomings. First of all, the light pseudoscalar mesons are quite badly off, except for the kaon. This may be due to the largely NR formalism without Dirac spin structure , the disregard of chiral symmetry, and possibly also the neglect of scalar-pseudoscalar two-meson channels. Moreover, for most processes, the experimental meson-meson phase shifts are only roughly and not very accurately reproduced, which is no wonder in view of the neglect of final-state interactions from direct t-channel meson exchange, and the absence of any "fitting freedom". Then there are problems with pseudothresholds in heavy-light systems, owing to the non-covariance of minimally relativistic equations in configuration space. Finally, too many, viz. 3, parameters — albeit a very modest number as compared to other approaches — are needed for the $^3P_0$ transition potential, due to the lack of a microscopic description in terms of quark exchange.

Nevertheless, we are convinced one can deal with these shortcomings by reformulating the model in momentum space, in a covariant three-dimensional (3D) quasipotential framework. Here, one could object why we do not attempt to tackle the full Bethe-Salpeter (BS) equation. The reason is twofold: first of all, the numerical effort to solve the resulting coupled 4D integro-differential equations with many channels would be enormous; but perhaps more importantly, the confinement mechanism employing potentials is an inherently 3D concept, possibly leading to a pathological, non-confining behavior when generalized to four dimensions (see e.g. Ref. [2], third paper). Anyhow, a covariant formulation would also allow to include, at a later stage, the Dirac spin structure of the quarks, and address the issue of dynamical chiral-symmetry breaking and generation of constituent mass. The resulting multichannel equations in momentum space we shall show to be explicitly solvable, provided we stick to harmonic confinement and work on a spline basis.

This paper is organized as follows. In Sec. 2 the good old harmonic oscillator (HO) is solved in momentum space using the spline technique, for two different 3D relativistic equations. In Sec. 3 a two-channel toy model, with one harmonically confined $q\bar{q}$ channel coupled via quark exchange to one free two-meson channel, is formally solved in closed form, employing again a spline expansion. Conclusions and an outlook on how to further develop the model are presented in Sec. 4.

2. RELATIVISTIC HARMONIC OSCILLATOR

In this section, we review the well-known 3D HO, first in coordinate space, and then in momentum space. In the latter representation, a 3D relativistic generalization is straightforward.

2.1. Non-Relativistic Harmonic Oscillator

The equal-mass radial Schrödinger equation (SE) for a local potential reads

$$\left\{\frac{1}{m}(-\frac{d^2}{dr^2} + \frac{l(l+1)}{r^2}) + V(r)\right\} u(r) = E u(r). \tag{1}$$

If we take the HO potential

$$V(r) = \frac{1}{2}\mu\omega^2 r^2, \quad \mu = \frac{1}{2}m, \tag{2}$$

define $x \equiv \sqrt{\mu\omega}\, r$, and assume for the moment $l = 0$, we get

$$\left\{-\frac{d^2}{dx^2} + x^2\right\} u(x) = \frac{E}{\frac{1}{2}\omega} u(x). \tag{3}$$

The spectrum of the 3D HO potential is very well known, viz. ($l = 0$)

$$E = \omega\left(2n + \frac{3}{2}\right), \quad n = 0, 1, 2, \ldots, \tag{4}$$

leading to the eigenvalues $3, 7, 11, \ldots$ for Eq. (3).

Alternatively, we can work in momentum space. The homogeneous Lippmann-Schwinger (LS) equation for the bound-state vertex function reads

$$\Gamma(\vec{p}) = \int \frac{d^3 p'}{(2\pi)^3} V(\vec{p}, \vec{p}') G_0(\vec{p}'; E) \Gamma(\vec{p}'), \tag{5}$$

where the free two-body Green's function is given by

$$G_0^{-1}(\vec{p}') = E - \frac{\vec{p}'^2}{m} + i\varepsilon. \tag{6}$$

Now define the wave function $\Psi \equiv G_0 \Gamma$. Then Eq. (5) becomes

$$G_0^{-1}(\vec{p})\Psi(\vec{p}) = \int \frac{d^3p'}{(2\pi)^3} V(\vec{p}, \vec{p}')\Psi(\vec{p}'),$$ (7)

or equivalently

$$\frac{\vec{p}^2}{m}\Psi(\vec{p}) + \int \frac{d^3p'}{(2\pi)^3} V(\vec{p}, \vec{p}')\Psi(\vec{p}') = E\Psi(\vec{p}).$$ (8)

This is just the SE in momentum space. In this representation, the HO potential is given by

$$V(\vec{p}, \vec{p}') = -\frac{1}{2}\mu\omega^2 \delta^{(3)}(\vec{p} - \vec{p}')\Delta_{\vec{p}},$$ (9)

which is local as in coordinate space. Substituting into (8) yields

$$\left\{\frac{\vec{p}^2}{m} - \frac{1}{2}\mu\omega^2 \Delta_{\vec{p}}\right\}\Psi(\vec{p}) = E\Psi(\vec{p}).$$ (10)

Taking again $l = 0$ and introducing $v(\mathrm{p}) = \mathrm{p}\Psi(\mathrm{p})$, with $\mathrm{p} \equiv |\vec{p}|$, we obtain

$$\left\{\frac{\mathrm{p}^2}{m} - \frac{1}{2}\mu\omega^2 \frac{d^2}{d\mathrm{p}^2}\right\}v(\mathrm{p}) = E v(\mathrm{p}).$$ (11)

If we define the dimensionless variable $y \equiv \mathrm{p}/\sqrt{\mu\omega}$ and write $m = 2\mu$, we get

$$\left\{y^2 - \frac{d^2}{dy^2}\right\}v(y) = \frac{E}{\frac{1}{2}\omega} v(y).$$ (12)

This is identical to Eq. (3). Note that now the roles of the kinetic and the potential term are interchanged.

2.2. Blankenbecler–Sugar–Logunov–Tavkhelidze (BSLT) Equation

The momentum-space equal-mass (scalar) BSLT [15] equation for the wave function has exactly the same form as in the LS case of Eq. (7), but now with center-of-mass (CM) Green's function

$$G_0^{-1}(\vec{p}) = \frac{E_p}{m^2}\left(\frac{s}{4} - E_p^2 + i\varepsilon\right),$$ (13)

where $E_p = \sqrt{\vec{p}^2 + m^2}$, and s is the total CM energy squared $s = P^2 = (p_1 + p_2)^2 = (2m + E)^2$. Note that G_0^{BSLT} can be derived in a completely covariant fashion, and written down explicitly for an arbitrary frame [16]. In the NR limit we have $E_p \approx m$ and $s \approx 4m^2 + 4mE$, recovering the LS Green's function (6). With the new G_0^{-1}, we get instead of Eq. (8)

$$\frac{\vec{p}^2}{m}\Psi(\vec{p}) + \frac{m}{E_p}\int \frac{d^3p'}{(2\pi)^3} V(\vec{p}, \vec{p}')\Psi(\vec{p}') = \frac{s - 4m^2}{4m}\Psi(\vec{p}).$$ (14)

370

In terms of the above-defined dimensionless variable y, the BSLT equation becomes

$$\left\{ y^2 - \frac{1}{\sqrt{1 + \frac{\omega}{2m} y^2}} \frac{d^2}{dy^2} \right\} v(y) = \frac{s - 4m^2}{2m\omega} v(y). \tag{15}$$

From this equation we readily see that deviations from the LS case are determined by the parameter $\omega/2m$. Furthermore, in the limit $\omega/2m \to 0$, the eigenvalues $3, 7, 11, \ldots$ get recovered. However, there are additional relativistic effects due to $s \neq 4m^2 + 4mE$.

2.3. Equal-Time (ET) Equation

The ET, also called Mandelzweig–Wallace [17], Green's function can be obtained from the BSLT (or Salpeter) equation by adding an extra term G_0^C to the two-body Green's function G_0 which approximates crossed-ladder contributions. This extra term becomes exact in the eikonal limit, or when one of the particles gets infinitely heavy. Therefore, the covariant ET equation manifestly has the correct one-body limit, i.e., one recovers either the Klein–Gordon or the Dirac equation, for scalar or spin-1/2 particles, respectively. In the scalar case, the CM ET Green's function is given by

$$G_0^{-1}(\vec{p}) = \frac{E_p^3}{m^2} \frac{\frac{s}{4} - E_p^2 + i\varepsilon}{2E_p^2 - \frac{s}{4}}. \tag{16}$$

The final form of the S-wave ET equation for the HO potential in momentum space reads

$$\left\{ y^2 - \frac{1}{\sqrt{1 + \frac{\omega}{2m} y^2}} \left[2 - \frac{1}{1 + \frac{\omega}{2m} y^2} \right] \frac{d^2}{dy^2} \right\} v(y) = \frac{s - 4m^2}{2m\omega} \left\{ 1 - \frac{\frac{\omega}{2m}}{\left(1 + \frac{\omega}{2m}\right)^{\frac{3}{2}}} \frac{d^2}{dy^2} \right\} v(y). \tag{17}$$

2.4. Non-Zero Orbital Angular Momentum

For $l \neq 0$, the dimensionless LS equation becomes

$$\left\{ -\frac{d^2}{dx^2} + \frac{l(l+1)}{x^2} + x^2 \right\} u(x) = \frac{E}{\frac{1}{2}\omega} u(x). \tag{18}$$

In order to numerically solve the LS and corresponding BSLT, ET equations, we substitute $u(x) \to x^l \tilde{u}(x)$. Then

$$\left\{ -\frac{d^2}{dx^2} + \frac{2l}{x^2}\left(1 - x\frac{d}{dx}\right) + x^2 \right\} \tilde{u}(x) = \frac{E}{\frac{1}{2}\omega} \tilde{u}(x). \tag{19}$$

The function $\tilde{u}(x)$ has the right boundary conditions to allow an easy numerical solution using an expansion on a spline basis (for details, see Ref. [18] and second paper of Ref. [2]). Thus, the NR HO spectrum is recovered with great accuracy, viz.

$$E = \omega\left(2n + l + \frac{3}{2}\right), \quad n, l = 0, 1, 2, \dots . \tag{20}$$

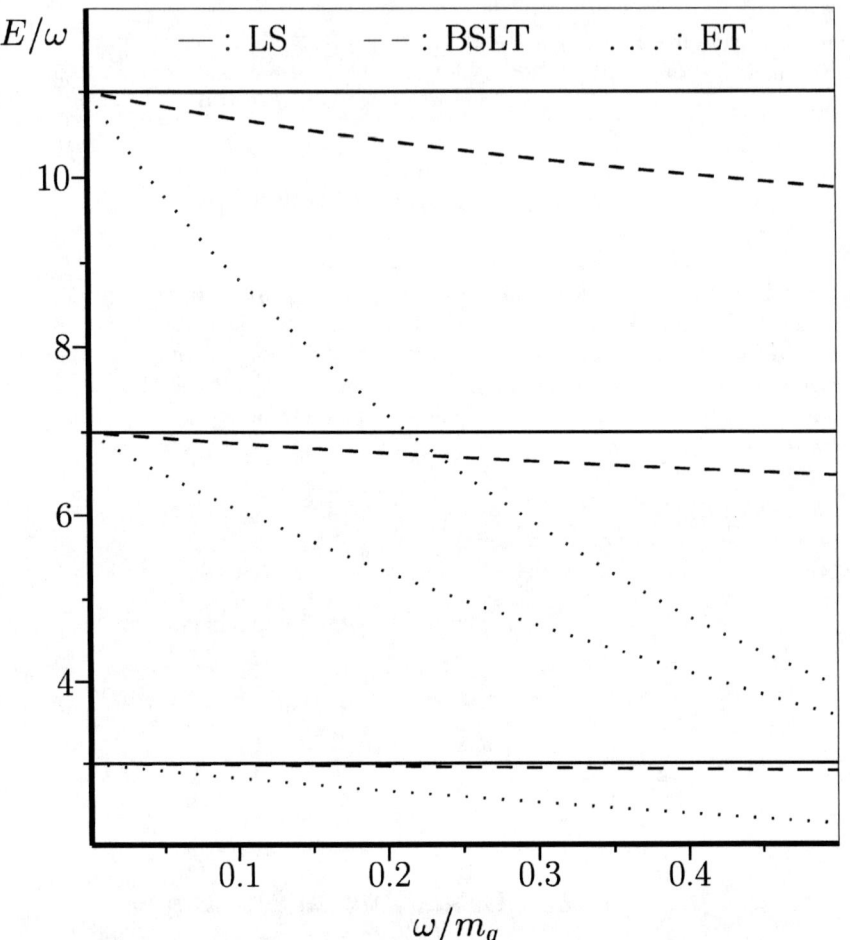

FIGURE 1. Dimensionless energy levels of the HO for the LS, BSLT, and ET equations, as a function of ω/m_q; first three radial levels are depicted ($l = 0$).

In the case of the BSLT and ET equations, the same can be done by replacing everywhere

$$-\frac{d^2}{dy^2} \longrightarrow \frac{2l}{y^2}\left(1 - y\frac{d}{dy}\right) - \frac{d^2}{dy^2} . \tag{21}$$

The LS, BSLT, and ET results for $l = 0$ are depicted in Fig. 1. We see that the relativistic corrections are relatively modest for the BSLT equation, but can become huge in the ET case, especially for excited states. The latter large effect has to do with the sign change of the ET propagator (16) as a function of the relative momentum, and has also been observed in the Dirac case (see Ref. [2], second paper). However, this overshooting of the ET equation will strongly depend on the chosen Dirac structure of the confining force in a realistic model, and hence should not cause any worries at this stage.

3. UNITARIZED MODEL IN MOMENTUM SPACE

Consider now a simple two-channel model of one confined $q\bar{q}$ system coupled to one free meson-meson channel. In a unitarized picture, this can formally be described by a 2×2 T matrix, with the dynamical equation (LS, BSLT, or ET)

$$\begin{pmatrix} T_{00} & T_{01} \\ T_{10} & T_{11} \end{pmatrix} = \begin{pmatrix} V_{00} & V_{01} \\ V_{10} & V_{11} \end{pmatrix} + \begin{pmatrix} V_{00} & V_{01} \\ V_{10} & V_{11} \end{pmatrix} \begin{pmatrix} G_0 & 0 \\ 0 & G_1 \end{pmatrix} \begin{pmatrix} T_{00} & T_{01} \\ T_{10} & T_{11} \end{pmatrix}. \quad (22)$$

Equation (22) is diagrammatically represented in Fig. 2. Dashed lines are mesons, solid lines with arrows are (anti)quarks, and curly exchanges symbolize confinement. Furthermore, open circles represent meson-quark-antiquark vertices, and dots stand for three-meson vertices. Note that the communication between the $q\bar{q}$ and two-meson channels takes place through quark exchange. This will give rise to a generally non-local $^3P_0$ transition potential in configuration space, as soon as the quarks and mesons have different masses. The precise form of such a potential will also depend on possible form factors to be attributed to the various vertices.

In order to solve Eq. (22), we must realize that no asymptotic scattering of quarks is possible, so T_{00}, T_{01}, and T_{10} must in the end be eliminated from the equations: only T_{11} can be on shell. If we now take again harmonic confinement, Eq. (22) becomes a set of coupled integro-differential equations. Omitting for clarity angular-momentum indices, we get e.g. for T_{01}

$$T_{01}(p,p') = V_{01}(p,p') - \frac{1}{2}\mu\omega^2\Delta_p G_0(p)T_{01}(p,p') + \int \frac{k^2dk}{4\pi^2}V_{01}(p,k)G_1(k)T_{11}(k,p'). \quad (23)$$

Note that p' is an essentially dummy variable in the equations, so we can choose a fixed value for it, say $\bar{p}$. If we now expand T_{01} and T_{11} on a spline basis, and discretize the initial momentum p, then T_{01} can be expressed in terms of T_{11} as follows.

$$T_{01}(p_i,\bar{p}) = \sum_{j=1}^{N} t_{01}^j s_{01}^j(p_i), \quad v_{01}^i \equiv V_{01}(p_i,\bar{p}); \quad (24)$$

$$T_{11}(k,\bar{p}) = \sum_{j=1}^{N} t_{11}^j s_{11}^j(k); \quad (25)$$

$$A_{01}^{ij} \equiv \left\{1 + \frac{1}{2}\mu\omega^2\Delta_{p_i} G_0(p_i)\right\} s_{01}^j(p_i); \quad (26)$$

$$A_{11}^{ij} \equiv \int \frac{k^2dk}{4\pi^2}V_{01}(p_i,k)G_1(k)s_{11}^j(k). \quad (27)$$

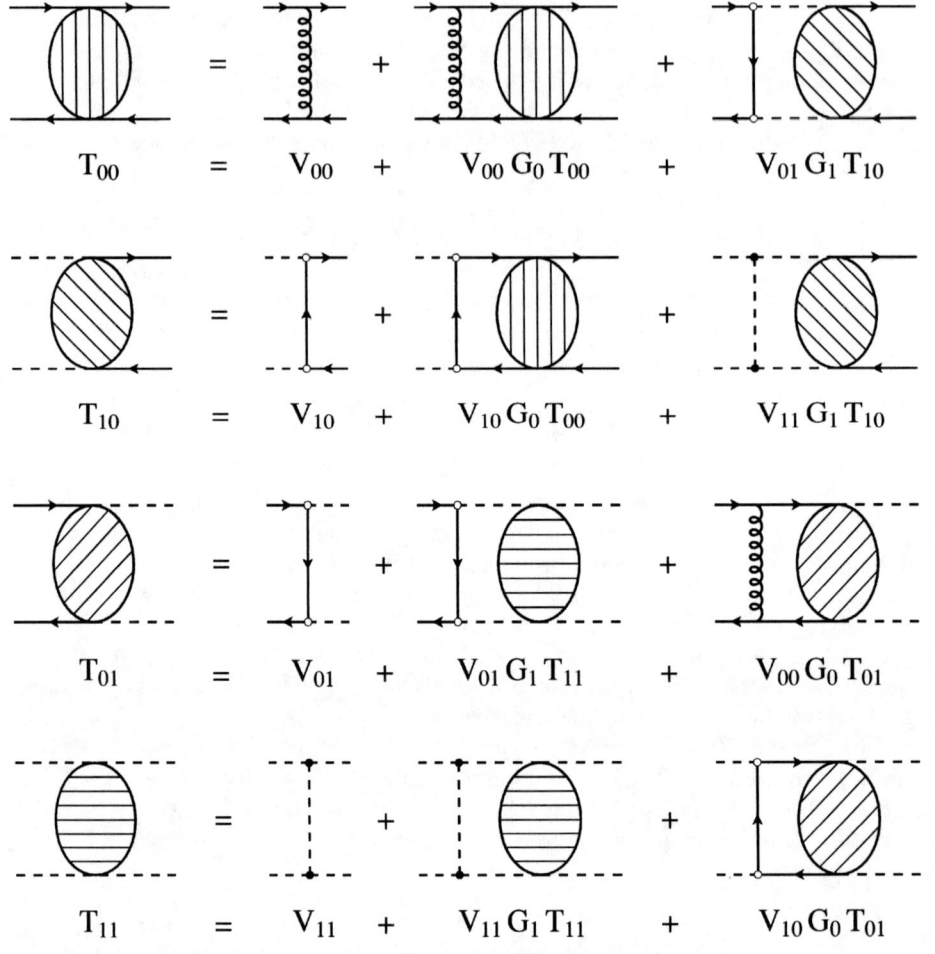

$$T_{00} \quad = \quad V_{00} \quad + \quad V_{00}\,G_0\,T_{00} \quad + \quad V_{01}\,G_1\,T_{10}$$

$$T_{10} \quad = \quad V_{10} \quad + \quad V_{10}\,G_0\,T_{00} \quad + \quad V_{11}\,G_1\,T_{10}$$

$$T_{01} \quad = \quad V_{01} \quad + \quad V_{01}\,G_1\,T_{11} \quad + \quad V_{00}\,G_0\,T_{01}$$

$$T_{11} \quad = \quad V_{11} \quad + \quad V_{11}\,G_1\,T_{11} \quad + \quad V_{10}\,G_0\,T_{01}$$

FIGURE 2. Graphical representation of Eq. (22); see text for explanation.

Combining Eqs. (24)–(27), we can write

$$\mathbf{A}_{01}\cdot\mathbf{t}_{01} = \mathbf{v}_{01} + \mathbf{A}_{11}\cdot\mathbf{t}_{11} \;\Rightarrow\; \mathbf{t}_{01} = \mathbf{A}_{01}^{-1}\cdot\{\mathbf{v}_{01} + \mathbf{A}_{11}\cdot\mathbf{t}_{11}\}\,, \tag{28}$$

where the boldface capitals and small letters are obvious shorthands for matrices and vectors, respectively. Turning next to T_{11}, we get the integral equation

$$T_{11}(\mathrm{p},\bar{\mathrm{p}}) - V_{11}(\mathrm{p},\bar{\mathrm{p}}) + \int \frac{k^2 dk}{4\pi^2} V_{11}(\mathrm{p},\mathrm{k})\,G_1(\mathrm{k})\,T_{11}(\mathrm{k},\bar{\mathrm{p}}) +$$

$$+ \int \frac{k^2 dk}{4\pi^2} V_{10}(\mathrm{p},\mathrm{k})\,G_0(\mathrm{k})\,T_{01}(\mathrm{k},\bar{\mathrm{p}})\,. \tag{29}$$

Similarly to Eqs. (24)–(27), we define now

$$B_{11}^{ij} \equiv s_{11}^j(p_i) - \int \frac{k^2 dk}{4\pi^2} V_{11}(p_i, k) G_1(k) s_{11}^j(k) ; \tag{30}$$

$$B_{01}^{ij} \equiv \int \frac{k^2 dk}{4\pi^2} V_{10}(p_i, k) G_0(k) s_{01}^j(k) , \quad v_{11}^i \equiv V_{11}(p_i, \bar{p}) . \tag{31}$$

Then we can express t_{11} in terms of itself, using Eq. (28):

$$\mathbf{B}_{11} \cdot \mathbf{t}_{11} = \mathbf{v}_{11} + \mathbf{B}_{01} \cdot \mathbf{t}_{01} \underset{(28)}{=} \mathbf{v}_{11} + \mathbf{B}_{01} \cdot \mathbf{A}_{01}^{-1} \cdot \{\mathbf{v}_{01} + \mathbf{A}_{11} \cdot \mathbf{t}_{11}\} . \tag{32}$$

This finally leads to

$$\{\mathbf{B}_{11} - \mathbf{B}_{01} \cdot \mathbf{A}_{01}^{-1} \cdot \mathbf{A}_{11}\} \cdot \mathbf{t}_{11} = \mathbf{v}_{11} + \mathbf{B}_{01} \cdot \mathbf{A}_{01}^{-1} \cdot \mathbf{v}_{01} , \tag{33}$$

$$\mathbf{t}_{11} = \{\mathbf{B}_{11} - \mathbf{B}_{01} \cdot \mathbf{A}_{01}^{-1} \cdot \mathbf{A}_{11}\}^{-1} \cdot \{\mathbf{v}_{11} + \mathbf{B}_{01} \cdot \mathbf{A}_{01}^{-1} \cdot \mathbf{v}_{01}\} . \tag{34}$$

Thus, we have obtained a closed-form expression for the T matrix (hence the S matrix), containing the *complete* information of the confinement-plus-scattering process. Needless to say that all these algebraic manipulations only make sense if the involved integrals actually exist and the spline expansions converge. Yet, we are confident this will be the case when appropriate boundary conditions are chosen, since the same spline expansion we are using here has been successfully applied to few-body scattering calculations in momentum space [19]. Nevertheless, in the Dirac case, to be dealt with in the future, form factors will have to be included for the meson-quark-antiquark vertices so as to make the integrals involving three quark propagators convergent. In any case, such form factors are physically meaningful, and should therefore already be considered in the approximation with spinless fermions, whenever doing phenomenology.

Bound states and resonances are given by zeroes in the determinant of the inverted-matrix factor right after the equal sign in Eq. (34). Resonance poles in the second Riemann sheet can be searched for by analytic continuation to complex energies, using e.g. contour-rotation methods. Moreover, from the structure of the equations one readily sees that the inclusion of final-state interactions, due to t-channel meson exchange in the meson-meson channel, simply amounts to taking a non-vanishing V_{11} in Eq. (29), which will not lead to any additional numerical effort. Finally, the generalization to many scattering channels and also more $q\bar{q}$ channels is straightforward, at the expense of bigger matrices.

4. CONCLUSIONS AND OUTLOOK

In the foregoing, we have demonstrated that the formulation of a unitarized quark/meson model with harmonic confinement is relatively easy in momentum space. Thus, the $^3P_0$ mechanism is dynamically described through P-wave quark exchange. Furthermore, the inclusion of final-state interactions via t-channel meson exchange is straightforward, and does not complicate the structure of the equations. The numerical resolution of these equations using splines looks very convenient, which should result in (approximate) analytic solutions, thereby facilitating an analytic continuation to complex energies in order to search for resonance poles.

Our first priority is, of course, to achieve a total control of the numerics, for real as well as complex energy. Then a simple toy model will be worked out so as to reproduce known coordinate-space results. Furthermore, cases problematic in the configuation-space formulation will be studied, such as heavy-light mesons. Also, the influence of t-channel meson exchange on meson-meson phase shifts will be investigated. In a somewhat more distant perspective, we intend to do a lot of phenomenology, and use the feedback to further refine the model.

ACKNOWLEDGMENTS

The authors thank F. Kleefeld and J. A. Tjon for valuable discussions, and P. Nogueira for expert graphical assistance. This work was partly supported by the *Fundação para a Ciência e a Tecnologia* (FCT) of the *Ministério do Ensino Superior, Ciência e Tecnologia* of Portugal, under contract number CERN/P/FIS/43697/2001.

REFERENCES

1. E. Eichten, K. Gottfried, T. Kinoshita, K. D. Lane, and T. M. Yan, Phys. Rev. D **21**, 203 (1980); J. L. Richardson, Phys. Lett. B **82**, 272 (1979).
2. S. Godfrey and N. Isgur, Phys. Rev. D **32** (1985) 189; P. C. Tiemeijer and J. A. Tjon, Phys. Rev. C **49**, 494 (1994); P. C. Tiemeijer, Ph.D. Thesis, Utrecht (1993).
3. A. Le Yaouanc, L. Oliver, S. Ono, O. Pene and J. C. Raynal, Phys. Rev. D **31**, 137 (1985); P. J. Bicudo and J. E. Ribeiro, Phys. Rev. D **42**, 1611 (1990); **42**, 1625 (1990); **42**, 1635 (1990).
4. N. Isgur and M. B. Wise, Phys. Rev. Lett. **66**, 1130 (1991); M. Neubert, Phys. Rept. **245**, 259 (1994) [hep-ph/9306320].
5. M. Gell-Mann and M. Lévy, Nuovo Cim. **16**, 705 (1960); R. Delbourgo and M. D. Scadron, Mod. Phys. Lett. A **10**, 251 (1995) [hep-ph/9910242]; Int. J. Mod. Phys. A **13**, 657 (1998) [hep-ph/9807504].
6. Y. Nambu and G. Jona-Lasinio, Phys. Rev. **122**, 345 (1961); V. Bernard, A. H. Blin, B. Hiller, Y. P. Ivanov, A. A. Osipov, and U. G. Meissner, Annals Phys. **249**, 499 (1996) [hep-ph/9506309].
7. J. Gasser and H. Leutwyler, Annals Phys. **158**, 142 (1984); J. A. Oller, E. Oset, and J. R. Peláez, Phys. Rev. **D59** (1999) 074001 [Erratum-ibid. **D60** (1999) 099906] [hep-ph/9804209].
8. D. Lohse, J. W. Durso, K. Holinde, and J. Speth, Nucl. Phys. A **516**, 513 (1990).
9. E. v. Beveren, G. Rupp, N. Petropoulos, and F. Kleefeld, hep-ph/0211411.
10. E. van Beveren, T. A. Rijken, K. Metzger, C. Dullemond, G. Rupp, and J. E. Ribeiro, Z. Phys. **C30**, 615 (1986).
11. E. van Beveren and G. Rupp, Eur. Phys. J. C **10**, 469 (1999) [hep-ph/9806246].
12. E. M. Aitala *et al.* [E791 Collaboration], Phys. Rev. Lett. **89**, 121801 (2002) [hep-ex/0204018].
13. E. van Beveren and G. Rupp, Eur. Phys. J. C **22**, 493 (2001) [hep-ex/0106077].
14. Nils A. Törnqvist and Matts Roos, Phys. Rev. Lett. **76** (1996) 1575 [hep-ph/9511210].
15. A. A. Logunov and A. N. Tavkhelidze, Nuovo Cim. **29** (1963) 380; R. Blankenbecler and R. Sugar, Phys. Rev. **142**, 1051 (1966).
16. R. M. Woloshyn and A. D. Jackson, Nucl. Phys. B **64**, 269 (1973).
17. S. J. Wallace and V. B. Mandelzweig, Nucl. Phys. A **503**, 673 (1989).
18. G. L. Payne, Lect. Notes Phys. **273**, 64 (1987).
19. M. T. Peña, H. Garcilazo, P. U. Sauer, and U. Oelfke, Phys. Rev. C **45**, 1487 (1992).

The Production and Decay of Heavy Dimesons

Mitja Rosina*, Damijan Janc†, Daniele Treleani** and
Alessio Del Fabbro**

*Faculty of Mathematics and Physics, University of Ljubljana, Jadranska 19, P.O. Box 2964, 1001
Ljubljana, Slovenia and Jožef Stefan Institute, 1000 Ljubljana, Slovenia
†Jožef Stefan Institute, 1000 Ljubljana, Slovenia
**Dipartimento di Fisica Teorica, Università di Trieste, Strada Costiera 11, Miramare-Grignano,
and INFN, Sezione di Trieste, I-34014 Trieste, Italy

Abstract. We show that it is necessary to go beyond a single hadron (beyond the quark-antiquark or three-quark systems) in order to distinguish the colour structure of the effective quark-quark interaction and the relevance of 3-body forces. We critically discuss the proposed models which suggest the dimeson $bb\bar{u}\bar{d}$ to be bound by ~ 100 MeV and the $cc\bar{u}\bar{d}$ dimeson to be unbound. Only experiment can judge. We estimate the probability of producing $bb\bar{u}\bar{d}$ at LHC by double gluon-gluon fusion and search for a characteristic decay.

MOTIVATION

There is a strong motivation to understand the effective interaction between heavy quarks (and antiquarks) since it is expected to be "cleaner" than between light quarks. For heavy particles the nonrelativistic constituent quark model is more acceptable, the perturbative QCD contributions (such as one-gluon-exchange) is more adequate and chiral fields are less important.

While the effective interaction between a heavy quark and a heavy antiquark has been reasonably well studied and fitted by the charmonium and bottomium spectra, there is no free diquark to study the effective interaction between two heavy quarks. One has to dress the diquark in order to obtain a colour singlet object. Therefore, the double-heavy baryons ccq, bcq, and bbq (q=u,d, or s), as well as the double-heavy dimesons $cc\bar{q}\bar{q}$, $bc\bar{q}\bar{q}$ and $bb\bar{q}\bar{q}$ (also called tetraquarks) can be considered as a laboratory to study the properties of the diquark. They have not yet been seen experimentally (except some tentative signals [1]). The purpose of our study is to demonstrate, how important information for quark models they would offer, and to stimulate the experimentalists to invest due efforts to discover them in near future .

It is straightforward to extrapolate the one-gluon-exchange (OGE) interaction from $Q\bar{Q}$ to QQ (Q= any quark). The charge conjugation changes the $\bar{Q}$ antitriplet to Q triplet. Then the colour factor $\lambda \cdot \lambda/4 = -4/3$ for the $Q\bar{Q}$ singlet changes to $-2/3$ for the QQ antitriplet (the "$V_{QQ} = \frac{1}{2}V_{Q\bar{Q}}$ rule").

On the other hand, it is questionable whether the (linear) confining potential should also possess such a colour factor and obey the $V_{QQ} = \frac{1}{2}V_{Q\bar{Q}}$ rule. The fact that the ground state energies and some excited states of light and heavy baryons are reasonably well

CP660, Hadron Physics: Effective Theories of Low Energy QCD, edited by A. H. Blin et al.
© 2003 American Institute of Physics 0-7354-0120-9/03/$20.00

reproduced with such a "universal" OGE + confining effective interaction is encouraging [2] but not conclusive. There may be other mechanisms for the $V_{QQ} = \frac{1}{2}V_{Q\bar{Q}}$ rule. For example, the flux tubes in a "Mercedes" configuration can be mimicked by twice weaker two-body flux lines since the length of the arms of the "Mercedes" is approximately half the length of the circumference of the triangle. The colour singlet 3-quark system is insensitive to the features of the *colour · colour* operator since it is just a constant in the 3-body singlet representation. To explore the colour structure of the effective interaction one has to go beyond mesons and baryons to dimesons and other exotics.

Moreover, there may be other contributions to the effective interaction. Regarding the short-range part, the important effect is the strong spin-spin splitting which can be due to OGE, pion exchange or some instanton effects, the relative importance of which is not yet clear. Light baryon spectra are better reproduced by the meson exchange interaction [3] than by the OGE interaction but the extension to the heavy baryons and to mesons is still uncertain.

The study of double-heavy baryons and of double-heavy dimesons are complementary. They both refer to the colour triplet heavy diquark and try to determine the strength of the interaction and test the "$V_{QQ} = \frac{1}{2}V_{Q\bar{Q}}$ rule". The dimeson, however, possesses also a $(6,\bar{6})$ configuration with a sextet heavy diquark and antisextet cloud of two light antiquarks; configuration mixing offers the opportunity to test the colour structure of the interaction.

Furthermore, there are theoretical reasons for three-body and four-body forces. Three-body forces have been shown to be welcome to improve the absolute position of baryons with respect to mesons [2]. The dimeson can give a better hint about the relative strength of the three-body interaction since there are 4 three-body contributions compared to just one in a baryon.

In next Section we give arguments why we expect the bb$\bar{u}\bar{d}$ dimeson to be strongly bound while the cc$\bar{u}\bar{d}$, bc$\bar{u}\bar{d}$ and others are most likely unbound. In the third Section we estimate the chances to produce the bb$\bar{u}\bar{d}$ dimeson in LHC, and in the final Section we call for new ideas how to detect it.

BINDING ENERGIES

With Two-Body Forces

We estimate the binding energy of the bb$\bar{u}\bar{d}$ dimeson assuming the $V_{QQ} = \frac{1}{2}V_{Q\bar{Q}}$ rule and no three-body forces. If this prediction is falsified by the experiment a revision of the two-body QQ interaction and/or introduction of many-quark forces will be needed.

For the sake of clarity we present here a simplified derivation which makes three further assumptions: (i) that the spin-dependent interactions between heavy quarks are not important, (ii) that the configuration with both b quarks bound in a compact diquark with spin 1 and antitriplet colour dominates, and (iii) that the two light antiquarks in the dimeson behave equally as in a Λ_b baryon. We have, however, performed also a careful four-body calculation without such unnecessary assumptions [4] and the results were practically the same.

We want the result to be as little model dependent as possible. Therefore we consider any model which reproduces exactly the masses of Λ_b baryon and several mesons. Then we can consider Nature as an "analogue computer" and express the mass of the dimeson in terms of those experimental masses. Now we compare the following hadrons

$$
\begin{aligned}
M_{bb\bar{u}\bar{d}} &= 2M_b + M_u + M_d + E_{bb} + E_{\bar{u}\bar{d}[bb]} \\
M_{\Upsilon} &= 2M_b + E_{b\bar{b}} \\
M_{\Lambda_b} &= M_b + M_u + M_d + E_{\bar{u}\bar{d}\bar{b}}
\end{aligned}
$$

where $E_{\bar{u}\bar{d}[bb]} \approx E_{\bar{u}\bar{d}\bar{b}}$ is the potential plus kinetic energy contribution of the two light quarks in the field of a heavy diquark or quark, respectively, and it cancels in the difference. This would be exactly true in the limit where the mass of the b quark goes to infinity and the heavy diquark is point-like so that we can neglect the size of the heavy diquark in the dimeson.

We can estimate the diquark binding energy using the following trick. The binding energy E_{meson} of a quark and antiquark in the meson is a function of the reduced mass only (neglecting spin forces):

$$
\left[\frac{p^2}{M_Q} + V_{Q\bar{Q}} \right] \psi = E_{\text{meson}}(M_Q) \psi
$$

For a diquark the Schrödinger equation is similar as for a meson, but with twice weaker interaction. To get the similarity, we mimic the kinetic energy with twice smaller reduced mass.

$$
\left[\frac{p^2}{M_Q} + V_{QQ} \right] \psi = \frac{1}{2} \left[\frac{p^2}{M_Q/2} + V_{Q\bar{Q}} \right] \psi = \frac{1}{2} E_{\text{meson}}(M_Q/2) \psi.
$$

We obtain $E_{\text{meson}}(M_b/2)$ by plotting the binding energies of mesons as a function of the reduced mass [4]. We do that for different choices of constituent quark masses in order to estimate the uncertainty due to quark masses. The plot is very smooth and we estimate the binding energy of the heavy bb diquark $E_{\text{diquark}}(M_b) = \frac{1}{2} E_{\text{meson}}(M_b/2) = -390 \pm 15$ MeV, whereas for bottomium $\frac{1}{2} E_{\text{meson}}(M_b) = -560 \pm 15$ MeV.

This gives us the phenomenological estimate for the binding energy of the dimeson with respect to the BB* threshold

$$
\begin{aligned}
\Delta E_{bb\bar{u}\bar{d}} &= M_{\Lambda_b} + [M_{\Upsilon} - E_{\text{meson}}(M_b) + E_{\text{meson}}(M_b/2)]/2 - M_B - M_{B^*} \\
&= -130 \pm 20 \, \text{MeV}.
\end{aligned}
$$

This agrees well with our detailed calculation [4] and with some previous four body calculations [2] [5] in the constituent quark model.

An analogous calculation using Λ_c and J/ψ instead of Λ_b and Υ gives for the mass difference between the $cc\bar{u}\bar{d}$ dimeson and the DD* threshold a value of $+97$ MeV which means a prediction, that this dimeson is definitely unbound.

With Two- and Three-Body Forces

Nothing is definite. If a future experiment finds the bb dimeson unbound or the cc dimeson bound we shall have to revise our general ideas about the effective quark-quark interaction, and/or introduce many-quark forces.

The recent CDF experiment on double-heavy baryons in Fermilab [1] has caused some confusion. The ccd candidate at 3520 MeV is somewhat low but still manageable with two-body interactions. One would need smaller constituent quark masses than [2]. Why? Smaller quark masses need less negative (kinetic + potential) binding energies to fit the heavy mesons. Therefore, going from $Q\bar{Q}$ to the QQ diquark using the the the $V_{QQ} = \frac{1}{2}V_{Q\bar{Q}}$ rule one loses less binding and the ccd baryon becomes better bound. (Note that a phenomenological estimate similar to the one for the dimesons would give 3537 to 3560 MeV). The ccd candidate at 3783 seems, however, too light for an excited state and too heavy for the ground state.

On the other hand, the ccu baryon candidate at 3460 MeV is not believable due to its 60 MeV isospin splitting. If, however, correct, it would require help from many-body forces or some other mechanisms which would lead to an almost bound cc dimeson (our phenomenological estimate can then be written as $\Delta E_{cc\bar{u}\bar{d}} = M_{\Xi_{cc}} + M_{\Lambda_c} - M_D - M_D - M_{D^*} = +3\,\text{MeV}$).

Some weak three-body terms have been introduced in ref.[2] in order to improve (lower) the absolute position of the baryon spectrum if the meson spectrum is fitted. We are exploring which colour structures could be used and whether the parametrisation could be stretched so as to bind cc dimeson or unbind the bb dimeson without spoiling the fit to mesons and baryons.

A MODEL FOR THE SYNTHESIS OF THE HEAVY DIMESON

The $bb\bar{u}\bar{d}$ dimesons have not been seen in the present machines. Anyway, estimates of the production and detection rate are very pessimistic and have for this reason not been published. Therefore we encourage to look for them at LHC. For the synthesis we propose a three-step model [8][9].

(i) First, two b-quarks are formed in the proton-proton collision by a double gluon-gluon fusion: $(g+g) + (g+g) \to (b+\bar{b})(b+\bar{b})$. This was shown to be the dominant production mechanism [6]. One might wonder why we need a TeV machine to produce GeV particles. The answer is simple. The two colliding protons can be considered as two packages of virtual gluons whose number is huge for low Bjorken-x. Only the number of gluons with $x < 0.001$ might be sufficient to make dimesons detectable.

The forward detector LHCb will cover the pseudorapidity region $1.8 < \eta < 4.9$ and will detect the B and $\bar{B}$ hadrons in the low p_T region. We are interested in double-b production in which the two b-quarks are close enough in phase space to synthesize a diquark. By requiring that the two b are produced with $|p_1(j) - p_2(j)| < \Delta$, $j = x, y, z$, we get the cross section $\sigma \approx 0.4(\Delta/\text{GeV})^3$ nb which is approximately proportional to the momentum volume up to 2 GeV: $d\sigma/d^3p \approx 0.4\,\text{nb}/\text{GeV}^3$. At the expected luminosity L=0.1 events/(second nb) this corresponds to 144 interesting bb pairs per hour per GeV^3.

(ii) In the second step, the two b *quarks join into a diquark.* We assume simultaneous production of two independent b quarks with momenta $\vec{p}_1, \vec{p}_2$. Since they appear wherever within the nucleon volume, we modulate their wavefunctions with a Gaussian profile with the "oscillator parameter" $B = \sqrt{2/3}\sqrt{<r^2>} = 0.69$ fm corresponding to the nucleon rms radius

$$\mathcal{N}_B \ \exp \ (-\vec{r}_1^2/2B^2 + i\vec{p}_1\vec{r}_1)\mathcal{N}_B\exp(-\vec{r}_2^2/2B^2 + i\vec{p}_2\vec{r}_2)$$
$$\equiv \mathcal{N}_{B/\sqrt{2}} \ \exp \ (-\vec{R}^2/2(B/\sqrt{2})^2 + i\vec{P}\vec{R})\mathcal{N}_{B\sqrt{2}}\exp(-\vec{r}^2/2(B\sqrt{2})^2 + i\vec{p}\vec{r})$$

where the normalization factor $\mathcal{N}_\beta = \pi^{-3/4}\beta^{-3/2}$.

We make an impulse approximation that this two-quark state is instantaneously transformed in any of the eigenstates of the two-quark Hamiltonian. Then the amplitude of the diquark formation M is equal to the overlap between the two free quarks and the diquark with the same centre-of-mass motion. By approximating the diquark wavefunction with a Gaussian with the oscillator parameter $\beta = 0.23$ fm we get

$$
\begin{aligned}
M(p) &= \int d^3r\, \mathcal{N}_{B\sqrt{2}}\exp(-\vec{r}^2/2(B\sqrt{2})^2 - i\,\vec{p}\vec{r})\mathcal{N}_\beta\exp(-\vec{r}^2/2\beta^2) \\
&= \sqrt{\frac{2\sqrt{2}B\beta}{2B^2+\beta^2}}^3 \ \exp[-(p^2/2)(2B^2\beta^2/(2B^2+\beta^2))]
\end{aligned}
$$

For the production cross section we take into account that $\beta << B$ and that $d\sigma/d^3p$ is practically constant and can be taken out of the integral

$$\sigma = \int d^3p \frac{d\sigma}{d^3p}M^2(p) \approx \frac{d\sigma}{d^3p}\left(\frac{\sqrt{2\pi\hbar}}{B}\right)^3 = 0.15\text{nb}$$

which corresponds to $L\sigma = 54$ diquarks/hour.

(iii) In the third step, the diquark gets dressed. It either acquires one light quark to become the doubly-heavy baryon bbu, bbd or bbs, or two light antiquarks to become a dimeson.

We estimate the probabilities of dressing f_{dress} using the analogy with dressing a single quark ("fragmentation of a quark into hadrons"). We make use of experimental data obtained at Fermilab and at LEP experiments [7]:

$$\bar{b} \to \bar{b}u, \bar{b}d, \bar{b}s, \bar{b}u\bar{d} = 0.37 \pm 0.02,\ 0.37 \pm 0.02,\ 0.16 \pm 0.03,\ 0.09 \pm 0.03.$$

Since a heavy diquark acts similarly as a heavy quark, we expect similar branching ratios:

$$\text{bb} \to \text{bbu, bbd, bbs, bb}u\bar{d} \approx 0.37,\ 0.37,\ 0.16,\ 0.09.$$

This yields the production rate of the dimeson $L\sigma f_{\text{dress}} \sim 5 - 6$ events/hour.

DISCUSSION ABOUT THE DETECTION OF DIMESONS

We expect that the $bb\bar{u}\bar{d}$ dimeson will be stable against strong and electromagnetic decay and will decay only weakly, with a lifetime of about 1 ps (corresponding to the width of 1 meV). The main channel would be the independent decay of each b quarks into c quark. This essentially means the independent decay of each B meson in the dimeson, for example $B \to D +$ anything. However, such a channel will be difficult to distinguish from the decay of two unbound B mesons which will be the main background contribution. There are no good two-body decay channels of B mesons to allow the reconstruction of the total energy of the dimeson; moreover, each separate exclusive decay channel has a low branching ratio of up to a few percent.

Much more characteristic would be the simple two-body decay channel $bb\bar{u}\bar{d} \to \Upsilon + \pi$ with the kinetic energy 876 MeV of both mesons together (in the c.m. system). Of course there would be a crowd of other Υ mesons, but few at this energy. The inspiration comes from the $B^0 \to \bar{B}^0$ oscillation which unfortunately is not feasible for bound B mesons because the BB and B$\bar{B}$ states are not degenerate. The weak transition $b\bar{u} \to u\bar{b}$ is negligible because of the low CKM amplitudes.

Some hope is offered by the angular correlation of the b and c quarks after the decay of the first b$\to$c, from which the original b-b correlation could be deduced. If a dimeson were formed, the correlation should be isotropic in the c.m. system of the dimeson since the two heavy quarks are expected to be in the relative 1s state. On the other hand, two independent b-quarks would tend to move more in the same direction.

We call for new ideas for the detection of doubly b dimesons!

REFERENCES

1. Mattson, M., et al. (SELEX Collaboration), *Phys. Rev. Lett.* **89**, 112001-1-5 (2002); Russ, J. S. (on behalf of the SELEX Collaboration), hep-ex/0209075.
2. Silvestre-Brac, B., and Semay, C., *Z. Phys.* **C57**, 273-282 (1993).
3. Glozman, L.Ya., Plessas, W., Varga, K., and Wagenbrunn, R.F., *Phys.Rev.D* **58**, 094030-1-8 (1998).
4. Janc, D., and Rosina, M., *Few-Body Systems* **31**, 1-11 (2001).
5. Brink, D. M., and Stancu, Fl., *Phys. Rev.* **D57**, 6778-6787 (1998).
6. Del Fabbro, A., and Treleani,D., *Phys. Rev.* **D61**, 077502 (2000); *Phys. Rev.* **D63**, 057901-1-4 (2001); *Nucl. Phys. B* **92**, 130 (2001).
7. Affolder, T., et al. (CDF Collaboration), *Phys. Rev. Lett.* **84**, 1663-1668 (2000).
8. Rosina, M., Janc, D., Treleani,D., and Del Fabbro, A., *Bled Workshops in Physics* **3**, No.3, 63-66 (2002), also available at http://www-f1.ijs.si/Bled/Pub.
9. Janc, D., Rosina, M., Treleani,D., and Del Fabbro, A., to appear in: *Few-Body Systems Suppl.*, (2003).

Scalar and Tensor Force Contribution to the Nucleon-Nucleon Interaction Within a Chiral Constituent Quark Model

D. Bartz

Institute of Physics B5, University of Liège, Sart Tilman, B-4000 Liège 1, Belgium

Abstract. The nucleon-nucleon problem is studied as a six-quark system in a nonrelativistic chiral constituent quark model where the Hamiltonian contains a linear confinement and a pseudoscalar meson (Goldstone boson) exchange interaction between the quarks [1, 2]. This hyperfine interaction has a long-range Yukawa-type part, depending on the mass of the exchanged meson and a short-range part, mainly responsible for the good description of the baryon spectra.

The nucleon-nucleon phases shifts are calculated using the resonating group method, convenient for treating the interaction between composite particles. The behavior of the phase shifts for the $^3S_1$ and $^1S_0$ channels show the presence of a strong repulsive core.

We show that the introduction of a scalar σ-meson exchange [3] and the tensor force [4] associated to the pseudoscalar meson exchange interaction lead to a satisfactory agreement with the experiencal phase shifts.

INTRODUCTION

The study of the nucleon-nucleon (NN) interaction in the framework of quark models has already some history. Twenty years ago Oka and Yazaki [5] published the first nucleon-nucleon $L = 0$ phase shifts, calculated within the resonating group method. Those results were obtained from a model based on the one-gluon exchange (OGE) interaction between quarks. Within such models one could explain the short-range repulsion of the NN potential as due to the chromomagnetic spin-spin interaction, combined with quark interchanges between 3q clusters. However, in order to describe the data, long- and medium-range interactions were added at the nucleon level. The challenge was then to describe both the baryon, as a three-quark system, and the NN interaction, as a six-quark problem at the quark level, within the same quark model. Relative successes were obtain as shown in several review papers [6, 7].

Here we use a constituent quark model where the short-range quark-quark interaction is entirely due to pseudoscalar meson exchange, instead of one-gluon exchange. This is the chiral constituent quark model of Ref. [1], parametrized in a nonrelativistic version in Ref. [2]. The origin of this model is thought to lie in the spontaneus breaking

CP660, *Hadron Physics: Effective Theories of Low Energy QCD*, edited by A. H. Blin et al.
© 2003 American Institute of Physics 0-7354-0120-9/03/$20.00

of chiral symmetry in quantum chromodynamics (QCD) which implies the existence of Goldstone bosons (pseudoscalar mesons) and constituent quarks with dynamical mass. If a quark-pseudoscalar meson coupling is assumed this generates a pseudoscalar meson exchange between quarks which is spin and flavour dependent. The spin-flavour structure is crucial in reproducing the baryon spectra. In the following this model will be referred to as the Goldstone boson exchange (GBE) model.

It is important to correctly describe both the baryon spectra and the baryon-baryon interaction with the same model. The GBE model gives a good description of the baryon spectra and in particular the correct order of positive and negative parity states, both in nonstrange and strange baryons, in contrast to the OGE model. The pseudoscalar exchange interaction has two parts : a repulsive Yukawa potential tail and an attractive contact interaction. When regularized, the latter generates the short-range part of the quark-quark interaction. This dominates over the Yukawa part in the description of baryon spectra. The present status of this model is presented in Ref. [8]. The whole interaction contains the main ingredients required in the calculation of the NN potential, and it is thus natural to study the NN problem within the GBE model. In addition, the two-meson exchange interaction between constituent quarks reinforces the effect of the flavour-spin part of the one-meson exchange and also provides a contribution of a σ-meson exchange type [9] required to describe the middle-range attraction. Besides the spin-spin term, the pseudoscalar meson exchange gives rise to a tensor term as well. By coupling of the $^3S_1$ and $^3D_1$ states, we shall show how this tensor force leads to the necessary attraction in the $^3S_1$ phase shift.

RESULTS

Here we report on dynamical calculations of the NN interaction obtained in the framework of the GBE model and based on the resonation group method. The chiral interaction parametrization [2] contains the following spin-spin part

$$
V_\gamma^{SS}(r_{ij}) = \frac{g_{\gamma q}^2}{4\pi} \frac{1}{12m_i m_j} \left\{ \mu_\gamma^2 \frac{e^{-\mu_\gamma r_{ij}}}{r_{ij}} - \Lambda_\gamma^2 \frac{e^{-\Lambda_\gamma r_{ij}}}{r_{ij}} \right\}, \tag{1}
$$

where $\Lambda_\gamma = \Lambda_0 + \kappa \mu_\gamma$ and with $\gamma = \pi, \eta, \eta'$. The parameters are

$$
\frac{g_{\pi q}^2}{4\pi} = \frac{g_{\eta q}^2}{4\pi} = 1.24, \frac{g_{\eta' q}^2}{4\pi} = 2.7652,
$$

$$
m_{u,d} = 340 \text{ MeV}, \ C = 0.77 \text{ fm}^{-2},
$$

$$
\mu_\pi = 139 \text{ MeV}, \ \mu_\eta = 547 \text{ MeV}, \ \mu_{\eta'} = 958 \text{ MeV},
$$

$$
\Lambda_0 = 5.82 \text{ fm}^{-1}, \ \kappa = 1.34, \ V_0 = -112 \text{ MeV}. \tag{2}
$$

In Ref. [10] we obtained the $^3S_1$ and $^1S_0$ phase shifts in single and three coupled channels calculations. We found that the coupling to the $\Delta\Delta$ and CC (hidden colour) channels change slightly the NN phase shift in the laboratory energy interval 0 - 350 MeV.

The behaviour of the phase shifts explicitly shows that the GBE model can explain the short-range repulsion, as due to both the flavour-spin quark-quark interaction and to the quark interchange between clusters. Moreover, the results indicate that the $^3S_1$ and $^1S_0$ phase shifts are quite close to each other when the interaction contains only a spin-spin part.

However, to describe the scattering data and the deuteron properties at the same time, intermediate- and long-range attraction potentials are necesary. In Ref. [3] a σ-meson exchange interaction has been added at the quark level to the six-quark Hamiltonian. This interaction has the form

$$V_\sigma = -\frac{g^2_{\sigma q}}{4\pi}\left(\frac{e^{-\mu_\sigma r}}{r} - \frac{e^{-\Lambda_\sigma r}}{r}\right),$$ (3)

The following set of the parametres entering this potential has been choosen [11]

$$\frac{g^2_{\sigma q}}{4\pi} = \frac{g^2_{\pi q}}{4\pi} = 1.24, \quad \mu_\sigma = 0.278 \text{ GeV}, \quad \Lambda_\sigma = 0.337 \text{ GeV}.$$ (4)

Note that the sigma mass, $\mu_\sigma = 2m_\pi$, is consistent with the findings of Ref. [12] in a linear σ-model.

As one can see from Fig. 1, with these values, the theoretical $^1S_0$ phase shift gets close to the experimental points without altering the good short-range behaviour, and in particular the change of sign of the phase shift at $E_{lab} \approx 260$ MeV. Without a scalar meson exchange, the phase shift would be negative everywhere. Thus the addition of a σ-meson exchange interaction alone leads to a good description of the phase shift in a large energy interval. Moreover the influence of the three coupled channels improve slightly the NN phase shift in the low energy domain. One can argue that the still existing discrepancy at low energies could possibly be removed by the coupling of the $^5D_0$ N-Δ channel. To achieve this coupling, as well as to describe the $^3S_1$ phase shift, the introduction of the tensor interaction is necessary.

As we have already mentioned, the pseudoscalar interaction potential between the constituent quarks contains both a spin-spin and a tensor part which, in the broken $SU_F(3)$, reads

$$V_\gamma(r_{ij}) = V^{SS}_\gamma(r_{ij})\vec{\sigma}_i \cdot \vec{\sigma}_j + V^T_\gamma(r_{ij})S^T_{ij}$$ (5)

where S^T_{ij} is given by

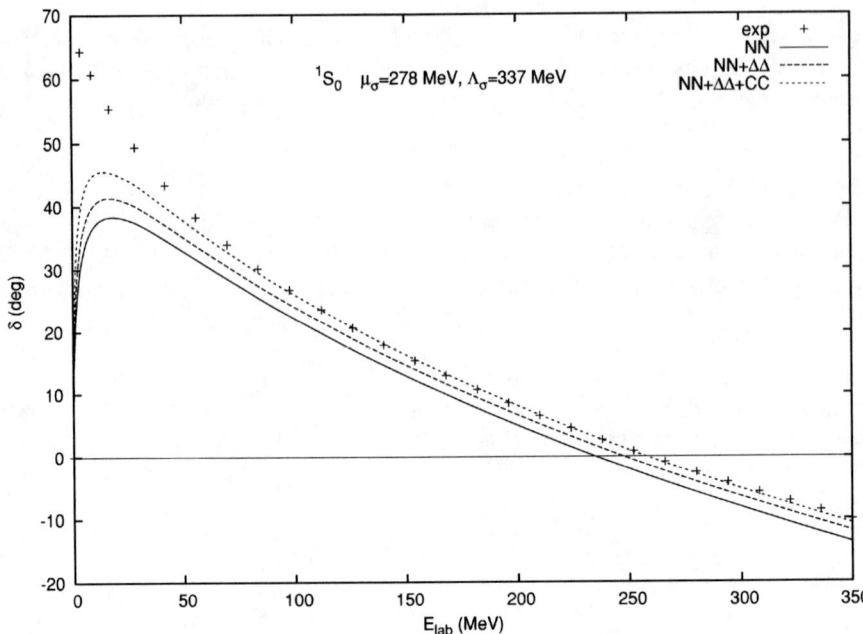

FIGURE 1. The $^1S_0$ NN scattering phase shift obtained in the GBE model as a function of E_{lab}. The coupled channel calculations are from Ref. [11]. Experimental data are from Ref. [13].

$$S_{ij}^T = \frac{3(\vec{r}_{ij} \cdot \vec{\sigma}_i)(\vec{r}_{ij} \cdot \vec{\sigma}_j)}{r^2} - \vec{\sigma}_i \cdot \vec{\sigma}_j . \qquad (6)$$

If in the derivation of the tensor potential we use the same form factor as in the spin-spin part, we obtain the following form for the tensor part of the pseudoscalar exchange potential

$$V_\gamma^T (r_{ij}) = T_1(r_{ij}) + T_2(r_{ij}) , \qquad (7)$$

where

$$T_1(r) = G_f \frac{g_{\gamma q}^2}{4\pi} \frac{1}{12m_i m_j} \left\{ \mu_\gamma^2 (1 + \frac{3}{\mu_\gamma r} + \frac{3}{\mu_\gamma^2 r^2}) \frac{e^{-\mu_\gamma r}}{r} \right\} , \qquad (8)$$

and the "regularized" part

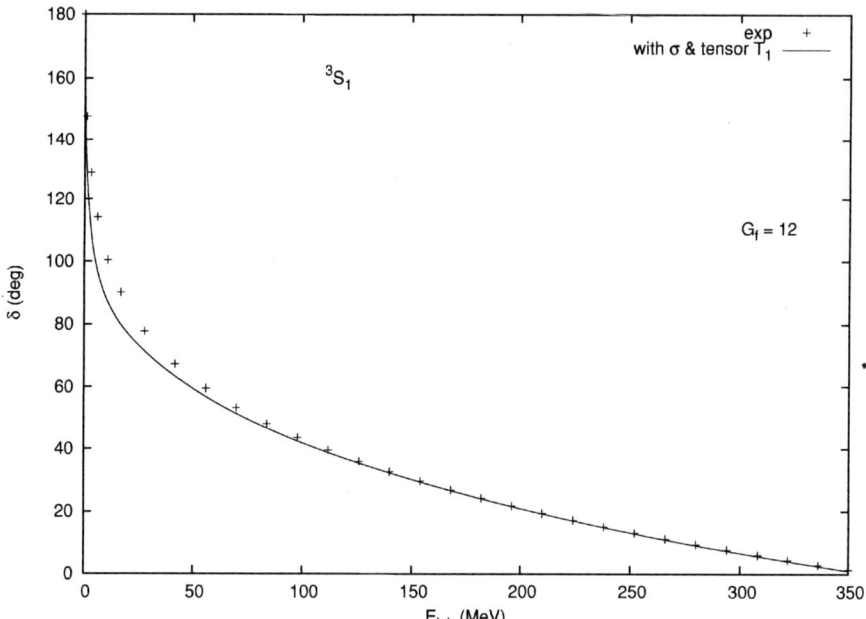

FIGURE 2. The $^3S_1$ scattering phase shift with the global factor $G_f = 12$. The "regularized" part (T_2, see text) of the tensor coulpling is equal to zero. Experimental data are from Ref. [13].

$$T_2(r) = G_f \frac{g_{\gamma q}^2}{4\pi} \frac{1}{12 m_i m_j} \left\{ -\Lambda_\gamma^2 (1 + \frac{3}{\Lambda_\gamma r} + \frac{3}{\Lambda_\gamma^2 r^2}) \frac{e^{-\Lambda_\gamma r}}{r} \right\} . \qquad (9)$$

A dimensionless global factor G_f has been introduced in order to allow an adjustment of the strength of this interaction such as to be as close as possible to the experiment [11].

Actually preliminary studies were first performed with a tensor interaction as given in Ref. [14]. Keeping the same parametres as in the above reference we found that the tensor force has only a negligible influence on the $^3S_1$ phase shift. In practice a much stronger tensor force is necessary to account for this phase shift. This is partially linked to the fact that the "regularized" part of (7) decreases the contribution of the main component of the tensor force, namely the T_1 term.

It is then interesting to analyze the contribution of the "regularized" part of (7) as compared to the T_1 term. In Fig. 2 we show the calculated phase shift with the tensor force as defined in Eq. (7) with the global factor $G_f = 12$ and without the "regularized" part of the tensor force. In this case good agreement is obtained at laboratory energies below 350 MeV by using the complete $SU_F(3)$ broken version of the model. If the T_2

term were introduced in the calculation, the attractive effect of the T_1 term would have been reduced. In this case, a global factor $G_f = 33$ (see Ref. [15]) would have been necessary to reproduce the experimental data of $^3S_1$ phase shift.

Naturally the question arises whether or not one can describe the N-spectrum and the NN interaction with the same quark model. The N-spectrum requires only a small tensor force while the NN interaction a large one. It would certainly be useful to search for other parametrizations of the GBE interaction and in particular to study the semi-relativistic version of the GBE model before giving a definite answer to this question.

ACKNOWLEDGMENTS

I would like to thank Floarea Stancu who help me to accomplish this research.

REFERENCES

1. L. Ya. Glozman and D. O. Riska, Phys. Rep. **268** 263 (1996)
2. L. Ya. Glozman, Z. Papp, W. Plessas, K. Varga and R. F. Wagenbrunn, Nucl. Phys. **A623** 90c (1997)
3. D. Bartz and Fl. Stancu, Nucl. Phys. **A688** 915 (2001)
4. D. Bartz and Fl. Stancu, Nucl. Phys. **A699** 316c (2002)
5. M. Oka and K. Yazaki, Phys. Lett. **90B** 41 (1980); Progr. Theor. Phys **66** 556 (1981); ibid **66** 572 (1981)
6. M. Oka and K. Yazaki, in *Quarks and nuclei*, Ed. Weise, World Scientific, Singapore 469 (1984)
7. K. Shimizu, Rep. Prog. Phys **52** 1 (1999); K. Shimizu, S. Takeuchi and A. Buchmann, Prog. Theor. Suppl **137** 43 (2000)
8. W. Plessas, Nucl. Phys. **A699** 316c (2002)
9. D. O. Riska and G. E. Brown, Nucl. Phys. **A653** 251 (1999)
10. D. Bartz and Fl. Stancu, Phys. Rev. **C63** 034001 (2001)
11. D. Bartz, PhD Thesis, University of Liège, 2002 hep-ph/0205138
12. B. Golli and M. Rosina, Phys. Lett. **B393** 161 (1997)
13. V. G. J. Stoks, R. A. M. Klomp, M. C. M. Rentmeester and J. J. de Swart, Phys. Rev. **C48** 792 (1993); V. G. J. Stoks, R. A. M. Klomp, C. P. F. Terheggen and J. J. de Swart, Phys. Rev. **C49** 2950 (1994)
14. W. Plessas et al., Proceedings of the Second International Conference on Perspectives in Hadronic Physics, Eds. S. Boffi, C. Ciofi degli Atti, M. Giannini, World Scientific, Singapore, (2000)
15. F. Stancu, Proceedings of XVIII European Conference on Few-Body Systems, Bled (Slovenia) Sept 8-14 2002, Few-Body Systems Suppl. (2003), to be published.

Summary talk

Georges Ripka

We owe this meeting to the efforts of A.H.Blin, B.Hiller, A.A.Osipov, M.C.Ruivo and E.van Beveren who went out of their way to make it successful and interesting. The talks were varied in kind, including reviews, reminders, development of models, advances in technical aspects, and more.

The $1/N_c$ expansion, and its implementation in terms of planar diagrams, now an old idea, continues to serve as an invaluable guide and it has been an *a posteriori* justification in choosing relevant processes in hadron physics where the coupling constants are large. We were shown how the $1/N_c$ expansion transpires in the values of the Gasser-Leutwyler coefficients which summarize low-energy data with no reference to QCD. I found it heartening to see useful results emerge from a less dogmatic view of chiral perturbation theory, when it is used in conjunction with coupled-channel Bethe-Salpeter amplitudes and other methods of unitarization. The small parameters of chiral perturbation theory are still treated as such, but they serve to expand different functions. In this context, we were impressed to see the full scalar nonet emerge in coupled channel calculations, which should be taken seriously by people who use Nambu-Jona Lasinio type models because they suggest that the $q\bar{q}$ threshold is closer to $1.3\,GeV$ than the $600-700\,MeV$ value, usually adopted at zero temperature. There is indeed a difference between a $q\bar{q}$ bound state and an observed resonance. We were also shown how chiral perturbation theory can be applied to study the propagation of pions in the thermal bath which might be produced during a heavy ion collision.

At low temperature but high density, that is, beyond the chiral phase transition, when constituent quark masses are small, we were shown a surprisingly rich and fascinating array of possible superconducting phases, as well as mixed phases, which are controlled by the prevailing chemical potentials. This is a nice example of how simple (Nambu-Jona Lasinio type) models can be used to explore and speculate on hitherto undetected phases and properties of hadronic matter, with obvious astrophysical implications. I say "explore and speculate" rather than "predict" because we should always keep in mind that models, used to describe hadronic matter in one phase, are expressed in terms of parameters which may have quite different values in another phase. And we don't really know what these values are since none of the models are really derived from QCD. This *caveat* also applies to the study of the η, η' masses presented at this workshop. Even at zero temperature and baryonic density, some of the models have not yet been completely explored: we were shown how a proper treatment of the cubic structure of the 't Hooft determinant can yield new solutions to the gap equation.

We were gratified by several exceptionally clear talks on particle production in heavy ions collisions, such as those produced at RHIC. The surprising success of a thermal-

CP660, *Hadron Physics: Effective Theories of Low Energy QCD*, edited by A. H. Blin et al.
© 2003 American Institute of Physics 0-7354-0120-9/03/$20.00

ization hypothesis is an important result which will possibly be better understood when new data will reveal under which conditions this hypothesis fails, that is, its domain of validity. Useful and careful calculations of the decay modes of J/ψ particles were presented.

We learned of many other developments: model-independent relations between the $I = 1$ states belonging to the $70'$ plet of the baryon spectrum, some failings of known approximation schemes in the many-body problem, the Nolen-Schiffer anomaly reconsidered in terms of a $\omega - \rho$ mixing term, evaluations of the Casimir energy, progress on difficult problems such as the $K \rightarrow \pi\pi$ amplitudes was reported as well as on nondegenerate large mass asymptotics of regularized quark loops, the realization of the Atiyah-Singer theorem on a lattice, the temperature dependence of the neutron electric dipole moment, and even a new field theory! I believe that we can be grateful to the organizers of this meeting for having brought all these subjects to our attention and provoked heated discussions.

Several factors have contributed to the success of this workshop.

- the kindness and open-mindedness of the organizers, characteristic of the theoretical physics division of the University of Coimbra;
- the fact that all participants presented their own work. This is in sharp contrast to so many other conferences in which heads of department are invited to summarize work done by others.
- we will not regret travelling to the Iberian peninsula, because we were gratified by an important participation of Spanish physicists from various laboratories and it was a pleasure to see them engaged in vivid discussions among each other, a feature I would like to see happening more often among my colleagues in France;
- we benefited from an important contingent of physicists from Portugal, and even from remote laboratories in Brazil. I felt ashamed to feel the need to consult the map, which Debora Perez Menezes gracefully displayed at the beginning of her talk in order to help us locate the Universidade Federal de Santa Catarina, where fine research is being performed.
- important contributions were also made by a number of Russian physicists who are dispersed throughout Europe. I think we should be aware of the great tragedy which befell on Russian physics after fall of the Iron Curtain, when the economic crisis destroyed so many centers of excellence of the former Soviet Union. This should serve as a useful reminder that our discipline and science depends on the willingness and the means of our governments, as it has in the past.

I had mixed feelings about the growing participation of laptops in spite of the impressive virtuosity of those who use them to display their talks on the screen. I suppose that sound will soon be added and this might then dispense the need to invite the speakers themselves...

I invite you all to join me in thanking the organizers for keeping us happy and physics alive and interesting.

Workshop Program

WEDNESDAY, SEPTEMBER 25

08:30 - Registration

09:30 - Opening Session

Effective Field Theory Methods and Applications

Chair: G. Ripka

10:00 - 10:40   A. Pich
Colourless mesons in a polychromatic world

10:40 - Coffee break

11:10 - 11:50   D. Espriu
Pions, strings and conformal invariance

11:50 - 12:30   G. Colangelo
Dispersion relation and soft pion theorems for $K \to \pi\pi$

12:30 - Lunch

Chair: A. Pich

14:30 - 15:10   E. Arriola
Meson-Baryon interaction in unitarized chiral perturbation theory

15:10 - 15:50   N. Scoccola
Excited baryon spectroscopy in the $1/N_c$ expansion

15:50 - Coffee break

16:20 - 17:00   A. Osipov
Large mass invariant asymptotics of the effective action

17:00 - 17:25   B. Hiller
Path integral bosonization of the 't Hooft determinant: fluctuations and multiple vacua

18:30 - Welcome Cocktail

THURSDAY, SEPTEMBER 26

Effective Field Theory Methods and Applications (cont.)

Chair: D. Espriu

09:00 - 09:40   G. Ripka
Phenomenological models of color confinement based on dual superconductors

09:40 - 10:20   H. Weigel
Casimir Energies in the light of renormalizable quantum field theories

10:20 - Coffee break

10:50 - 11:30   J. Peláez
Chiral perturbation theory, unitarity and light mesons

12:10 - Lunch

Chair: E. Arriola

14:10 - 14:50   J. Oller
Nucleon-Nucleon interaction in a new chiral expansion

14:50 - 15:15   P. Bicudo
Criterium for the index theorem on the lattice

15:15 - Coffee break

Hot and dense matter

Chair: M. Buballa

15:45 - 16:25   B. Krippa
Effective field theory for nuclear matter

16:25 - 17:05   A. Gómez Nicola
Thermal meson properties within chiral perturbation theory

17:05 - 17:45   M. Chabab
The Theta term in QCD sum rules at finite temperature and the neutron electric dipole moment

19:30 - Departure to Palácio de São Marcos

20:00 - Dinner at Palácio de São Marcos

FRIDAY, SEPTEMBER 27

Hot and dense matter (cont.)

Chair: D. Blaschke

09:00 - 09:40   W. Florkowski
Thermal model for RHIC, part I: particle ratios and spectra

09:40 - 10:20   W. Broniowski
Thermal model for RHIC, part II: elliptic flow and HBT radii

10:20 - Coffee break

Chair: Y. Kalinovsky

10:50 - 11:30   M. Buballa
Color superconductivity in electrically and color neutral quark matter

11:30 - 12:10   D. Blaschke
Effects of quark matter and color superconductivity in compact stars

12:10 - Lunch

Chair: W. Broniowski

14:00 - 14:40   D. Menezes
Effects of the $\rho - \omega$ mixing interaction in relativistic models

14:40 - 15:05   C. Providência
EOS of nuclear matter within a generalized NJL model

15:05 - 15:30   P. Costa
η-η' in a hot and dense medium

15:30 - Coffee break

J/Ψ Physics

Chair: M. Nielsen

16:00 - 16:40   J. Hüfner
Anomalous suppression in J/Ψ and Ψ' production in Pb-Pb collisions at the SPS of CERN

16:40 - 17:20   J. Dias de Deus
Percolation and the J/Ψ suppression

17:20 - 18:00   Y. Kalinovsky
J/Ψ dissociation and hadron formfactors

SATURDAY, SEPTEMBER 28

J/Ψ Physics (cont.)

Chair: J. Hüfner

09:00 - 09:40   Y. Ivanov
Photoproduction of Charmonia off Protons and Nuclei

09:40 - 10:20   M. Nielsen
J/Ψ dissociation: a QCD sum rule approach

10:20 - Coffee break

Quark models

Chair: D. Menezes

10:50 - 11:30   M. Scadron
Meson form factors and the quark-level linear sigma model (LσM)

11:30 - 12:10   F. Kleefeld
Consistent relativistic quantum theory for systems/particles described by non-hermitian Hamiltonians and Lagrangians

12:10 - 12:35   J.P. Lansberg
A model for the pion structure function

12:35 - Lunch

Chair: M. Rosina

14:35 - 14:55   E. van Beveren
Classification of scalar mesons

14:55 - 15:25   G. Rupp
Relativistic unitarised quark/meson model in momentum space

15:25 - Coffee break

Chair: G. Rupp

15:55 - 16:35   M. Rosina
The production and decay of heavy dimesons

16:35 - 17:00   D. Bartz
Tensor force contribution to the nucleon-nucleon interaction within a chiral constituent quark model

17:00 - 17:20   G. Ripka
Summary

List of Participants

Pedro Alberto
Centro de Física Computacional,
Departamento de Física,
Universidade de Coimbra,
P-3004-516 Coimbra, Portugal
pedro@teor.fis.uc.pt

Enrique Ruiz Arriola
Departamento de Fisica Moderna,
Facultad de Ciencias,
Universidad de Granada,
Campus de Fuentenueva,
Granada 18071, Spain
earriola@ugr.es

Daniel Bartz
Physics Department B5a,
University of Liège,
allée du 6 Aôut, 17,
B-4000 Liège 1, Belgium
d.bartz@ulg.ac.be

Eef van Beveren
Centro de Física Teórica,
Departamento de Física,
Universidade de Coimbra,
P-3004-516 Coimbra, Portugal
eef@teor.fis.uc.pt

Pedro Bicudo
Centro de Física das Interacções Fundamentais,
Instituto Superior Técnico,
Edifício Ciência, Piso 3,
Avenida Rovisco Pais,
P-1049-001, Lisboa Codex, Portugal
bicudo@ist.utl.pt

David Blaschke
Fachbereich Physick,
Universitat Rostock,
D-18051 Rostock, Germany
david@darss.mpg.uni-rostock.de

Alex H. Blin
Centro de Física Teórica,
Departamento de Física,
Universidade de Coimbra,
P-3004-516 Coimbra, Portugal
alex@teor.fis.uc.pt

Lucília Brito
Centro de Física Teórica,
Departamento de Física,
Universidade de Coimbra,
P-3004-516 Coimbra, Portugal
lucilia@teor.fis.uc.pt

Wojciech Broniowski
The H. Niewodniczański Institute of Nuclear Physics,
ul. Radzikowskiego 152
PL-31342 Cracow, Poland
b4bronio@cyf-kr.edu.pl

Michael Buballa
Institut für Kernphysik,
Technische Universität Darmstadt,
Schloßgartenstraße 9,
D-64289, Darmstadt, Germany
Michael.Buballa@physik.tu-darmstadt.de

Mohamed Chabab
LPHEA, Physics Department,
Faculty of Science-Semlalia,
Cadi Ayyad University,
P.O. Box 2390, Marrakesh, Morocco
mchabab@ucam.ac.ma

Gilberto Colangelo
Institut für Theoretische Physik,
Universität Bern,
Sindlerstraße 5,
CH-3012, Bern, Switzerland
gilberto@itp.unibe.ch

Pedro Costa
Centro de Física Teórica,
Departamento de Física,
Universidade de Coimbra,
P-3004-516 Coimbra, Portugal
pfcosta@portugalmail.pt

Jorge Dias de Deus
CENTRA, Instituto Superior Técnico,
Edifício Ciência, Piso 3 Av. Rovisco Pais,
11049-001, Lisboa, Portugal
jdd@ist.utl.pt

Domenec Espriu
Departament d'Estructura i Constituents de la Matèria,
Universitat de Barcelona,
Diagonal 647,
08028 Barcelona, Spain
espriu@ecm.ub.es

Manuel Fiolhais
Centro de Física Computacional,
Departamento de Física,
Universidade de Coimbra,
P-3004-516 Coimbra, Portugal
tmanuel@teor.fis.uc.pt

Wojciech Florkowski
The H. Niewodniczański Institute of Nuclear Physics,
ul. Radzikowskiego 152
PL-31342 Cracow, Poland
florkows@amun.ifj.edu.pl

Angel Gómez Nicola
Departamento de Física Teórica II,
Facultad de Ciencias Fisicas,
Universidad Complutense,
28040 Madrid, Spain
gomez@fis.ucm.es

Brigitte Hiller
Centro de Física Teórica,
Departamento de Física,
Universidade de Coimbra,
P-3004-516 Coimbra, Portugal
brigitte@teor.fis.uc.pt

Jörg Hüfner
Institut für Theoretische Physik,
Universität Heildelberg,
Philosophenweg 19,
D 69120 Heidelberg, Germany
huefner@tphys.uni-heidelberg.de

Yuri P. Ivanov
Institut für Theoretische Physik,
Universität Heildelberg,
Philosophenweg 19,
D 69120 Heidelberg, Germany
yupi@tphys.uni-heidelberg.de

Yuri L. Kalinovsky
Laboratory of Information Technologies,
Joint Institute for Nuclear Research,
Joliot Curie 6, Moscow Region,
141980 Dubna, Russia
kalinov@cv.jinr.ru

Frieder Kleefeld
Centro de Física das Interacções Fundamentais,
Instituto Superior Técnico,
Edifício Ciência, Piso 3,
Avenida Rovisco Pais,
P-1049-001, Lisboa Codex, Portugal
kleefeld@cfif.ist.utl.pt

Boris V. Krippa
Theoretical Physics Group,
Department of Physics and Astronomy,
University of Manchester,
Manchester M13 9PL, UK
krippa@dirac.phy.umist.ac.uk

J. P. Lansberg
Université de Liège,
Dèpartement de Physique B5,
Sart Tilman,
B-4000 Liège 1, Belgium
JPH.Lansberg@ulg.ac.be

Débora Peres Menezes
Departamento de Física, CFM,
Universidade Federal de Santa Catarina,
Florianópolis,
SC - CP. 476 - CEP 88.040 - 900, Brazil
debora@server.fsc.ufsc.br

Marina Nielsen
Instituto de Fisica,
Universidade de São Paulo,
Caixa Postal 66318, 05315-970 - São Paulo, Brasil
mnielsen@if.usp.br

Orlando Oliveira
Centro de Física Computacional,
Departamento de Física,
Universidade de Coimbra,
P-3004-516 Coimbra, Portugal
orlando@teor.fis.uc.pt

José Antonio Oller
Departamento de Física,
Campus Universitario de Espinardo,
Universidad de Murcia,
E-30071, Murcia, Spain
oller@um.es

Alexandre A. Osipov
Centro de Física Teórica,
Departamento de Física,
Universidade de Coimbra,
P-3004-516 Coimbra, Portugal
alexguest@teor.fis.uc.pt

Humberto Pascoal
Centro de Física Teórica,
Departamento de Física,
Universidade de Coimbra,
P-3004-516 Coimbra, Portugal
pascoal@teor.fis.uc.pt

José R. Peláez
Dipartimento di Fisica,
Universitá degli Studi di Firenze,
INFN, Sede di Sesto Fiorentino,
Via G.Sansone,1,
50019 Sesto Fiorentino (FI), Italia
pelaez@fi.infn.it

Nicholas Petropoulos
Centro de Física Teórica,
Departamento de Física,
Universidade de Coimbra,
P-3004-516 Coimbra, Portugal
nicholas@teor.fis.uc.pt

Antonio Pich
Departament de Física Teòrica, IFIC
Universitat de València - CSIC,
Edifici d'Instituts de Paterna,
Apartat Oficial 22085
E-46071 València, Spain
Pich@ific.uv.es

Constança Providência
Centro de Física Teórica,
Departamento de Física,
Universidade de Coimbra,
P-3004-516 Coimbra, Portugal
constanca@teor.fis.uc.pt

João da Providência
Centro de Física Teórica,
Departamento de Física,
Universidade de Coimbra,
P-3004-516 Coimbra, Portugal
providencia@teor.fis.uc.pt

Georges Ripka
Service de Physique Theorique,
Centre d'Études de Saclay,
F - 91191 Gif-sur-Yvette Cedex, France
ripka@spht.saclay.cea.fr

Mitja Rosina
Faculty of Mathematics and Physics,
University of Ljubljana,
Jadranska 19,
P.O. Box 2964, 1001 Ljubljana, Slovenia
Mitja.Rosina@ijs.si

Maria Conceição Ruivo
Centro de Física Teórica,
Departamento de Física,
Universidade de Coimbra,
P-3004-516 Coimbra, Portugal
maria@teor.fis.uc.pt

George Rupp
Centro de Física das Interacções Fundamentais,
Instituto Superior Técnico,
Edifício Ciência, Piso 3,
Avenida Rovisco Pais,
P-1049-001, Lisboa Codex, Portugal
george@ajax.ist.utl.pt

Michael D. Scadron
Physics Department,
University of Arizona,
Tucson AZ 85721, USA
scadron@physics.arizona.edu

Norberto N. Scoccola
Physics Department,
Comisión Nacional de Energía Atómica,
Av.Libertador 8250,
(1429) Buenos Aires, Argentina.
scoccola@tandar.cnea.gov.ar

Célia A. de Sousa
Centro de Física Teórica,
Departamento de Física,
Universidade de Coimbra,
P-3004-516 Coimbra, Portugal
celia@teor.fis.uc.pt

José N. Urbano
Centro de Física Computacional,
Departamento de Física,
Universidade de Coimbra,
P-3004-516 Coimbra, Portugal
urbano@teor.fis.uc.pt

Herbert Weigel
Institute for Theoretical Physics,
Tübingen University,
Auf der Morgenstelle 14,
D-72076 Tübingen, Germany
herbert.weigel@uni-tuebingen.de

AUTHOR INDEX

A

Aguilera, D. N., 209
Andrianov, A. A., 17

B

Baran, A., 185
Bartz, D., 383
Bicudo, P., 130
Birse, M. C., 147
Bissey, F., 339
Blaschke, D., 209, 272
Bracco, M. E., 296
Broniowski, W., 177, 185
Buballa, M., 196
Burau, G., 272

C

Chabab, M., 170
Colangelo, G., 25
Costa, P., 242
Cudell, J. R., 339
Cugnon, J., 339

D

da Providência, J., 231
Del Fabbro, A., 377
Dias de Deus, J., 266
Dobado, A., 156

E

Espriu, D., 17

F

Florkowski, W., 177, 185

G

García Recio, J., 34
Gómez Nicola, A., 102, 156
Grigorian, H., 209

H

Hiller, B., 62, 69
Hošek, J., 77
Hüfner, J., 259, 283

I

Ivanov, Y. P., 283

J

Jaminon, M., 339
Janc, D., 377

K

Kalinovsky, Y. L., 242, 272
Kleefeld, F., 311, 325, 353
Kopeliovich, B. Z., 283
Krippa, B. V., 147

L

Lansberg, J. P., 339
Llanes-Estrada, F. J., 156

M

Matheus, R. D., 296
McGovern, J. A., 147
Menezes, D. P., 223
Moreira, J. M., 231
Moszkowski, S. A., 231

N

Navarra, F. S., 296
Neumann, F., 196
Nielsen, M., 296
Nieves, J., 34

O

Oertel, M., 196
Oller, J. A., 116
Osipov, A. A., 62, 69

P

Peláez, J. R., 102, 156
Petropoulos, N., 353
Pich, A., 3
Providência, C., 223, 231

R

Ripka, G., 77, 389
Rodrigues da Silva, R., 296
Rosina, M., 377
Ruivo, M. C., 242
Ruiz Arriola, E., 34
Rupp, G., 311, 353, 367

S

Scadron, M. D., 311
Scoccola, N. N., 48
Stassart, P., 339

T

Tarasov, A. V., 283
Toki, H., 209
Treleani, D., 377

V

van Beveren, E., 311, 353, 367
Vicente Vacas, M., 34

W

Walet, N. R., 147
Weigel, H., 88

Y

Yasui, S., 209

Z

Zhuang, P., 259